Advanced Drug Delivery Systems in Management of Chronic Obstructive Pulmonary Disease

This book offers a comprehensive overview of the epidemiology, etiology, and pathophysiology of chronic obstructive pulmonary disease (COPD). It addresses the limitations of existing drug delivery methods and explores advanced delivery systems to overcome these challenges, providing an exhaustive account of their intricate mechanisms. The introductory chapters elucidate pathways responsible for COPD progression, followed by a detailed analysis of established biomarkers and potential targets in contemporary COPD therapy. Subsequent chapters provide insights into ongoing treatment modalities, their efficacy, drawbacks, and prospective solutions to counter the setbacks of COPD therapy. The subsequent section covers state-of-the-art drug delivery technologies and novel drug formulations designed to enhance drug deposition and absorption in COPD lungs. It further explores a methodical yet coordinated explanation of targeted personalized therapies and emerging approaches, including nanoparticles, polymeric carriers, and vesicular delivery systems. Toward the end, the book discusses ongoing and completed clinical trials encompassing the management of COPD through advanced drug delivery approaches. It serves as a valuable resource for professionals, scientists, academicians, and clinicians specializing in respiratory health.

Advanced Drug Delivery Systems in Management of Chronic Obstructive Pulmonary Disease

Edited by
Parteek Prasher, Mousmee Sharma, Gang Liu,
Amlan Chakraborty, and Kamal Dua

CRC Press
Taylor & Francis Group
Boca Raton London New York

CRC Press is an imprint of the
Taylor & Francis Group, an **informa** business

Designed cover image: Shutterstock
First edition published 2025
by CRC Press
2385 NW Executive Center Drive, Suite 320, Boca Raton FL 33431

and by CRC Press
4 Park Square, Milton Park, Abingdon, Oxon, OX14 4RN

CRC Press is an imprint of Taylor & Francis Group, LLC

ISBN: 978-1-032-81259-5 (hbk)
ISBN: 978-1-032-81260-1 (pbk)
ISBN: 978-1-003-49892-6 (ebk)

DOI: 10.1201/9781003498926

Typeset in Times
by KnowledgeWorks Global Ltd.

Contents

About the Editors

Dr. Parteek Prasher, an Associate Professor in the Department of Chemistry at the University of Petroleum & Energy Studies, possesses eight years of research experience in medicinal chemistry and biomaterials. Leading a research team, he focuses on developing rationally designed molecules to target various disorders, including inflammation, cancer, and antimicrobial multidrug resistance. Dr. Prasher has received prestigious fellowships and awards, including the Junior/Senior Research Fellowship by the Council of Scientific and Industrial Research (2012–2015) and the TARE Fellowship by the Department of Science and Technology-Science and Engineering Research Board (DST-SERB) for three years.

Dr. Mousmee Sharma, an Assistant professor in the Department of Chemistry at Uttaranchal University, possesses more than five years of research experience in physical chemistry and biomaterials, with a focus on bio membranes. Her recent research centres on developing mucoadhesive materials for drug delivery across physiological mucus barriers. Dr. Mousmee, a "Gold Medallist" in post-graduation, has authored 80 research and review articles and 20 book chapters in reputable journals.

Dr. Gang Liu completed his Bachelor of Science in Biotechnology at Tianjin University of Commerce in 2008, followed by a Master of Science in Biotechnology at the University of Wollongong in 2011. He earned his PhD in Immunology and Microbiology at the University of Newcastle (Australia) in 2016, focusing on airway remodeling in respiratory diseases such as chronic obstructive pulmonary disease (COPD), severe asthma, and idiopathic pulmonary fibrosis (IPF). Dr. Liu conducted postdoctoral research from 2016–2017, investigating inflammation's role in respiratory diseases, and later studied tissue structure changes and remodeling in the gut-lung crosstalk at Hunter Medical Research Institute (HMRI) from 2017–2019. Since 2019, he has held a tenure position as a lecturer at the University of Technology (UTS), leading the fibrosis program at the Center for Inflammation in the UTS/Centenary Institute. Dr. Liu was awarded a CREATE Hope Fellowship from Lung Foundation Australia (2022–2023) and was promoted to senior lecturer at UTS in early 2023. Currently, he teaches and coordinates immunology and pathology subjects at UTS while supervising three PhD students. His ongoing research focuses on understanding the mechanisms of lung fibrosis in various lung diseases and identifying new therapeutic options.

Dr. Amlan Chakraborty specializes in unraveling the mechanisms of innate immune cells in airway inflammation and remodeling linked to chronic allergic airway diseases, particularly asthma. He holds a research master's degree and a PhD, focusing on developing immunomodulators to target airway inflammation in acute lung injury and asthma. As a postdoctoral research associate, he investigated peptide-conjugated nanoparticles in asthma-related inflammation. Joining the Fibrosis Group at Monash Biomedicine Discovery Institute in 2020, he secured two industry-funded fellowships to explore induced pluripotent stem cells' role in pulmonary fibrosis and understand epithelial cell repair in chronic allergic airway disease. Dr. Chakraborty has more than 20 publications, 7 pending patents, and has attracted $200K in commercial funding. His post-PhD research concentrates on using nanoparticles, including serelaxin-conjugated nanoparticles, to attenuate airway inflammation and target airway remodelling in pulmonary fibrosis and asthma. He also investigates the role of pulmonary myeloid cells in asthma-associated airway remodeling using functionalized nanoparticles, traceable through innovative imaging techniques. Recently, he has expanded his research interest to explore circadian mechanisms in asthma.

Dr Kamal Dua, is an Associate Professor in the Discipline of Pharmacy at UTS. Dr Dua has been recognised and named as one of Australia's Top 250 researchers in 2025 and 2024 in Research Magazine by the Australian in the research discipline of Health and Medical sciences, and #1/research leader in the area of Pharmacology and Pharmacy (2025) and Toxicology (2024) respectively. With over 17 years of research experience, he has specialised in drug delivery targeting inflammatory diseases. Additionally, Dr Dua also leads drug delivery research at the Woolcock Institute of Medical Research. In this role, he focuses on advancing targets identified in research projects to develop innovative formulations, taking the first steps toward clinical translation. Dr. Dua's research encompasses two complementary areas: drug delivery and immunology. His work explores how these disciplines can mutually benefit, contributing to the goal of promoting longer and healthier lives for the community. His commitment to synergy is reflected in his extensive publication record in reputable journals. Dr Dua's research interests are centered on harnessing the pharmaceutical potential of modulating critical regulators, such as interleukins and microRNAs. He also specialises in developing new and effective drug delivery formulations for managing inflammation in chronic airway diseases and cancer.

Contributors

Mohit Agrawal
School of Medical & Allied Sciences
K.R. Mangalam University
Gurgaon, Haryana, India

Abhilasha Ahlawat
Faculty of Pharmaceutical Sciences
Maharshi Dayanand University
Rohtak, Haryana, India

Waseem Ahmad
Department of Chemistry
Graphic Era (Deemed to Be University)
Dehradun, Uttarakhand, India

Shvetank Bhatt
School of Health Sciences and Technology
Dr. Vishwanath Karad MIT World Peace University
Pune, Maharashtra, India

Gaurav Bisht
Department of Chemistry
School of Applied and Life Sciences
Uttaranchal University
Dehradun, Uttarakhand, India

Rachelle Catanzariti
Discipline of Pharmacy
Graduate School of Health, University of Technology Sydney
Ultimo, NSW, Australia

Sarath Chandran C
College of Pharmaceutical Sciences
Govt. Medical College Kannur

Gopika Nath C
College of Pharmaceutical Sciences
Govt. Medical College Kannur

Prarit Chandel
School of Pharmacy
Chitkara University
Himachal Pradesh, India

Shraddha Chugh
Department of Chemistry
University of Petroleum & Energy Studies
Dehradun, Uttarakhand, India

Chinmay Devidas Deshmukh
School of Health Sciences and Technology
Dr. Vishwanath Karad MIT World Peace University
Pune, Maharashtra, India

Kamal Dua
Discipline of Pharmacy
Graduate School of Health
Faculty of Health
University of Technology Sydney
Ultimo, NSW 2007, Australia

Harish Dureja
Faculty of Pharmaceutical Sciences
Maharshi Dayanand University
Rohtak, Haryana, India
Faculty of Health
Australian Research Centre in Complementary and Integrative Medicine
University of Technology Sydney
Ultimo, NSW, Australia

Abhishek Dwivedi
Department of Pharmaceutics
Maharishi Arvind College of Pharmacy
Ambabari, Jaipur, Rajasthan, India

Rabab Fatima
Department of Chemistry
University of Petroleum & Energy Studies
Dehradun, Uttarakhand, India

Nisha Gulati
Faculty of Pharmaceutical Sciences
Maharshi Dayanand University
Rohtak, Haryana, India

Tarif Hussian
Faculty of Pharmaceutical Sciences
Maharshi Dayanand University
Rohtak, Haryana, India

Hem C. Joshi
Department of Botany
D.S.B. Campus, Kumaun University
Nainital, Uttarakhand, India

Richa Joshi
Department of Botany
D.S.B. Campus, Kumaun University
Nainital, Uttarakhand, India

Hiba. K
College of Pharmaceutical Sciences
Govt. Medical College Kannur

Rahul. K
College of Pharmaceutical Sciences
Govt. Medical College Kannur

Reshma. K
College of Pharmaceutical Sciences
Govt. Medical College Kannur

Kanika
School of Pharmaceutical Sciences
MVN University
Palwal, Haryana, India

Neha Kanojia
School of Pharmacy
Chitkara University
Himachal Pradesh, India

Shijina Kappally
College of Pharmacy
Sharjah University
UAE

Mourvi Sachin Kulkarni
School of Health Sciences and Technology
Dr. Vishwanath Karad MIT World Peace University
Pune, Maharashtra, India

Chetan Kumar
School of Pharmaceutical and Populations Health Informatics
Faculty of Pharmacy
DIT University
Dehradun, Uttarakhand, India

Dileep Kumar
Poona College of Pharmacy
Department of Pharmaceutical Chemistry
Bharati Vidyapeeth (Deemed to be University)
Pune, Maharashtra, India
UC Davis Comprehensive Cancer Center
Department of Entomology and Nematology,
University of California Davis,
Davis, CA, USA

Girish Kumar
School of Pharmaceutical Sciences
MVN University
Palwal, Haryana, India

Tanisqa Mall
Department of Chemistry
University of Petroleum & Energy Studies
Dehradun, Uttarakhand, India

Anurag Mishra
NIMS Institute of Pharmacy
NIMS University Rajasthan
Jaipur, Rajasthan, India

Lokesh Nagar
Faculty of Pharmaceutical Sciences
Maharshi Dayanand University
Rohtak, Haryana, India

Anirudhha Pravin Nilkari
School of Health Sciences and Technology
Dr. Vishwanath Karad MIT World Peace University
Pune, Maharashtra, India

Ritu Painuli
Department of Chemistry
School of Applied and Life Sciences
Uttaranchal University
Dehradun, Uttarakhand, India

Shivam Pandey
Department of Chemistry
School of Applied and Life Sciences
Uttaranchal University
Dehradun, Uttarakhand, India

Kamla Pathak
School of Pharmaceutical Sciences
MVN University
Palwal, Haryana, India
Faculty of Pharmacy
Uttar Pradesh
University of Medical Sciences
Etawah, Uttar Pradesh, India

Parteek Prasher
Department of Chemistry
University of Petroleum & Energy Studies
Dehradun, Uttarakhand, India

Sobhit Singh Rajput
Department of Pharmaceutics
Aligarh College of Pharmacy
Aligarh, Uttar Pradesh, India

Annu Saini
Faculty of Pharmaceutical Sciences
Maharshi Dayanand University
Rohtak, Haryana, India

Prachi Saini
School of Pharmaceutical Sciences
MVN University
Palwal, Haryana, India

Ayeh Bani Saeid
Discipline of Pharmacy
Graduate School of Health
University of Technology Sydney
Ultimo, NSW, Australia
Faculty of Health
Australian Research Centre in Complementary and Integrative Medicine
University of Technology Sydney
Ultimo, NSW, Australia

Abhijith. K. Shaji
College of Pharmaceutical Sciences
Govt. Medical College Kannur

Mousmee Sharma
Department of Chemistry
Uttaranchal University
Dehradun, Uttarakhand, India

Ajay Kumar Shukla
Institute of Pharmacy
Dr Rammanohar Lohia Avadh University
Ayodhya, Uttar Pradesh, India

Ajay Singh
Department of Chemistry
School of Applied and Life Sciences
Uttaranchal University
Dehradun, Uttarakhand, India

Lubhan Singh
Faculty of Pharmacy
Swami Vivekanand Subharti University
Meerut, Uttar Pradesh, India

Sachin Kumar Singh
School of Pharmaceutical Sciences
Lovely Professional University
Phagwara, Punjab, India
Faculty of Health
Australian Research Centre in Complementary and Integrative Medicine
University of Technology Sydney
Ultimo, NSW, Australia

Yogendra Singh
Department of Pharmacology
Maharishi Arvind College of Pharmacy
Ambabari, Jaipur, Rajasthan, India

Neeta Solanki
Faculty of Pharmaceutical Sciences
Maharshi Dayanand University
Rohtak, Haryana, India

Komal Thapa
School of Pharmacy
Chitkara University
Himachal Pradesh, India

Nitish Kumar Tiwary
School of Health Sciences and Technology
Dr. Vishwanath Karad MIT World Peace University
Pune, Maharashtra, India

Garima Verma
Faculty of Pharmacy
Swami Vivekanand Subharti University
Meerut, Uttar Pradesh, India

Nitin Verma
School of Pharmacy
Chitkara University
Himachal Pradesh, India

Tarun Virmani
School of Pharmaceutical Sciences
MVN University
Palwal, Haryana, India

Viswapriya Viswalingam
Poona College of Pharmacy
Department of Pharmaceutical Chemistry
Bharati Vidyapeeth (Deemed to be University)
Pune, Maharashtra, India

Stewart Yeung
Discipline of Pharmacy
Graduate School of Health
University of Technology Sydney
Ultimo, NSW, Australia
Faculty of Health
Australian Research Centre in Complementary and Integrative Medicine
University of Technology Sydney
Ultimo, NSW, Australia

Flavia C. Zacconi
Facultad de Química y de Farmacia
Pontificia Universidad Católica de Chile
Santiago, Chile
Institute for Biological and Medical Engineering
Schools of Engineering
Medicine and Biological Sciences
Pontificia Universidad Católica de Chile
Santiago, Chile

1 Introduction to COPD

Epidemiology, Etiology, and Pathophysiology

Mohit Agarwal and Anurag Mishra
NIMS Institute of Pharmacy, NIMS University Rajasthan
Jaipur, Rajasthan, India

Yogendra Singh
Department of Pharmacology, Maharishi Arvind College
of Pharmacy, Ambabari, Jaipur, Rajasthan, India

1.1 INTRODUCTION

Chronic obstructive pulmonary disease (COPD) is defined by the Global Initiative for Chronic Obstructive Lung Disease (GOLD) initiative as a respiratory illness that causes impaired airflow and respiratory symptoms due to anomalies in the airways and/or the alveoli (Vogelmeier et al., 2017). Contrary to previous societal recommendations, GOLD 2020 now considers asthma and COPD to be distinct illnesses and does not acknowledge the overlap between the two as a unique entity (Sam & Kraft, 2022). More than 10 million Americans are impacted by COPD, making it the fourth leading cause of death in the country. COPD is gaining more and more attention in public health circles all around the globe. Approximately 250 million people throughout the world suffer from COPD. Pulmonary function decline due to adult respiratory infections is debatable, although severe, long-lasting impairments in lung function are seldom observed after a single bout of pneumonia or acute bronchitis. Loss of lung function is a common side effect of smoking, and it worsens with time and increasing frequency. In contrast to the 30 mL and 22 mL annual declines in forced expiratory volume in one second (FEV1) in men and women, respectively, who were able to maintain quitting smoking, a large study of current smokers found that these values dropped by 66 mL and 54 mL per year for men and women, respectively. Fifteen percent of those who smoke start to have increasingly incapacitating symptoms in their 40s and 50s. Nearly two-thirds of COPD patients report heavy cigarette use in their past. The link between cigarette smoking and COPD is obvious. It would indicate that new nicotine delivery products are less dangerous than their predecessors. Electronic cigarettes, often known as e-cigarettes, and heated tobacco products (HTPs) are alternatives to traditional cigarettes or combustible cigarettes. Personal vaporizers that heat a nicotine-containing liquid to produce an aerosol are known as electronic cigarettes. Tobacco is heated in a regulated fashion using a battery-operated device, which is how HTPs work. When compared to the 850 °C combustion temperature at the cigarette tip, the temperature produced by an HTP—330–350 °C—is significantly lower. The rationale behind this approach is that the resulting aerosol should be far less harmful to the lungs than cigarette smoke (Auer et al., 2017; Baker, 1974).

The other one-third of COPD patients could be subjected to a mixture of outside cigarette smoke, dusts and chemicals from work, and interior contaminants in the air from biomass fuel used for heating and cooking in houses without enough ventilation. Other potential causes include air pollution, allergies, environmental factors, airway infections, and genetics (specifically, alpha-1-antiprotease deficiency). Two potential major risk factors are allergic reactions and bronchoconstriction caused

DOI: 10.1201/9781003498926-1

by nonspecific airway stimulation. Research suggests that prenatal and early childhood exposure to allergens and pollutants might stunt a child's lung development, decrease expiratory airflow, and cause spirometric values to be lower than expected in middle age (Abbas et al., 2024). Asthma symptoms include a higher propensity for bronchoconstriction in reaction to a range of external stimuli, such as histamine and methacholine. Airway hyperresponsiveness is a common symptom among COPD patients as well. When it comes to airway responsiveness, airflow restriction, and pulmonary symptoms, older people with a history of persistent asthma and smokers with COPD have many similarities. The majority believe that asthma is an allergic illness and that COPD is mostly caused by inflammation and damage caused by smoking. Nevertheless, it is possible that these two conditions have comparable environmental and genetic variables, and that the chronic form in older subjects might manifest identically. For subjects with asthmatic infancy who develop a smoking habit, this is of utmost importance.

The majority of research focuses on the onset of COPD. It is believed that in genetically vulnerable hosts, the "progression" and "consolidation" stages of COPD patients take place across decades, even after smoking cessation (Rochester et al., 2022). Exacerbations of COPD are a common complication of respiratory infections. New evidence from the COPDGene and ECLIPSE trials shows that people with better lung function levels to begin with lose more lung function over the course of a COPD exacerbation. Although there is an inadequate amount of sufficient longitudinal data, new research indicates that pneumonia early on may raise the possibility of COPD in adulthood. This finding adds to the growing body of evidence linking respiratory infections in infancy to a greater chance of COPD in later life.

COPD includes emphysema, chronic bronchitis, and chronic obstructive airway syndrome (a kind of asthma with airway blockage that is not commonly reversible) (Lynch et al., 2015). The outward appearances of these entities vary and have distinct clinical and physiological effects. Therefore, COPD is a diverse collection of interconnected disorders, each with a distinct set of characteristics. The manifestations of COPD are influenced by several risk factors, with tobacco smoking being the most prevalent among them (Friedlander et al., 2007).

1.2 ETIOLOGY

COPD is a chronic airflow obstruction condition. Airflow blockage frequently develops and is linked to aberrant lung inflammation to toxic particles or gasses. The consumption of tobacco products a significant determinant of COPD (Kohansal et al., 2009). The four interconnected events that make up the current prevailing model for the pathophysiology of emphysema are: (1) In genetically predisposed people, of small and large the respiratory tract, as well as the outermost air spaces of the lung, are recruited by inflammatory and immunological cells under prolonged exposure to cigarette smoke. (2) The extracellular matrix that supports the lung's the airway walls, vasculature, and gas exchange surfaces is damaged when inflammatory cells produce proteinases. (3) Structural cell death, triggered by oxidant-driven deterioration, senescence of cells, and biochemical digestion of cell-matrix materials leads to alveolar destruction, vascular pruning, and the loss of smaller airways. (4) Air space expansion and emphysema are both caused by disrupted repair of elastin and other components of the extracellular matrix.

Smoking can impact the alveoli, small airways (≤2 mm diameter), and upper respiratory tract. When the alveoli and major airways undergo remodeling, it results in biological alterations including coughing and formation of sputum. Most people with COPD experience airway inflammation, damage, and emphysema formation; however, these seem to be separate processes, and the extent to which each contributes to blockage in any one individual varies.

Infections: Although the exact nature of the host–pathogen interaction is still a mystery, researchers are looking at the potential significance of persistent and frequent infections in the development of COPD, which is marked by persistent inflammation. It has been found

TABLE 1.1
Investigate the Causal Variables Linked to COPD

Host-Based Variables	Environmental Variables
Airway hyperreactivity	Indoor air pollution
Genetic	Smoking
Infection	Occupation
	Low socioeconomic status

that some diseases, such HIV and tuberculosis (TB), might raise the risk of COPD. There is evidence that lower respiratory infections in children have a negative impact on lung health as adults (Table 1.1).

Infection with the Mycobacterium TB: An estimated 1.3 million people died from TB in 2019, making it the leading infectious killer on a global scale. Even after a microbiologic cure, the most prevalent kind of TB, pulmonary TB, can leave a person with serious chronic pulmonary sequelae (Jin et al., 2018). Whether the airflow blockage is due to scarring or other postfibrotic alterations is unknown, although tuberculosis is a significant determinant of risk for COPD (Jin et al., 2018; Yakar et al., 2017).

Infection with the HIV: COPD linked to HIV was first detected in 1989 and has a prevalence rate of 6–15% (Fitzpatrick et al., 2018; Kuhlman et al., 1989). Individuals with HIV who engage in smoking have an increased likelihood of developing COPD, a condition that can ultimately result in emphysema (Crothers et al., 2011; Gingo et al., 2014). Endothelial dysfunction, systemic inflammation due to prolonged immunological activation, and elevated oxidative stress all play a role in regulating the risk (Fitzpatrick et al., 2018). Whether these processes are a direct result of HIV infection or a reaction to altered innate immune responses is not yet apparent. The incidence of COPD and pulmonary hypertension among smokers with HIV seems to be higher in patients with higher viral loads and lower CD4 cell counts, rather than being an unforeseen consequence of medication for HIV (Almodovar et al., 2011; Morris et al., 2012). Emphysema in HIV patients is irreversible regardless of viral load or CD4 cell count recovery, despite the correlation between HIV disease management and worse COPD outcomes.

Infections of the lower respiratory tract in early age: Particularly important in children is the complicated interplay between host-based variables and environmental susceptibility to illness. Peak lung function is affected by a variety of events, including those that occur during pregnancy and subsequent exposures, illnesses in early infancy, and other moments in time (Lange et al., 2015). A delayed lung maturity raises the chance of acquiring COPD in old age. The FEV1 continues to decrease in adults who smoke (Agusti & Faner, 2019). It is critical to encourage general lung health as the state of the lungs in early infancy is a predictor of ventilatory performance in later years. Efforts to reduce tobacco use, lessen contact with allergens, and improve healthcare access and management of respiratory illnesses can all contribute to this goal (Stern et al., 2007).

1.3 PATHOGENESIS AND PATHOPHYSIOLOGY

COPD comprises emphysema, a pathologically identified pulmonary problem with enlarged air spaces and destroyed alveoli; prolonged bronchitis, a medically specified lung functional condition with persistent wheezing and sputum; and/or small airway ailments, a disorder with constricted and decreased numbers of delicate bronchioles. Airway inflammation and lung parenchyma are the primary sites of pathogenic alterations in COPD. Paraenchyma consists of terminal airways and the acinus, which comprises the respiratory bronchiole, alveolar ducts, alveolar sacs, and alveoli grouped together (Rennard, 1998). Although severe COPD exacerbations hold significant clinical and economic significance, their cause and mechanisms remain little understood. However, lower airway inflammation is a hallmark of COPD. There is a protein cascade that becomes activated during an inflammatory response. Chemokines, cytokines, enzymes, growth factors, receptors, and adhesion molecules are all part of this course of study.

Inflammation often causes a cell-specific induction of gene transcription, which in turn increases protein expression. Many of these genes are not expressed in normal cells when at rest, but they are stimulated during inflammation. Airway inflammation pathophysiology is heavily influenced by transcription factors, which control the expression of several pro-inflammatory genes (Caramori et al., 2004). The existence of both type 1 and 3 immune responses is strongly associated with inflammation in COPD. Group 1 innate lymphoid cells, the transcription factor T-bet, which regulates the production of IFNγ, and Th1 are all components of the antimicrobial type 1 that work together. Macrophages that promote inflammation are also more active. In the type 3 response, which primarily targets fungus, neutrophil inflammation is caused by the secretion of interleukin (IL)-17 and IL-22 by group 3 innate lymphoid cells and the expression of RORγt by type 17 helper T cells (Th17) (Barnes, 2018). The lung's structural stability depends on elastin, a very durable component of the matrix that surrounds cells called extracellular matrix (ECM) along the primary molecule of elastic fibers (Vij et al., 2018).

Later on, however, new molecular pathways that play a role in emphysema, as well as a complicated network of inflammatory and immunological cells, were discovered. Lung macrophages and epithelial cells are activated when exposed to oxidants from cigarette smoke. The airways that allow gases to exchange hands—the alveoli, bronchioles, and respiratory ducts—are damaged or destroyed in emphysema. Smokers nearly universally have an accumulation of macrophages in their respiratory bronchioles. Another finding in smokers' alveolar spaces is an increase in neutrophils, B lymphocytes, and T lymphocytes, with a concentration of CD8+ cells observed in particular. Coalescence of the fragile alveolar structure into huge emphysematous air gaps causes the alveolar walls to become punctured and eventually demolished. Furthermore, they produce proteinases and chemokines, which attract other immunological and inflammatory cells. Animal models have linked the transcription factor NRF2, which regulates the oxidant-antioxidant balance, and superoxide dismutase 3, a powerful antioxidant, to the formation of emphysema, highlighting the critical nature of oxidative stress in COPD (Ma, 2013). In COPD, malfunctioning mitochondria may amplify cellular oxidative stress. Mitochondria serve as the main energy generators in the majority of eukaryotic cells, producing adenosine triphosphate (ATP) by oxidative phosphorylation. They play a crucial role in detecting nutrients and oxygen. Nevertheless, mounting evidence indicates that mitochondria play a vital role in the development of lung diseases by controlling essential cellular processes such as calcium balance, generation of reactive oxygen species (ROS) in mitochondria, release of factors that promote cell death, changes in cellular metabolism, cell demise, and inflammation. One way that macrophages can be activated is by the oxidant-induced deactivation of histone deacetylase-2 (HDAC2) (Dasgupta et al., 2020; Dromparis & Michelakis, 2013). This causes a change in the balance toward acetylated or open chromatin, which in turn exposes nuclear factor-κB sites. As a result, matrix metalloproteinases and pro-inflammatory cytokines like IL-8 and tumor necrosis factor-alpha (TNF-α) are transcribed. Ultimately, this leads to the recruitment of neutrophils. Cigarette smoke also triggers the recruitment of CD8+ T lymphocytes, which secrete CXCL-7 (Kim et al., 2015). This stimulation causes macrophages to produce macrophage elastase (matrix metalloproteinase12 [MMP-12]). Tobacco smoking hinders the immune system's capacity to eliminate dead cells, which impedes the body's ability to recover. Figure 1.1 shows the progression of COPD by alteration in the alveoli shape in bronchial tree of respiratory airways.

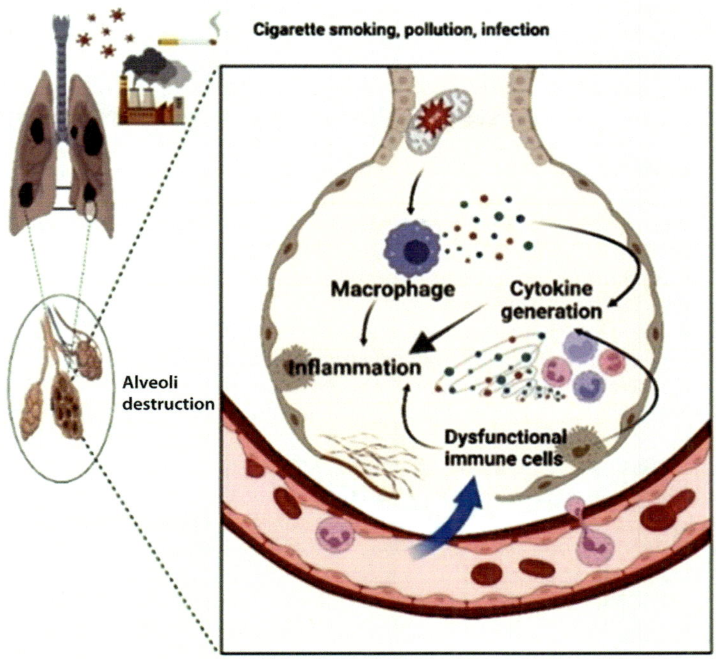

FIGURE 1.1 Inflammation associated bronchitis and emphysema triggers for chronic obstructive pulmonary disease.

1.4 CONCLUSIONS

COPD is a major cause for alarm since it is a chronic, progressive, non-infectious respiratory condition. Non-smokers account for about 25–45% of COPD cases, despite smoking being a major risk factor for the disease. Therefore, more research into non-smoking risk factors is necessary to ascertain the ways in which these variables impact the development and symptoms of COPD. COPD and asthma are separate conditions with their own unique pathologies, risk factors, and outcomes; nevertheless, the same drugs may be able to effectively treat both conditions, depending on the drug choice. To alleviate symptoms and lessen the likelihood of exacerbations, bronchodilators are often prescribed to the vast majority of COPD patients. When compared to systemic pharmaceutical administration, the inhaled route has fewer adverse effects, making it the preferred method of medication delivery. Patients experiencing symptoms should take both long-acting drugs on a regular basis and short-acting medications as needed.

REFERENCES

Abbas, R., Ali, M. E., Fitzpatrick, M. E., & Rivera-Lebron, B. (2024). Chronic Obstructive Pulmonary Disease. In M. A. Papadakis, S. J. McPhee, M. W. Rabow, K. R. McQuaid, & M. Gandhi (Eds.), *Current Medical Diagnosis & Treatment 2024*. McGraw-Hill Education. https://accessmedicine.mhmedical.com/content.aspx?aid=1202062851

Agusti, A., & Faner, R. (2019). Lung function trajectories in health and disease. *Lancet Respir Med*, *7*(4), 358–364. https://doi.org/10.1016/s2213-2600(18)30529-0

Almodovar, S., Hsue, P. Y., Morelli, J., Huang, L., & Flores, S. C. (2011). Pathogenesis of HIV-associated pulmonary hypertension: Potential role of HIV-1 Nef. *Proc Am Thorac Soc*, *8*(3), 308–312. https://doi.org/10.1513/pats.201006-046WR

Auer, R., Concha-Lozano, N., Jacot-Sadowski, I., Cornuz, J., & Berthet, A. (2017). Heat-not-burn tobacco cigarettes: Smoke by any other name. *JAMA Intern Med*, *177*(7), 1050–1052. https://doi.org/10.1001/jamainternmed.2017.1419

Baker, R. (1974). Temperature distribution inside a burning cigarette. *Nature*, *247*(5440), 405–406.

Barnes, P. J. (2018). Targeting cytokines to treat asthma and chronic obstructive pulmonary disease. *Nat Rev Immunol*, *18*(7), 454–466. https://doi.org/10.1038/s41577-018-0006-6

Caramori, G., Ito, K., & Adcock, I. M. (2004). Transcription factors in asthma and COPD. *IDrugs*, *7*(8), 764–770.

Crothers, K., Huang, L., Goulet, J. L., Goetz, M. B., Brown, S. T., Rodriguez-Barradas, M. C., Oursler, K. K., Rimland, D., Gibert, C. L., Butt, A., & Justice, A. (2011). HIV infection and risk for incident pulmonary diseases in the combination antiretroviral therapy era. *Am J Respir Crit Care Med*, *183*(3), 388–395. https://doi.org/10.1164/rccm.201006-0836OC

Dasgupta, A., Wu, D., Tian, L., Xiong, P. Y., Dunham-Snary, K. J., Chen, K.-H., Alizadeh, E., Motamed, M., Potus, F., & Hindmarch, C. C. (2020). Mitochondria in the pulmonary vasculature in health and disease: Oxygen-sensing, metabolism, and dynamics. *Comprehensive Physiology*, *10*(2), 713–765.

Dromparis, P., & Michelakis, E. D. (2013). Mitochondria in vascular health and disease. *Annu Rev Physiol*, *75*, 95–126.

Fitzpatrick, M. E., Kunisaki, K. M., & Morris, A. (2018). Pulmonary disease in HIV-infected adults in the era of antiretroviral therapy. *Aids*, *32*(3), 277–292. https://doi.org/10.1097/qad.0000000000001712

Friedlander, A. L., Lynch, D., Dyar, L. A., & Bowler, R. P. (2007). Phenotypes of chronic obstructive pulmonary disease. *COPD: J Chron Obstruct Pulmon Dis*, *4*(4), 355–384.

Gingo, M. R., He, J., Wittman, C., Fuhrman, C., Leader, J. K., Kessinger, C., Lucht, L., Slivka, W. A., Zhang, Y., McMahon, D. K., Sciurba, F. C., & Morris, A. (2014). Contributors to diffusion impairment in HIV-infected persons. *Eur Respir J*, *43*(1), 195–203. https://doi.org/10.1183/09031936.00157712

Jin, J., Li, S., Yu, W., Liu, X., & Sun, Y. (2018). Emphysema and bronchiectasis in COPD patients with previous pulmonary tuberculosis: Computed tomography features and clinical implications. *Int J Chron Obstruct Pulmon Dis*, *13*, 375–384. https://doi.org/10.2147/copd.S152447

Kim, V., Cornwell, W. D., Oros, M., Durra, H., Criner, G. J., & Rogers, T. J. (2015). Plasma chemokine signature correlates with lung goblet cell hyperplasia in smokers with and without chronic obstructive pulmonary disease. *BMC Pulm Med*, *15*, 1–10.

Kohansal, R., Martinez-Camblor, P., Agustí, A., Buist, A. S., Mannino, D. M., & Soriano, J. B. (2009). The natural history of chronic airflow obstruction revisited: An Analysis of the Framingham offspring cohort. *Am J Respir Crit Care Med*, *180*(1), 3–10. https://doi.org/10.1164/rccm.200901-0047OC

Kuhlman, J. E., Knowles, M. C., Fishman, E. K., & Siegelman, S. S. (1989). Premature bullous pulmonary damage in AIDS: CT diagnosis. *Radiology*, *173*(1), 23–26. https://doi.org/10.1148/radiology.173.1.2781013

Lange, P., Celli, B., Agustí, A., Boje Jensen, G., Divo, M., Faner, R., Guerra, S., Marott, J. L., Martinez, F. D., Martinez-Camblor, P., Meek, P., Owen, C. A., Petersen, H., Pinto-Plata, V., Schnohr, P., Sood, A., Soriano, J. B., Tesfaigzi, Y., & Vestbo, J. (2015). Lung-function trajectories leading to chronic obstructive pulmonary disease. *N Engl J Med*, *373*(2), 111–122. https://doi.org/10.1056/NEJMoa1411532

Lynch, D. A., Austin, J. H., Hogg, J. C., Grenier, P. A., Kauczor, H.-U., Bankier, A. A., Barr, R. G., Colby, T. V., Galvin, J. R., & Gevenois, P. A. (2015). CT-definable subtypes of chronic obstructive pulmonary disease: A statement of the Fleischner society. *Radiology*, *277*(1), 192–205.

Ma, Q. (2013). Role of nrf2 in oxidative stress and toxicity. *Annu Rev Pharmacol Toxicol*, *53*, 401–426.

Morris, A., Gingo, M. R., George, M. P., Lucht, L., Kessinger, C., Singh, V., Hillenbrand, M., Busch, M., McMahon, D., Norris, K. A., Champion, H. C., Gladwin, M. T., Zhang, Y., Steele, C., & Sciurba, F. C. (2012). Cardiopulmonary function in individuals with HIV infection in the antiretroviral therapy era. *Aids*, *26*(6), 731–740. https://doi.org/10.1097/QAD.0b013e32835099ae

Rennard, S. I. (1998). COPD: Overview of definitions, epidemiology, and factors influencing its development. *Chest*, *113*(4), 235S–241S.

Rochester, C. L., Akgün, K. M., Possick, J. D., Kapo, J. M., & Lee, P. J. (2022). Chronic Obstructive Pulmonary Disease. In J. B. Halter, J. G. Ouslander, S. Studenski, K. P. High, S. Asthana, M. A. Supiano, C. S. Ritchie, & K. Schmader (Eds.), *Hazzard's Geriatric Medicine and Gerontology* (8th ed.). McGraw-Hill Education. accessmedicine.mhmedical.com/content.aspx?aid=1190911509

Sam, A., & Kraft, M. (2022). Asthma-COPD overlap. *Curr Pulmonol Rep*, *11*(1), 1–14. https://doi.org/10.1007/s13665-021-00284-0

Stern, D. A., Morgan, W. J., Wright, A. L., Guerra, S., & Martinez, F. D. (2007). Poor airway function in early infancy and lung function by age 22 years: A non-selective longitudinal cohort study. *Lancet*, *370*(9589), 758–764. https://doi.org/10.1016/s0140-6736(07)61379-8

Vij, N., Chandramani-Shivalingappa, P., Van Westphal, C., Hole, R., & Bodas, M. (2018). Cigarette smoke-induced autophagy impairment accelerates lung aging, COPD-emphysema exacerbations and pathogenesis. *Am J Physiol Cell Physiol*, *314*(1), C73–C87.

Vogelmeier, C. F., Criner, G. J., Martinez, F. J., Anzueto, A., Barnes, P. J., Bourbeau, J., Celli, B. R., Chen, R., Decramer, M., & Fabbri, L. M. (2017). Global strategy for the diagnosis, management, and prevention of chronic obstructive lung disease 2017 report. GOLD executive summary. *Am J Respir Crit Care Med*, *195*(5), 557–582.

Yakar, H. I., Gunen, H., Pehlivan, E., & Aydogan, S. (2017). The role of tuberculosis in COPD. *Int J Chron Obstruct Pulmon Dis*, *12*, 323–329. https://doi.org/10.2147/copd.S116086

2 Biomarkers for the Management of Chronic Obstructive Pulmonary Disease

Anurag Mishra
NIMS Institute of Pharmacy, NIMS University Rajasthan
Jaipur, Rajasthan, India

Sobhit Singh Rajput
MIPER, Mangalayatan University,
Beswan, Aligarh, Uttar Pradesh, India

Abhishek Dwivedi
Department of Pharmaceutics, Maharishi Arvind College of Pharmacy
Ambabari, Jaipur, Rajasthan, India

Yogendra Singh
Department of Pharmacology, Maharishi Arvind College of Pharmacy,
Ambabari, Jaipur, Rajasthan, India

2.1 INTRODUCTION

"Biomarkers are characteristics that are objectively assessed and analyzed as indicators for standard life processes, diseases, or drug responses to a treatment response," according to the National Institutes of Health (NIH) (De Gruttola et al., 2001). The majority of molecular biomarkers come from blood. Biomarker measurement has made use of buffy coat, serum, plasma, and the separation of certain kinds of blood cells. Other bodily fluids for which biomarkers have been created include urine, with a focus on acute kidney injury (AKI) biomarkers. When it comes to diagnosing diseases of the brain, spinal cord, and meninges, clinically valuable biomarkers can be found in cerebrospinal fluid. For decades, clinicians have used ascites and pleural effusions as samples for potential new biomarkers associated with these organ systems. Biomarkers and DNA may be found in saliva, which is an intriguing and underused source (Lindsay & Costello, 2017). It is also non-invasive. Biomarker assessment has historically made use of virtually any biological sample that is readily available, including synovial fluid, surgical drain fluid, tissue biopsies, surgical specimens, and pretty much anything else.

Hospitalizations are significantly increased due to severe symptoms of chronic obstructive pulmonary disease (COPD). However, the choice to admit patients to the ICU is determined by their clinical presentation and the use of empiric medication. Establishing reliable biomarkers is essential for COPD screening and intervention. As a result of biomarkers, not only will diseases be more efficiently diagnosed, but phenotypic differences can be better understood, disease severity can be better stratified, and COPD prognosis can be better monitored (Zemans et al., 2017).

 DOI: 10.1201/9781003498926-2

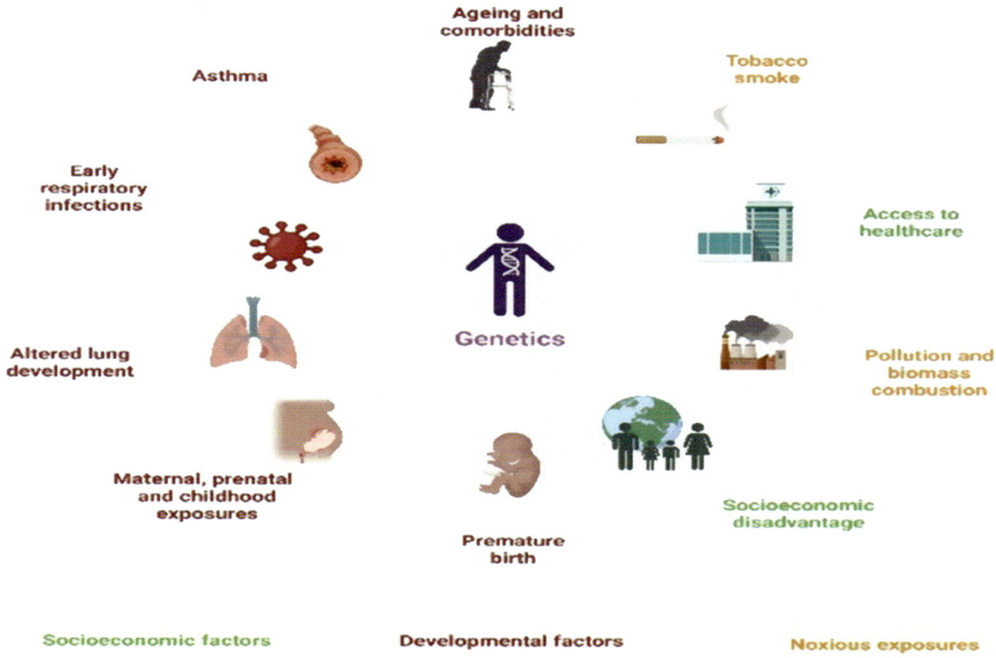

FIGURE 2.1 Chronic obstructive pulmonary disease diagnostic variables.

Many COPD biomarkers have been studied by researchers to better understand the disease's physiology and to treat patients (Figure 2.1). Blood or plasma fibrinogen, hypercapnia, sputum purulence, and blood C-reactive protein (CRP) (Ramakrishnan & Bafadhel, 2022), interferons (INFs), tumor necrosis factor-alpha (TNF-α), interleukins (ILs), eosinophil cationic protein, leukotrienes (LTs), eosinophil peroxidase, myeloperoxidase, and exhaled nitric oxide (eNO), leptin, resistin, chemerin and fetuin A (Małujło-Balcerska et al., 2023), and Phospholipase A2 group VII (PLA2G7) (Zhao et al., 2023) are some of the robust validated biomarkers that are now available. Furthermore, the lack of sensitivity and specificity in COPD and asthma biomarkers has prevented them from being used in clinical treatment. Having said that, there isn't a single biomarker that can capture all of COPD's facets. Instead, the focus should be on the validity, reliability, and evaluation circumstances of possible biomarkers.

Prior research in COPD patient cohorts has evaluated the diagnostic, phenotypic, and prognostic utility of a large number of blood biomarkers. COPD patients have been reported to have higher levels of many inflammatory markers, the majority of which are CRP and IL-6 (Fabbri & Rabe, 2007; Gan et al., 2004; Heidari, 2012). Table 2.1 summarizes the most important findings from this research, which are a number of connections. Because these results could point to clinical factors helpful for the integrated evaluation of the COPD patient, they should be considered. There does not appear to be a strong correlation between the observed connections and important clinical outcomes of the condition, and these relationships are inconsistent and not always repeatable (Hollander et al., 2017). Therefore, there is currently no blood biomarker that demonstrates performance features that enable definitive clinical translation and pharmacological recommendations in patients with COPD. For this reason, there is still a need for COPD-specific serum/plasma biomarkers.

TABLE 2.1
Showcasing the Several Bloodstream Biomarkers for Diagnosis and Treatment of COPD.

Biomarkers	Main Findings
C-reactive protein (CRP)	CRP has been employed to anticipate biomarkers and measure the level of inflammation among those in COPD. Elevated levels of CRP were seen in both patients with acute exacerbation-COPD and those with stable COPD (Heidari, 2012).
MicroRNAs	In lung cells, miR-233 regulates HDAC2 expression and activity (Houshmandfar et al., 2021). MiR-203 reduces smokers' immunological responses, which may cause COPD (Shi et al., 2015).
Extracellular vesicles (EVs)	Patients with emphysema showed increased levels of CD31+ EVs, indicating endothelial cell death. Elevated levels of CD62E+ EVs, which are a sign of endothelial activation, were seen in individuals with severe COPD and hyperinflation (Thomashow et al., 2013). Elevated levels of CD62E+ EVs at baseline may suggest that individuals with COPD are more prone to experiencing exacerbations (Takahashi et al., 2012).
Fibrinogen	COPD mortality was substantially linked with fibrinogen levels (Danesh et al., 2005).
PLA2G7	The level of PLA2G7 mRNA were found to be increased in individuals with COPD (Zhao et al., 2023).
Ratios of white blood cell count and its precursor cells	In COPD and exacerbated COPD, there is an elevated ratio of neutrophils to lymphocytes, platelets to lymphocytes, and monocytes to lymphocytes (Cai et al., 2024; Faria et al., 2016; Xiang et al., 2017).
Thioredoxin	Mitigates COPD lung damage by regulating both inflammation and oxidative stress through transcription elements such as PI3-APK and NF-kB, among others (Kansal et al., 2023).

2.2 CURRENT STATUS OF BIOMARKERS IN COPD

Numerous COPD biomarkers have been the subject of extensive study in the past few decades, and they are approaching readiness for application in healthcare. The more rigorously validated biomarkers are mentioned in the next section. It is important to mention that the authors specifically concentrate on biomarkers that can be easily tested in the systemic circulation due to their convenient accessibility. Other potentially relevant samples, such as those found in sputum or inhaled air, are not addressed in this discussion.

2.3 A HIGHLY INTRIGUING BIOMARKER: FIBRINOGEN

The Food and Drug Administration (FDA) of the United States is actively evaluating fibrinogen, a particularly advanced and anticipated blood biological marker in COPD, towards credentials (Duvoix et al., 2013). In addition with age, smoking, and lung health, plasma fibrinogen has been linked to COPD vulnerability, progressing disease, and mortality (total and disease specific) (Duvoix et al., 2013). The correlation with death is substantial and noticeable in both COPD and normal population groups. For instance, the largest meta-analysis to date found a 3.7-fold increase in the likelihood of COPD-specific mortality. This study included 154,211 patients from 31 prospective research projects with a total of 1.38 million person-years of follow-up (Danesh et al., 2005). Compared to current or past smokers, this meta-analysis discovered a statistically significant association between plasma fibrinogen levels and COPD fatality between individuals who had never smoked (hazard ratio of 5.5 vs. 3.7 for all participants).

The Copenhagen City Heart Study also showed a higher likelihood of hospitalization for COPD in those with elevated levels of plasma fibrinogen (Dahl et al., 2001). As with other diagnostic biomarkers, the results on plasma fibrinogen levels for exacerbation have been inconsistent and contradictory. There is conflicting evidence about the relationship between plasma fibrinogen levels and acute exacerbations; some studies have found an increase during these episodes (Koutsokera et al., 2009; Polatli et al., 2008), while others have failed to detect a change (Valipour et al., 2008). Both positive and negative studies have been hindered by small sample numbers and muddled by the inclusion of therapeutic therapies such systemic corticosteroids and antibiotics. To further validate or disprove fibrinogen as a diagnostic biomarker for exacerbation, more research is required.

2.4 IMMUNE-RELATED BIOMARKER IN COPD

Several inflammatory disorders, including COPD, are thought to be influenced by PLA2G7, a powerful molecule with both pro- and anti-inflammatory properties (Deng et al., 2021). Furthermore, the same study explored notable associations seen among Lp-PLA2 levels and BMI, forced expiratory volume in one second (FEV1)/forced vital capacity (FVC), FFMI, CAT and mMRC score, and a six-minute walk distance in individuals with COPD. Consistent with prior studies, a separate study discovered that the levels of Lp-PLA2 in the blood serum and the levels of PLA2G7 mRNA were noticeably increased in patients with COPD when compared to healthy individuals (Zhao et al., 2023). Nevertheless, there was no discernible disparity between smokers with normal lung function and non-smokers who were in good health. This finding suggests that the levels of PLA2G7 are not influenced by cigarette smoke. The processes that were previously documented included the recruitment and induction of macrophages. Our findings point to PLA2G7 as a potential immunological biomarker that promotes COPD development by activating myeloid-derived suppressor cells expansion and suppression. Possible effective treatment for COPD could involve targeting PLA2G7.

2.5 RATIOS OF NEUTROPHILS TO LYMPHOCYTES, PLATELETS TO LYMPHOCYTES, AND MONOCYTES TO LYMPHOCYTES AS BIOMARKERS IN COPD

Lymphocytes and neutrophils are important markers seen in complete blood count (CBC) when it comes to inflammation. In several disorders, CBC levels can represent a systemic inflammatory response (Song et al., 2021). An indicator of systemic inflammation is the existence of neutrophil-lymphocyte ratio (NLR) (Faria et al., 2016). Also, certain systemic inflammatory diseases have been found to be predictors of the platelet-lymphocyte ratio (PLR) (Cai et al., 2024). In addition, the monocyte-lymphocyte ratio (MLR) has recently emerged as a useful indicator for a variety of cancer patients (Xiang et al., 2017). Recent research has shown that three clinically useful indices obtained from CBC tests can aid in the diagnosis of infection and the progression of inflammation, and they can serve as independent risk factors (Luo et al., 2022; Martini et al., 2020; Wang et al., 2022). While several studies have looked at how well certain CBC parameters may diagnose COPD, there hasn't been much study into how CBC indicators relate to clinical diagnosis evaluations of COPD patients. The study indicated that the levels of neutrophils, monocytes and leukocytes were considerably greater in the COPD and exacerbated-COPD groups when contrasted with the control group. On the other hand, COPD patients' immune systems are compromised due to the long-term use of steroid medicines like budesonide. In light of their immunosuppressed status, individuals with exacerbated-COPD may not have a enough of or a substantial boost in the total amount of leukocytes, neutrophils, and monocytes to adequately clear infections.

When comparing stable-COPD with exacerbated-COPD in terms of leukocytes, neutrophils, and monocytes, the one-way ANOVA test revealed no statistically significant difference (Cai et al., 2024).

2.6 CRP AND COPD EXACERBATION

Significant and intermediate airflow restriction was linked to a higher incidence of ischemia alterations on electrocardiograms in COPD patients who were older than 50 according to the TNHNE Survey. The circulating CRP and fibrinogen levels were strongly linked with the extent of airflow obstruction (Bircan et al., 2008). Thus, serum CRP levels may serve as a predictor of lung function volumes, such as FEV1 and other related metrics (Wu et al., 2005). Measuring serum CRP adds more evidence when it comes to verifying a COPD exacerbation because of how sensitive it is to changes brought on by the condition. An elevated CRP level was associated with a 72.5% sensitivity and a 100% specificity for a serious flare-up in a COPD patient during a study by Bircan et al. (2008). Leukocytosis and purulent sputum were associated with elevated serum CRP levels, whereas mucoid sputum and leukocytosis-free patients had lower levels, respectively.

2.7 MicroRNAs IN COPD

MicroRNAs (miRNAs) have recently gained attention as a promising diagnostic tool for a range of lung disorders, including asthma (Soccio et al., 2023), pulmonary tuberculosis (TB) (Li et al., 2023), lung cancer (Mohamed et al., 2024), and other similar conditions. New evidence suggests a strong link between miRNA and COPD. Recent research has highlighted the critical involvement of miRNAs in COPD pathogenesis. These miRNAs regulate important pathways for communications such Wnt, JAK-STAT, MAPK and NF-κB, which in turn aid in the formation, maturation, and maintenance of pulmonary function. Bronchial epithelial cell death and proliferation brought on by persistent inflammation is an important step in COPD development (Ahmadi et al., 2023; Kim et al., 2022; Purohit et al., 2023). Key participants in airway inflammation include several miRNAs that target genes associated with inflammation and enhance pathways such as TLR and NOD-like receptor pathways (Dang et al., 2017). These miRNAs include miR-93-5p, miR-1273-3p, miR-320a/b, and miR-24-3p. Additionally, studies have shown that miR-233 regulates the levels of activity and expression of HDAC2 in lung cells, which affects airway inflammation in COPD (Houshmandfar et al., 2021). Izzotti et al. investigated 126 miRNAs that were down-regulated in cigarette smoke-exposed rats' lungs. These miRNAs mainly control things like cell proliferation, angiogenesis, stress response, and apoptosis, so they could be relevant to how COPD damage and repair work (Izzotti et al., 2009). Pathogenic factors in COPD also include immune dysregulation. Through its targeting of the TAK1 and PIK3CA genes, Shi et al. showed that miR-203 inhibits the immunological response. They found evidence that miR-203 suppresses the immune response in smokers, which may lead to COPD development and worsening (Shi et al., 2015). Reports have indicated that regulatory T cells derived from COPD patients have notable changes in miRNA expression patterns. One such regulator of the Th1–Th17 balance and a potential contributor to COPD is miR-199a-5p, which is exclusive to regulatory T cells. Other important factors in COPD development include autophagy and cell death (Chatila et al., 2014). Tobacco smoke causes human bronchial epithelial cells to undergo autophagy and apoptosis, and miR-21 plays a role in this process (Zeng et al., 2018). To add insult to injury, human pulmonary microvascular endothelial cells undergo a dramatic upregulation of cell death with miR-34a overexpression. Importantly, human pulmonary microvascular endothelial cells are protected against apoptosis caused by increased miR-34a when Notch1 is overexpressed (Song et al., 2021).

2.8 EXTRACELLULAR VESICLES AS POTENTIAL COPD BIOMARKERS

Recent research has looked at extracellular vesicles (EVs) as a possible diagnostic and prognostic biomarker for COPD, with applications including defining the COPD phenotype and differentiating between stable and exacerbation states of the disease (Reid et al., 2021). When endothelial cells undergo death or activation, they release membrane particles called EVs into the systemic circulation. Isolation of EVs has been accomplished using a variety of bodily fluids, such as urine, blood, and bronchoalveolar lavage. EVs produce a number of endothelial cell markers that are stimulus-specific. To be more precise, endothelial cell activation is associated with CD62E expression and endothelial cell death with CD31+ EVs. In addition, chronic damage is linked to CD51 expression (Thomashow et al., 2013). Research has shown a correlation between CD31+ EVs and reduced diffusing capacity of the lungs (DLCO), or carbon monoxide diffusing capability. A negative correlation between FEV1 and EVs expressing CD31+ is also seen. The severity of COPD was associated with elevated levels of CD31+ EVs in COPD patients. Imaging studies have linked CD31+ EVs to the emphysematous COPD phenotype. On the other hand, lung hyperinflation was associated with an increase in EVs expressing CD62E+, which was seen exclusively in patients with severe COPD. Regular acute exacerbations of COPD episodes and exacerbation proneness were associated with higher CD62E+ EVs (E-selectin). High quantities of CD51+ EVs were found in COPD patients. Nevertheless, DLCO was not associated with CD51+ or CD62E+ EVs (Thomashow et al., 2013). In conclusion, EVs as biomarkers for COPD diagnosis and treatment need more investigation.

2.9 CONCLUSIONS

Notably, biomarkers are not utilized nearly as frequently in healthcare as they formerly were. But there is optimism that our expanding dataset will result in improved decision-making and outcomes, particularly given the exponential growth of biomarker research. It is possible to enhance patient care by ongoing practice review and the implementation of these new ideas and studies.

Modern COPD treatments primarily aim to alleviate the symptoms experienced by patients. This highlights the need for more study into the etiology of COPD to identify and perhaps treat the underlying mechanisms. More individualized care for people with COPD may be possible if we better understand the disease heterogeneity across COPD patients, particularly with regard to COPD severity, progression, comorbidities, and clinical status. In particular, in the future, it may be quite practical to measure and evaluate each patient's own biomarker panel.

Additional study into the biological reasons of COPD exacerbations and worsenings, as well as the development of highly effective therapies with low on- and off-target side effects, is urgently needed in view of the rising number of COPD cases and deaths. Hence, new, safer, and more successful treatment techniques need to be developed. Table 2.1 highlights that thioredoxin (TRX) reduces lung damage in COPD via a number of pathways, including modulating oxidative stress and inflammation via transcription factors including PI3-APK and NF-kB (Kansal et al., 2023). Given the limitations of existing COPD medications and the potential adverse effects, including as immunosuppression, further investigation into these pathways involving TRX is necessary before its use as a therapy. Contrarily, TRX has shown in several studies that it can reduce CS-induced oxidative stress and inflammation, thereby attenuating emphysematous changes in COPD animal models. This effect is observed even in animals that corticosteroids are unable to manage. Consequently, future research can test the efficacy of TRX during acute exacerbations. Thus, future prospects may arise from studies investigating TRX as a therapy for COPD. Furthermore, it is possible to reduce variability and increase benefit through the creation and implementation of biomarker-based recommendations and procedures in the future. The biomarker age will undoubtedly revolutionize and mold the course of medical history via well-planned research and teamwork.

REFERENCES

Ahmadi, A., Ahrari, S., Salimian, J., Salehi, Z., Karimi, M., Emamvirdizadeh, A., Jamalkandi, S. A., & Ghanei, M. (2023). p38 MAPK signaling in chronic obstructive pulmonary disease pathogenesis and inhibitor therapeutics. *Cell Commun Signal*, *21*(1), 314. https://doi.org/10.1186/s12964-023-01337-4

Bircan, A., Gokirmak, M., Kilic, O., Ozturk, O., & Akkaya, A. (2008). C-reactive protein levels in patients with chronic obstructive pulmonary disease: Role of infection. *Med Princ Pract*, *17*(3), 202–208. https://doi.org/10.1159/000117793

Cai, C., Zeng, W., Wang, H., & Ren, S. (2024). Neutrophil-to-lymphocyte ratio (NLR), platelet-to-lymphocyte ratio (PLR) and monocyte-to-lymphocyte ratio (MLR) as biomarkers in diagnosis evaluation of acute exacerbation of chronic obstructive pulmonary disease: A retrospective, observational study. *Int J Chron Obstruct Pulmon Dis*, *19*, 933–943. https://doi.org/10.2147/copd.S452444

Chatila, W. M., Criner, G. J., Hancock, W. W., Akimova, T., Moldover, B., Chang, J. K., Cornwell, W., Santerre, M., & Rogers, T. J. (2014). Blunted expression of miR-199a-5p in regulatory T cells of patients with chronic obstructive pulmonary disease compared to unaffected smokers. *Clin Exp Immunol*, *177*(1), 341–352. https://doi.org/10.1111/cei.12325

Dahl, M., Tybjaerg-Hansen, A., Vestbo, J., Lange, P., & Nordestgaard, B. G. (2001). Elevated plasma fibrinogen associated with reduced pulmonary function and increased risk of chronic obstructive pulmonary disease. *Am J Respir Crit Care Med*, *164*(6), 1008–1011. https://doi.org/10.1164/ajrccm.164.6.2010067

Danesh, J., Lewington, S., Thompson, S., Lowe, G., Collins, R., Kostis, J., Wilson, A., Folsom, A., Wu, K., & Benderly, M. (2005). Fibrinogen studies collaboration: Plasma fibrinogen level and the risk of major cardiovascular diseases and nonvascular mortality: An individual participant meta-analysis. *JAMA*, *294*(14), 1799–1809.

Dang, X., Qu, X., Wang, W., Liao, C., Li, Y., Zhang, X., Xu, D., Baglole, C. J., Shang, D., & Chang, Y. (2017). Bioinformatic analysis of microRNA and mRNA regulation in peripheral blood mononuclear cells of patients with chronic obstructive pulmonary disease. *Respir Res*, *18*(1), 4. https://doi.org/10.1186/s12931-016-0486-5

De Gruttola, V. G., Clax, P., DeMets, D. L., Downing, G. J., Ellenberg, S. S., Friedman, L., Gail, M. H., Prentice, R., Wittes, J., & Zeger, S. L. (2001). Considerations in the evaluation of surrogate endpoints in clinical trials: Summary of a National Institutes of Health workshop. *Control Clin Trials*, *22*(5), 485–502. https://doi.org/10.1016/s0197-2456(01)00153-2

Deng, M., Yin, Y., Zhang, Q., Zhou, X., & Hou, G. (2021). Identification of inflammation-related biomarker Lp-PLA2 for patients with COPD by comprehensive analysis. *Front Immunol*, *12*. https://doi.org/10.3389/fimmu.2021.670971

Duvoix, A., Dickens, J., Haq, I., Mannino, D., Miller, B., Tal-Singer, R., & Lomas, D. A. (2013). Blood fibrinogen as a biomarker of chronic obstructive pulmonary disease. *Thorax*, *68*(7), 670–676.

Fabbri, L. M., & Rabe, K. F. (2007). From COPD to chronic systemic inflammatory syndrome? *Lancet*, *370*(9589), 797–799. https://doi.org/10.1016/s0140-6736(07)61383-x

Faria, S. S., Fernandes, P. C., Jr., Silva, M. J., Lima, V. C., Fontes, W., Freitas-Junior, R., Eterovic, A. K., & Forget, P. (2016). The neutrophil-to-lymphocyte ratio: A narrative review. *Ecancermedicalscience*, *10*, 702. https://doi.org/10.3332/ecancer.2016.702

Gan, W. Q., Man, S. F., Senthilselvan, A., & Sin, D. D. (2004). Association between chronic obstructive pulmonary disease and systemic inflammation: A systematic review and a meta-analysis. *Thorax*, *59*(7), 574–580. https://doi.org/10.1136/thx.2003.019588

Heidari, B. (2012). The importance of C-reactive protein and other inflammatory markers in patients with chronic obstructive pulmonary disease. *Caspian J Intern Med*, *3*(2), 428–435.

Hollander, Z., DeMarco, M. L., Sadatsafavi, M., McManus, B. M., Ng, R. T., & Sin, D. D. (2017). Biomarker development in COPD: Moving from P values to products to impact patient care. *Chest*, *151*(2), 455–467. https://doi.org/10.1016/j.chest.2016.09.012

Houshmandfar, S., Saeedi-Boroujeni, A., Rashno, M., Khodadadi, A., & Mahmoudian-Sani, M.-R. (2021). miRNA-223 as a regulator of inflammation and NLRP3 inflammasome, the main fragments in the puzzle of immunopathogenesis of different inflammatory diseases and COVID-19. *Naunyn Schmiedebergs Arch Pharmacol*, *394*(11), 2187–2195. https://doi.org/10.1007/s00210-021-02163-6

Izzotti, A., Calin, G. A., Arrigo, P., Steele, V. E., Croce, C. M., & De Flora, S. (2009). Downregulation of microRNA expression in the lungs of rats exposed to cigarette smoke. *FASEB J*, *23*(3), 806–812. https://doi.org/10.1096/fj.08-121384

Kansal, H., Chopra, V., Garg, K., & Sharma, S. (2023). Role of thioredoxin in chronic obstructive pulmonary disease (COPD): A promising future target. *Respir Res*, *24*(1), 295. https://doi.org/10.1186/s12931-023-02574-4

Kim, H.-T., Panza, P., Kikhi, K., Nakamichi, Y., Atzberger, A., Guenther, S., Ruppert, C., Guenther, A., & Stainier, D. Y. R. (2022). WNT/RYK signaling functions as an antiinflammatory modulator in the lung mesenchyme. *Proc Natl Acad Sci U S A*, *119*(24), e2201707119. https://doi.org/10.1073/pnas.2201707119

Koutsokera, A., Kiropoulos, T. S., Nikoulis, D. J., Daniil, Z. D., Tsolaki, V., Tanou, K., Papaioannou, A. I., Germenis, A., Gourgoulianis, K. I., & Kostikas, K. (2009). Clinical, functional and biochemical changes during recovery from COPD exacerbations. *Respir Med*, *103*(6), 919–926. https://doi.org/10.1016/j.rmed.2008.12.006

Li, X., Xu, Y., & Liao, P. (2023). Diagnostic performance of microRNA-29a in active pulmonary tuberculosis: A systematic review and meta-analysis. *Clinics (Sao Paulo)*, *78*, 100290. https://doi.org/10.1016/j.clinsp.2023.100290

Lindsay, A., & Costello, J. T. (2017). Realising the potential of urine and saliva as diagnostic tools in sport and exercise medicine. *Sports Med*, *47*(1), 11–31. https://doi.org/10.1007/s40279-016-0558-1

Luo, S., Yang, W. S., Shen, Y. Q., Chen, P., Zhang, S. Q., Jia, Z., Li, Q., Zhao, J. T., & Xie, P. (2022). The clinical value of neutrophil-to-lymphocyte ratio, platelet-to-lymphocyte ratio, and D-dimer-to-fibrinogen ratio for predicting pneumonia and poor outcomes in patients with acute intracerebral hemorrhage. *Front Immunol*, *13*, 1037255. https://doi.org/10.3389/fimmu.2022.1037255

Małujło-Balcerska, E., Kumor-Kisielewska, A., & Śmigielski, W. (2023). Leptin, resistin and fetuin a concentration as the potential useful biomarkers in stable COPD – An exploratory study. *Cytokine*, *169*, 156275. https://doi.org/10.1016/j.cyto.2023.156275

Martini, D. J., Liu, Y., Shabto, J. M., Carthon, B. C., Hitron, E. E., Russler, G. A., Caulfield, S., Kissick, H. T., Harris, W. B., Kucuk, O., Master, V. A., & Bilen, M. A. (2020). Novel risk scoring system for patients with metastatic renal cell carcinoma treated with immune checkpoint inhibitors. *Oncologist*, *25*(3), e484–e491. https://doi.org/10.1634/theoncologist.2019-0578

Mohamed, E., García Martínez, D. J., Hosseini, M. S., Yoong, S. Q., Fletcher, D., Hart, S., & Guinn, B. A. (2024). Identification of biomarkers for the early detection of non-small cell lung cancer: A systematic review and meta-analysis. *Carcinogenesis*, *45*(1–2), 1–22. https://doi.org/10.1093/carcin/bgad091

Polatli, M., Çakir, A., Cildag, O., Bolaman, A. Z., Yenisey, C., & Yenicerioglu, Y. (2008). Microalbuminuria, von Willebrand factor and fibrinogen levels as markers of the severity in COPD exacerbation. *J Thromb Thrombolysis*, *26*(2), 97–102. https://doi.org/10.1007/s11239-007-0073-1

Purohit, M., Gupta, G., Afzal, O., Altamimi, A. S. A., Alzarea, S. I., Kazmi, I., Almalki, W. H., Gulati, M., Kaur, I. P., Singh, S. K., & Dua, K. (2023). Janus kinase/signal transducers and activator of transcription (JAK/STAT) and its role in lung inflammatory disease. *Chem Biol Interact*, *371*, 110334. https://doi.org/10.1016/j.cbi.2023.110334

Ramakrishnan, S., & Bafadhel, M. (2022). Biomarkers in COPD. In S. M. Janes (Ed.), *Encyclopedia of Respiratory Medicine* (2nd ed., pp. 559–572). Academic Press. https://doi.org/10.1016/B978-0-08-102723-3.00257-2

Reid, L. V., Spalluto, C. M., Watson, A., Staples, K. J., & Wilkinson, T. M. A. (2021). The role of extracellular vesicles as a shared disease mechanism contributing to multimorbidity in patients with COPD. *Front Immunol*, *12*, 754004. https://doi.org/10.3389/fimmu.2021.754004

Shi, L., Xin, Q., Chai, R., Liu, L., & Ma, Z. (2015). Ectopic expressed miR-203 contributes to chronic obstructive pulmonary disease via targeting TAK1 and PIK3CA. *Int J Clin Exp Pathol*, *8*(9), 10662–10670.

Soccio, P., Moriondo, G., Lacedonia, D., Tondo, P., Pescatore, D., Quarato, C. M. I., Carone, M., Foschino Barbaro, M. P., & Scioscia, G. (2023). MiRNA and exosomal miRNA as new biomarkers useful to phenotyping severe asthma. *Biomolecules*, *13*(10). https://doi.org/10.3390/biom13101542

Song, Q., Chen, P., & Liu, X. M. (2021). The role of cigarette smoke-induced pulmonary vascular endothelial cell apoptosis in COPD. *Respir Res*, *22*(1), 39. https://doi.org/10.1186/s12931-021-01630-1

Takahashi, T., Kobayashi, S., Fujino, N., Suzuki, T., Ota, C., He, M., Yamada, M., Suzuki, S., Yanai, M., Kurosawa, S., Yamaya, M., & Kubo, H. (2012). Increased circulating endothelial microparticles in COPD patients: A potential biomarker for COPD exacerbation susceptibility. *Thorax*, *67*(12), 1067–1074. https://doi.org/10.1136/thoraxjnl-2011-201395

Thomashow, M. A., Shimbo, D., Parikh, M. A., Hoffman, E. A., Vogel-Claussen, J., Hueper, K., Fu, J., Liu, C. Y., Bluemke, D. A., Ventetuolo, C. E., Doyle, M. F., & Barr, R. G. (2013). Endothelial microparticles in mild chronic obstructive pulmonary disease and emphysema. The multi-ethnic study of atherosclerosis chronic obstructive pulmonary disease study. *Am J Respir Crit Care Med*, *188*(1), 60–68. https://doi.org/10.1164/rccm.201209-1697OC

Valipour, A., Schreder, M., Wolzt, M., Saliba, S., Kapiotis, S., Eickhoff, P., & Burghuber, O. C. (2008). Circulating vascular endothelial growth factor and systemic inflammatory markers in patients with stable and exacerbated chronic obstructive pulmonary disease. *Clin Sci*, *115*(7), 225–232. https://doi.org/10.1042/cs20070382

Wang, X., Ni, Q., Wang, J., Wu, S., Chen, P., & Xing, D. (2022). Systemic inflammation response index is a promising prognostic marker in elderly patients with heart failure: A retrospective cohort study. *Front Cardiovasc Med*, *9*, 871031. https://doi.org/10.3389/fcvm.2022.871031

Wu, S. J., Chen, P., Jiang, X. N., & Liu, Z. J. (2005). C-reactive protein level and the correlation between lung function and CRP levels in patients with chronic obstructive pulmonary diseases. *Zhong Nan Da Xue Xue Bao Yi Xue Ban*, *30*(4), 444–446.

Xiang, J., Zhou, L., Li, X., Bao, W., Chen, T., Xi, X., He, Y., & Wan, X. (2017). Preoperative monocyte-to-lymphocyte ratio in peripheral blood predicts stages, metastasis, and histological grades in patients with ovarian cancer. *Transl Oncol*, *10*(1), 33–39. https://doi.org/10.1016/j.tranon.2016.10.006

Zemans, R. L., Jacobson, S., Keene, J., Kechris, K., Miller, B. E., Tal-Singer, R., & Bowler, R. P. (2017). Multiple biomarkers predict disease severity, progression and mortality in COPD. *Respir Res*, *18*(1), 117. https://doi.org/10.1186/s12931-017-0597-7

Zeng, Z., He, S., Lu, J., Liu, C., Lin, H., Xu, C., Xie, L., & Sun, S. (2018). MicroRNA-21 aggravates chronic obstructive pulmonary disease by promoting autophagy. *Exp Lung Res*, *44*(2), 89–97. https://doi.org/10.1080/01902148.2018.1439548

Zhao, X., Yue, Y., Wang, X., & Zhang, Q. (2023). Bioinformatics analysis of PLA2G7 as an immune-related biomarker in COPD by promoting expansion and suppressive functions of MDSCs. *Int Immunopharmacol*, *120*, 110399. https://doi.org/10.1016/j.intimp.2023.110399

3 Current Treatment Modalities for Chronic Obstructive Pulmonary Disease

Challenges and Limitations

Mourvi Sachin Kulkarni, Nitish Kumar Tiwary, Anirudhha Pravin Nilkari, Chinmay Devidas Deshmukh, and Shvetank Bhatt
School of Health Sciences and Technology,
Dr. Vishwanath Karad MIT World Peace University
Pune, Maharashtra, India

3.1 INTRODUCTION

Chronic obstructive pulmonary disease COPD is a progressive lung condition that involves narrowing of bronchial tubes in the lungs, making it difficult to breathe. When this condition occurs, it is usually long lasting (chronic) and worsens over time. COPD is an umbrella term that typically involves emphysema, chronic bronchitis and bronchial asthma (Toelle et al., 2013). More precisely, emphysema affects the air sacs in the lungs and the walls between them, damaging them and making them less elastic. In bronchitis, the lining of the airway is constantly irritated, inflamed and swollen with an excess production of mucus, whereas asthma results in the tightening and swelling of the muscles in the airway causing it to narrow. COPD is an important public health challenge and is a major cause of chronic morbidity and mortality throughout the world, affecting both men and women. It is the fourth leading cause of death in the world currently (Ho et al., 2019).

COPD is characterized by persistent respiratory symptoms with airflow limitations caused by exposure to a noxious stimulus or gas. Tobacco smoking, outdoor, indoor and occupational air pollution, family history, genetics, lack of lung functions, airway hyper responsiveness and poor lung growth are the risk factors for COPD. Non-smokers may also develop COPD due to due to a variety of such risk factors (Ortiz-Quintero et al., 2022). The etiology and pathophysiology of COPD are shown in Figure 3.1.

When a patient presents in a hospital with the COPD symptoms, the case is further confirmed by diagnostic tests. The pulmonologist assess the patient depending on medical history, risk factors, addiction, occupation or any abnormalities in the airway followed by spirometry and chest X-ray. Airflow limitation is typically revealed by such tests (Johns et al., 2014; Vestbo et al., 2013). The severity of COPD is also assessed by measuring the forced expiratory volume (FEV), Forced Vital Capacity (FVC) or Medical Research Council (MRC) dyspnea scale (Bailey, 2012). In addition, certain typical features are observed in COPD, asthma, emphysema and bronchiectasis, which can be differentiated as shown in Table 3.1.

Management of COPD is based on removal of the underlying risk factors that contribute substantially to the progression of COPD. They are cessation of smoking, use of tobacco, and awareness of these risk factors among the patients (Andreas et al., 2009). The patients are also counseled, and called for follow-up training sessions. Pharmacological therapy is used to handle mild to moderate

DOI: 10.1201/9781003498926-3

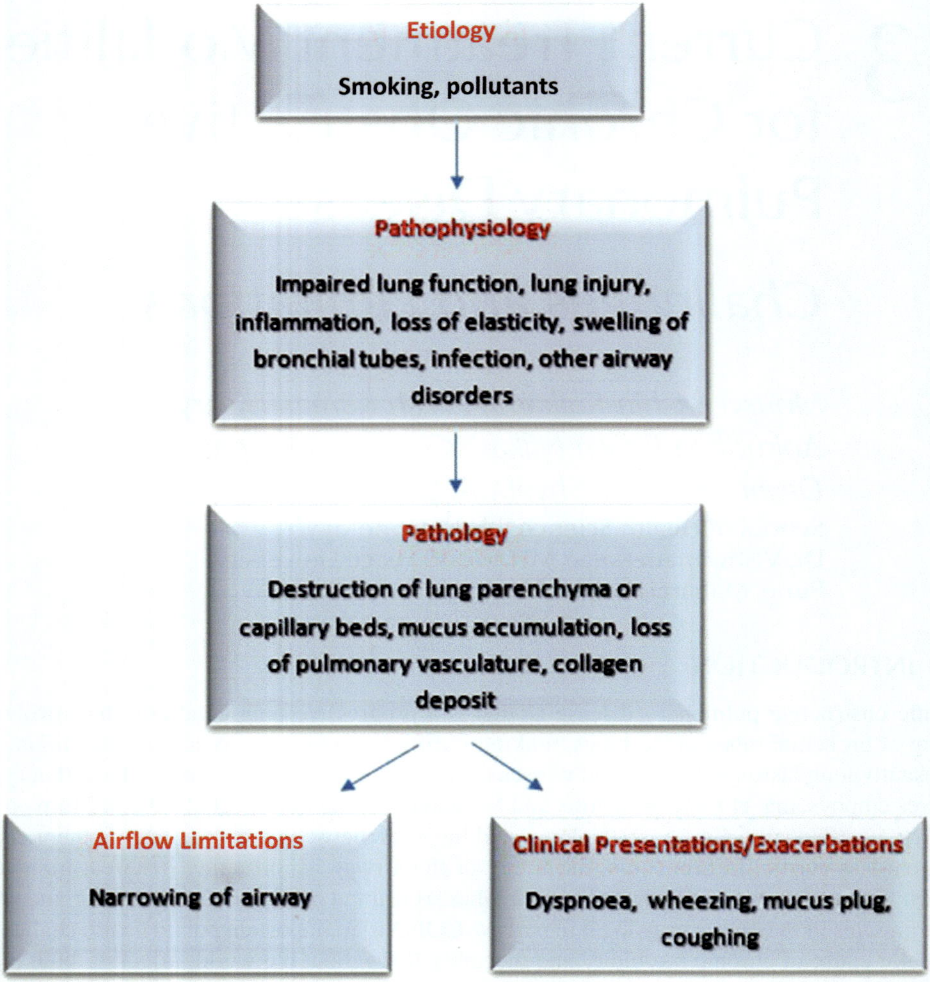

FIGURE 3.1 Etiology and pathophysiology of chronic obstructive pulmonary disease leading to airflow limitations.

patients by helping to enhance health status and activity tolerance while reducing symptoms and the frequency and intensity of exacerbations. For each patient, the pharmacological treatment regime is designed considering age, risk factors, co-morbidities, drug availabilities, patient response and delivery devices and methods. Solid dosage forms and inhalers are the most commonly prescribed drugs for management of COPD (Price et al., 2018). Oxygen therapy, non-invasive ventilation and surgical or bronchoscopic interventional treatments benefit patients with severe forms of these diseases that are further complicated by chronic hypoxemia, hypercapnia, loss of natural respiration or advanced emphysema. These treatments will reduce future hospitalization and mortality among such patients. Additionally, pulmonary rehabilitation enhances quality of life, participation in daily activities both physically and emotionally, and also improves the signs and symptoms of patient. Supportive therapy is also given, which includes vaccinations that reduce the changes of respiratory tract infections (Moreno et al., 2015).

This chapter discusses the advantages and disadvantages of current available pharmacotherapy and their combinations for management and treatment of COPD.

TABLE 3.1
Differential Diagnosis of COPD Patients

Diagnosis	Clinical Manifestations
COPD	• Onset takes place in mid-life. • Symptoms are slowly progressive. • Tobacco smoking history.
Asthma	• Early onset mostly during childhood. • Variations in symptoms. These symptoms vary day by day. • Worsening of symptoms during night and early morning. • May present with allergies, rhinitis and/or eczema. • Asthma may worsen and can become COPD. Particularly in those with a family history. • Obesity as co-morbid factor
Bronchiectasis	• Purulent sputum with more volume. Linked more commonly with bacterial infections. • Chest X-ray/CT demonstrates dilatation of bronchi, thickening of bronchial wall.
Obliterative bronchiolitis	• Begins earlier in life; non-smokers. • May have a history of acute fume exposure or rheumatoid arthritis. • Observed following bone marrow or lung transplantation. CT shows areas that are hypodense during expiration.
Diffuse panbronchiolitis	• Greater in patients who are Asian. • Nearly all patients have chronic sinusitis, and the majority are male and non-smokers. • Diffuse tiny centrilobular nodular opacities and hyperinflation are seen on the chest X-ray and HRCT.

3.2 PATHOPHYSIOLOGY OF COPD

Airway restriction is the main feature of COPD, which is not reversible. It also has a specific feature known as inflammation of the respiratory tract. This inflammatory response has both innate and adaptive immunity components, which is triggered more in the presence of poisonous gas substances, specifically cigarette smoke (Rovina et al., 2013). The inflammatory component is more predominant and decides the severity of the disease and airway obstruction. It is observed that all cigarette smokers have certain types of lung inflammation, but the individuals who develop COPD have an unusual response to inhaling toxic substances (Laniado-Laborín, 2009). The respiratory system's hyper-responsiveness causes increased mucus output, emphysema (destroyed alveolar tissue), bronchitis (chronic, progressive inflammation of the bronchi), inflammation of the tiny airways and fibrosis (bronchiolitis). Furthermore, one of the distinctive characteristics of COPD is an increased resistance to airflow (Scichilone et al., 2006). The distinct characteristic of COPD is an increased quantity of cells, including neutrophils, macrophages and CD8+ T lymphocytes. These inflammatory cells are responsible for the growth of different inflammatory components that significantly contribute to airway blockage and inflammation (Barnes et al., 2003). The inflammatory pattern of COPD is markedly different from asthma. In addition, epithelial cells also have a significant role in the pathogenesis of COPD. Epithelial cells of airways and alveoli are activated by components of smoke and produce mediators of inflammation such as tumor necrosis factor-alpha (TNF-α), interleukin (IL)-1, granulocyte-macrophage colony-stimulating factor (GM-CSF) and IL-8. These mediators are important sources for the production of transforming growth factor-β (TGF- β) and may result in carcinoma of the bronchi and squamous metaplasia (Rovina et al., 2013). In addition, dendritic cells or antigen-presenting cells have an important role in the starting of the innate and adaptive immune response. These dendritic cells are activated after entry of foreign particles or antigens and the initiate activation and proliferation of various inflammatory and immune mediators such as macrophages, neutrophils and T and B lymphocytes. In addition, prostaglandins, leukotrienes, thromboxane, bradykinin and chemokines also have important roles in the pathological progression of COPD (MacNee, 2013; Yan et al., 2018).

3.2.1 Imbalance of Protease and anti-Protease

Imbalance in the activity or levels of proteases and anti-proteases is also considered an important factor in the progression of COPD. The imbalance occurs because of the decreased production or inactivation of anti-proteases and the increased activity of proteases (Pandey et al., 2017). The increased oxidative stress stimulates the inflammatory signaling cascade and obligates several inflammatory cells to release a protease combination and leads to inactivation of anti-proteases (Wang et al., 2020). Serine proteases, neutrophil elastase, MMP are the examples of some proteases (Twigg et al., 2015).

3.2.2 Oxidative Stress

Imbalance between oxidative stress parameters and anti-oxidant enzymes along with endogenous anti-oxidants is a hallmark for the pathological progression of COPD (Fischer et al., 2015). Excessive production of reactive oxygen species (ROS) leads to irreversible destruction of macromolecules such as DNA, lipids, proteins, etc. ROS further cause the activation of epithelial cells and macrophages to produce chemotactic molecules that recruit other important inflammatory cells such as monocytes, neutrophils and lymphocytes inside the lungs. Increased oxidative stress also causes an imbalance between proteases and anti-proteases. An increase in the level of the oxidative stress marker malondialdehyde (MDA) after the peroxidation of lipids is responsible for inducing pulmonary inflammation (Mittal et al., 2014). This iNOS pathway's upregulation is thought to be the primary cause of pulmonary hypertension and emphysema, both of which are associated with cigarette smoking. Patients with COPD have lower plasma levels of anti-oxidant enzymes, including glutathione peroxidase, catalase and superoxide dismutase. Furthermore, lowered glutathione levels are also caused by cigarette smoke (Domej et al., 2014).

3.3 TREATMENT OF COPD

3.3.1 Quit Smoking

Quitting smoking is an important non-pharmacological approach for the treatment of COPD. Cigarette smoke is one of the contributing factors of COPD. Various chemicals present in cigarettes, e-cigarettes and cigars can damage the lungs irreversibly and decrease lung function to a significant extent. Bupropion, a noradrenaline dopamine reuptake inhibitor, and varenicline, a partial agonist-antagonist that is highly selective for the predominant brain nicotinic acetylcholine receptor 42 subtype that mediates the reinforcing effects of nicotine, can be used to stop active smoking (Fagerström & Hughes, 2008; Mansvelder et al., 2007). In addition, one should avoid the triggers of COPD exacerbations, i.e., should avoid passive smoking, occupational hazards, etc.

3.3.2 Challenges and Limitations of Quitting Smoking and Nicotine Replacement Therapy

The effectiveness of nicotine replacement therapy is questionable. Alone, these therapies without self-motivation or help from family members and doctors are less effective, with effects seen in less than 5% of patients. Insomnia, headache, heart palpitations and difficulty in speech are the common adverse effects that limit sudden smoking cessation. In addition to these side effects, nicotine replacement therapy increases the chances of oral ulcers, skin irritation, taste impairment and hyperkeratosis (Devi et al., 2020).

3.4 BRONCHODILATORS

Bronchodilators work through beta-2 adrenergic receptors and increase the level of cyclic AMP, thereby causing bronchodilation. Salbutamol and terbutaline are two drugs that belong to this class of short-acting bronchodilators (SABA). Generally, these drugs are given with metered dose inhalers for precise dosing and targeted delivery to the affected area, i.e., bronchi and lungs. SABA work within minutes, but last only 4–6 hours. Another category of bronchodilator is the long-acting bronchodilator (LABA) that works through a similar mechanism to SABA, but these are long-acting and have more half-life. They may be slow at onset, but their duration of action is between 12 and 24 hours. Some important drugs in this category are salmeterol and formoterol (Maltais et al., 2020).

The main treatment for respiratory conditions like asthma and COPD is still bronchodilators. This explains the growing interest in finding new classes of bronchodilator medications as well as in determining how to best utilize the ones we have. Yet, it has been difficult to create new bronchodilator classes since many of them have unsatisfactory safety profiles and are not more effective than the current bronchodilator classes. According to recent research, optimal bronchodilation occurs when the autonomic nervous system is activated in two ways: by mimicking the sympathetic nervous system by stimulating β 2-adrenoceptors (β 2-ARs) on airway smooth muscle with β 2-AR-selective agonists, and by antagonizing muscarinic receptors to reduce the effect of parasympathetic innervation of the lung (Deeney et al., 2022).

3.5 ANTI-CHOLINERGIC DRUGS

Selective M_3-muscarinic receptor inhibition causes bronchodilation as well as a reduction in the production of mucus. To treat COPD, inhalational anti-cholinergics (with anti-muscarinic activity) such as ipratropium bromide, tiotropium bromide and oxitropium bromide are prescribed (Shah et al., 2023). The U.S. Food and Drug Administration (FDA) recently reviewed and authorized the new twice-daily anti-cholinergic medication aclidinium bromide for the treatment of stable chronic obstructive pulmonary disease (SPD) (Jones, 2013). The quaternary ammonium, presents in all currently prescribed drugs, restricts their ability to penetrate the blood–brain barrier and be absorbed systemically. In addition, it acts for a longer period of time than tertiary amines, which are their predecessors. Atropine shares structural similarities with ipratropium, tiotropium and aclidinium, but oxitropium is a congener of scopolamine (Moulton & Fryer, 2011). Because these drugs have less of an effect on the mucociliary clearance of the airway than the belladonna alkaloids, they avoid the potentially dangerous reduction of airway secretions (and subsequent inspissation of viscid material that leads to worsening obstruction and infection) seen with atropine and scopolamine. All of these drugs are delivered by inhalation.

Compared to ipratropium, oxitropium acts a little longer and is taken two to three times a day at a dose of 200 mcg (or 100 mcg every puff). It was not made commercially available in the United States, although it was marketed in Europe in a metered-dose inhaler (MDI) form. In 2004, oxitropium was stopped. Exercise capacity and FEV1 have been demonstrated to be enhanced by oxitropium, either by itself or in conjunction with an inhalational beta agonist (Zheng et al., 2006).

Tiotropium is taken at a dose of 18 mcg per day using a dry powder inhaler (DPI). Patients on tiotropium who have moderate to severe renal impairment (creatinine clearance $\leq$ 50 mL/min) should be continuously monitored for signs of anti-cholinergic side effects, as tiotropium is mostly eliminated via the kidneys (Anzueto & Miravitlles, 2020). Tiotropium has been thoroughly studied in numerous clinical trials. Tiotropium was found to enhance exercise capacity, lessen exacerbations of COPD, enhance lung function and enhance quality of life in the trials. Tiotropium and ipratropium did not slow the deterioration of lung function in long-term investigations (Barr et al., 2005).

3.6 COMBINATION OF BRONCHODILATORS AND MUSCARINIC ANTAGONIST

For the treatment of COPD, a combination of short-acting muscarinic antagonists (SAMA), ipratropium bromide, levalbuterol and salbutamol, is commonly utilized. Other combinations of SABA and SAMA, such as fenoterol/ipratropium and salbutamol/ipratropium, are also used for therapeutic purposes. Conversely, long-acting muscarinic antagonist (LAMA) is aclidinium bromide, glycopyrronium bromide and tiotropium, and LABA is arformoterol, formoterol, indacaterol and salmeterol. Formoterol/aclidinium, formoterol/glycopyrronium, indacaterol/glycopyrronium, vilanterol/umeclidinium and olodaterol/tiotropium are some other combinations of LABA and LAMA that are also commonly utilized (van Geffen et al., 2023).

3.7 CHALLENGES AND LIMITATIONS OF BRONCHODILATORS AND ANTI-MUSCARINIC DRUGS

The effectiveness and safety profile of inhalational anti-cholinergics are significantly influenced by their pharmacodynamics and pharmacokinetic properties. Compared to medicines administered orally or intravenously, all inhaled agents cause reduced systemic exposure. This results in a superior tolerance profile and a broader therapeutic index for all inhalational drugs. Since all the anticholinergics on the market now dissolve in water, their ability to cross biological membranes is restricted. They have a lower rate of systemic absorption and a lower likelihood of passing through the blood–brain barrier when inhaled.

Although LABAs and LAMAs play a crucial role in the treatment of COPD, there have been concerns expressed that exposure to these medications may result in unfavorable cardiovascular events. The pharmacological effects of LABAs and LAMAs are shown to extend themselves beyond the site of pulmonary application, particularly in the heart, suggesting a biological plausibility of adverse cardiovascular events from their use. Clinical trials have produced conflicting results despite the fact that numerous observational studies have demonstrated an elevated risk of adverse cardiovascular events associated with the use of LABAs and LAMAs. In addition, these bronchodilators alone are not very effective and must be used in combination with anti-cholinergic drugs and steroids (Wang et al., 2019).

3.8 STEROID

Depending on the stage of the disease, different guidelines apply to the administration of systemic steroids in COPD. Based on the results of multiple placebo-controlled trials conducted in secondary-care settings, oral steroids are recommended in moderate and severe acute exacerbations (Abroug et al., 2014). These studies demonstrate that systemic steroids enhance gas exchange, dyspnea and lung function. Steroid use also led to shorter hospital stays, a lower relapse rate and fewer treatment failures. Additionally, administering steroids orally is not worse than administering them intravenously, and an eight-week course of treatment is not better than a two-week treatment. It has even been verified recently that a five-day course is not inferior than a 14-day course.

However, there is limited evidence of systemic steroids having positive effects on stable COPD. While higher doses of prednisolone (≥ 30 mg/day) may temporarily improve lung function, doses of less than 10–15 mg/day did not show the same effects. The efficacy of 5 mg prednisolone plus 1600 μg budesonide per day in a two-year experiment did not show a significant improvement over budesonide alone in terms of lung function decline (Tashkin et al., 2019). There are also significant adverse effect issues. It is particularly well known that there are effects on respiratory and peripheral muscular strength. This may result in a lower survival rate and is brought on by a myopathy that exhibits atrophy and necrosis. Additionally, data suggest that a risk factor for the long-term use of systemic steroids is an increased risk of osteoporosis in COPD (Dekhuijzen & Decramer, 1992).

Large prospective trials have demonstrated some benefits for the use of corticosteroids in COPD, but patient-to-patient variability in efficacy appears to be significant. The usage of inhaled corticosteroid therapy (ICS) is justified by a variety of factors. It makes it possible to provide a medication directly to the intended organ, avoid systemic absorption and utilize less cumulative corticosteroid doses (Pandya et al., 2014). While it would be ideal if there was no systemic absorption of ICS, this is not the reality. Nearly none of the ICS often used to treat COPD, such as fluticasone propionate and budesonide, are absorbed through the gastrointestinal system due to first-pass metabolism of the liver. Thus, the lungs are the primary route of systemic absorption of corticosteroids. When it comes to fluticasone, individuals with poor lungs absorb the drug less systemically than those with healthy lungs (Pandya et al., 2014). Although, a reduction in total systemic corticosteroid levels may be advantageous in theory.

3.9 CHALLENGES AND LIMITATIONS OF THE USE OF STEROIDS

One of the major challenges of steroid therapy is immunosuppression that leads to fungal infection or other opportunistic infection if inhalers are not used properly. One should gargle properly to avoid these opportunistic infections. The chance of infection is less with the use of MDIs, as very precise amount of drug is administered. In severe cases, systemic use of these steroids increases the chance of opportunistic infections. There are known adverse biological effects of using ICS, such as adrenal suppression, loss of bone mass density and an increased risk of fracture, glaucoma and skin bruising (McEvoy & Niewoehner, 1997). Also, it has been observed that patients with stable COPD using systemic corticosteroids are more likely to experience glucose intolerance, reduced osteocalcin levels in their serum and adrenal insufficiency (Falk et al., 2008). Additionally, it has been found that the length and dosage of systemic steroid usage in the long-term management of COPD correlate with the degree of respiratory and peripheral muscle weakening. Systemic steroid courses for exacerbations have been associated with effects like weight gain, depression, anxiety, insomnia and hyperglycemia (Wood-Baker et al., 2005).

3.10 PHOSPHODIESTERASE (PDE) INHIBITORS

Another family of medications that has demonstrated efficacy in treating COPD is the phosphodiesterase (PDE) inhibitor, which works by inhibiting different isoforms of PDE enzymes. The release of different inflammatory mediators, bronchial relaxation and cardiovascular activity are all impacted by this enzyme. The amount of cyclic guanosine monophosphate (GMP) increases when inhibitors block this enzyme. PDE3 plays a role in bronchoconstriction, and bronchodilation results from inhibition of this enzyme (Bardin et al., 1998). The PDE4 enzyme is found in the majority of inflammatory cell types and is thought to be a key target for newly developed COPD treatments. The FDA authorized roflumilast, a PDE4 inhibitor, in 2011 for the treatment of COPD (Rabe, 2011). The number of neutrophils and eosinophils in induced sputum was decreased by roflumilast therapy. Additionally, roflumilast dramatically decreased the levels of soluble IL-8, neutrophil elastase, eosinophil cationic protein and α2-macroglobulin. Pfizer stopped using tofimilast after failing to show its efficacy (Grootendorst et al., 2007). The drug UK-500001, developed by Pfizer, also failed to show the efficacy and was discontinued in COPD patients (Phillips, 2020). However, some companies are still working on targeting PDE enzyme such as AstraZeneca, Glied, Eisai, etc.

3.11 CHALLENGES AND LIMITATIONS OF PDE4 INHIBITORS

Roflumilast is poorly tolerable and has significant gastrointestinal adverse events. Diarrhea, weight loss, nausea, headache, insomnia, decreased appetite and dizziness are common adverse effects of PDE4 inhibitors (Rabe, 2011).

3.11.1 Anti-oxidants

Cigarette smoke has high concentrations of gaseous and soluble oxidants that can directly cause cell injury and tissue necrosis. In addition, various components of cigarette smoke can activate the macrophages in the lungs and attract neutrophils. This leads to the initiation of inflammatory cascade. Various preclinical studies demonstrated usefulness of anti-oxidants by inhibiting inflammation and other morphological changes. However, oxidants are also essential for reducing inflammation and killing pathogens, though oxidants can inactivate alpha-1 antitrypsin and other protective proteins, damage lung tissue and increase mucus production (Vézina & Cantin, 2018). The level of various anti-oxidant enzymes such as superoxide dismutase (SOD), catalase, glutathione peroxidase is decreased in COPD patients. Consumption of anti-oxidant vitamins such as vitamin C and E can reduce oxidative stress and lipid peroxidation of cells. Furthermore, in smokers with a high body mass index, these vitamins lower plasma F2-isoprostane levels, indicating that these dietary supplements may be used to modulate lung oxidative stress. N-acetylcysteine, carbocysteine, N-isobutyrylcysteine, erdosteine, sulforaphane and diet anti-oxidants (multiple) are evaluated clinically for their effectiveness in COPD (Rahman, 2006). However, various anti-depressants showed their effectiveness in preclinical models, but further, proof of concept studies are warranted to understand the effect and signaling mechanism of these drugs. More monotherapy trials are required. We need to see if these drugs are effective as monotherapy or can only be used as adjuvant treatment in COPD patients.

3.11.2 Challenges and Limitations of Anti-oxidants

These anti-oxidants have the limitation of poor penetrability to the lung cells. Their absorption is variable and, in a similar way, these anti-oxidants have shown their effectiveness in various preclinical models; in clinical front, however, their efficacy, particularly as monotherapy, needs to be studied more (Schwedhelm et al., 2003).

3.12 ANTI-INFLAMMATORY DRUGS

Inflammation is an important hallmark of COPD, and targeting inflammation can be considered a useful strategy for the treatment of COPD. Various anti-inflammatory drugs are in preclinical and clinical phases for the treatment of COPD. Leukotriene modifiers and anti-histaminic drugs are useful and helpful, as these drugs significantly reduce inflammation (Cazzola et al., 2023). Various CXCR2 inhibitors, TNF-α antibodies and IL-1β monoclonal antibodies are in various phases of clinical trials.

3.12.1 Challenges and Limitations of Anti-Inflammatory Drugs

Lack of promising results have been seen in various phases of clinical trials. These drugs are not able to show promising results in moderate-to-severe COPD.

3.13 CONCLUSION

Available treatment strategies are helpful to reduce the progression of COPD to some extent. However, newer therapies with better efficacy and fewer side effects are warranted. None of the drugs used for COPD is effective as a monotherapy treatment of COPD. The drug that could hamper the progression of COPD completely would be a gold standard for treatment. Tolerance develops to the initial SABA treatment quickly due to the progression of the disease. Anti-cholinergic drugs are effective in combination. Steroid drugs are useful to avoid worsening of the condition. All the previously mentioned major drugs, including PDE4 inhibitors and anti-inflammatory drugs, have limitations. The use of alternative treatments such as herbal drugs, or siddha or Unani system drugs may be beneficial approaches for the treatment of COPD. In addition, the use of effective anti-oxidant therapy as an adjuvant to other medications can also produce beneficial effects in the treatment of COPD. If we look into the non-pharmacological treatment approach of quitting smoking, passive and active are both mandatory. We should avoid often sitting next to a friend in our circle who smokes continuously. Yoga, meditation and pranayama have additional benefits and can improve the symptoms in COPD patients to a significant extent. These three techniques improve lung function, lung capacity and breathing efficiency. In addition, regular walking to maintain a healthy body weight is also mandatory, as obesity is one of the risk factors for the progression of COPD.

REFERENCES

Abroug, F., Ouanes, I., Abroug, S., Dachraoui, F., Abdallah, S., Ben, Hammouda, Z., & Ouanes-Besbes, L. (2014). Systemic corticosteroids in acute exacerbation of COPD: A meta-analysis of controlled studies with emphasis on ICU patients. *Annals of Intensive Care*, *4*, 32. https://doi.org/10.1186/s13613-014-0032-x

Andreas, S., Hering, T., Mühlig, S., Nowak, D., Raupach, T., & Worth, H. (2009). Smoking cessation in chronic obstructive pulmonary disease: An effective medical intervention. *Deutsches Arzteblatt International*, *106*(16), 276–282. https://doi.org/10.3238/arztebl.2009.0276

Anzueto, A., & Miravitlles, M. (2020). Tiotropium in chronic obstructive pulmonary disease - a review of clinical development. *Respiratory Research*, *21*(1), 199. https://doi.org/10.1186/s12931-020-01407-y

Bailey, K. L. (2012). The importance of the assessment of pulmonary function in COPD. *The Medical Clinics of North America*, *96*(4), 745–752. https://doi.org/10.1016/j.mcna.2012.04.011

Bardin, P. G., Dorward, M. A., Lampe, F. C., Franke, B., & Holgate, S. T. (1998). Effect of selective phosphodiesterase 3 inhibition on the early and late asthmatic responses to inhaled allergen. *British Journal of Clinical Pharmacology*, *45*(4), 387–391. https://doi.org/10.1046/j.1365-2125.1998.t01-1-00689.x

Barnes, P. J., Shapiro, S. D., & Pauwels, R. A. (2003). Chronic obstructive pulmonary disease: Molecular and cellular mechanisms. *The European Respiratory Journal*, *22*(4), 672–688. https://doi.org/10.1183/0903 1936.03.00040703

Barr, R. G., Bourbeau, J., Camargo, C. A., & Ram, F. S. F. (2005). Inhaled tiotropium for stable chronic obstructive pulmonary disease. *The Cochrane Database of Systematic Reviews*, *2005*(2), CD002876. https://doi.org/10.1002/14651858.CD002876.pub2

Cazzola, M., Hanania, N. A., Page, C. P., & Matera, M. G. (2023). Novel anti-inflammatory approaches to COPD. *International Journal of Chronic Obstructive Pulmonary Disease*, *18*, 1333–1352. https://doi.org/10.2147/COPD.S419056

Deeney, B. T., Cao, G., Orfanos, S., Lee, J., Kan, M., Himes, B. E., Parikh, V., Koziol-White, C. J., An, S. S., & Panettieri, R. A. J. (2022). Epinephrine evokes shortening of human airway smooth muscle cells following β(2) adrenergic receptor desensitization. *American Journal of Physiology. Lung Cellular and Molecular Physiology*, *323*(2), L142–L151. https://doi.org/10.1152/ajplung.00444.2021

Dekhuijzen, P. N., & Decramer, M. (1992). Steroid-induced myopathy and its significance to respiratory disease: A known disease rediscovered. *The European Respiratory Journal*, *5*(8), 997–1003.

Devi, R. E., Barman, D., Sinha, S., Hazarika, S. J., & Das, S. (2020). Nicotine replacement therapy: A friend or foe. *Journal of Family Medicine and Primary Care*, *9*(6), 2615–2620. https://doi.org/10.4103/jfmpc.jfmpc_313_20

Domej, W., Oettl, K., & Renner, W. (2014). Oxidative stress and free radicals in COPD–implications and relevance for treatment. *International Journal of Chronic Obstructive Pulmonary Disease*, *9*, 1207–1224. https://doi.org/10.2147/COPD.S51226

Fagerström, K., & Hughes, J. (2008). Varenicline in the treatment of tobacco dependence. *Neuropsychiatric Disease and Treatment*, *4*(2), 353–363. https://doi.org/10.2147/ndt.s927

Falk, J. A., Minai, O. A., & Mosenifar, Z. (2008). Inhaled and systemic corticosteroids in chronic obstructive pulmonary disease. *Proceedings of the American Thoracic Society*, *5*(4), 506–512. https://doi.org/10.1513/pats.200707-096ET

Fischer, B. M., Voynow, J. A., & Ghio, A. J. (2015). COPD: Balancing oxidants and antioxidants. *International Journal of Chronic Obstructive Pulmonary Disease*, *10*, 261–276. https://doi.org/10.2147/COPD.S42414

Grootendorst, D. C., Gauw, S. A., Verhoosel, R. M., Sterk, P. J., Hospers, J. J., Bredenbröker, D., Bethke, T. D., Hiemstra, P. S., & Rabe, K. F. (2007). Reduction in sputum neutrophil and eosinophil numbers by the PDE4 inhibitor roflumilast in patients with COPD. *Thorax*, *62*(12), 1081–1087. https://doi.org/10.1136/thx.2006.075937

Ho, T., Cusack, R. P., Chaudhary, N., Satia, I., & Kurmi, O. P. (2019). Under- and over-diagnosis of COPD: A global perspective. *Breathe (Sheffield, England)*, *15*(1), 24–35. https://doi.org/10.1183/20734735.0346-2018

Johns, D. P., Walters, J. A. E., & Walters, E. H. (2014). Diagnosis and early detection of COPD using spirometry. *Journal of Thoracic Disease*, *6*(11), 1557–1569. https://doi.org/10.3978/j.issn.2072-1439.2014.08.18

Jones, P. (2013). Aclidinium bromide twice daily for the treatment of chronic obstructive pulmonary disease: A review. *Advances in Therapy*, *30*(4), 354–368. https://doi.org/10.1007/s12325-013-0019-2

Laniado-Laborín, R. (2009). Smoking and chronic obstructive pulmonary disease (COPD). Parallel epidemics of the 21 century. *International Journal of Environmental Research and Public Health*, *6*(1), 209–224. https://doi.org/10.3390/ijerph6010209

MacNee, W. (2013). Systemic inflammatory biomarkers and co-morbidities of chronic obstructive pulmonary disease. *Annals of Medicine*, *45*(3), 291–300. https://doi.org/10.3109/07853890.2012.732703

Maltais, F., Naya, I. P., Vogelmeier, C. F., Boucot, I. H., Jones, P. W., Bjermer, L., Tombs, L., Compton, C., Lipson, D. A., & Kerwin, E. M. (2020). Salbutamol use in relation to maintenance bronchodilator efficacy in COPD: A prospective subgroup analysis of the EMAX trial. *Respiratory Research*, *21*(1), 280. https://doi.org/10.1186/s12931-020-01451-8

Mansvelder, H. D., Fagen, Z. M., Chang, B., Mitchum, R., & McGehee, D. S. (2007). Bupropion inhibits the cellular effects of nicotine in the ventral tegmental area. *Biochemical Pharmacology*, *74*(8), 1283–1291. https://doi.org/10.1016/j.bcp.2007.07.034

McEvoy, C. E., & Niewoehner, D. E. (1997). Adverse effects of corticosteroid therapy for COPD. A critical review. *Chest*, *111*(3), 732–743. https://doi.org/10.1378/chest.111.3.732

Mittal, M., Siddiqui, M. R., Tran, K., Reddy, S. P., & Malik, A. B. (2014). Reactive oxygen species in inflammation and tissue injury. *Antioxidants & Redox Signaling*, *20*(7), 1126–1167. https://doi.org/10.1089/ars.2012.5149

Moreno, D., Barroso, J., & Garcia, A. (2015). Vaccines for patients with COPD. *Recent Patents on Inflammation & Allergy Drug Discovery*, *9*(1), 23–30. https://doi.org/10.2174/1872213x09666150223114958

Moulton, B. C., & Fryer, A. D. (2011). Muscarinic receptor antagonists, from folklore to pharmacology: Finding drugs that actually work in asthma and COPD. *British Journal of Pharmacology*, *163*(1), 44–52. https://doi.org/10.1111/j.1476-5381.2010.01190.x

Ortiz-Quintero, B., Martínez-Espinosa, I., & Pérez-Padilla, R. (2022). Mechanisms of lung damage and development of COPD due to household biomass-smoke exposure: Inflammation, oxidative stress, microRNAs, and Gene polymorphisms. *Cells*, *12*(1). https://doi.org/10.3390/cells12010067

Pandey, K. C., De, S., & Mishra, P. K. (2017). Role of proteases in chronic obstructive pulmonary disease. *Frontiers in Pharmacology*, *8*, 512. https://doi.org/10.3389/fphar.2017.00512

Pandya, D., Puttanna, A., & Balagopal, V. (2014). Systemic effects of inhaled corticosteroids: An overview. *The Open Respiratory Medicine Journal*, *8*, 59–65. https://doi.org/10.2174/1874306401408010059

Phillips, J. E. (2020). Inhaled phosphodiesterase 4 (PDE4) inhibitors for inflammatory respiratory diseases. *Frontiers in Pharmacology*, *11*, 259. https://doi.org/10.3389/fphar.2020.00259

Price, D., Keininger, D. L., Viswanad, B., Gasser, M., Walda, S., & Gutzwiller, F. S. (2018). Factors associated with appropriate inhaler use in patients with COPD - lessons from the REAL survey. *International Journal of Chronic Obstructive Pulmonary Disease*, *13*, 695–702. https://doi.org/10.2147/COPD.S149404

Rabe, K. F. (2011). Update on roflumilast, a phosphodiesterase 4 inhibitor for the treatment of chronic obstructive pulmonary disease. *British Journal of Pharmacology*, *163*(1), 53–67. https://doi.org/10.1111/j.1476-5381.2011.01218.x

Rahman, I. (2006). Antioxidant therapies in COPD. *International Journal of Chronic Obstructive Pulmonary Disease*, *1*(1), 15–29. https://doi.org/10.2147/copd.2006.1.1.15

Rovina, N., Koutsoukou, A., & Koulouris, N. G. (2013). Inflammation and immune response in COPD: Where do we stand? *Mediators of Inflammation*, *2013*, 413735. https://doi.org/10.1155/2013/413735

Schwedhelm, E., Maas, R., Troost, R., & Böger, R. H. (2003). Clinical pharmacokinetics of antioxidants and their impact on systemic oxidative stress. *Clinical Pharmacokinetics*, *42*(5), 437–459. https://doi.org/10.2165/00003088-200342050-00003

Scichilone, N., Battaglia, S., La Sala, A., & Bellia, V. (2006). Clinical implications of airway hyperresponsiveness in COPD. *International Journal of Chronic Obstructive Pulmonary Disease*, *1*(1), 49–60. https://doi.org/10.2147/copd.2006.1.1.49

Shah, B. K., Singh, B., Wang, Y., Xie, S., & Wang, C. (2023). Mucus hypersecretion in chronic obstructive pulmonary disease and its treatment. *Mediators of Inflammation*, *2023*, 8840594. https://doi.org/10.1155/2023/8840594

Tashkin, D. P., Lipworth, B., & Brattsand, R. (2019). Benefit:Risk profile of budesonide in obstructive airways disease. *Drugs*, *79*(16), 1757–1775. https://doi.org/10.1007/s40265-019-01198-7

Toelle, B. G., Xuan, W., Bird, T. E., Abramson, M. J., Atkinson, D. N., Burton, D. L., James, A. L., Jenkins, C. R., Johns, D. P., Maguire, G. P., Musk, A. W. B., Walters, E. H., Wood-Baker, R., Hunter, M. L., Graham, B. J., Southwell, P. J., Vollmer, W. M., Buist, A. S., & Marks, G. B. (2013). Respiratory symptoms and illness in older Australians: The burden of obstructive lung disease (BOLD) study. *The Medical Journal of Australia*, *198*(3), 144–148. https://doi.org/10.5694/mja11.11640

Twigg, M. S., Brockbank, S., Lowry, P., FitzGerald, S. P., Taggart, C., & Weldon, S. (2015). The role of serine proteases and antiproteases in the cystic fibrosis lung. *Mediators of Inflammation*, *2015*, 293053. https://doi.org/10.1155/2015/293053

van Geffen, W. H., Tan, D. J., Walters, J. A., & Walters, E. H. (2023). Inhaled corticosteroids with combination inhaled long-acting beta2-agonists and long-acting muscarinic antagonists for chronic obstructive pulmonary disease. *The Cochrane Database of Systematic Reviews*, *12*(12), CD011600. https://doi.org/10.1002/14651858.CD011600.pub3

Vestbo, J., Hurd, S. S., Agustí, A. G., Jones, P. W., Vogelmeier, C., Anzueto, A., Barnes, P. J., Fabbri, L. M., Martinez, F. J., Nishimura, M., Stockley, R. A., Sin, D. D., & Rodriguez-Roisin, R. (2013). Global strategy for the diagnosis, management, and prevention of chronic obstructive pulmonary disease: GOLD executive summary. *American Journal of Respiratory and Critical Care Medicine*, *187*(4), 347–365. https://doi.org/10.1164/rccm.201204-0596PP

Vézina, F.-A., & Cantin, A. M. (2018). Antioxidants and chronic obstructive pulmonary disease. *Chronic Obstructive Pulmonary Diseases (Miami, Fla.)*, *5*(4), 277–288. https://doi.org/10.15326/jcopdf.5.4.2018.0133

Wang, C., Zhou, J., Wang, J., Li, S., Fukunaga, A., Yodoi, J., & Tian, H. (2020). Progress in the mechanism and targeted drug therapy for COPD. *Signal Transduction and Targeted Therapy*, *5*(1), 248. https://doi.org/10.1038/s41392-020-00345-x

Wang, M.-T., Lai, J.-H., Tsai, C.-L., & Liou, J.-T. (2019). Risk of adverse cardiovascular events with use of inhaled long-acting bronchodilators in management of chronic obstructive pulmonary disease. *Journal of Food and Drug Analysis*, *27*(3), 657–670. https://doi.org/10.1016/j.jfda.2018.12.006

Wood-Baker, R. R., Gibson, P. G., Hannay, M., Walters, E. H., & Walters, J. A. E. (2005). Systemic corticosteroids for acute exacerbations of chronic obstructive pulmonary disease. *The Cochrane Database of Systematic Reviews*, *1*, CD001288. https://doi.org/10.1002/14651858.CD001288.pub2

Yan, F., Gao, H., Zhao, H., Bhatia, M., & Zeng, Y. (2018). Roles of airway smooth muscle dysfunction in chronic obstructive pulmonary disease. *Journal of Translational Medicine*, *16*(1), 262. https://doi.org/10.1186/s12967-018-1635-z

Zheng, J., Kang, J., Cai, B., Zhou, X., Cao, Z., Bai, C., & Zhong, N. (2006). [Comparison of tiotropium inhalation capsules and ipratropium metered dose inhaler in a randomized, double-blind, double-dummy, efficacy and safety study in patients with chronic obstructive pulmonary disease]. *Zhonghua jie He He Hu Xi Za Zhi = Chinese Journal of Tuberculosis and Respiratory Diseases*, *29*(6), 363–367.

4 Inhalation Drug Delivery Systems for the Management of Chronic Obstructive Pulmonary Disease

Garima Verma
Faculty of Pharmacy, Swami Vivekanand Subharti University
Meerut, Uttar Pradesh, India

Ajay Kumar Shukla
Institute of Pharmacy, Dr Rammanohar Lohia Avadh University
Ayodhya, Uttar Pradesh, India

Lubhan Singh
Faculty of Pharmacy, Swami Vivekanand Subharti University
Meerut, Uttar Pradesh, India

4.1 INTRODUCTION

A respiratory condition known as chronic obstructive pulmonary disease (COPD) is characterized by ongoing inflammation and airflow obstruction [Saxena et al., 2022]. Dyspnea, or difficulty breathing, is caused by airway obstruction, slower expiration rates, and increased levels of neutrophils and macrophages, which are indicative of an inflammatory response in COPD [Vij et al., 2016]. According to the World Health Organization, tobacco use causes more than 70% of COPD cases in high-income countries and between 30% and 40% of cases in low- to middle-income nations [WHO, 2023].

Chronic airway inflammation, tissue remodeling, and irreversible lung parenchyma damage are among the clinical characteristics of COPD that lead to mucus accumulation and limited airflow. This results in symptoms like wheezing, dyspnea, prolonged coughing, and dilated chest walls. Hospitalization is frequently necessary due to recurrent exacerbations, which increases the financial burden on patients and healthcare systems. Cigarette smoking is the main risk factor for COPD because it contains a number of pro-inflammatory and pro-oxidant chemicals that cause irreversible tissue damage, oxidative stress, and airway inflammation. According to research conducted both *in vivo* and *in vitro*, smoking cigarettes induces inflammation in the lungs by interfering with the signaling pathways of macrophages and bronchoepithelial cells, which in turn causes a pro-inflammatory and pro-oxidant state [De Rubis et al., 2023]. Cigarette smoke causes inflammation because it releases pro-inflammatory cytokines such interleukin (IL)-1α, IL-1β, IL-8, IL-18, and GDF-15 and inhibits the anti-inflammatory cytokine IL-10 [De Rubis et al., 2023]. Glutathione plays a crucial role as a reactive oxygen species (ROS) scavenger and detoxifying agent against tobacco's carcinogenic chemicals. Oxidative stress in COPD is caused by an imbalance between the generation and clearance of the ROS [De Rubis et al., 2023].

 DOI: 10.1201/9781003498926-4

The majority of people affected with COPD are middle-aged and older, and the condition is more common as people age. In Australians 45 years of age and older, 4.8% have COPD, according to the 2017–2018 National Health Survey (Australian Health Survey), with comparable rates among men (4.5%) and women (5.1%). Among Aboriginal and Torres Strait Islanders (ATSI) 45 years of age and above, the prevalence of COPD was 10% in 2018–2019; this means that ATSI are 2.3 times more likely to develop COPD than the overall population. In 2019, COPD accounted for 3.23 million deaths worldwide, 90% of which were in low- to middle-income countries and individuals under the age of 70. In the world, COPD ranks as the sixth most common cause of poor health.

The pharmacological treatments for COPD that are currently available can be used alone or in combination, depending on the severity of the disease. These medicines include inhaled corticosteroids (ICS), long-acting beta-2 agonists (LABA), short-acting beta-2 agonists (SABA), and long-acting muscarinic antagonists (LAMA). These therapies have substantial adverse effects even when they reduce symptoms. There are several non-pharmacological methods to reduce lung volume, such as exercise, pulmonary rehabilitation, immunizations, vaccines, and healthy eating. Due to systemic medication exposure, conventional pharmaceutical therapies frequently result in side effects in addition to their primary purpose of symptom reduction. A prospective substitute is the targeted drug delivery system, which may lessen side effects by administering medications to the specific tissues that are impacted. Delivery techniques based on nanoparticles (NPs) have great promise because of their unique physicochemical characteristics, which improve medication selectivity and bioavailability. For example, organic NPs have demonstrated promise in the treatment of a number of respiratory conditions, including COPD.

Advanced and NP-based drug delivery methods for the treatment of COPD are examined in this chapter. Using keywords such as "nanoparticles as advanced drug delivery systems COPD," we performed a PUBMED search that produced 27 pertinent papers published in the past ten years. Twelve more papers were found for analysis by conducting additional searches with keywords including "COPD," "PLGA," "COPD pulmonary delivery limitations," "phytoceuticals," "COPD nanotherapeutics," and "liposomes." The purpose is to demonstrate how these cutting-edge therapy approaches can effectively cure COPD (Figure 4.1).

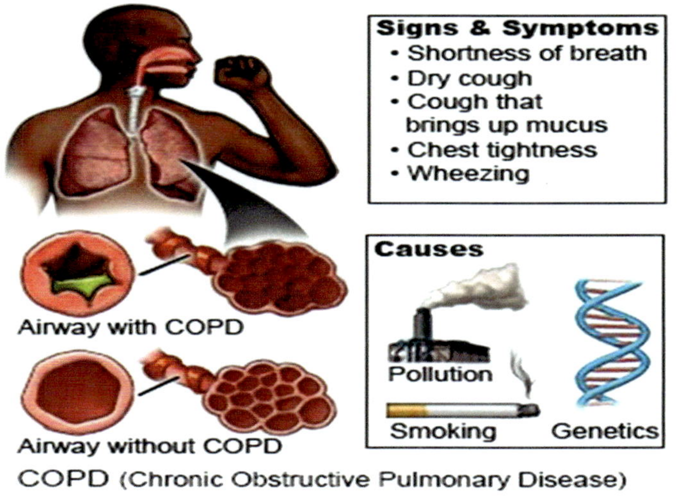

FIGURE 4.1 Sign, symptoms, and causes of chronic obstructive pulmonary disease.

4.2 LIMITATIONS OF CURRENT COPD THERAPIES

Current COPD therapies have made significant strides in managing symptoms and improving quality of life for patients. However, they still come with several limitations:

1. ***Inhaler technique***: Many COPD medications are delivered via inhalers, and incorrect inhaler technique is common among patients. This can lead to inadequate drug delivery to the lungs, reducing the effectiveness of the treatment.
2. ***Adherence***: Patient adherence to COPD medication regimens can be challenging due to the complexity of treatment regimens, side effects, and the progressive nature of the disease. Missing doses or not following the prescribed schedule can lead to exacerbations and disease progression.
3. ***Limited effectiveness***: Present treatments cannot provide a cure for COPD, but they can help control symptoms and reduce the disease's progression. Despite receiving medication, patients may continue to have exacerbations, a steady reduction in lung function, and a lower quality of life.
4. ***Side effects***: Some COPD medications can cause side effects such as dry mouth, sore throat, increased heart rate, and increased risk of pneumonia. These side effects can impact patient adherence and quality of life.
5. ***Cost***: COPD medications can be expensive, especially for patients without insurance or with limited access to healthcare. Cost barriers may prevent some patients from accessing or adhering to prescribed treatments.
6. ***Limited options for severe cases***: In advanced stages of COPD, when airflow obstruction becomes severe, treatment options may be limited. Invasive interventions such as lung transplantation or lung volume reduction surgery may be considered, but they come with significant risks and are not suitable for all patients.
7. ***Non-pharmacological approaches***: While medications are a cornerstone of COPD management, non-pharmacological approaches such as pulmonary rehabilitation, oxygen therapy, and smoking cessation programs are also important. However, access to these services may be limited in some regions, and their effectiveness can vary depending on individual patient factors.

Addressing these limitations requires ongoing research and development efforts to improve existing therapies, develop new treatment modalities, and enhance patient education and support programs. Additionally, efforts to improve access to healthcare services and reduce disparities in COPD care are crucial for optimizing patient outcomes [David and Edwards, 2022; Rodrigo et al., 2017; Skolnik et al., 2020].

4.3 BENEFITS AND DRAWBACKS OF TRADITIONAL MEDICINE

A chronic inflammatory lung illness that is not entirely reversible, COPD is typified by a restriction in airflow. Conventional therapies for COPD aim to manage symptoms, improve lung function, and prevent exacerbations. Here are some advantages and disadvantages of these therapies:

Advantages:

1. ***Bronchodilators***: These drugs, which include anticholinergics and beta-agonists, serve to relax the muscles surrounding the airways, which facilitates breathing. They offer prompt alleviation of symptoms like wheezing and dyspnea.
2. ***Corticosteroids***: ICSs reduce airway inflammation, helping to prevent exacerbations and improve lung function over time. They are often used in combination with bronchodilators for moderate to severe COPD.

3. ***Oxygen therapy***: Supplemental oxygen therapy raises blood oxygen levels, which lessens the strain on the heart and lungs and relieves exhaustion and shortness of breath symptoms.
4. ***Pulmonary rehabilitation***: Exercise training, information, and support are all used in pulmonary rehabilitation programs to help patients with COPD increase their ability to exercise, manage their symptoms, and live better.
5. ***Vaccinations***: Annual influenza vaccines and pneumococcal vaccines help reduce the risk of respiratory infections, which can exacerbate COPD symptoms.

Disadvantages:

1. ***Side effects***: Conventional COPD medications can cause side effects such as dry mouth, sore throat, increased heart rate, and increased risk of pneumonia (especially with long-term use of corticosteroids).
2. ***Tolerance development***: Over time, some individuals may develop tolerance to bronchodilators, requiring higher doses or additional medications for symptom control.
3. ***Risk of exacerbations***: Despite treatment, COPD exacerbations can still occur, leading to hospitalizations and decline in lung function.
4. ***Cost***: Some COPD medications, particularly newer formulations or combination therapies, can be expensive, placing financial strain on patients, especially those without adequate insurance coverage.
5. ***Complexity of treatment***: Managing COPD often requires a combination of medications, oxygen therapy, pulmonary rehabilitation, and lifestyle changes, which can be challenging for patients to adhere to consistently.
6. ***Limited disease modification***: While conventional therapies can help manage symptoms and reduce exacerbations, they do not modify the underlying progression of COPD or reverse existing lung damage.
7. ***Inhaler technique***: The efficacy of COPD drugs depends on proper inhaler technique, yet many patients have trouble using it correctly, which results in less-than-ideal treatment outcomes.

Conventional therapies for COPD offer significant benefits in symptom management, exacerbation prevention, and improving quality of life. However, they also have limitations such as side effects, tolerance development, and the inability to modify disease progression. Treatment decisions should be individualized based on the severity of COPD, patient preferences, and healthcare provider guidance [David and Edwards, 2022; Rodrigo et al., 2017; Skolnik et al., 2020].

4.4 COPD AND SYSTEMATIC DRUG DELIVERY

4.4.1 Systemic Drug Delivery Systems

NP-based systemic drug delivery systems (NP-SDDS), as seen in Figure 4.2, are formulations that encapsulate the therapeutic agent, convey it to the intended place, and subsequently release the active molecules into the intended tissues. The controlled delivery of the medication to the therapeutic target in real time is one of the many benefits of these technologies' dispersion processes. In most NP-SDDS, the active ingredient is transported and diffused by use of additional carriers and drug molecules. Additionally, the various ways that drug delivery systems are used can affect pharmacokinetics, delivery ease, toxicity, and infection risk, among other factors.

NP-SDDS are being used more widely in drug delivery because of their ability to penetrate the lower respiratory tract and enhance drug permeability and bioavailability. They are an attractive therapeutic option because of their biocompatibility, low toxicity, and surface modification capability. In a study, Mohamed et al., showed that Poly (glycerol adipate-co-pentadecalactone)

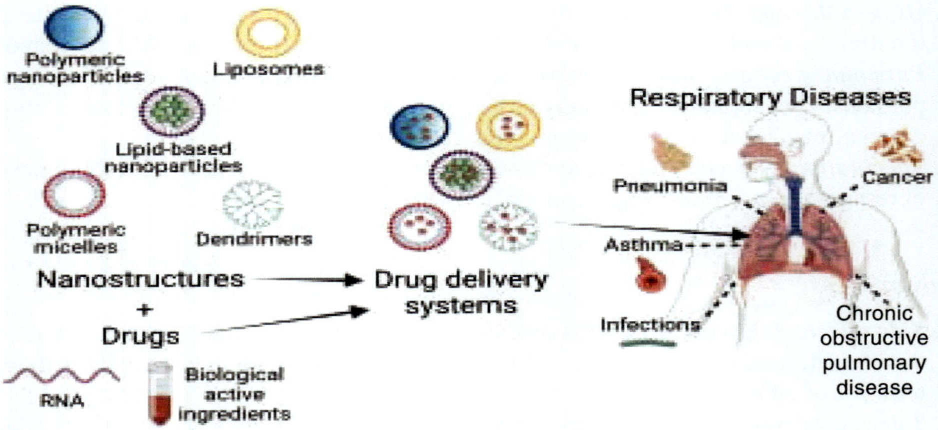

FIGURE 4.2 Chronic obstructive pulmonary disease and systematic drug delivery.

NPs loaded with miRNA146 and the cationic lipid DOTAP might control inflammation. At a dose of 0.625 mg/mL, the oil-in-water emulsion solvent evaporation method used to generate the NPs resulted in a dose-dependent 40% reduction of IRAK1 gene expression. Through the IL-1 beta signaling route, miR-146a successfully decreased IL-8 promoter activity while retaining its structural integrity, indicating promise for COPD treatment [Mohamed et al., 2019].

The internal organ environment presents hurdles for conventional drug delivery methods, and they frequently suffer from insufficient mucosal penetration, poor patient adherence, and suboptimal biodistribution. Utilizing NP and microparticle systems can help to address these problems by prolonging the drug's half-life, which lowers toxicity and permits sustained release [Guo, 2023].

It has been found that extracellular vesicles (EVs) contribute to COPD pathogenesis, originate from various cell types, exhibit altered miRNA levels, and vary according to the disease's pathophysiological status. This suggests EVs' potential utility in diagnosing and monitoring COPD. Despite the current research being of high quality, more studies are needed to further elucidate the functional characteristics of EVs in COPD pathogenesis [Gomez, 2022].

4.5 THE ROLE OF EXTRACELLULAR VESICLES IN COPD

A variety of cells, including neutrophils, lymphocytes, monocytes, macrophages, and chemokines, are involved in the intricate pathophysiology of COPD. Three essential elements are cellular senescence, lung aging, and chronic inflammation. EVs are found in bodily fluids and have become important COPD indicators by facilitating intercellular communication. Their functions and potential as therapeutic targets are covered in this review [Neri et al., 2022].

4.5.1 Microvesicles in COPD Diagnosis

Bronchoalveolar lavage (BAL) fluid contains microvesicles (MVs), which may be used as COPD biomarkers. According to this study, smokers with COPD had larger concentrations of MVs made from alveolar macrophages than did non-smokers or individuals without the illness. These levels showed a correlation with both airway obstruction and smoking history, suggesting MVs' involvement in inflammation and the advancement of COPD. Consequently, MVs may be useful indicators for diagnosing COPD and comprehending its processes [Bazzan et al., 2021].

These studies collectively highlight the emerging importance of EVs and MVs in diagnosing, understanding, and potentially treating COPD and related pulmonary conditions.

4.5.2 Roles in Respiratory Disease Pathogenesis

Alveoli, critical lung structures composed of various parenchymal and bone marrow-derived cells like alveolar macrophages, perform essential functions vital for lung health. The interactions among these cells are crucial for maintaining homeostasis and influencing disease progression. Recent studies have isolated individual lung cell types to investigate changes in gene expression, humoral factors, and adhesion molecules, thereby revealing intercellular communication dynamics. Particularly significant are EVs, lipid bilayer-bound structures released from cells that transport substances such as microRNAs (miRNAs), mRNAs, and proteins, serving as intercellular messengers. EVs, including apoptotic bodies, exosomes, and microparticles, have gained attention for their potential in understanding disease mechanisms and aiding diagnostics. Studies have examined the functions of exosomes and microparticles in preserving lung homeostasis and their roles in idiopathic PF, COVID-19, acute respiratory distress syndrome (ARDS), and COPD [Yamada, 2021].

4.5.3 Therapeutic Strategies for Bacterial Lung Infections

Bacterial lung infections, often following respiratory viral illnesses such as influenza and COVID-19, as well as chronic conditions like COPD and cystic fibrosis (CF), are traditionally treated with antibiotics. However, antibiotics can disrupt the body's natural balance and promote drug-resistant bacteria. Biofilms, protective layers formed by bacteria, complicate treatment by shielding bacteria from the immune system and drugs. Alternative treatments like bacteriophages and antimicrobial peptides exist but face challenges in effectively reaching infection sites and maintaining efficacy. Advanced delivery methods using micro/nanoparticles to transport and gradually release antimicrobial agents offer promise in overcoming these obstacles, potentially improving recovery from bacterial infections [Chee & García, 2023].

4.5.4 Intra-Alveolar Neutrophil-Derived Microvesicles in COPD

COPD management is hindered by the lack of biochemical markers, despite progress in understanding the disease. MVs, involved in inflammatory diseases like COPD, have not been clearly linked to disease severity. In a study involving 62 COPD patients, various MV subtypes were identified in plasma and bronchoalveolar lavage fluid (BALF), with BALF neutrophil MVs significantly correlating with clinical evaluation scores, including lung function and exercise tests. These MVs also showed a strong correlation with the BODE index, suggesting their potential as biomarkers for COPD severity and pathophysiology [Soni et al., 2021].

4.5.5 MSC-Derived Extracellular Vesicles in Lung Diseases

Mesenchymal stromal cells (MSCs) are one type of cell that releases EVs, which are microscopic particles made from the plasma membrane. These particles can be released spontaneously, in reaction to external stimuli, or during interactions between cells. EVs are divided into three categories based on their origin, size, and mode of release: apoptotic bodies, MVs, and exosomes. Different components, including proteins, lipids, nucleic acids, and organelles, are present in MSC-derived EVs that can deliver anti-inflammatory and reparative signals to injured cells. They have a great deal of therapeutic potential in regenerative medicine, particularly for lung conditions such as CF, ARDS, idiopathic PF, COPD, and COVID-19. Nonetheless, a number of obstacles need to be overcome before EV-based treatments can be used in clinical settings [Abreu et al., 2021].

4.5.6 METTL16-Mediated m6A Modification of PM2.5 Causes Pulmonary Microvascular Injury in COPD Patients

COPD is markedly worsened by exposure to fine particulate matter (PM2.5), while the exact processes are still unknown. To accelerate the onset of COPD, this study created COPD-induced rat models and found that PM2.5 promoted pulmonary microvascular injury by enhancing endothelial apoptosis, increasing vascular permeability, and decreasing angiogenesis. Serum VEGF levels and plasma endothelial microparticles (EMPs) were used to confirm microvascular damage in COPD patients. N6-methyladenosine (m6A) modification is altered by PM2.5 exposure, according to m6A sequencing, and in COPD rats, the m6A regulator METTL16 is upregulated. When m6A levels in METTL16-knockdown pulmonary microvascular endothelial cells (PMVECs) were measured, two METTL16-regulated genes linked to vascular damage were found: sulfatase 2 (Sulf2) and cytohesin-1 (Cyth1). These genes were downregulated and m6A methylation increased in response to PM2.5 exposure, indicating a potential role for METTL16-mediated m6A modification of Sulf2 in PM2.5-induced microvascular damage. This finding sheds light on the role of m6A modification in COPD and identifies new treatment targets [Guo, 2023].

4.5.7 Systemic Endothelial and Platelet Derived Microparticles

Microparticles of endothelium and platelets (eMPs and pMPs) are indicators of cellular activity, malfunction, or apoptosis associated with cardiovascular diseases. This study looked at the relationship between COPD outcomes and platelet/endothelial dysfunction and cardiovascular consequences. Fifty-eight COPD patients and 19 controls were included in the study, and they were followed for an average of 1.17 years. Lung function, BODE scores, quality of life, exacerbations, comorbidities, and mortality were all evaluated by researchers. During acute exacerbations and at enrollment, 17 patients had plasma samples taken for eMP and pMP values, which were then subjected to flow cytometry analysis. According to the findings, there were no discernible variations in MP levels between COPD patients and controls, nor were there any relationships with diffusion ability, quality of life, or the presence of exacerbations. In contrast to controls, GOLD stage 3 COPD patients had greater levels of eMP ULEX and pMP CD62P+. In contrast to GOLD stage 2/3 patients, controls and GOLD stage 1 patients had a lower CD62E+/CD31+ ratio, suggesting more apoptosis. During acute exacerbations, pMP41+31+ levels increased with higher BODE scores, but eMP ULEX lectin+ levels declined. Following comorbidity adjustment, the majority of eMPs and pMPs showed no discernible relationship with the severity or status of COPD [Lascano et al., 2021].

4.5.8 Endothelial Microparticles and miRNA-126 in COPD

EVs are important intercellular communication mediators that are implicated in a number of disorders. Among these are endothelium microparticles (EMPs). This research focuses on how EMPs from primary pulmonary microvascular endothelial cells contribute to the development of COPD, specifically by transferring microRNA-126 (miR-126).

To look at this, Western blotting, flow cytometry, and transmission electron microscopy were used to characterize the EMPs that were isolated via gradient centrifugation. After that, these EMPs were used in COPD models created in vivo and in vitro using cigarette smoke extract (CSE). Additionally, EMPs were transfected with miR-126 mimics or inhibitors using calcium chloride to assess their effects on lung inflammation. The study found that inflammation-related mRNA and protein levels were significantly elevated in the COPD models. However, treatment with EMPs notably reduced these levels. The inflammatory marker HMGB1 was further downregulated by the overexpression of miR-126 in EMPs, augmenting the anti-inflammatory properties of EMPs. These results suggest that EMPs, particularly those transferring miR-126, can mitigate lung inflammation in COPD, highlighting a potential therapeutic avenue for this disease [Ma et al., 2022].

4.5.9 Apoptosis and PECAM EMPs in Pulmonary Microvascular Endothelial Cells Associated with COPD

In particular, PECAM EMPs, a type of EMP, is important for the development of COPD. In COPD patients and human pulmonary microvascular endothelial cells (HPMECs) exposed to CSE, the quantities and effects of extracellular matrix proteins were investigated in a study by Zeng et al. (2022). According to the study, COPD patients had higher amounts of PECAM, VE-cadherin, and E-selectin EMPs than non-COPD volunteers did. PECAM EMPs were also observed to be higher in HPMECs subjected to CSE. When HPMECs were exposed to PECAM EMP, there was a time-independent rise in apoptosis and cleaved caspase 3/9 expression associated with a fall in phospho-Akt (p-Akt) expression. The antioxidant Eukarion-134 (EUK-134), which raised p-Akt expression and decreased apoptosis in the presence of PECAM EMPs, may be able to partially offset these apoptotic effects. According to this, PECAM EMPs in COPD cause HPMECs to undergo apoptosis by blocking Akt signaling, and antioxidants may have a preventive role [Zeng et al., 2022].

4.6 INNOVATIVE THERAPIES

Inhalation therapy is being explored for PF, leveraging innovative methods such as microparticles and NPs. While traditionally used for asthma and COPD, these new approaches hold promise for PF by targeting specific stages of the disease process. Despite current limitations, these therapies offer substantial benefits and warrant further research [Zheng, 2024].

Lipid-polymer hybrid inhalable nanoparticles for the purpose of delivering roflumilast to lung macrophages in a targeted manner were developed to improve COPD treatment. These NPs had a core-shell shape with a considerable phospholipid content and surface PEG density. They were created by mixing lipid vesicles with polymeric NPs loaded with roflumilast. This design ensured slow drug release, improved mucus penetration, and efficient macrophage uptake via mannose-mediated targeting. The formulation was adapted for inhalation using a nano-into-micro (NiM) approach, encapsulating the NPs in poly(vinyl alcohol)/leucine microparticles through spray drying, facilitating efficient delivery to the lungs [Craparo, 2022a].

Craparo (2022) developed an innovative rapamycin (Rapa) formulation to treat lung inflammation associated with COPD, asthma, and CF. This novel formulation employs NiM particles, created by encapsulating Rapa in pegylated copolymer NPs via nanoprecipitation. These NPs, which demonstrate high drug loading, encapsulation efficiency, and sustained release in lung-simulated fluids, are further processed into mannitol-based microparticles through spray drying. These microparticles are suitable for breath-activated dry powder inhalers (DPIs), facilitating effective mucosal penetration and prolonged drug release, thereby offering a promising therapeutic strategy for lung inflammation [Craparo, 2022b].

Knap et al. (2023) developed poly(sebacic acid) microparticles loaded with azithromycin to address bacterial infections in CF and COPD patients. These microparticles, characterized by their spherical shape and optimal size range of 1–5 μm, exhibit nearly 100% encapsulation efficiency and rapid degradation, with about 50% mass loss after 24 hours. Antibacterial efficacy against *Staphylococcus aureus* and cytotoxicity tests in lung epithelial cells confirm the microparticles' potential for safe and effective pulmonary drug delivery, highlighting their promise for treating lung infections [Knap et al., 2023].

Alogna et al. (2024) introduced an inhalable technology called LungToHeartNIM for targeting heart failure (HF) treatment, alongside a novel mimetic peptide (MP). In a pig model of chronic HF, MP-loaded NPs delivered via a dry powder significantly improved heart function, reversed cardiac remodeling, and reduced pulmonary congestion without adverse effects. These findings suggest LungToHeartNIM as a groundbreaking approach for delivering peptide-based therapies to the heart, potentially revolutionizing HF treatment [Alogna et al., 2024].

Kim et al. (2021) investigated the impact of omega-3 polyunsaturated fatty acid (PUFA) supplementation on endothelial function in COPD patients. In a randomized controlled trial, participants received either high-dose fish oil or a placebo for six months. While there was no significant improvement in flow-mediated dilation (FMD), the fish oil group showed increased EMPs and better respiratory function as per the St. George's Respiratory Questionnaire (SGRQ), indicating potential benefits of omega-3 PUFAs in COPD management [Kim et al., 2021].

Banat et al. (2024) developed a DPI containing nanosized ketoprofen-embedded mannitol-coated microparticles for treating pulmonary inflammations like COPD and CF. This formulation, achieved through advanced homogenization and co-spray drying techniques, displayed optimal pulmonary delivery characteristics, significant anti-inflammatory effects, and low cytotoxicity, showcasing its potential for effective lung inflammation treatment [Banat et al., 2024].

Fischer et al. (2021) designed a cylindrical microparticulate system for small interfering RNA (siRNA) delivery to macrophages, addressing the challenges of oligonucleotide delivery in pulmonary diseases. The system demonstrated effective lung deposition, high siRNA loading capacity, and sustained release. *In vitro* studies showed a significant reduction in tumor necrosis factor-alpha (TNF-α) release from human macrophages, indicating its efficacy in delivering and protecting siRNA for potential asthma, COPD, and CF treatments [Fischer et al., 2021]. Developed and was found that the inhalable dry powder microparticles of azithromycin using electrospray. These microparticles, sized at 2.5 μm, were suitable for respiratory deposition. Physicochemical analysis confirmed their efficacy against bacteria *in vitro*, and biocompatibility with alveolar and bronchial cell lines was demonstrated [Arauzo et al., 2021]. Craparo et al. (2024) devised a rapamycin-based therapy for COPD-related inflammation by creating aerosolizable NiM particles. Rapamycin-loaded lipid/polymer hybrid NPs (Rapa@Man-LPHNPs) were synthesized, measuring approximately 128 nm with a slightly negative ζ potential. These NPs exhibited a slow drug release rate, with 74 wt% intact after 24 hours. To enable inhalation, Rapa@Man-LPHNPs were encapsulated within PVA:LEU microparticles via spray drying, resulting in particle sizes suitable for effective nebulization (FPF 27.5% and MMAD 4.3 μm).

4.6.1 Polymeric Nanoparticles for miRNA Delivery in COPD Treatment

Therapeutics based on RNA interference (RNAi) present a viable method of controlling gene expression. In particular, it is becoming clear that miRNAs may be used as therapeutic targets for COPD. To downregulate IRAK1, this study focused on miR-146a adsorbed onto poly(glycerol adipate-co-ω-pentadecalactone) (PGA-co-PDL) NPs. Oil-in-water single emulsion solvent evaporation was used to create these NPs, and the cationic lipid dioleoyltrimethylammoniumpropane (DOTAP) was one of the ingredients. MiR-146a-loaded NPs were proven to have internalized in A549 cell lines by fluorescence and confocal imaging. The effects of the miR-146a-NPs were dose-dependent, with IRAK1 expression being reduced by 40% at the highest NP concentrations (0.321 mg/ml and 0.625 mg/ml). Furthermore, the IL-8 promoter activity was inhibited by the miR-146a-NPs. These findings demonstrate the potential of PGA-co-PDL-DOTAP NPs as an miRNA delivery method for the treatment of COPD by indicating that miR-146a retains its biological activity when adsorbed onto these NPs [Mohamed et al., 2019].

4.6.2 Targeted Neutrophil Nano-Drug Delivery

The airway's defense mechanisms make it difficult to target airway cells with medications. This work explored the possibility of using PEGylated immuno-conjugated poly(lactic-co-glycolic) acid (PLGA) nanoparticles (PINPs) to target neutrophils and other inflammatory cells with targeted medication delivery. Using transmission electron microscopy, dynamic laser scattering, and flow cytometry, the study determined the size, shape, surface characteristics, and targeting potential of the NPs. In animal models of obstructive lung disorders, the results demonstrated that PINPs could successfully travel the airway, attach to neutrophils, and release the anti-inflammatory medication ibuprofen. Neutrophil inflammation and associated pulmonary disorders were effectively managed by this targeted approach [Vij et al., 2016].

4.6.3 Nanoparticles as Cutting Edge COPD Medication Delivery Systems

The symptoms of COPD include airway blockage, inflammation, and the generation of free radicals. Significant side effects from current therapies, which include systemic drugs, include decreased bone density and an elevated risk of pneumonia. Patient adherence to these medications is further complicated by the need for regular dosing. NP-based drug delivery systems offer high selectivity, enhanced bioavailability, and biocompatibility, in contrast to conventional medicines that treat symptoms only and do not stop the disease's progression. Side effect reduction and decreased dosage frequency are the goals of these systems. To show how NP-based drug delivery systems can transform the treatment of COPD, this study addresses the drawbacks of conventional COPD therapies and emphasizes their cutting-edge uses [Jessamine et al., 2024].

4.6.4 Airway-targeted PLGA Nanoparticle Synthesis and Assessment for COPD

Chronic airway inflammation is a defining feature of several persistent obstructive airway diseases, including asthma, CF, and COPD. As recently highlighted, inflammation poses a significant challenge to the successful delivery and therapeutic efficacy of NP-based delivery systems for a variety of chronic airway illnesses since the NPs need to navigate and overcome the airway's defensive mechanisms. NPs that are able to successfully elude these defenses may facilitate the delivery of tailored medications to sick cells. Over the past ten years, there has been an increase in interest in creating targeted NPs for the treatment of cancer; nevertheless, the development of innovative systems for the management of chronic inflammatory and obstructive airway disorders has received relatively less attention. This research describes the methods for producing drug-loaded, multifunctional NPs that are specifically aimed at specific types of airway cells. The formulations and selective drug delivery methods displayed here are currently undergoing pre-clinical development in our laboratory to treat chronic airway diseases such as COPD, CF, and asthma [Vij et al., 2016].

4.7 OLIGONUCLEOTIDE THERAPY

Oligonucleotide therapy (ONT), originating from Zamecnik and Stephenson in 1978, aims to modulate gene expression by inhibiting translation. Research has explored RNA-targeting approaches such as siRNA, ribozymes, and miRNA for therapeutic potential, particularly in diseases like COPD. Antisense oligonucleotides (ASO) bind to target mRNA, inducing gene knockdown via RNase H-mediated cleavage without altering the ASO itself. In contrast, siRNA triggers gene silencing.

MiRNA regulates gene expression, impacting various cellular processes. Optimizing oligonucleotide structure and concentration is crucial for efficacy. ONTs have gained U.S. Food and Drug Administration (FDA) approval for diseases like homozygous familial dyslipidemia (e.g., Kynamro). However, clinical trials for COPD and asthma utilizing ONTs have faced challenges.

MiRNA-based nanotherapeutics offer potential in respiratory diseases. Dysregulated miRNAs contribute to COPD pathogenesis. Antagomirs, inhibiting specific miRNAs, show promise in restoring normal cellular functions. However, their limited cellular membrane penetration requires polymer-based carriers.

For instance, inhibiting miRNA-570-3p in COPD airways can restore sirtuin-1 expression, aiding lung function and reducing cellular senescence. ONTs hold promise for COPD treatment but face challenges in clinical translation and delivery optimization. Furthermore, because pulmonary surfactants, or zwitterionic lipids, are naturally found in the respiratory system and help ONTs bind to target cells, inhaled drug delivery is the best option for ONT [Liao, 2017; Séguin and Ferrari, 2009]. Additionally, by allowing for slower complex breakdown and direct access to the target region, this method prolongs the duration of the therapy in the lungs. By doing so, systemic toxicity may be decreased by lowering the dose frequency. Nevertheless, additional investigation is required to evaluate the therapy's long-term clinical safety. However, there is also a desire for chemical alterations to boost oligonucleotide efficacy [Séguin and Ferrari, 2009].

4.7.1 The Promise of ONTs for PF

Fibrotic lung diseases pose significant clinical challenges, driving the search for innovative treatments. ONTs emerge as a promising avenue by targeting elusive disease pathways. Direct delivery to the lungs offers advantages by minimizing systemic side effects and enhancing tolerability. Chemical designs of ONTs tailored for fibrotic lung diseases are explored, alongside discussions on delivery methods, physiological barriers, and toxicity concerns. Despite their efficacy in targeted delivery and reduced side effects, no ONTs have yet received clinical approval, underscoring challenges for future therapies [Paul et al., 2023].

4.7.2 IL-6 Trans-signaling and AIM2 Inflammasome in Emphysema Pathogenesis

Emphysema, characterized by lung inflammation and tissue damage, lacks clear molecular understandings. This study elucidates the role of innate immune inflammasome complexes in driving emphysema progression. Elevated expression of AIM2, a DNA sensor, along with increased IL-1β levels, contributes to alveolar cell apoptosis. Blocking AIM2 and IL-1 receptor effectively mitigates emphysema progression. Furthermore, IL-6 trans-signaling enhances AIM2 expression, suggesting a potential therapeutic target for emphysema treatment [Ruwanpura et al., 2022].

4.7.3 CELA1 Inhibition for Emphysema Treatment

CELA1, a serine protease, plays a crucial role in emphysema progression. Inhibition of CELA1 with the antibody KF4 prevents emphysema by inhibiting stretch-induced lung elastase. This intervention preserves lung elastin without increasing lung immune cells, indicating its potential as a therapy for stabilizing lung matrix in emphysema [Ojha et al., 2024].

4.7.4 AntimiRs Targeting microRNA-34a for Lung Disease Therapy

AntimiRs, which inhibit miRNAs, show promise for treating lung diseases like COPD and IPF. Targeting miR-34a, a miRNA linked to chronic lung ailments, these compounds effectively lower its levels in bronchial epithelial cells. Testing in 3D cultured cells mimicking lung conditions reveals varying compound properties affecting uptake and potency, suggesting potential for inhalation delivery in lung-related therapies [Weidner et al., 2023].

4.8 GENE DELIVERY IN ACUTE LUNG INJURY

Lung illnesses may benefit from gene therapy. Curcumin was incorporated into micelles and given to afflicted mouse models in a study aimed at treating acute lung injury (ALI), together with the heme oxygenase-1 (HO-1) gene. This compound demonstrated the effectiveness of gene delivery in pulmonary illnesses by efficiently reducing hemolysis and pro-inflammatory cytokines in edema [Kim et al., 2019].

4.9 APPLICATIONS OF NANOPARTICLES IN COPD

4.9.1 Nanoparticles Made of Polymers

Made of lactic acid and glycolic acid joined by ester bonds, PLGA NPs are biodegradable and present a promising option for drug administration [Awasthi et al., 2020]. With sizes typically smaller than 100 nm [Saxena et al., 2022], PLGA NPs exhibit enhanced mucosal penetration and reduced steric hindrance, particularly advantageous for airway applications [Di Francesco et al., 2020]. Variations in NP shape, such as elongated spheroids or honeycomb structures, can influence penetration efficiency and drug-release kinetics [Di Francesco et al., 2020]. Encapsulation within PLGA NPs shields therapeutic agents from degradation, facilitating targeted delivery across cellular barriers [Alsaab et al., 2022].

In addressing inflammatory conditions like COPD, PLGA-based delivery systems, such as targeting neutrophil-mediated inflammation, ibuprofen-loaded PLGA-polyethylene glycol nanoparticles (PLGA-PEG-NP) have demonstrated potential [Di Francesco et al., 2020]. Vij et al. (2016) reported that characterization tests using methods such as transmission electron microscopy and dynamic laser scattering validated the PLGA NPs' physical characteristics and their ability to distribute ibuprofen efficiently in mouse models of obstructive lung disorders. Significant therapeutic efficacy in reducing cigarette smoke-induced inflammation was shown using real-time imaging [Alsaab et al., 2022; Vij et al., 2016].

Despite their demonstrated efficacy and safety, further pre-clinical research is warranted to optimize PLGA-drug conjugates, given their reported effectiveness and low toxicity levels [Alsaab et al., 2022]. MiRNAs can be delivered by polymeric NPs, which have shown promise in treating COPD. Using PGA-co-PDL NPs loaded with miR146a to target the expression of the IRAK1 gene was the subject of a recent study. These NPs showed promising properties when made with the cationic lipid DOTAP and the oil-in-water single emulsion solvent evaporation method. Significantly, following exposure, more than 75% of A549 cells were still alive, indicating that these NPs showed remarkable biocompatibility. Moreover, miR-146a loaded NPs showed long-term release and efficient uptake by A549 cells, resulting in a 40% drop in IRAK1 expression and a reduction in IL-8 promoter activity. This work highlights the promise of miRNA-loaded polymeric NPs as a therapy strategy for COPD [Gao et al., 2023].

The effects of acute calcium channel blocker nifedipine on exercise-induced pulmonary hemodynamics and oxygen delivery were studied in patients with COPD who had various angiotensin-converting enzyme (ACE) gene polymorphisms. Thirty-three COPD patients with the II, ID, and DD genotypes participated in the trial and were given a placebo or nifedipine. Exercise-induced right cardiac catheterization was used to examine pulmonary and systemic hemodynamic parameters. The findings showed that, in comparison to the II genotype, the DD genotype was linked to higher levels of pulmonary hypertension and decreased tissue oxygenation during exercise. On the other hand, nifedipine treatment had no significant effect on oxygen delivery or pulmonary vascular resistance (PVR) but did significantly lower mean pulmonary arterial pressure (mPAP) following exercise in all genotypes. These results imply that nifedipine may not significantly improve poor oxygen supply in COPD patients with distinct ACE genotypes, even while it may reduce exercise-induced pulmonary hypertension [Kanazawa et al., 2004].

A promising therapeutic target for the treatment of COPD has been identified as histone deacetylase 2 (HDAC2). Histone acetylation is inhibited by HDAC2, which is essential for controlling genes associated with inflammation. Reduced HDAC2 activity is linked to corticosteroid resistance and exacerbates inflammation in diseases such as COPD. Antioxidants, nitric oxide (NO) synthase inhibitors, and low-dose theophylline are some of the treatment approaches that try to increase HDAC activity. Additionally, effective modulation of inflammatory pathways in COPD and other chronic inflammatory illnesses may be possible through direct targeting of HDACs with siRNA or small-molecule activators [Barnes, 2005].

Targeting pulmonary conditions like COPD, asthma, and lung cancer presents a great opportunity for the use of siRNA-based nano-drug delivery systems. Since protein misfolding is a major cause of respiratory illnesses, siRNA-based therapeutics is essential for gene targeting. However, the effective delivery of siRNA poses challenges, necessitating protection for efficient gene therapy. Nano-based delivery systems offer a solution to these challenges by enabling targeted delivery of siRNA. This perspective underscores the significance of employing siRNA-loaded NPs for addressing major pulmonary disorders, thus opening new avenues for effective therapeutic interventions [Krishnaswami and Kandasamy, 2022].

Prior studies using mouse models showed that exposure to cigarette smoke raised levels of a particular non-coding RNA, mito-ncR-LDL805, which improved mitochondrial function by turning on important genes. However, following extended smoking exposure, this response was reduced in COPD mouse models. Researchers set out to find its human equivalent, monitor its reaction in human cells that have been exposed to smoke, validate its influence on mitochondrial function, and measure its levels in patients with COPD to investigate its significance in people. The human form of the RNA temporarily increased, moved to the nucleus, and improved mitochondrial function in human lung cells exposed to smoking. On the other hand, lower quantities of this RNA were seen in COPD patients, indicating a possible treatment target to stop the disease's progression [Mathuram et al., 2024].

Obstructive pulmonary diseases like CF often require repetitive antibiotic treatment, leading to potential drug interactions with medications like Orkambi® (lumacaftor/ivacaftor). Aerosol NPs aim to overcome barriers posed by bacteria and biofilm in airways for effective drug delivery. This review explores aerosol formulations tailored for the respiratory tract, current clinical applications, and innovative approaches to reduce drug interactions by incorporating antibacterial activity into nanocarriers while ensuring efficient transfection [Mottais, 2017]. Improved hydrofluoroalkane-based pressurized metered-dose inhalers (MDIs) have been developed, along with innovations such as dry powder and soft mist inhalers. The development of effective drug carriers, safety demonstrations for non-lung targets, and improved delivery to newborns for early intervention are all critical to the future of aerosol therapy [Laube, 2005].

Dendrimers are a type of synthetic branching polymeric macromolecule that is polyfunctional because it has an outside functional unit, an interior repeating unit layer, and a core. They are generally sized between 10 nm and 100 nm [Mishra & Singh, 2020]. They are synthesized chemically through polymerization, resulting in globular structures with highly modifiable surfaces, enhancing biocompatibility and degradation ease [Mishra et al., 2021].

Khan and colleagues explored the encapsulation of siRNA into poly(amidoamine) (PAMAM) and poly(propylene imine) (PPI) dendrimers, integrating lipid modifications as a means of treating inflammatory lung disorders, including asthma and COPD. Tyrosine-protein kinase receptor-targeting Tie-2 endothelial cells in the lung were the synergistic benefits of this strategy. Notably, great efficacy was revealed by PAMAM-conjugated dendrimers with C15 cholesterol tails and polyphenylene ethynylene (PPE) dendrimers with C15 and C14 cholesterol chains without raising proinflammatory cytokines or inducing weight loss in mice [Khan et al., 2015]. Vasile et al. (2014) showed that ascorbic acid and gold-cored PPI dendrimers do not have the same antioxidant effects as gold-PAMAM dendrimer nanocomposites. Ascorbic acid's rate constant was 85 times lower than that of gold-PAMAM dendrimers, indicating the latter's promise as a targeted medication delivery mechanism for oxidative stress [Vasile et al., 2014].

1. Studies the use of dendrimers for autophagy enhancement and the treatment of obstructive lung disorders, stressing the significance of preserving cellular homeostasis via processes such as autophagy–lysosomal systems, especially in COPD and CF. Dendrimer-based approaches targeting autophagy aim to correct cellular dysfunctions, offering promising avenues for treating lung diseases and promoting healthy aging. Clinical development is ongoing in this area [Vij, 2020].

2. A promising new approach to improving the efficacy of treatments for inflammatory lung illnesses like COPD, PF, and asthma is nanotechnology. Traditional drugs used for these conditions often face challenges related to low solubility and stability, limiting their efficacy. Current research trends aim to solve these issues by producing innovative medications like siRNA, mRNA, and corticosteroids as well as biomacromolecules like DNA and long-acting β-agonists. A solution provided by nanotechnology is the effective drug delivery of NPs, which are made to be smaller than 1000 nm. Various carriers like liposomes and polymer-based NPs are being investigated for inhalation, ingestion, or injection delivery methods. These NPs enhance drug solubility, stability, and targeted delivery, thereby improving treatment outcomes [Loo & Lee, 2022].
3. For instance, MDIs are widely used for pulmonary drug delivery due to their efficiency. Nanotechnology-assisted MDIs offer controlled and targeted drug delivery systems, particularly advantageous in bypassing hepatic metabolism and reducing side effects. Polymeric NPs, liposomes, and solid lipid nanostructures are among the nanocarriers explored for drug delivery via MDIs [Kumar et al., 2022].
4. Furthermore, defective autophagy in chronic obstructive lung disorders such as CF and COPD may be salvaged by nanotechnology. In individuals with CF, medications such as cysteamine administered via specialized nanocarriers such as PLGA-PEG or dendrimers may help improve lung function and autophagy [Vij, 2017].
5. Oxidative stress exacerbates chronic respiratory diseases, necessitating innovative drug delivery systems to address inflammation effectively. Microparticles, NPs, and liposomes offer avenues for managing oxidative stress by enhancing drug effectiveness and targeting specific sites within the lungs. In summary, nanotechnology-based therapeutics show considerable potential in revolutionizing the treatment of inflammatory lung diseases by overcoming traditional drug delivery challenges and improving drug efficacy [Dua et al., 2019].

Recent advancements in NP applications in respiratory disorders have been extensively reviewed, shedding light on their transformative potential in treating such conditions. The comprehensive review integrates both fundamental and clinical research findings on NP-based drugs, demonstrating their efficacy in addressing the limitations of traditional medications. This review, conducted using databases like PubMed, Web of Knowledge, and Scopus (until August 2022), found that asthma, COPD, PF, lung cancer, and infections are just a few of the respiratory conditions for which nanomedicine shows promise. Different kinds of NPs have shown therapeutic benefits, including metal NPs, carbon nanotubes, dendrimers, and liposomes. These provide new therapy options with lower side effects and increased efficacy [Taghavizadeh Yazdi et al., 2023].

Enhancing drug delivery to the lungs using lung targeting mediated by micro- and nanocarriers has shown promise. By making use of the alveoli's large surface area, these carriers present opportunities to enhance systemic and localized medication therapy. For diseases including lung cancer, COPD, tuberculosis, and asthma, direct delivery to the lungs is especially important. This review explores the utilization of nano- and microcarriers, including microspheres, NPs, liposomes, niosomes, and dendrimers, for precise delivery of bioactive substances to the lungs. These carriers have the potential to prolong drug action duration, reduce dosage, enhance patient adherence, and minimize adverse effects. However, further research is needed to identify specific lung receptors for targeted delivery to precise lung regions such as bronchioles and alveoli [Kurmi et al., 2010].

A cutting-edge approach to medication administration for long-term inflammatory respiratory conditions like COPD and asthma is oligonucleotide treatment. Because these novel treatments work by RNAi or gene silence, more effective delivery strategies are required for accurate tissue and cell targeting. NPs play a pivotal role in facilitating this targeted delivery, ensuring efficient drug transport to the intended site. Oligonucleotides are also employed for gene transfer using carriers like polymers, liposomes, and inorganic materials. This review explores advanced dosage forms such as NPs and liposomes for delivering oligonucleotides like siRNA and miRNA. Additionally, it discusses future prospects and goals for ONTs in managing respiratory diseases [Mehta et al., 2019].

Researchers have developed specialized nanomaterials inspired by dendrimers for delivering RNA to lung endothelial cells, offering potential treatment avenues for inflammatory conditions like asthma and obstructive pulmonary disease. Modification of dendrimer molecules has optimized the delivery of siRNAs to lung blood vessels. By fine-tuning the chemical structure, materials have been identified that effectively target specific lung endothelial cells without inducing harmful inflammation or weight loss in animal models. These findings highlight the promising potential of these materials for delivering RNA therapies directly to the lungs [Khan et al., 2015].

Asthma, COPD, and acute lung damage are just a few of the lung disorders where inflammation is crucial. In an acute lung damage model, the goal of a study was to develop efficient carriers for siRNA targeting TNF-α. The improved biocompatibility of phosphorus-based dendrimers with protonated amino groups at the pyrrolidinium or morpholinium terminal was the deciding factor in their selection. Comparing dendriplexes with pyrrolidinium surface groups to those with morpholinium, the former showed improved siRNA complexation, cellular uptake, and TNF-α silencing effectiveness *in vitro*. The enhanced cellular absorption of pyrrolidinium dendriplexes was ascribed to their greater pKa value, which led to stronger siRNA complexation and better defense against enzymatic destruction. *In vivo* experiments in a mouse model of short-term ALI showed the superior silencing effect of pyrrolidinium dendriplexes when supplied via nasal injection. These results point to the possibility of using phosphorus dendriplexes to deliver siRNA in the treatment of inflammatory lung diseases [Bohr et al., 2017].

4.9.2 Inorganic Nanoparticles

Since inorganic NPs have special properties, like fewer peripheral contacts during transportation, they have a lot to offer in terms of medication delivery [Mishra and Singh, 2020]. For example, ambient magnetic fields can affect the magnetic NPs' ability to reach target areas, with the particles using either an active or passive approach depending on the characteristics of their ligand. The biodegradability of these NPs can vary based on their coating, taking into account factors like toxicity and excretion routes [Mulhall and Criner, 2016].

However, inorganic NPs also pose risks, particularly cardiotoxic effects, as evidenced by their penetration through the alveolar capillary barrier and accumulation in systemic organs, including the heart [Guerrero-Beltrán et al., 2017]. Metal toxicity, harm to the heart's membranes, oxidative stress, inflammation, and DNA damage are examples of toxicity mechanisms.

Exploring the advantages and disadvantages of inorganic NPs, they are crucial in ONT for chronic inflammatory respiratory diseases [Mehta et al., 2019]. Oligonucleotide-based therapies, like siRNA and miRNA, rely on efficient delivery methods for gene silencing or interference. NPs offer targeted delivery, enhancing the efficacy of these therapies. Moreover, recent advancements, such as LungToHeartNIM technology, show promise in treating chronic HF [Alogna et al., 2024]. By delivering mimetic peptides targeting impaired L-type calcium channel trafficking directly to the heart, this inhalable technology significantly improves heart function and reverses cardiac remodeling in a porcine HF model. The study underscores the potential of NP-based delivery systems in revolutionizing HF therapy.

4.10 PLANT-BASED NANOPARTICLES LOADED WITH BIOACTIVE COMPOUNDS

Medicinal plants offer significant therapeutic potential for COPD treatment due to their rich phytochemical content. These compounds have garnered research interest for their diverse structures and historical medicinal uses worldwide [Paudel et al., 2022]. Many medicinal plants have demonstrated efficacy in treating respiratory conditions like COPD, asthma, tuberculosis, and bronchitis, owing to their anti-inflammatory and antimicrobial properties [Paudel et al., 2022]. Plant extracts from species such as loquat, lotus, pomegranate, licorice root, and green tea exhibit antioxidant and anti-inflammatory effects, potentially aiding in lung inflammation management [Liu et al., 2022; Paudel et al., 2022].

The quick disintegration, low bioavailability, and poor solubility of phytochemicals limit their clinical utility even with their potential for medicinal use. To address these drawbacks, new developments in drug delivery techniques such as polymeric NPs, liposomal delivery, liquid crystalline NPs, micelles, and nanoemulsions have been made [Devkota et al., 2022; Liu et al., 2022; Paudel et al., 2022].

Liu et al. (2022) used an asthma mouse model to study the anti-inflammatory effects of 18β-glycyrrhetinic acid (18β-GA) *in vivo*. It was discovered that 18β-GA considerably decreased inflammation and enhanced lung function by blocking nuclear factor kappa B (NF-κB) phosphorylation and increasing the production of HO-1 and nuclear factor erythroid 2-related factor 2 (Nrf2) [Liu, 2022]. Unfortunately, the low water solubility and poor bioavailability of 18β-GA make it unsuitable for clinical usage; therefore, to improve its therapeutic efficacy and bioavailability, it must be encapsulated into nanocarriers. There is potential to enhance the therapeutic viability of 18β-GA by encapsulating it in sophisticated delivery systems and treating inflammatory respiratory illnesses such as asthma [Liu et al., 2022]. This underscores the importance of leveraging nanotechnology to harness the full therapeutic potential of phytochemicals in clinical practice.

Natural compounds, particularly marine bioactive compounds, have been explored for their potential utility in both COPD and cognitive impairment. These compounds offer a promising avenue for addressing both conditions simultaneously. Additionally, studies have investigated the therapeutic potential of plant-derived compounds in COPD management. For instance, Kaur (2023) focused on the protective effects of the oleo-gum resin of *Commiphora wightii* (guggul) against COPD-linked lung inflammation and emphysema, isolating key bioactive compounds that demonstrated anti-inflammatory activity. Moreover, Nicikowski and Reguła (2021) discussed how bioactive compounds in food of plant origin can serve as natural immunomodulators in asthma and COPD, potentially offering relief from inflammation and infections. Furthermore, Ben-Zaken Cohen et al., (2007) examined the increasing burden of COPD and lung cancer in women, attributing it partly to biological differences in cigarette smoke metabolism. Estrogen and related compounds may enhance the expression of cytochrome P450 (CYP) enzymes, which play a role in metabolizing cigarette smoke components, potentially contributing to the higher susceptibility of women to smoking-related lung diseases.

These studies collectively underscore the multifaceted approach needed to address COPD and its associated complications, including cognitive impairment, inflammation, and immune dysregulation, by exploring the therapeutic potential of natural compounds and understanding biological differences that may influence disease progression. In the study by Deng et al. (2021), the researchers explored the efficacy of Qiwei Putao powder (Uzhumu-7), a traditional Mongolian remedy, in treating COPD. Through experimentation on mice with induced COPD, they identified 61 bioactive compounds within the powder and conducted network pharmacology analysis to unveil 30 potential COPD-related targets. Notably, certain ingredients in the powder were found to activate Nrf2, offering promise in protecting lung cells and shedding light on its potential mechanisms in COPD treatment.

To comprehend the molecular mechanism of Jianpiyifei II (JPYF II) granules in treating COPD, Xie et al. (2024) performed a study. By using molecular docking simulations, network pharmacology analysis, and experimental evaluations, they discovered that JPYF II has 88 bioactive components that target 342 proteins, of which 284 are associated with the therapy of COPD. Their results demonstrated how JPYF II, which targets the IL-17/MAPK/AP-1 signaling axis within the IL-17 pathway, effectively improves COPD. The Bufei Yishen formula (BYF) was the subject of Li et al.'s (2020) investigation as a conventional herbal remedy for COPD. As part of BYF I's effective-constituent compatibility (ECC), they merged the essential effective components from BYF. They showed that ECC of BYF I exhibited bioactive equivalency to BYF using rat models of COPD, offering insights into the effectiveness of traditional Chinese medicine (TCM) formulae in treating COPD.

Chen et al. (2016) reviewed marine natural product inhibitors of neutrophil-associated inflammation, emphasizing the potential of marine-derived compounds in treating inflammatory disorders like COPD. Researchers examined the efficacy of Chuan Bei Pi Pa dropping tablets (CBPP), a TCM formula, in the treatment of COPD in a different study conducted by Ma et al. (2019). Peiminine (PEI), which targets the epidermal growth factor receptor (EGFR) pathway and shows potential in enhancing lung function and reducing PF to prevent COPD exacerbation, was discovered during a phosphoproteomics study. Targeting inflammation and oxidative stress, TCMs have shown promise in treating COPD. Zhou et al. (2016) conducted a study that screened 196 TCMs for their ability to combat oxidative stress and inflammation. Among these, 38 extracts increased QR activity, while 55 inhibited NO production. Notably, 20 extracts exhibited dual inhibition of oxidative stress and inflammation, suggesting promising efficacy against COPD. This research not only identifies potential lead compounds for COPD treatment but also opens avenues for discovering new therapeutic agents from active TCMs [Zhou et al., 2016].

Furthermore, Feng et al. (2024) investigated the effects of the Tibetan medicine Longdan zhike tablet (LDZK) on COPD. The study identified 45 key compounds in LDZK and demonstrated its efficacy in improving respiratory function, reducing inflammation, and protecting against cellular damage. LDZK primarily acted through the MAPK pathway, indicating its potential as a supplementary treatment for COPD inflammation (Feng et al., 2024). Moreover, Astragaloside IV (AST), as discussed by Zhang (2023), shows promise in treating COPD-related PF. AST targets the RAS/RAF/FoxO signaling pathway, inhibiting fibroblast transformation, reducing inflammation, and blocking epithelial-mesenchymal transition (EMT). These findings highlight AST as a natural compound with potential for COPD-related PF treatment [Zhang et al., 2023].

In summary, these studies underscore the potential of TCMs and natural compounds like LDZK and AST in managing COPD by targeting oxidative stress, inflammation, and fibrosis, offering promising avenues for further research and clinical application. The distribution of jaboticabin, an anti-inflammatory component present in different portions of the plant, was investigated in the study on *Myrciaria cauliflora*, also known as jaboticaba. The fruit peel contained significant amounts of jaboticabin, while the wood contained polyphenols, such as the antioxidant-rich 3,3′-dimethylellagic acid-4-O-sulfate. These substances showed encouraging anti-inflammatory properties, pointing to possible therapeutic use in the treatment of COPD [Zhao et al., 2019].

Antioxidant enzymes play a vital role in defending the vascular system against oxidative stress, a key factor in vascular diseases like diabetes, atherosclerosis, and COPD. Recent research highlights the positive impact of phytochemicals found in everyday foods on enhancing antioxidant enzyme function to prevent such diseases [Lee and Park, 2021].

Understanding the mechanisms and treatments for lung conditions like IPF, COPD, and asthma is crucial. Bleomycin (BLM) is a common inducer used in *in vivo* models to simulate lung injury. However, there's variation in its induction methods, necessitating further research into its molecular mechanisms [Ayilya et al., 2023].

Inflammatory responses in skeletal muscles are associated with diseases like COPD and inflammatory myopathies. Medicinal plants offer potential anti-inflammatory compounds, though research in this area is challenging. Exploring phytoconstituents for treating inflammation in skeletal muscles provides molecular insights into potential treatments [Karati et al., 2022].

Targeting drug delivery to the lungs using micro- and nanocarriers offers promising advantages for conditions like asthma, tuberculosis, COPD, and lung cancer. These carriers can improve drug release, prolong action duration, lower doses, enhance patient adherence, and mitigate adverse effects, though further research is needed to target specific lung regions (Kurmi, 2010).

Houpo Mahuang Decoction (HPMHD), a TCM, shows promise in treating chronic bronchitis (CB) by reducing inflammation via the NF-κB pathway. Chemical analysis identified several bioactive compounds in HPMHD effective against CB, offering potential for CB treatment [Gao et al., 2023].

The Bu-Fei formula (BFF) has shown potential in treating COPD, yet its precise mechanisms remain unclear. Zhang et al. (2022) investigated its effects on COPD using rat models. Results demonstrated significant improvements in lung function, reduced inflammation, and positive effects on specific molecular targets associated with COPD. Molecular docking analysis suggested that BFF influences apoptotic pathways, shedding light on its therapeutic mechanism for COPD.

Cecropia pachystachya Trécul, commonly known as embaúba, has been traditionally used for various medicinal purposes. Sousa Mourão et al. (2022) identified and isolated bioactive compounds from its roots and investigated their interaction with enzymes 5-lipoxygenase (5-LOX) and α-1-antitrypsin through computational tests. Tormentic acid and sitostenone exhibited significant affinity for 5-LOX and α-1-antitrypsin enzymes, respectively, suggesting the potential of embaúba as an alternative treatment for chronic lung diseases.

In China, BFHX, or bufei-Huoxue capsules, are used to treat COPD. A dependable UHPLC-MS/MS method was created by Men et al. (2021) to measure eight bioactive chemicals in rat plasma after BFHX consumption. The technique showed stability, accuracy, and precision, which made it easier to study the pharmacokinetics of the bioactive components in BFHX after oral administration.

Wang et al., (2023) investigated the metabolic regularity of bioactive compounds in Bufei Jianpi granule (BJG) using UHPLC coupled with triple quadrupole mass spectrometry. Ten components in rat plasma after BJG administration were analyzed, revealing pharmacokinetic parameters indicative of their biological activities, which contribute to improving lung function and treating COPD. This study provides insights for the clinical application of BJG and the identification of quality markers for its control. Newly synthesized compounds derived from tetrahydroquinoline and tetrahydroisoquinoline, featuring a 2-phenyl-5-furan unit, have shown promise in inhibiting phosphodiesterase type 4 (PDE4), particularly PDE4B. These compounds demonstrated significant inhibition of TNF-α release induced by lipopolysaccharide (LPS), suggesting potential for treating conditions like asthma/COPD and LPS-induced sepsis. Among them, tetrahydroquinoline derivatives showed superior activity compared to tetrahydroisoquinoline derivatives. Compound 4 m, containing a 4-methoxybenzene group, exhibited promising selective activity against PDE4B. Structure-activity relationship studies revealed the importance of the tetrahydroquinoline moiety in interacting with PDE4B. Further research is warranted to explore compound 4 m's potential as a PDE4B inhibitor [Li et al., 2018].

Traditional medicinal plant *Vitex negundo* L. has been explored for treating COPD. Researchers identified 86 active compounds from the plant and assessed their potential targets, suggesting effectiveness against COPD through molecular docking studies. While this offers hope for new treatments, additional studies and clinical trials are necessary for validation [Adarshan et al., 2022].

Phenolic chemicals in Jaboticaba and fake Jaboticaba fruits were shown to differ from one another by comparative study. While cyanidin-3-O-galactoside and cyanidin-3-O-arabinose were prevalent in false jaboticaba, key chemicals such as delphinidin-3-O-glucoside were identified in jaboticaba. Furthermore, jaboticabin, a possible COPD treatment molecule, was found [Wu et al., 2013].

Mosla scabra, a medicinal plant, has shown potential in reducing lung inflammation caused by influenza. Network pharmacology identified active compounds and targets related to inflammation and infection pathways, which were validated through animal tests. This study highlights the effectiveness of TCM in treating viral pneumonia [Cai and Zhang, 2022].

A compound from *Curcuma zedoaria* (Christm.) Roscoe, known as HMDD, exhibits dual action as a β2 adrenoreceptor agonist and NLRP3 inhibitor. This compound showed promising results in downregulating NLR signaling, reducing NLRP3 receptor expression, and suppressing inflammation markers, suggesting its potential as a non-steroidal anti-inflammatory drug for COPD and related conditions [Chen et al., 2023].

Unlocking herbal medicines to fight cancer-related muscle loss: Cachexia, a severe condition causing muscle and fat tissue loss, is often linked to chronic illnesses like cancer. Despite being overlooked, herbal medicines are being explored for potential treatments. Through advanced techniques like network pharmacology and transcriptomics, researchers identified 34 bioactive compounds from 12 herbs targeting 32 genes associated with cachexia pathways, showing promise in combating cancer-related muscle loss [Muthamil et al., 2023].

Rhamnus pentapomica *R. Parker – a possible agent to fight cancer:* Using Gas Chromatography-Mass Spectrometry (GC-MS), a recent study examined the stem bark of *Rhamnus pentapomica* R. Parker for its anti-glioblastoma and antioxidant qualities. Significant antioxidant and anti-glioblastoma activity was demonstrated by several extract fractions, indicating that this extract may be worth more research as a natural cancer treatment [Rehman et al., 2024].

A promising inhibitor for lung diseases: Elevated levels of human neutrophil elastase (HNE) are linked to various lung conditions. Scientists have developed a potent inhibitor, BAY 85-8501, which tightly binds to HNE's pockets, showing efficacy in rodent models of ALI and undergoing clinical trials for pulmonary diseases [von Nussbaum et al., 2015].

Understanding cigarette smoke's impact on biomolecules: Cigarette smoke contains harmful chemicals contributing to diseases like COPD and cancer. A study revealed how cigarette smoke interacts with biomolecules, forming derivatives like N-(3-oxobutyl)-Tyr, shedding light on its impact and providing methods for further research [Takahashi et al., 2013].

Unraveling the secrets of Bufei Jianpi formula: The Bufei Jianpi formula (BJF), a Chinese herbal remedy for COPD, was investigated using a systems pharmacology approach. It identified 145 bioactive ingredients and 175 potential targets, shedding light on its mechanisms and offering insights into complex prescriptions for COPD treatment [Zhao et al., 2015].

Researchers synthesized new compounds based on 3,5-dimethylpyrazole derivatives and tested them as potential inhibitors of PDE4, a key enzyme involved in inflammatory responses. One compound, called If, showed significant inhibition of PDE4B and reduced TNF-α release induced by LPS. It demonstrated effectiveness in models of asthma/COPD and LPS-induced sepsis [Hu et al., 2018].

A study explored *Rhododendron micranthum*, a traditional Chinese remedy for CB, and found eight new compounds and eight known compounds in its twigs and leaves. Three of these compounds displayed anti-inflammatory properties by suppressing NF-κB-dependent gene transcription in laboratory cells [Zhang et al., 2019]. Soluble epoxide hydrolase (sEH) is a therapeutic target for diseases like COPD. Despite lacking a crystal structure with bound substrate, new mimics of sEH inhibitors and epoxyeicosatrienoic acid (EET) regioisomers, along with computational modeling, revealed the orientation of EET regioisomers and enantiomers in sEH's catalytic cavity. This sheds light on the role of epoxide stereochemistry in substrate binding [Lee et al., 2017]. Licorice's 18β-GA and glycyrrhizic acid (GA) have the ability to improve glucocorticoid sensitivity, which is crucial for the treatment of inflammatory diseases like COPD and asthma. They activate signaling pathways involving PI3K, AP1, CRE, and NFAT, and reduce ROS levels, thus enhancing the effectiveness of glucocorticoid treatment [Kao et al., 2013].

Native to northern South America, *Myrciaria vexator* fruits are rich in bioactive substances such as delphinidin-3-O-glucoside, cyanidin-3-O-glucoside, and ellagic acid. Ellagic acid may play a part in the treatment of COPD as it demonstrated promise in reducing the expression of MMP-1 that is produced by cigarette smoke. This work adds to our understanding of *M. vexator's* chemical makeup [Dastmalchi et al., 2012].

Throughout history, respiratory illnesses have been treated with medicinal plants. Examining botanicals such as tulsi, ginger, and aloe, this study investigates their efficacy. More than 35,000 plant species worldwide have healing properties. Some, like echinacea and ginger, show promise in treating COPD and asthma. Others, like acacia and mint, target pneumonia and colds. This review suggests that medicinal plants could be valuable in developing new respiratory treatments [Firdaus et al., 2024].

A compound called S-allylmercapto-l-cysteine (SAMC) found in garlic shows promise in treating COPD. In a study, SAMC reduced MUC5AC secretion and increased AQP5 expression in airway cells. It works by inhibiting certain pathways involved in COPD progression [Yang et al., 2016].

Tanshinones from *Salvia miltiorrhiza* have anti-inflammatory properties. Two compounds, dihydrotanshinone I and cryptotanshinone, were studied. Cryptotanshinone showed better effectiveness, suggesting it could be a potent anti-inflammatory agent [Ruan et al., 2019].

Amanita muscaria, known as fly agaric, was historically used by shamans. Its compounds were later discovered to affect acetylcholine, leading to the development of drugs for lung diseases [Lee et al., 2018].

Brusatol derivatives were created to release NO, which reduces inflammation. One compound, 75, showed strong anti-inflammatory effects with low toxicity, making it a promising candidate for treating diseases like COPD [Tang et al., 2014].

An investigation was conducted on the impact of *Nigella sativa* (L.) seed essential oil on HNE activity. They discovered that the oil suppressed HNE activity, particularly its primary component, carvacrol. This suggests that it could be used as a natural antielastase agent to treat respiratory diseases such as COPD [Kacem and Meraihi, 2006].

There is little study on COPD, despite ongoing FDA investigations on medicine delivery using NPs for respiratory conditions such lung cancer. Nonetheless, patents point to the possibility of using NPs to treat COPD. Vij et al. (2016), for example, targeted inflammatory cells with NPs to reduce inflammation and start lung tissue healing. In a similar NP that carried medicinal compounds to treat lung inflammation, Castellani et al. (2018) used drugs encapsulated in NPs to address the inflammatory and oxidative aspects of respiratory diseases. Smart inhalers, or digital inhalers, are gaining popularity as a way to control COPD and asthma. They gather information on how inhalers are used, enhancing adherence, technique, and lung function. Based on this data, healthcare professionals can customize treatment; nevertheless, additional research is required for broader adoption and cost-effectiveness [Chan et al., 2021]. Adherium, Respiro, and Hailie® are a few commercial examples.

4.11 FUTURE DIRECTIONS

Compared to traditional COPD treatments, NP-based drug delivery systems have demonstrated promise in managing and possibly treating COPD with improved levels of safety and accuracy. Although using NPs for medicinal purposes has many positive aspects, dose-dependent toxicity and related adverse effects provide a challenge. Reductions in particle size and an increase in reactive surface area may cause this. To guarantee prospective patients' safety, future studies should be specifically designed to evaluate how various NPs respond to various environmental situations. Moreover, physical changes like surface-area modifications that raise chemical reactivity may also raise the risk of oxidative stress, inflammation, and genetic material damage, all of which can lead to toxicity. Furthermore, since kidney function may be harmed by the removal of the agent, it is crucial to assess the level of toxicity in the case of non-biodegradable metal NPs. Therefore, more investigation is needed to evaluate the long-term danger associated with these advanced therapeutic applications and determine which NP drug delivery systems are suitable for which demographic groups with differing degrees of COPD severity.

4.12 CONCLUSION

Potential COPD treatments have entered a new and inventive era thanks to nanomedicine. The capacity of nanotechnology to develop cutting-edge drug delivery systems that allow for a higher degree of precision when delivering the medication to the target site has shown it to be a highly desirable technology. Additionally, nanotechnology enables a precise and regulated dosage, potentially resulting in a lower frequency of administration. COPD patients' quality of life will undoubtedly be enhanced by NP-based medication delivery systems, which have fewer side effects than the majority of today's medications. Further clinical research is necessary to fully grasp the therapeutic potential of NPs in the treatment of COPD.

The use of medication delivery systems based on NPs has attracted a lot of interest in tackling the treatment issues related to COPD and lung cancer. Notably, oxidative stress and inflammation both of which are frequently made worse by cigarette smoking are common pathophysiological traits shared by lung cancer and COPD. Low survival rates in lung cancer and the incurability of COPD are caused by the poor efficacy and side effects of current treatment options for these disorders. Recent advancements in therapeutic strategies, including plant-derived compounds (phytoceuticals) and nucleic acid-based treatments, exhibit promise but are hindered by issues like poor solubility and enzymatic degradation. NP-based drug delivery systems offer a viable solution by circumventing these limitations. By encapsulating therapeutic agents within NPs, these systems facilitate improved delivery, enhancing the pharmacokinetic profiles of nucleic acids and phytoceuticals. Such advancements underscore the pivotal role of NP-based delivery systems in enabling the clinical translation of innovative treatments for lung cancer and COPD.

Furthermore, chronic pulmonary diseases, encompassing a spectrum from COPD to lung cancers, affect millions globally, necessitating efficacious therapeutic interventions. Traditional treatments, including bronchodilators and corticosteroids, have demonstrated limited efficacy in managing these conditions. Nanotechnology presents a promising avenue by augmenting targeted drug delivery to the pulmonary system. The unique properties of NPs enhance drug pharmacokinetics, offering a potential solution to the challenges posed by conventional therapies. Exploring NP-based approaches for treating chronic pulmonary diseases reveals a range of preparation techniques. Notably, nanosuspensions administered via nebulization and dry powder carriers have emerged as effective delivery systems for targeting the lungs. Moreover, considerations regarding NP toxicity, patented formulations, and commercial pulmonary drug delivery devices underscore the comprehensive approach required in advancing treatment modalities for chronic pulmonary diseases.

Addressing bacterial biofilm-associated lung infections presents another critical challenge in respiratory medicine. Airway mucus dysfunction and compromised immune defenses exacerbate these infections, leading to significant morbidity and mortality. NP-based drug delivery systems offer a potential solution by overcoming barriers to biofilm accessibility within the respiratory system. Recent research highlights the diverse array of NPs utilized to combat biofilms, emphasizing the need for innovative strategies to tackle chronic lung infections effectively. Targeted therapeutics and accurate drug delivery systems are essential for the best possible treatment outcomes in the field of chronic inflammatory respiratory disorders, which includes asthma, COPD, and CF. Combining NP-based delivery methods with additional approaches like ICSs and innovative biologicals shows promise in treating chronic airway inflammation while reducing side effects. By elucidating recent advancements in drug delivery innovations, this comprehensive review aims to enhance understanding and improve treatment outcomes in complex respiratory conditions.

In summary, the integration of nanotechnology into drug delivery systems represents a transformative approach in respiratory medicine. NP-based strategies offer enhanced precision, efficacy, and targeted delivery, thereby revolutionizing therapeutic interventions for lung cancer, COPD, bacterial biofilm-associated lung infections, and chronic inflammatory respiratory diseases. These advancements underscore the burgeoning role of nanomedicine in reshaping the landscape of respiratory disease management.

REFERENCES

Abreu SC, Lopes-Pacheco M, Weiss DJ, Rocco PRM. Mesenchymal stromal cell-derived extracellular vesicles in lung diseases: Current status and perspectives. Front Cell Dev Biol. 2021;9:600711. doi: 10.3389/fcell.2021.600711.

Adarshan S, Muthuramalingam P, Jeyasri R, Lakshmi MA, Sathishkumar R, Pandian SK, Shin H, Chen JT, Ramesh M. *Vitex negundo* L. derived specialized molecules unveil the multi-targeted therapeutic avenues against COPD: A systems pharmacology approach. Front Biosci (Landmark Ed). 2022;27(3):87. doi: 10.31083/j.fbl2703087.

Alogna A, Berboth L, Faragli A, Ötvös J, Lo Muzio FP, di Mauro V, Modica J, Quarta E, Semmler L, Deißler PM, Berger YW, Tran KL, de Marchi B, Longinotti-Buitoni G, Degli Esposti L, Guillot E, Bazile D, Iafisco M, Dotti A, Bang ML, de Luca C, Brandenberger C, Benazzi L, di Silvestre D, de Palma A, Primeßnig U, Hohendanner F, Perna S, Buttini F, Colombo P, Mühlfeld C, Steendijk P, Mauri P, Tschöpe C, Borlaug B, Pieske BM, Attanasio P, Post H, Heinzel FR, Catalucci D. Lung-to-heart nano-in-micro peptide promotes cardiac recovery in a pig model of chronic heart failure. J Am Coll Cardiol. 2024;83(1):47–59. doi: 10.1016/j.jacc.2023.10.029.

Alsaab H.O., Alharbi F.D., Alhibs A.S., Alanazi N.B., Alshehri B.Y., Saleh M.A., Alshehri F.S., Algarni M.A., Almugaiteeb T., Uddin M.N., Alzhrani R.M., PLGAbased nanomedicine: History of advancement and development in clinical applications of multiple diseases, Pharmaceutics. 2022;14(12):2728. https://doi.org/10.3390/pharmaceutics14122728.

Arauzo B, Lopez-Mendez TB, Lobera MP, Calzada-Funes J, Pedraz JL, Santamaria J. Excipient-free inhalable microparticles of azithromycin produced by electrospray: A novel approach to direct pulmonary delivery of antibiotics. Pharmaceutics. 2021;13(12):1988. doi: 10.3390/pharmaceutics13121988.

Australian Institute of Health and Welfare, Chronic Respiratory Conditions, 2023. https://www.aihw.gov.au/reports/chronic-respiratory-conditions/chronic-respiratory-conditions/contents/chronic-obstructive-pulmonary-disease.

Awasthi R., Bhushan B., Kulkarni G.T. Chapter 9 - Concepts of nanotechnology in nanomedicine: From discovery to applications, in: Dua K., Hansbro P.M., Wadhwa R., Haghi M., Pont L.G., Williams K.A. (Eds.), Targeting Chronic Inflammatory Lung Diseases Using Advanced Drug Delivery Systems, Academic Press, 2020, pp. 171–209. https://doi.org/10.1016/B978-0-12-820658-4.00009-1.

Ayilya BL, Balde A, Ramya M, Benjakul S, Kim SK, Nazeer RA. Insights on the mechanism of bleomycin to induce lung injury and associated in vivo models: A review. Int Immunopharmacol. 2023;121:110493. doi: 10.1016/j.intimp.2023.110493.

Banat H, Csóka I, Paróczai D, Burian K, Farkas Á, Ambrus R. A novel combined dry powder inhaler comprising nanosized ketoprofen-embedded mannitol-coated microparticles for pulmonary inflammations: Development, in vitro-in silico characterization, and cell line evaluation. Pharmaceuticals (Basel). 2024;17(1):75. doi: 10.3390/ph17010075.

Barnes PJ. Targeting histone deacetylase 2 in chronic obstructive pulmonary disease treatment. Expert Opin Ther Targets. 2005;9(6):1111–21. doi: 10.1517/14728222.9.6.1111.

Bazzan E, Radu CM, Tinè M, Neri T, Biondini D, Semenzato U, Casara A, Balestro E, Simioni P, Celi A, Cosio MG, Saetta M. Microvesicles in bronchoalveolar lavage as a potential biomarker of COPD. Am J Physiol Lung Cell Mol Physiol. 2021;320(2):L241–L245. doi: 10.1152/ajplung.00362.2020.

Ben-Zaken Cohen S, Paré PD, Man SF, Sin DD. The growing burden of chronic obstructive pulmonary disease and lung cancer in women: Examining sex differences in cigarette smoke metabolism. Am J Respir Crit Care Med. 2007;176(2):113–20. doi: 10.1164/rccm.200611-1655PP.

Bohr A, Tsapis N, Andreana I, Chamarat A, Foged C, Delomenie C, Noiray M, El Brahmi N, Majoral JP, Mignani S, Fattal E. Anti-inflammatory effect of anti-TNF-α SiRNA cationic phosphorus dendrimer nanocomplexes administered intranasally in a murine acute lung injury model. Biomacromolecules. 2017;18(8):2379–2388. doi: 10.1021/acs.biomac.7b00572.

Cai W, Zhang SL. Anti-inflammatory mechanisms of total flavonoids from Mosla scabra against influenza a virus-induced pneumonia by integrating network pharmacology and experimental verification. Evid Based Complement Alternat Med. 2022;2022:2154485. doi: 10.1155/2022/2154485.

Castellani S., Trapani A., Spagnoletta A., di Toma L, Magrone T., Di Gioia S., Mandracchia D., Trapani G., Jirillo E., Conese M., Nanoparticle delivery of grape seed-derived proanthocyanidins to airway epithelial cells dampens oxidative stress and inflammation. J. Transl. Med. 2018;16(1):140. https://doi.org/10.1186/s12967-018-1509-4.

Chee E, García AJ. Biomaterial therapeutic strategies for treatment of bacterial lung infections. Biofilm. 2023;5:100111. doi: 10.1016/j.bioflm.2023.100111.

Chen CY, Tsai YF, Chang WY, Yang SC, Hwang TL. Marine natural product inhibitors of neutrophil-associated inflammation. Mar Drugs. 2016;14(8):141. doi: 10.3390/md14080141.

Chen X, Zhou H, Hou T, Lu J, Wang J, Zhou L, Zhao Y, Liu Y, Wang J, Liang X, Chen C. The dual-targeting mechanism of an anti-inflammatory diarylheptanoid from *Curcuma zedoaria* (Christm.) Roscoe with the capacity for β2-adrenoreceptor agonism and NLRP3 inhibition. Chem Biol Interact. 2023;386:110771. doi: 10.1016/j.cbi.2023.110771.

Craparo EF, Cabibbo M, Scialabba C, Giammona G, Cavallaro G. Inhalable formulation based on lipid-polymer hybrid nanoparticles for the macrophage targeted delivery of roflumilast. Biomacromolecules. 2022;23(8):3439–3451. doi: 10.1021/acs.biomac.2c00576.

Craparo EF, Cabibbo M, Scialabba C, Casula L, Lai F, Cavallaro G. Rapamycin-based inhaled therapy for potential treatment of COPD-related inflammation: Production and characterization of aerosolizable nano into micro (NiM) particles. Biomater Sci. 2024;12(2):387–401. doi: 10.1039/d3bm01210g.

Craparo EF, Drago SE, Quaglia F, Ungaro F, Cavallaro G. Development of a novel rapamycin loaded nano-into micro-formulation for treatment of lung inflammation. Drug Deliv Transl Res. 2022;12(8):1859–1872. doi: 10.1007/s13346-021-01102-5.

Dastmalchi K, Flores G, Wu SB, Ma C, Dabo AJ, Whalen K, Reynertson KA, Foronjy RF, D Armiento JM, Kennelly EJ. Edible *Myrciaria vexator* fruits: Bioactive phenolics for potential COPD therapy. Bioorg Med Chem. 2012;20(14):4549–4555. doi: 10.1016/j.bmc.2012.05.013.

David S., Edwards C.W. Forced Expiratory Volume, StatPearls Publishing, 2022. https://www.ncbi.nlm.nih.gov/books/NBK540970/.

De Rubis G., Paudel K.R., Manandhar B., Singh S.K., Gupta G., Malik R., Shen J., Chami A., MacLoughlin R., Chellappan D.K., Oliver B.G., Hansbro P.M., Dua K., Agarwood oil nanoemulsion attenuates cigarette smoke-induced inflammation and oxidative stress markers in BCi-NS1.1 airway epithelial cells. Nutrients. 2023;15(4):1019.

Devkota H.P., Adhikari-Devkota A., Paudel K.R., Panth N., Gupta G., Chellappan D.K., Hansbro P.M., Dua K., in: D.K. Chellappan, K. Pabreja, M. Faiyazuddin (Eds.), Phytochemicals and Their Nanoformulations Targeted for Pulmonary Diseases, 2022.

Di Francesco M., Primavera R., Summa M., Pannuzzo M., Di Francesco V., Di Mascolo D., Bertorelli R., Decuzzi P. Engineering shape-defined PLGA microPlates for the sustained release of anti-inflammatory molecules. J. Contr. Release. 2020;319:201–212. https://doi.org/10.1016/j.jconrel.2019.12.039.

Dua K, Malyla V, Singhvi G, Wadhwa R, Krishna RV, Shukla SD, Shastri MD, Chellappan DK, Maurya PK, Satija S, Mehta M, Gulati M, Hansbro N, Collet T, Awasthi R, Gupta G, Hsu A, Hansbro PM. Increasing complexity and interactions of oxidative stress in chronic respiratory diseases: An emerging need for novel drug delivery systems. Chem Biol Interact. 2019;299:168–178. doi: 10.1016/j.cbi.2018.12.009.

Feng Y, Qin P, Wang R, Mi Y, Li Y, Feng J, Shen W, Dong H, Duo J, Ma L, Yao X, Hu X, Xiong F, Shi X, Wang H. Effects of Tibetan medicine Longdan zhike tablet on chronic obstructive pulmonary disease through MAPK pathway. J Ethnopharmacol. 2024;328:118082. doi: 10.1016/j.jep.2024.118082.

Firdaus A, Yunus MH, Izhar SK, Afaq U. Medicinal plants in the treatment of respiratory diseases and their future aspects. Recent Pat Biotechnol. 2024. doi: 10.2174/0118722083278561231212072408.

Fischer T, Tschernig T, Drews F, Brix K, Meier C, Simon M, Kautenburger R, Schneider M. siRNA delivery to macrophages using aspherical, nanostructured microparticles as delivery system for pulmonary administration. Eur J Pharm Biopharm. 2021;158:284–293. doi: 10.1016/j.ejpb.2020.11.024.

Gao F, Zhang T, Zhang H, Dai Z, Gu Y, Lu M, Zhang Z, Zeng Q, Shang B, Gao S, Wang N, Xu B, Lei H. Explore bioactive ingredients and potential mechanism of Houpo Mahuang decoction for chronic bronchitis based on UHPLC-Q exactive orbitrap HRMS, network pharmacology, and experiment verification. J Ethnopharmacol. 2023;303:115924. doi: 10.1016/j.jep.2022.115924.

Guerrero-Beltrán C.E., Bernal-Ramírez J., Lozano O., Oropeza-Almaz´an Y., Castillo E.C., Garza J.R., García N., Vela J., García-García A., Ortega E., Torre-Amione G., Ornelas-Soto N., García-Rivas G., Silica nanoparticles induce cardiotoxicity interfering with energetic status and Ca2+ handling in adult rat cardiomyocytes. Am. J. Physiol. Heart Circ. Physiol. 2017; 312(4):H645–H661. https://doi.org/10.1152/ajpheart.00564.2016

Guo Y., Mao Z., Ran F., Sun J., Zhang J., Chai G., Wang J. Nanotechnology-based drug delivery systems to control bacterial-biofilm-associated lung infections. Pharmaceutics. 2023;15(11):2582. https://doi.org/10.3390/pharmaceutics15112582

Guo X., Zuo X., Zhou Z., Gu Y., Zheng H., Wang X., Wang G., Xu C., Wang F, PLGA-based micro/nanoparticles: An overview of their applications in respiratory diseases. Int. J. Mol. Sci. 2023;24(5). https://doi.org/10.3390/ijms24054333

Hu DK, Zhao DS, He M, Jin HW, Tang YM, Zhang LH, Song GP, Cui ZN. Synthesis and bioactivity of 3,5-dimethylpyrazole derivatives as potential PDE4 inhibitors. Bioorg Med Chem Lett. 2018;28(19):3276–3280. doi: 10.1016/j.bmcl.2018.03.031.

Jessamine Victoria, Mehndiratta Samir, De Rubis Gabriele, Paudel Keshav Raj, Shetty Saritha, Suares Divya, Chellappan Dinesh Kumar, Oliver Brian G., Hansbro Phillip M., Dua Kamal. The application of nanoparticles as advanced drug delivery systems in attenuating COPD. Heliyon. 2024;10:1–15.

Kacem R, Meraihi Z. Effects of essential oil extracted from *Nigella sativa* (L.) Seeds and its main components on human neutrophil elastase activity. Yakugaku Zasshi. 2006;126(4):301–5. doi: 10.1248/yakushi.126.301.

Kanazawa H, Tateishi Y, Yoshikawa J. Acute effects of nifedipine administration in pulmonary haemodynamics and oxygen delivery during exercise in patients with chronic obstructive pulmonary disease: Implication of the angiotensin-converting enzyme gene polymorphisms. Clin Physiol Funct Imaging. 2004;24(4):224–8. doi: 10.1111/j.1475-097X.2004.00552.x.

Kao TC, Wu CH, Yen GC. Glycyrrhizic acid and 18β-glycyrrhetinic acid recover glucocorticoid resistance via PI3K-induced AP1, CRE and NFAT activation. Phytomedicine. 2013;20(3-4):295–302. doi: 10.1016/j.phymed.2012.10.013.

Karati D, Varghese R, Mahadik KR, Sharma R, Kumar D. Plant bioactives in the treatment of inflammation of skeletal muscles: A molecular perspective. Evid Based Complement Alternat Med. 2022;2022:4295802. doi: 10.1155/2022/4295802.

Khan O.F., Zaia E.W., Jhunjhunwala S., Xue W., Cai W., Yun D.S., Barnes C.M., Dahlman J.E., Dong Y., Pelet J.M., Webber M.J., Tsosie J.K., Jacks T.E., Langer R., Anderson D.G., Dendrimer-inspired nanomaterials for the in vivo delivery of siRNA to lung vasculature. Nano Lett. 2015;15(5):3008–3016. https://doi.org/10.1021/nl5048972.

Kim G., Piao C., Oh J., Lee M. Combined delivery of curcumin and the heme oxygenase-1 gene using cholesterol-conjugated polyamidoamine for anti-inflammatory therapy in acute lung injury. Phytomedicine. 2019;56:165–174. https://doi.org/10.1016/j.phymed.2018.09.240.

Kim JS, Thomashow MA, Yip NH, Burkart KM, Lo Cascio CM, Shimbo D, Barr RG. Randomization to Omega-3 fatty acid supplementation and endothelial function in COPD: The COD-fish randomized controlled trial. Chronic Obstr Pulm Dis. 2021;8(1):41–53. doi: 10.15326/jcopdf.8.1.2020.0132.

Knap K, Reczyńska-Kolman K, Kwiecień K, Niewolik D, Płonka J, Ochońska D, Jeleń P, Mielczarek P, Kazek-Kęsik A, Jaszcz K, Brzychczy-Włoch M, Pamuła E. Poly(sebacic acid) microparticles loaded with azithromycin as potential pulmonary drug delivery system: Physicochemical properties, antibacterial behavior, and cytocompatibility studies. Biomater Adv. 2023;153:213540. doi: 10.1016/j.bioadv.2023.213540.

Krishnaswami V, Kandasamy R. Current drug targeting using siRNA-based nano therapeutics for pulmonary diseases. Curr Pharm Des. 2022;28(28):2279–2282. doi: 10.2174/13816128286662207271 44223.

Kumar R, Mehta P, Shankar KR, Rajora MAK, Mishra YK, Mostafavi E, Kaushik A. Nanotechnology-assisted metered-dose inhalers (MDIs) for high-performance pulmonary drug delivery applications. Pharm Res. 2022;39(11):2831–2855. doi: 10.1007/s11095-022-03286-y.

Kurmi BD, Kayat J, Gajbhiye V, Tekade RK, Jain NK. Micro- and nanocarrier-mediated lung targeting. Expert Opin Drug Deliv. 2010;7(7):781–794. doi: 10.1517/17425247.2010.492212.

Lascano J, Katz J, Cearras M, Campos M. Association of systemic endothelial-derived and platelet-derived microparticles with clinical outcomes in chronic obstructive pulmonary disease. Chronic Obstr Pulm Dis. 2021;8(3):382–395. doi: 10.15326/jcopdf.2021.0211.

Laube BL. The expanding role of aerosols in systemic drug delivery, gene therapy, and vaccination. Respir Care. 2005;50(9):1161–76.

Lee KS, Henriksen NM, Ng CJ, Yang J, Jia W, Morisseau C, Andaya A, Gilson MK, Hammock BD. Probing the orientation of inhibitor and epoxy-eicosatrienoic acid binding in the active site of soluble epoxide hydrolase. Arch Biochem Biophys. 2017;613:1–11. doi: 10.1016/j.abb.2016.10.017.

Lee MR, Dukan E, Milne I. Amanita muscaria (fly agaric): From a shamanistic hallucinogen to the search for acetylcholine. J R Coll Physicians Edinb. 2018;48(1):85–91. doi: 10.4997/JRCPE.2018.119.

Lee SE, Park YS. The emerging roles of antioxidant enzymes by dietary phytochemicals in vascular diseases. Life (Basel). 2021;11(3):199. doi: 10.3390/life11030199.

Li LY, Zhang CT, Zhu FY, Zheng G, Liu YF, Liu K, Zhang CH, Zhang H. Potential natural small molecular compounds for the treatment of chronic obstructive pulmonary disease: An overview. Front Pharmacol. 2022;13:821941. doi: 10.3389/fphar.2022.821941.

Li YS, Liu XY, Zhao DS, Liao YX, Zhang LH, Zhang FZ, Song GP, Cui ZN. Tetrahydroquinoline and tetrahydroisoquinoline derivatives as potential selective PDE4B inhibitors. Bioorg Med Chem Lett. 2018;28(19):3271–3275. doi: 10.1016/j.bmcl.2018.04.068.

Loo CY, Lee WH. Nanotechnology-based therapeutics for targeting inflammatory lung diseases. Nanomedicine (Lond). 2022;17(12):865–879. doi: 10.2217/nnm-2021-0447.

Ma Y, He X, Liu X, Long Y, Chen Y. Endothelial microparticles derived from primary pulmonary microvascular endothelial cells mediate lung inflammation in chronic obstructive pulmonary disease by transferring microRNA-126. J Inflamm Res. 2022;15:1399–1411. doi: 10.2147/JIR.S349818.

Mathuram TL, Su Y, Bard JE, Perry NA, Chen CW, Warren MT, Linden PA, Perry Y, Hatzoglou M, Wu Y, Blumental-Perry A. Mitochondrial ncRNA LDL-805 declines in alveolar epithelial type 2 cells of chronic obstructive pulmonary disease patients. bioRxiv [Preprint]. 2024;1(14):575591. doi: 10.1101/2024.01.14.575591.

Mehta M, Deeksha, Tewari D, Gupta G, Awasthi R, Singh H, Pandey P, Chellappan DK, Wadhwa R, Collet T, Hansbro PM, Kumar SR, Thangavelu L, Negi P, Dua K, Satija S. Oligonucleotide therapy: An emerging focus area for drug delivery in chronic inflammatory respiratory diseases. Chem Biol Interact. 2019;308:206–215. doi: 10.1016/j.cbi.2019.05.028.

Men WJ, Cheng LY, Chen MY, Zhang XY, Zhang Y, Zhou K. Study on pharmacokinetics of eight active compounds from Bufei-Huoxue capsule based on UHPLC-MS/MS. J Chromatogr B Analyt Technol Biomed Life Sci. 2021;1184:122974. doi: 10.1016/j.jchromb.2021.122974.

Mishra B., Singh J., Chapter 4 - Novel drug delivery systems and significance in respiratory diseases, in: K. Dua, P.M. Hansbro, R. Wadhwa, M. Haghi, L.G. Pont, K.A. Williams (Eds.), Targeting Chronic Inflammatory Lung Diseases Using Advanced Drug Delivery Systems, Academic Press, 2020, pp. 57–95. https://doi.org/10.1016/B978-0-12-820658-4.00004-2.

Mishra A.S., Sugumaran A., Damodharan N., Nanoparticulate delivery systems for the effective treatment of COPD. Annals of the Romanian Society for Cell Biology. 2021;25(4):13892–13904.

Mohamed A., Kunda N.K., Ross K., Hutcheon G.A., Saleem I.Y. Polymeric nanoparticles for the delivery of miRNA to treat chronic obstructive pulmonary disease (COPD). Eur. J. Pharm. Biopharm. 2019;136:1–8. https://doi.org/10.1016/j.ejpb.2019.01.002.

Mulhall P., Criner G., Non-pharmacological treatments for COPD, Respirology. 2016;21(5):791–809. https://doi.org/10.1111/resp.12782.

Muthamil S, Muthuramalingam P, Kim HY, Jang HJ, Lyu JH, Shin UC, Go Y, Park SH, Lee HG, Shin H, Park JH. Unlocking prognostic genes and multi-targeted therapeutic bioactives from herbal medicines to combat cancer-associated cachexia: A transcriptomics and network pharmacology approach. Int J Mol Sci. 2023;25(1):156. doi: 10.3390/ijms25010156.

Neri T, Celi A, Tinè M, Bernardinello N, Cosio MG, Saetta M, Nieri D, Bazzan E. The emerging role of extracellular vesicles detected in different biological fluids in COPD. Int J Mol Sci. 2022;23(9):5136. doi: 10.3390/ijms23095136.

Nicikowski J, Reguła J. Selected bioactive compounds in food of plant origin as natural immunomodulators in asthma and chronic obstructive pulmonary disease. Acta Sci Pol Technol Aliment. 2021;20(4):383–397. doi: 10.17306/J.AFS.0992.

Ojha M, Smith NJ, Devine AJ, Joshi R, Goodman EM, Fan Q, Schuman R, Porollo A, Wells JM, Tiwary E, Batie MR, Gray J, Deshmukh H, Borchers MT, Ammerman SA, Varisco BM. Anti-CELA1 antibody KF4 prevents emphysema by inhibiting stretch-mediated remodeling. JCI Insight. 2024;9(1):e169189. doi: 10.1172/jci.insight.169189.

Paudel K.R., Panth N., Manandhar B., Singh S.K., Gupta G., Wich P.R., Nammi S., MacLoughlin R., Adams J., Warkiani M.E., Chellappan D.K., Oliver B.G., Hansbro P. M., Dua K., Attenuation of cigarette-smoke-induced oxidative stress, senescence, and inflammation by berberine-loaded liquid crystalline nanoparticles: In vitro study in 16HBE and RAW264.7 cells. Antioxidants. 2022;11(5). https://doi.org/10.3390/antiox11050873.

Paul D, Miller MH, Born J, Samaddar S, Ni H, Avila H, Krishnamurthy VR, Thirunavukkarasu K. The promising therapeutic potential of oligonucleotides for pulmonary fibrotic diseases. Expert Opin Drug Discov. 2023;18(2):193–206. doi: 10.1080/17460441.2023.2160439.

Rehman YU, Iqbal A, Ali G, Alotaibi G, Ahmed A, Ayaz M. Phytochemical analysis, radical scavenging and glioblastoma U87 cells toxicity studies of stem bark of buckthorn (Rhamnus pentapomica R. Parker). BMC Complement Med Ther. 2024;24(1):12. doi: 10.1186/s12906-023-04309-w.

Rodrigo G.J., Price D., Anzueto A., Singh D., Altman P., Bader G., Patalano F., Fogel R., Kostikas K. LABA/LAMA combinations versus LAMA monotherapy or LABA/ICS in COPD: A systematic review and meta-analysis. Int J Chron Obstruct Pulmon Dis. 2017;12:907–922. https://doi.org/10.2147/copd.S130482.

Ruan Z, Niu L, Han L, Ren R, Xu Z, Dong W, Jiang L. In silico comparative molecular docking analysis and analysis of the anti-inflammatory mechanisms of action of tanshinone from Salvia miltiorrhiza. Exp Ther Med. 2019;18(2):1131–1140. doi: 10.3892/etm.2019.7679.

Ruwanpura SM, McLeod L, Dousha LF, Seow HJ, West AC, West AJ, Weng T, Alanazi M, MacDonald M, King PT, Bardin PG, Gabay C, Klinman DM, Bozinovski S, Vlahos R, Anderson GP, Rose-John S, Saad MI, Jenkins BJ. Cross-talk between IL-6 trans-signaling and AIM2 inflammasome/IL-1β axes bridge innate immunity and epithelial apoptosis to promote emphysema. Proc Natl Acad Sci U S A. 2022;119(36):e2201494119. doi: 10.1073/pnas.2201494119.

Saxena J., Bisen M., Misra A., Srivastava V.K., Kaushik S., Siddiqui A.J., Mishra N., Singh A., Jyoti A. Targeting COPD with PLGA-based nanoparticles: Current status and prospects. BioMed Res. Int. 2022;2022:5058121. https://doi.org/10.1155/2022/5058121.

Séguin R.M., Ferrari N., Emerging oligonucleotide therapies for asthma and chronic obstructive pulmonary disease. Expet Opin. Invest. Drugs. 2009;18(10):1505–1517. https://doi.org/10.1517/13543780903179294.

Skolnik N.S., Nguyen T.S., Shrestha A., Ray R., Corbridge T.C., Brunton S.A., Current evidence for COPD management with dual long-acting muscarinic antagonist/long-acting β2-agonist bronchodilators. Postgrad. Med. 2020;132(2):198–205. https://doi.org/10.1080/00325481.2019.1702834.

Soni S, Garner JL, O'Dea KP, Koh M, Finney L, Tirlapur N, Srikanthan K, Tenda ED, Aboelhassan AM, Singh S, Wilson MR, Wedzicha JA, Kemp SV, Usmani OS, Shah PL, Takata M. Intra-alveolar neutrophil-derived microvesicles are associated with disease severity in COPD. Am J Physiol Lung Cell Mol Physiol. 2021;320(1):L73–L83. doi: 10.1152/ajplung.00099.2020.

Sousa Mourão P, de Oliveira Gomes R, Crisóstomo Bezerra Costa CA, da Silva Moura OF, Sousa HG, Lemos Martins Júnior GR, Cabral Leão Ferreira D, Martins Maia Filho AL, Duarte de Freitas J, Rai M, Das Chagas Alves Lima F, Gourlart Santana AE, Chaves MH, Dos Santos Alves W, Uchôa VT. Cecropia pachystachya Trécul: Identification, isolation of secondary metabolites, in silico study of toxicological evaluation and interaction with the enzymes 5-LOX and α-1-antitrypsin. J Toxicol Environ Health A. 2022;85(20):827–849. doi: 10.1080/15287394.2022.2095546.

Taghavizadeh Yazdi ME, Qayoomian M, Beigoli S, Boskabady MH. Recent advances in nanoparticle applications in respiratory disorders: A review. Front Pharmacol. 2023;14:1059343. doi: 10.3389/fphar.2023.1059343.

Takahashi Y, Horiyama S, Honda C, Suwa K, Nakamura K, Kunitomo M, Shimma S, Toyoda M, Sato H, Shizuma M, Takayama M. A chemical approach to searching for bioactive ingredients in cigarette smoke. Chem Pharm Bull (Tokyo). 2013;61(1):85–89. doi: 10.1248/cpb.c12-00539.

Tang W, Xie J, Xu S, Lv H, Lin M, Yuan S, Bai J, Hou Q, Yu S. Novel nitric oxide-releasing derivatives of brusatol as anti-inflammatory agents: Design, synthesis, biological evaluation, and nitric oxide release studies. J Med Chem. 2014;57(18):7600–7612. doi: 10.1021/jm5007534.

Therapeutic Guidelines, Stable Chronic Obstructive Pulmonary Disease (COPD), 2022. https://tgldcdp-tg-org-au.ezproxy.lib.uts.edu.au/viewTopic?.

Vasile E., Serafim A., Petre D., Giol D., Dubruel P., Iovu H., Stancu I.C., Direct synthesis and morphological characterization of gold-dendrimer nanocomposites prepared using PAMAM succinamic acid dendrimers: Preliminary study of the calcification potential. Sci World J. 2014:103462. https://doi.org/10.1155/2014/103462.

Vij N. Nano-based rescue of dysfunctional autophagy in chronic obstructive lung diseases. Expert Opin Drug Deliv. 2017;14(4):483–489. doi: 10.1080/17425247.2016.1223040.

Vij N. Synthesis and evaluation of dendrimers for autophagy augmentation and alleviation of obstructive lung diseases. Methods Mol Biol. 2020;2118:155–164.

Vij N., Min T., Bodas M., Gorde A., Roy I., Neutrophil targeted nano-drug delivery system for chronic obstructive lung diseases. Nanomed. Nanotechnol. Biol. Med. 2016;12(8):2415–2427. https://doi.org/10.1016/j.nano.2016.06.008.

von Nussbaum F, Li VM, Allerheiligen S, Anlauf S, Bärfacker L, Bechem M, Delbeck M, Fitzgerald MF, Gerisch M, Gielen-Haertwig H, Haning H, Karthaus D, Lang D, Lustig K, Meibom D, Mittendorf J, Rosentreter U, Schäfer M, Schäfer S, Schamberger J, Telan LA, Tersteegen A. Freezing the bioactive conformation to boost potency: The identification of BAY 85-8501, a selective and potent inhibitor of human neutrophil elastase for pulmonary diseases. ChemMedChem. 2015;10(7):1163–1173. doi: 10.1002/cmdc.201500131.

Wang S, Yang XX, Li TJ, Tian XM, Wang YL, Bai G, Bao YR, Meng XS. Metabolic regularity of bioactive compounds in Bufei Jianpi granule in rats using ultra-high-performance liquid chromatography coupled with triple quadrupole mass spectrometry analysis technology. Biomed Chromatogr. 2023;37(12):e5740. doi: 10.1002/bmc.5740.

Weidner J, Kolosionek E, Holmila R, Ax E, Garreau M, Gnerlich F, Olsson H, Czechtizky W, Vollmer S, Rydzik AM. Gymnotic uptake of AntimiRs alter microRNA-34a levels in 2D and 3D epithelial cell culture. Mol Ther Nucleic Acids. 2023;33:898–907. doi: 10.1016/j.omtn.2023.08.014.

World Health Organization [WHO]. Chronic Obstructive Pulmonary Disease (COPD), 2023. https://www.who.int/news-room/fact-sheets/detail/chronic-obstructive-pulmonary-disease-(copd)#:~:text=Chronic%20obstructive%20pulmonary%20disease%20(COPD)%20is%20the%20third%20leading%20cause,%2Dincome%20countries%20(LMIC).&text=Tobacco%20smoking%20accounts%20for%20over,cases%20in%20high%2Dincome%20countries.

Wu SB, Wu J, Yin Z, Zhang J, Long C, Kennelly EJ, Zheng S. Bioactive and marker compounds from two edible dark-colored Myrciaria fruits and the synthesis of jaboticabin. J Agric Food Chem. 2013;61(17):4035–4043. doi: 10.1021/jf400487g.

Yamada M. Extracellular vesicles: Their emerging roles in the pathogenesis of respiratory diseases. Respir Investig. 2021;59(3):302–311. doi: 10.1016/j.resinv.2021.02.006.

Yang M, Wang Y, Zhang Y, Zhang F, Zhao Z, Li S, Zhang J, Cao X, Zhang D. S-allylmercapto-l-cysteine modulates MUC5AC and AQP5 secretions in a COPD model via NF-κB signaling pathway. Int Immunopharmacol. 2016;39:307–313. doi: 10.1016/j.intimp.2016.08.002.

Yao Min, Zhang Lijuan, Wang Lin. Astragaloside IV: A promising natural neuroprotective agent for neurological disorders. Biomed Pharmacother. 2023;159:114229. doi: 10.1016/j.biopha.2023.114229.

Zeng Y, Zhao Y, Chen Y, Cai S, Chen P. PECAM EMPs regulate apoptosis in pulmonary microvascular endothelial cells in COPD by activating the Akt signaling pathway. Tob Induc Dis. 2022;20:40. doi: 10.18332/tid/146959.

Zhang LX, Tian YG, Zhao P, Feng SX, Han XX, Li JS. Network pharmacology analysis uncovers the effect on apoptotic pathway by Bu-Fei formula for COPD treatment. J Ethnopharmacol. 2022;289:115022. doi: 10.1016/j.jep.2022.115022.

Zhang Z, Yan H, Zhu Y, Zhang H, Chai L, Li L, Wang X, Liu Y, Li Y. New lignans, sesquiterpenes and other constituents from twigs and leaves of Rhododendron micranthum. Fitoterapia. 2019;135:15–21. doi: 10.1016/j.fitote.2019.03.025.

Zhao DK, Shi YN, Petrova V, Yue GGL, Negrin A, Wu SB, D'Armiento JM, Lau CBS, Kennelly EJ. Jaboticabin and related polyphenols from Jaboticaba (*Myrciaria cauliflora*) with anti-inflammatory activity for chronic obstructive pulmonary disease. J Agric Food Chem. 2019;67(5):1513–1520. doi: 10.1021/acs.jafc.8b05814.

Zhao P, Li J, Li Y, Tian Y, Wang Y, Zheng C. Systems pharmacology-based approach for dissecting the active ingredients and potential targets of the Chinese herbal Bufei Jianpi formula for the treatment of COPD. Int J Chron Obstruct Pulmon Dis. 2015;10:2633–56. doi: 10.2147/COPD.S94043.

Zhou MX, Wei X, Li AL, Wang AM, Lu LZ, Yang Y, Ren DM, Wang XN, Wen XS, Lou HX, Shen T. Screening of traditional Chinese medicines with therapeutic potential on chronic obstructive pulmonary disease through inhibiting oxidative stress and inflammatory response. BMC Complement Altern Med. 2016;16(1):360. doi: 10.1186/s12906-016-1347-y.

5 Advancements in Smart Inhaler Technology for Enhancing Chronic Obstructive Pulmonary Disease Care

Viswapriya Viswalingam
Poona College of Pharmacy, Department of Pharmaceutical Chemistry, Bharati Vidyapeeth (Deemed to be University) Pune, Maharashtra, India

Dileep Kumar
Department of Pharm Chemistry, Manipal College of Pharmaceutical Sciences, Manipal Academy of Higher Education Manipal, Karnataka, India

5.1 INTRODUCTION

Chronic obstructive pulmonary disease (COPD) stands as the foremost cause of mortality across the globe, characterized by persistent respiratory and airflow blockage, yet it is both treatable and preventable (Agarwal et al., 2024). It ranked as the third most prevalent cause of death worldwide around 2016 and the fourth in the United States by 2017 (Kochanek, 2017; Rana et al., 2021). The incidence of COPD is anticipated to escalate further due to population growth and heightened exposure to risk factors like tobacco use, environmental pollutants, and biomass fuel combustion (Ryu et al., 2024; Tan & Ng, 2008).

COPD manifests in various forms, with emphysema and chronic bronchitis being the most common. These conditions often coexist and may vary in severity among patients (Cavaillès et al., 2013; Khalid et al., n.d.). Emphysema occurs when the air sacs at the ends of the lungs' airways become damaged, leading to their rupture, while chronic bronchitis entails persistent inflammation of the airways, resulting in a chronic cough with phlegm production (Pahal et al., 2024). Although asthma is generally distinct from COPD, some individuals may experience a combination known as asthma-COPD overlap syndrome (ACOS) (Alsayed et al., 2023; Louie et al., 2013).

The progression of COPD is typically categorized into four stages. In the initial stage, symptoms are mild and often unnoticed except during physical exertion, presenting as slight shortness of breath and a dry cough (Burkhardt & Pankow, 2014; Tsiligianni & Kocks, 2020). Subsequently, stage two exhibits worsened symptoms, including increased breathlessness, persistent cough, and phlegm production, with changes in phlegm color during exacerbations (Watase et al., 2023). Stage three signifies further aggravation, with more frequent flare-ups, pronounced morning symptoms, and potential peripheral edema (Goyal et al., 2024). The final stage, stage four, denotes the most severe condition, marked by pronounced breathlessness, rapid or irregular heartbeat, cognitive impairment, weight loss, and pulmonary hypertension (Yuan et al., n.d.).

DOI: 10.1201/9781003498926-5

Multiple factors contribute to the development of COPD, with long-term tobacco use serving as the primary risk factor. Exposure to indoor air pollutants from burning fuels, occupational hazards like dust and fumes, and environmental pollutants such as air pollution also play significant roles (Lu et al., 2022; Mishra et al., n.d.). Cigar, pipe, and secondhand smoke exposure further exacerbate the risk, emphasizing the detrimental impact of airborne contaminants on respiratory health. Aerosol therapies are vital for treating obstructive lung diseases, which affect about 15% of the world's population. These treatments are quite complex for patients with chronic diseases to manage on their own (Laube et al., 2011; McCarthy et al., 2020). To get the best results, patients need to coordinate specific breathing techniques with the release of aerosol from their inhalers. However, many patients struggle to stick to their treatment plans, which can lead to worse health outcomes and more hospital visits (Lavorini, 2013; Virchow et al., 2008).

The term "smart inhaler," lacking a universally agreed definition, refers to inhaler-based electronic monitoring systems designed to track medication usage, involving evaluating device technique (Hei et al., 2022; Zabczyk & Blakey, 2021). This definition excludes nebulizers and software systems that solely utilize separate electronic devices, like image detection technology on phones. Over the past decade, most smart inhalers have taken the form of standard inhalers with added electronic components (Himes et al., 2019; Mosnaim et al., 2021). Electronic monitoring of inhaler adherence has been made available for more than 30 years, with the Nebuliser Chronolog being the first widely used device (Craven et al., 2018). However, only in the past decade has electronic technology become compact, reliable, and affordable enough for widespread use. These advancements coincide with improvements in wireless technology and mobile computing, facilitating rapid transfer of inhaler data for remote analysis and user-friendly presentation (Javaid et al., 2022).

Current technological capabilities include recording inhaler usage times, reminding patients of upcoming doses, assessing technique by measuring inspiratory flow rate, and providing feedback on technique and timing (Dhruve & Jackson, 2022; Taylor et al., 2018). Data collected by these devices can be transmitted securely to healthcare professionals, offering real-world insights into patient adherence and medication competence (Khatiwada et al., 2024; Li et al., 2024). This additional layer of information encourages patient compliance, as users are aware of their activities being recorded, while healthcare providers can tailor consultations based on individual patient data to promote consistent and appropriate medication usage (Aremu et al., 2022; Jimmy & Jose, 2011). Moreover, knowing the adherence levels of each doctor's or team's patients enables peer comparison and benchmarking.

In recent years, digital inhaler devices, also known as electronic adherence monitors, have started to become more important in managing COPD (Eikholt et al., 2023; Hoyte et al., 2022). These devices, along with other digital tools like mobile apps, can collect data on how well patients stick to their medication schedules and how they're feeling. Newer versions of these devices can even track where the inhaler is used and give feedback in real time (Pérez-Jover et al., 2019).

5.2 INHALERS IN RESPIRATORY DISEASES

Throughout history, the development of inhalation therapies has played a pivotal role in managing respiratory diseases (Borghardt et al., 2018). Dating back more than 4000 years, ancient civilizations in Egypt and India utilized inhalation of substances like scopolamine and atropine from *Datura* species for their bronchodilator properties (Anderson, 2005; Soni et al., 2012). Fast forward to the late 18th century, where the English physician John Mudge crafted the first defined "inhaler," marking a milestone in respiratory treatment (Stein & Thiel, 2017). In the mid-19th century, nebulizers emerged, offering a method to nebulize liquid medicine for inhalation, albeit with unpredictable dosages and potential irritants (Anderson, 2005).

The 20th century saw significant advancements in inhalation therapy. Hand-bulb nebulizers and electric nebulizers became available, providing a means to deliver active principles, such as epinephrine, directly to the airways of asthmatic patients (Alharbi et al., 2021; Dhanani et al., 2016).

In 1956, the introduction of the pressurized metered-dose inhaler (pMDI) revolutionized treatment delivery, offering precise doses of medication in a convenient, portable device. The innovation continued with the introduction of dry powder inhalers (DPIs) in 1967, providing an alternative method of delivering powdered drugs directly to the airways (de Boer et al., 2022; Roche & Dekhuijzen, 2016).

In the early 2000s, a new generation of inhalers, known as soft mist inhalers, emerged, offering propellant-free, multiple-dose devices capable of delivering medication with minimal coordination required from the user (Biddiscombe & Usmani, 2018; Xiroudaki et al., 2021). These advancements in inhalation technology have significantly improved the management of respiratory diseases such as COPD (Chan et al., 2014; Chow et al., 2023).

The importance of inhalers in respiratory diseases cannot be overstated. Inhalation therapy offers several advantages over traditional oral or injectable routes, including faster onset of action, lower doses, and reduced side effects in non-targeted areas (Jardim & Nascimento, 2019; Newman, 2014). Furthermore, delivering medication directly to the airways ensures optimal efficacy, particularly in treating diseases affecting the bronchial tree and small airways (Scichilone et al., 2013; Usmani, 2014).

In inhalation therapy for localized treatment, the primary aim is to alleviate pulmonary symptoms by reducing airway inflammation and constriction (Borghardt et al., 2018; Parray et al., 2021). This is typically achieved through the inhalation of drugs like corticosteroids, beta-sympathomimetics, muscarinic antagonists, and antibiotics. Inhaled drugs offer significant advantages over systemic administration methods because they allow for high concentrations of the drug directly at the target organ, the lung (Czechtizky et al., 2022). Consequently, lower doses of inhaled medication can be as effective as or even more effective than higher doses administered systemically. This high concentration of medication in the lungs leads to effective treatment with minimal risk of systemic side effects (Guo et al., 2021; Labiris & Dolovich, 2003). Inhalation therapy also provides a rapid onset of action in the lungs compared to other routes of administration.

However, it's important to recognize that the lung isn't a uniform organ but rather a complex structure with various specialized parts, including conducting airways, bronchioles, and alveoli. Because of this complexity, the pharmacokinetic processes involved in delivering drugs to the lungs are unique and more complex than those for drugs administered via other routes (He et al., 2022; Man et al., 2023). Drug concentrations in the lungs depend on several pulmonary kinetic processes, which are influenced by factors such as the physicochemical properties of the drug, its formulation, the inhalation device used, and patient factors like inhalation technique proficiency (Borghardt et al., 2018; Newman, 2017). Therefore, designing inhaled drugs and inhalation devices requires a thorough understanding of these pulmonary kinetic processes (Labiris & Dolovich, 2003; Rinderknecht et al., 2024). Additionally, while pulmonary concentrations are critical for efficacy, it's also essential to consider systemic drug concentrations to evaluate the overall safety profile of an inhaled drug. Systemic concentrations depend on factors like gastrointestinal absorption, pulmonary absorption, and systemic distribution and elimination properties (Eriksson et al., 2020).

Therefore, inhalers have revolutionized the treatment landscape for respiratory diseases, providing patients with effective and convenient means of managing their conditions (Baryakova et al., 2023; Pleasants et al., 2022). From ancient remedies to modern inhalation devices, the evolution of inhalation therapy underscores its critical role in improving respiratory health and quality of life.

5.3 NEBULISERS

Nebulizers, falling into three main categories—jet (or pneumatic), ultrasonic, and mesh nebulizers—have become popular for their ease of use, requiring only regular breathing (Anderson et al., 2020; Dampage et al., 2022; Pleasants et al., 2022). This makes them suitable for emergencies and accessible to patients of varying ages and conditions, such as older adults and children, who may struggle with other inhalation devices (Hagmeyer et al., 2023; Volerman et al., 2021). However, nebulizers tend to be large, costly, and demand regular upkeep, leading to extended treatment times and imprecise drug dosages with low lung deposition (DiBlasi, 2015; Terry & Dhand, 2020). Consequently, higher medication doses are often necessary compared to alternative inhalers. While ultrasonic nebulizers can alter certain medications' properties and are unsuitable for thick drugs or suspensions, mesh nebulizers offer quiet operation, shorter treatment durations, and improved lung deposition (Ari, 2014; Hou et al., 2015). Despite advancements in atomizing aerosol technologies like AERxTM and MysticTM, designed to simplify inhalation by eliminating the need for precise hand-breath coordination, they remain unreleased commercially, primarily due to cost constraints (Lange & Finlay, 2006; Morton et al., 2020; Sosnowski, 2024).

5.4 PRESSURIZED METERED-DOSE INHALERS

Nebulizers and pMDIs have revolutionized inhaled therapy, offering efficient delivery of medication to the airways (Dhand, 2017; Holland et al., 2013). The first widely used multi-dose portable devices, pMDIs are capable of dispensing precise doses of various drugs, including glucocorticoids, beta-agonists, and antimuscarinic bronchodilators (Ibrahim et al., 2015; Pleasants & Hess, 2018). A typical pMDI comprises a metal canister containing medication and a liquefied gas propellant, a metering valve for dose accuracy, and a plastic mouthpiece for particle dispersion into the airways (Buttini et al., 2023; Javadzadeh & Yaqoubi, 2017). Activation of the device by pressing down on the canister releases the medication, aerosolizing it with the propellant's energy.

Early pMDIs produced a brief, high-speed plume, resulting in low lung deposition and significant oropharyngeal impact. Lung deposition was improved with the introduction of spacers and reservoirs (Vincken et al., 2018; Wachtel et al., 2024). Another challenge was the "cold-Freon effect," causing discomfort due to the rapid evaporation of the propellant. The transition from chlorofluorocarbon (CFC) to hydrofluoroalkane (HFA) propellants, mandated by the Montreal Protocol, enhanced hand-breath coordination and lung deposition by slowing the plume's speed. Modern pMDIs, like those with Modulite® technology, offer even greater pulmonary deposition, exceeding 50% of the nominal dose. Additionally, special unidirectional valve fittings enable pMDI use during mechanical ventilation, expanding their utility in clinical settings (Myrdal et al., 2014).

5.5 SOFT MIST INHALERS

Soft mist inhalers, currently exemplified by the Respimat®, were developed to address limitations seen in other inhalation systems. Unlike traditional inhalers, Respimat® operates without propellant or batteries, relying instead on a spring mechanism to propel a solution through a fine nozzle known as the uniblock (Anderson, 2006; Dalby et al., 2011). Upon activation, the solution is atomized and dispersed as two fine jets of liquid that converge at a predetermined angle. This collision generates a slow-moving cloud of inhalable particles, known as the soft mist, which patients inhale deeply and slowly (Wachtel et al., 2008). The soft mist consists primarily of fine particles (65–80%) and exits the nozzle at a velocity approximately one-tenth that of aerosol clouds delivered by pMDIs (Ge et al., 2021; Kumar et al., 2022). Additionally, the soft mist has a longer generation time of around 1.5 seconds, aiding in coordination between actuation and inhalation. These features help reduce oropharyngeal impact and improve lung deposition to over 50% of the delivered dose, regardless of the patient's inspiratory flow (Biddiscombe & Usmani, 2018; Iwanaga et al., 2019).

5.6 DRY POWDER INHALERS

DPIs are devices that deliver medication into the airways by the means of energy generated by the patient's inhalation. They contain a mixture of micronized drug particles and large carrier molecules, typically lactose (Levy et al., 2019; Newman, 2017). To ensure optimal deposition of the medication in the lungs, the drug-carrier blend must be dispersed during forceful inhalation against the internal resistance of the inhaler. Unlike pMDIs, DPIs do not require coordination between inhaler actuation and inhalation (de Boer et al., 2022; Rogliani et al., 2017). The patient's inhalation flow may vary depending on their abilities and clinical condition, while the resistance of the device varies between different models. Many DPIs can effectively deliver medication with an inspiratory flow rate of 30 liters per minute or higher, but some devices can accommodate flows as high as 90–100 liters per minute, resulting in a reproducible dose (Mahler, 2020; Ohar et al., 2022). Some DPIs come with premetered single doses contained in capsules that are pierced before inhalation, while others have multi-unit doses, including reservoir-based mechanisms or preloaded doses in blister packs (Chrystyn, 2007; Mohan et al., 2022). Recent advancements in DPI design aim to simplify device usage, with features such as simultaneous loading of doses upon device opening or breath-activated mechanisms. Additionally, certain DPIs may include feedback mechanisms to indicate successful drug dispensing (Berkenfeld et al., 2015; Ye et al., 2022).

5.7 ABOUT SMART INHALERS

Smart inhaler health platforms utilize advanced technology to transform traditional inhalers into smart devices capable of tracking inhaler usage. These platforms consist of special sensors and tiny electronic components integrated into the inhaler (Greene & Costello, 2019; Kaplan & Milstein, 2019). These sensors can detect when the inhaler is used and record the time and date of each usage.

To make use of this technology, patients need to connect their inhaler to their smartphone using a dedicated mobile application. Through this app, the data collected by the inhaler is sent to a secure online server (Chan et al., 2021; Dey et al., 2020). Both patients and their healthcare providers can access this data through their respective portals. The sensors in these smart inhalers are designed to work with popular smartphone operating systems like iOS and Android. They can also provide information about the inhaler's location by using the location data from the patient's smartphone (Kladovasilakis et al., 2023; Mosnaim et al., 2021).

Several smart inhaler devices are available commercially, such as Propeller Health and Teva Digihaler®. Some of these devices have sensors attached to the inhaler, while others have sensors built directly into the inhaler itself. The choice between these options can affect the regulatory approval process (Brennan et al., 2021; Teva, 2020).

These smart inhalers not only track when the inhaler is used but can also provide additional information. For example, they can measure the flow of air during inhalation, ensuring that the patient is using the inhaler correctly (Levy et al., 2019). Some devices can even remind patients to take their medication and alert them if they are using their rescue inhaler too frequently. Patients can easily set up their smart inhaler by downloading the manufacturer's app and syncing it with their smartphone via Bluetooth (Eikholt et al., 2023; Hui et al., 2022). They can then monitor their inhaler usage and other health-related information through the app. If patients do not have a smartphone, they can still access their inhaler data by bringing it to a clinic for downloading (Dhadge et al., 2020; Genes et al., 2018).

For healthcare providers, these smart inhalers offer a convenient way to track patient adherence to their medication regimen and assess their inhalation technique (Chrystyn et al., 2019; Turégano-Yedro et al., 2023). By remotely accessing the data generated by these devices, clinicians can provide timely feedback and support to their patients, ultimately leading to better management of respiratory conditions like COPD (Pinnock et al., 2023; Yang et al., 2022).

The initial generation of DPIs, considered the pioneering smart inhalers, utilized more sophisticated designs compared to older metered-dose inhalers (MDIs). When employed correctly, they enable better drug deposition and consistent dosing, thus supplanting MDIs in frontline maintenance therapy for COPD (Mohan et al., 2022; Ramadan & Sarkis, 2017). However, achieving the high inspiratory flow rates is necessary for certain DPIs poses challenges for patients with compromised lung function, resulting in diminished lung deposition and heightened systemic absorption (Baloira et al., 2021; Weers, 2022). This issue is particularly pertinent in COPD, where disease advancement correlates with reduced inspiratory flow rates. Various studies have highlighted frequent technique errors linked to commonly used DPI models, with research using acoustic devices to estimate inspiratory flow rates during inhaler usage revealing that insufficient inspiratory flow affects around 25% of inhaler uses in moderate to severe COPD (Kocks et al., 2023; Usmani et al., 2018).

Despite advancements in inhaler design, the clinical challenges of poor adherence and technique persist. The emergence of low-cost sensors and data transmission technologies has spurred the development of devices affixed to or integrated within inhalers to address this issue by detecting inhaler technique and assessing adherence (AL-Jahdali et al., 2013; Bosnic-Anticevich et al., 2023). The term "smart inhaler" encompasses various technologies, from basic tracking devices that record inhaler usage to devices that remind patients to take their medication, provide feedback on technique and timing, or analyze the interplay between environmental factors and individual behaviors prompting inhaler use (Poot et al., 2023; Raikar et al., 2023). Some devices incorporate sensors with advanced signal processing to analyze inspiratory flow rates and characterize inhaler technique. While considerable attention is given to the selection and capabilities of sensors in these devices and their data communication methods, their greatest potential lies in the wealth of data generated, which contains intricate information about individual patient behaviors and medication interactions (Raghupathi & Raghupathi, 2014; Sivarajah et al., 2017).

Recognizing low compliance with inhaled medication usage, timing, and technique as a significant barrier to mitigating harm from COPD in higher-income countries, the potential impact of smart inhalers on managing airways diseases remains underexplored despite their promise (George, 2018). In studies involving smart inhalers, primarily deployed to assess therapeutic interventions rather than systematically investigated, a scarcity of research exists, particularly with sizable sample sizes.

While publications in this domain are limited, evidence suggests a notable increase in the number of doses or preventer inhalers taken with the use of smart inhalers, observed across various demographics, care settings, and disease types (Hei et al., 2022; Zabczyk & Blakey, 2021). Such enhanced compliance could potentially mitigate severe outcomes like COPD-related fatalities if replicated on a population scale. Moreover, smart inhalers show promise in improving adherence to prescribed therapy, although current data primarily pertain to specific device types, necessitating further research to validate initial findings (Kanyongo & Ezugwu, 2023; Taylor et al., 2018). Long-term prospects indicate that connected devices may facilitate adherence to medical advice, including action plans, although empirical evidence is lacking. However, beyond compliance, several critical research questions persist regarding the acceptability of these devices to users, their environmental impact due to extensive disposable electronics production, and the ramifications of the lack of system interoperability (Abalansa et al., 2021; Andeobu et al., 2021).

Despite available randomized trial evidence, smart-inhaler use lacks clear recommendations in widely followed COPD guidelines. Notably, the latest Global Initiative for Asthma (GINA) guidelines offer limited recommendations, underscoring the need for further exploration and integration of smart inhalers to enhance adherence and reduce harm in airways diseases (George, 2018; Rogliani et al., 2017).

5.8 DIGITAL INHALERS IN COPD

Upgrading inhaled medications via digital technology can tackle enduring issues of non-adherence and inadequate inhaler technique in individuals with obstructive lung diseases (Bosnic-Anticevich et al., 2023; Volerman et al., 2021). With a mounting body of adjacent clinical evidence, advancements in digital inhaler sensors and platforms, alongside increased backing from healthcare organizations, substantial growth is anticipated with these technologies (Junaid et al., 2022; Pleasants et al., 2022). While all digital inhalers document adherence, their distinctions lie in their compatibility with commercial inhalers, their ability to guide inhaler technique, utilization of patient-reported outcomes (PROs), and ease of use for both professionals and patients (Hoyte et al., 2022). Given the intricacy of employing digital inhalers, collaboration with various entities within health systems is imperative, necessitating a well-planned integration process (Mehta, 2020). Cybersecurity and privacy are pivotal considerations for healthcare officials and patients, mandating review by each healthcare organization. In the United States, some payers reimburse for remote monitoring utilizing digital inhalers, yet reimbursement remains unavailable in other countries (Alanazi, n.d.; Vukotich, 2023). Various models for remote patient care exist, including ongoing digital interface between the professionals and patient to review data solely during clinical encounters. Personalization of therapies and feedback are pivotal for success (Mosnaim et al., 2021). While occurrences of digital inhaler malfunction are rare, patient attrition over a year remains noteworthy. Certain patients may encounter challenges in utilizing digital platforms or lack access to necessary technology (Pleasants et al., 2022). Further research is warranted to address cost-effectiveness, accuracy of inspiratory measurement-capable devices *in vivo*, efficacy in teaching inhaler technique, applicability for monitoring lung function, and ultimately real-world adoption and implementation in routine clinical practice (Klijn et al., 2017).

Medication adherence in individuals with COPD ranks among the lowest for prevalent chronic diseases, significantly impacting health outcomes and costs. Poor adherence in both conditions correlates with heightened exacerbations, inferior symptom control, and increased mortality risk (Homętowska et al., 2022; Zhang et al., 2022). Typically, long-term adherence to maintenance therapies hovers around 50%, although rates may be lower in COPD due to the absence of immediate consequences for non-adherence (Humenberger et al., 2018). Current data indicate that regular use of daily maintenance medications often falls below 50% in both conditions. The resultant poor adherence and clinical outcomes translate into escalated direct and indirect costs, with annual expenditures ranging from $7–17 billion USD due to suboptimal inhaler use (Iuga & McGuire, 2014). Conversely, patients adherent to their inhaled medications experience reduced healthcare utilization, enhanced PROs, and decreased mortality (Burnier & Egan, 2019; Cutler et al., 2018). Studies advocate for regular use of maintenance inhalers, emphasizing the importance of clinicians promoting daily controller use. Clinicians frequently overestimate adherence, and digital adherence monitoring is likely to uncover higher rates of non-adherence and uncontrolled disease than currently perceived based on patient self-reports (Poot et al., 2023).

Utilization of digital inhalers enhances medication adherence, defined as consistent daily use of controllers, and has been linked to decreased rescue inhaler utilization in multiple studies (Chan et al., 2021; Hei et al., 2022). Digital inhalers offer two monitoring approaches: passive monitoring, where the device tracks adherence without feedback to the patient, and active monitoring, where patients receive biofeedback (BF) or clinicians access the adherence data to discuss with the patient (Chan et al., 2021; Mosnaim et al., 2021). Active adherence monitoring involves real-time or periodic review of inhaler use, typically during clinical encounters. A randomized controlled trial involving 437 adults with uncontrolled asthma found that providing biofeedback through a smartphone application significantly increased adherence to fluticasone furoate/vilanterol DPI compared to passive monitoring (mean [SD] percentage adherence: 82% [17%] vs. 71% [27%]; difference: 12%, 95% CI: 5–19%; $p < 0.001$). Clinician-led active monitoring led to an additional 10% improvement. Similarly, another study demonstrated less decline in adherence to inhaled corticosteroids (ICS) over 14 weeks with digital inhaler use compared to controls (–2% vs. –17%; difference: 15%, 95% CI: 4–25%; $p < 0.01$), alongside a significant increase in SABA-free days (19% vs. 6%; $p < 0.01$) (Patel et al., 2013) (Figure 5.1).

Digital inhalers, also applicable in clinical drug trials, serve to quantify inhaler adherence and mitigate the influence of non-adherence on study outcomes. Comparable to clinical settings, drug trials heavily rely on patient self-reporting before and during the study, often utilizing daily diaries to record medication usage (Garin et al., 2023; van de Hei et al., 2023). However, self-reported measures are susceptible to overestimation of adherence due to social desirability bias. Although dose-counters on inhalers are employed to gauge adherence in trials, they tend to overestimate actual usage when compared to digital inhalers in real-world practice (Cachim et al., 2023; Jácome et al., 2021).

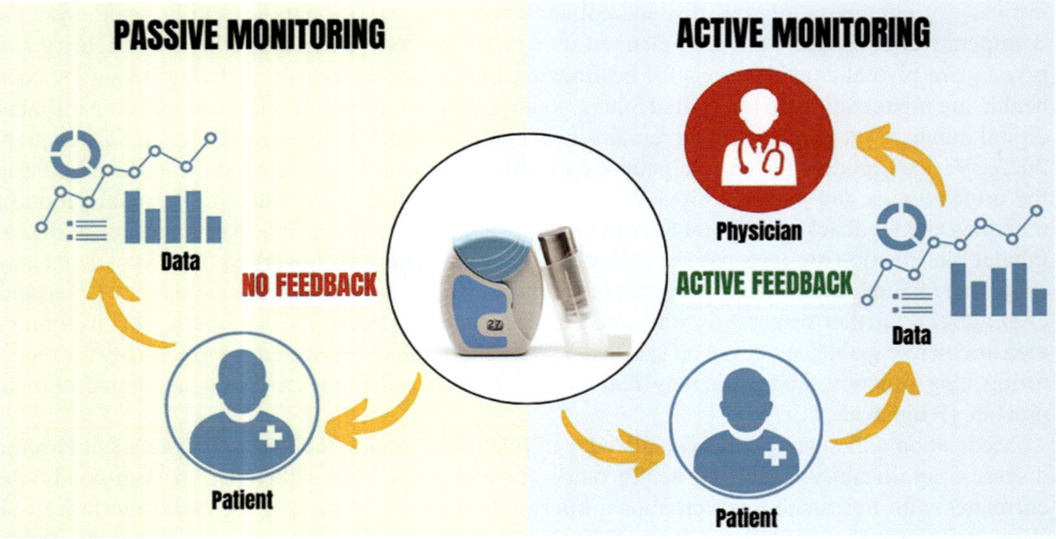

FIGURE 5.1 Smart inhalers are potential in encouraging active monitoring in chronic obstructive pulmonary disease patients.

While run-in periods and study design may mitigate the impact of inhaler non-adherence on outcomes, the true impact remains unclear without the continuous use of digital devices throughout the study (van de Hei et al., 2023). These devices hold significant potential in elucidating the genuine relationship between medication dose and efficacy, such as quantifying how various usage patterns (different doses and frequencies) correlate with reported outcomes (Longuespée et al., 2021; Mason et al., 2022). Despite studies funded by pharmaceutical companies exploring the use of digital inhalers and their impact on adherence and outcomes, the focus has primarily been on evaluating the digital device itself as an adherence intervention (Adejumo et al., 2022). There remains a gap in leveraging digital inhalers to investigate drug dose efficacy relationships within clinical drug trials. A recent systematic review examined how prior adherence to inhalers was assessed in published clinical trials of asthma add-on step 4/5 therapy (Rafi et al., 2022).

5.9 POTENTIAL HINDERANCES IN DIGITAL INHALERS

Examining the efficacy of smart inhalers on medication adherence and inhaler technique, research has delved into the effectiveness of audio-visual reminder feedback (AVRF) and Healthcare profesionals feedback (Perumal et al., 2020). Charles et al., in a 2007 study, enrolled individuals with asthma, revealing significantly higher medication usage in the group receiving AVRF compared to controls (Charles et al., 2007). Subsequent trials by supported these findings, indicating elevated adherence rates with AVRF and professional feedback. The "STudy of Asthma Adherence Reminders" (STAAR) further emphasized the advantages of combining AVRF with professional feedback, demonstrating improved adherence among pediatric asthma patients (Bittner et al., 2019). Expanding on these findings, Foster et al., extended the research to primary care settings, unveiling higher adherence rates with AVRF and feedback interventions (Foster et al., 2014). Nonetheless, further large-scale studies are necessary to validate these observations and ascertain the efficacy of smart inhalers in enhancing medication adherence.

Despite advancements in assessing medication adherence, challenges persist in determining concordance with prescribed doses (Harsha et al., 2019). When studied among individuals with COPD, unveiled disparities between dose usage and adherence to prescribed timing intervals.

Similarly, the INhaler Compliance Assessment (INCA) devices exhibited discrepancies between dose usage and adherence to prescribed dosing windows in severe asthma patients. These findings underscore the necessity for further research to accurately assess the relationship between smart inhalers and concordance with prescribed doses (Hesso et al., 2020; López-Campos et al., 2019).

Smart inhaler technology also facilitates feedback on inhaler technique, potentially enhancing inhaler competence (D'Arcy et al., 2014). Sulaiman et al., discovered that smart inhaler feedback led to higher rates of appropriate and timely dose administration compared to intensive education alone. Nevertheless, more multi-center studies are warranted to validate these findings on a larger scale (Agustí et al., 2023).

Furthermore, smart devices hold promise in automating COPD action plans, enhancing the speed and appropriateness of care escalation. Although promising research is underway, no studies specifically address this aspect, highlighting the need for future investigations to elucidate the relationship between smart devices and effective action plan deployment (Irish Thoracic Society, 2023; Watson & Wilkinson, 2022).

Innovative technology, aimed at personalizing management strategies based on real-time individualized data, is being developed. Various digital strategies, intended to monitor and enhance inhaled medication adherence and technique, have emerged (Blakey et al., 2018; Hein et al., 2020). A recent Cochrane review by Chan et al. explored the array of digital interventions available to tackle medication non-adherence, emphasizing their role in enhancing clinical outcomes in patients suffering from respiratory diseases compared to non-digital interventions and usual care. These interventions range from simple text-message reminders and self-management tools to more intricate digital inhalers equipped with add-on or integrated electromechanical sensors and associated microelectronics. While most of these devices can detect the time and date of inhaler actuation, more advanced ones can even capture clinical-grade inhalation parameters such as peak inspiratory flow (PIF), inhaled volume, and inhalation duration, distinguishing between a mere actuation and an optimal inhalation (Nieuwlaat et al., 2014).

Digital inhalers offer improved connectivity between patients and clinicians by facilitating the exchange of personalized data. Data on medication adherence can be transmitted between connected devices such as inhalers and smartphones via mobile apps (Adejumo et al., 2022; Kaplan et al., 2023). By providing accurate and objective data to all stakeholders, digital technologies hold potential to overcome recurring barriers such as non-adherence in patients with uncontrolled asthma, thus forming the cornerstone of data-driven, personalized patient care (Chrystyn et al., 2019).

To fully realize the capabilty of digital inhalers, further innovation is required to advance research and deepen our understanding of respiratory diseases and the patient's role in managing them (Honkoop et al., 2022; Slevin et al., 2019). Moreover, efforts are needed to inform clinical practice and establish appropriate infrastructure for seamlessly integrating data from digital devices into routine care (Hoffmann et al., 2023; Pannunzio et al., 2023). Monitoring alone cannot improve patient outcomes unless complemented by effective feedback and interventions. While substantial evidence exists on optimal respiratory disease management, there's a need for innovation in how digital inhalers, with their unique features, are integrated into clinical care, tailored to individual patients, and implemented at the right time (Pinnock et al., 2023). This presents our greatest challenge moving forward.

Technological innovation has empowered many patients with chronic conditions to take better control of their disease, leading to improved outcomes (Samal et al., 2021; Siminerio, 2010). However, despite the increasing availability of these products, there is a lack of infrastructure capable of interpreting the vast volume of patient data and supporting the widespread use of digital devices within health services. Understanding other therapy areas offer valuable insights into the facilitators and barriers to integrating digital technologies, such as inhalers, into clinical practice (Alowais et al., 2023).

For instance, in diabetes care, the considerable benefits of digital glucose monitoring were tempered by concerns about increased provider workload and responsibilities due to the data generated (Iyengar et al., 2016). Similar concerns have been raised in asthma care, where healthcare providers anticipate additional workload with the use of digital device data in routine practice. Moreover, sound cybersecurity within digital systems is essential, as highlighted in diabetes care with the development of DTSec, a cybersecurity standard for connected diabetes devices (Arafa et al., 2023). Streamlining data from various digital systems and integrating it into electronic medical records could enhance the efficiency of medical appointments for both patients and clinicians.

Despite the potential of digital solutions to offer advanced care, a consensus framework is necessary for implementation. While digital devices may improve patient adherence to inhaler medication and technique, conflicting reports exist on whether these changes translate into clinical and economic outcomes (Damschroder et al., 2022; Plekhanov et al., 2023).

Ongoing studies, such as the MAGNIFY trial in COPD, aim to investigate the effects of adherence support on disease outcomes. Cost-effectiveness assessments suggest that identifying patients most likely to benefit from adherence-enhancing interventions is crucial for maintaining treatment cost-effectiveness (Price et al., 2021).

Utilizing digital tools to improve outcomes in COPD is explored in this feasibility randomized controlled trial. Recent evidence suggests significant potential for clinical innovation through digital transformation, aiding healthcare systems and empowering patients in self-management (Kraus et al., 2021). However, clinical evidence is limited, necessitating better research to ascertain reliable outcomes of enhanced self-management, particularly in high-risk patient populations discharged from acute care (Paukkonen et al., 2022; Pollack et al., 2016). This study has demonstrated the feasibility of using a digital platform like the myCOPD app to support patients. Despite experiencing significant clinical events such as exacerbation, most subjects randomized to the app engaged with the system to an extent, enabling detection of clinically important benefits associated with the intervention (North et al., 2020).

The use of the app in this vulnerable patient population correlates with detectable improvements in disease control, measured by the validated CAT, reductions in exacerbations, and a signal of decreased risk of hospital readmission (Hei et al., 2022). Additionally, improved inhaler technique was observed in the digital arm compared to usual care. However, patient acceptance of digital tools remains a challenge, with some declining participation due to concerns about privacy, security, and technological efficacy, particularly in older adults. Nonetheless, ongoing implementation experiments aim to ensure access to digital health interventions across all patient demographics (North et al., 2020).

Engagement factors and technology accessibility are pivotal in the debate surrounding digital health interventions, especially in the context of recently acutely unwell patients, unlike stable patients in most published trials (Barello et al., 2016; Nahum-Shani et al., 2022). While the use of mobile technologies has surpassed its height in the past decade, equitable access remains an issue, especially among older and lower socioeconomic status individuals (Graetz et al., 2018; Lythreatis et al., 2022). Despite challenges, the potential benefits of digital interventions in COPD management are considerable, underscored by this trial's early signals of efficacy. Further research, including larger studies, is warranted to elucidate optimal usage patterns and long-term effects of digital interventions in COPD management, addressing an unmet clinical need in the field (Janjua et al., 2021).

However, translating these findings into real-world implementation faces challenges, particularly the lack of reimbursement incentives for healthcare providers to engage with digital technologies. Addressing this issue requires concerted efforts to reach a solution.

5.10 FUTURE PROSPECTS

While recent advancements in COPD device therapy offer numerous benefits, there are also potential drawbacks and challenges associated with their use. Understanding these challenges is crucial for a comprehensive grasp of the impact of these innovations. Several challenges have been outlined below.

First, many advanced COPD treatment devices, such as smart inhalers and remote monitoring systems, may come with higher costs compared to traditional inhalers or non-connected devices (Vachon et al., 2022). These expenses could pose a barrier to access for certain patients, especially those without adequate insurance coverage or limited financial resources. Additionally, technological barriers present another challenge (Biddell et al., 2023). COPD primarily affects older adults, and some patients may not be familiar with or comfortable using complex technologies. This can hinder the effective use of smart inhalers, mobile apps, or wearables (Bao et al., 2023).

Furthermore, remote monitoring systems and smart devices collect and share health data, raising concerns about the security and privacy of this information. Ensuring data privacy and compliance with security regulations is essential. Another challenge is the reliance on technology (Shojaei et al., 2024). While advanced devices offer various benefits, many patients are becoming overly reliant on technology to manage their illnesses. Over-dependence on technology may lead to neglect of other important aspects of COPD care, such as lifestyle changes and medication adherence (Coombs et al., 2022; Jha & Chandi, n.d.). Additionally, maintenance of these devices is necessary. If these devices experience glitches, software issues, or connectivity problems, it could lead to patient dissatisfaction and impact the effectiveness of therapy (Kim et al., 2017).

Moreover, increased reliance on remote monitoring and virtual communications may result in decreased personal interactions between patients and healthcare providers (Morag et al., 2023). Special interactions and empathetic communication are crucial in healthcare, and the lack of in-person discussions could affect the quality of care provided (Moudatsou et al., 2020).

Despite these challenges, continuous refinement, improvement of potential difficulties, and enhanced transparency could further improve their benefits and contribute to better COPD management and patient outcomes (Jha & Chandi, n.d.). Healthcare providers, manufacturers, and others must overcome these challenges and ensure that these advancements are accessible, effective, and user-friendly for all COPD patients (Rogliani et al., 2017).

Examining the potential of digital tools in enhancing outcomes for individuals with COPD, researchers delve into addressing the prevalent issue of medication non-adherence. This phenomenon, common in both COPD often leads to suboptimal health results (Bosnic-Anticevich et al., 2023). Furthermore, despite patients' belief in proper usage, many struggle with correct inhaler techniques, diminishing the efficacy of their prescribed medications (Figure 5.2).

To confront these challenges, researchers explore the utilization of digital solutions, such as smartphone applications and inhaler-attached devices, to monitor patients' medication intake (Kasoju et al., 2023). These tools offer more precise data compared to self-reported information, providing insights into medication usage frequency and adherence to prescribed techniques. This enables healthcare providers to make well-informed decisions regarding patient care (Walquist et al., 2022).

Various forms of medication non-adherence exist, including unintentional forgetfulness and intentional avoidance. Digital tools offer tailored solutions, such as reminders and interactive features, to address both types effectively (Hugtenburg et al., 2013). Additionally, they provide valuable feedback on inhaler usage, aiding patients and professionals in ensuring proper administration of medications (Stewart et al., 2023).

The potential of digital tools extends beyond improving adherence; they also hold promise in enhancing patient engagement and disease management. By offering personalized support and guidance, these tools empower individuals to take control of their health and improve their overall well-being (Georgantzinos et al., 2021; Zhang & Kamel Boulos, 2023).

Digital tools present a promising avenue for enhancing medication adherence and inhaler technique among COPD patients. With their ability to provide accurate data and personalized support, these tools have the potential to revolutionize the management of respiratory conditions and optimize patient outcomes.

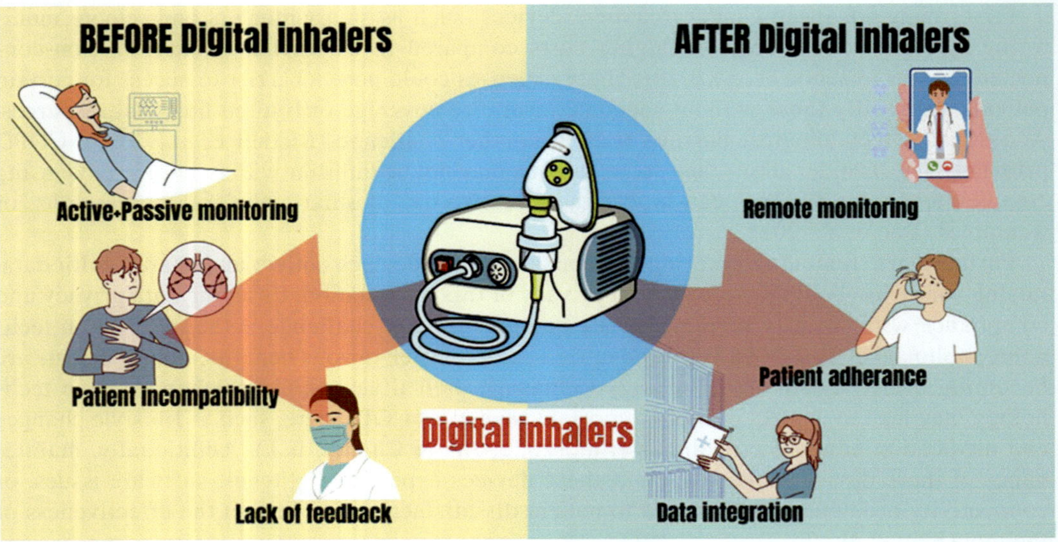

FIGURE 5.2 Digital inhalers have been a game changer in the treatment of respiratory diseases like chronic obstructive pulmonary disease, as they integrate different aspects of healthcare from medication adherence to data management.

5.11 CONCLUSION

Utilized widely in discussions surrounding inhaler use among patients with COPD, the terms "adherence" and "inhaler technique" veil intricate and personalized patient behaviors, attitudes, and disease management approaches that have historically lacked satisfactory attention. Incorporating significant technological advancements, inhaler devices now offer a range of digital solutions poised to transform disease management and outcomes. Linked to mobile applications, these digital inhaler devices hold the potential to instigate shifts in patient behaviors and attitudes toward managing their COPD, particularly when integrating interactive and gamified elements. Furthermore, they can facilitate education of ongoing techniques, empowering patients and aiding service providers in maximizing the value of consultations, informed by objective insights into patient inhaler usage patterns and respiratory health. Additionally, when integrated with innovative techniques like machine learning, they may forecast exacerbations and prompt preemptive intervention. Ultimately, digital inhalers stand to support an advanced precision-medicine approach to COPD care and shared decision-making. Nevertheless, there exists a risk that consumers and healthcare providers may feel overwhelmed by the plethora of choices and capabilities offered by these technologies. Overcoming the hurdle and realizing the potential of digital inhalers requires an approach that is informed, strategic, and respectful of user preferences. Integrating into management guidelines, providing suitable reports and strategies, and supporting real-world uptake are imperative measures. Only by taking such measures can the data captured by digital devices be harnessed to deliver personalized disease management.

REFERENCES

Abalansa, S., El Mahrad, B., Icely, J., & Newton, A. (2021). Electronic waste, an environmental problem exported to developing countries: The GOOD, the BAD and the UGLY. *Sustainability*, *13*(9), 5302. https://doi.org/10.3390/su13095302

Adejumo, I., Patel, M., McKeever, T. M., Shaw, D. E., & Bains, M. (2022). Qualitative study of user perspectives and experiences of digital inhaler technology. *NPJ Primary Care Respiratory Medicine*, *32*, 57. https://doi.org/10.1038/s41533-022-00320-9

Agarwal, A. K., Raja, A., & Brown, B. D. (2024). Chronic Obstructive Pulmonary Disease. In *StatPearls*. StatPearls Publishing. http://www.ncbi.nlm.nih.gov/books/NBK559281/

Agustí, A., Celli, B. R., Criner, G. J., Halpin, D., Anzueto, A., Barnes, P., Bourbeau, J., Han, M. K., Martinez, F. J., Montes de Oca, M., Mortimer, K., Papi, A., Pavord, I., Roche, N., Salvi, S., Sin, D. D., Singh, D., Stockley, R., López Varela, M. V., … Vogelmeier, C. F. (2023). Global initiative for chronic obstructive lung disease 2023 report: GOLD executive summary. *The European Respiratory Journal*, *61*(4), 2300239. https://doi.org/10.1183/13993003.00239-2023

Alanazi, A. T. (n.d). Clinicians' perspectives on healthcare cybersecurity and cyber threats. *Cureus*, *15*(10), e47026. https://doi.org/10.7759/cureus.47026

Alharbi, A. S., Yousef, A. A., Alharbi, S. A., Al-Shamrani, A., Alqwaiee, M. M., Almeziny, M., Said, Y. S., Alshehri, S. A., Alotaibi, F. N., Mosalli, R., Alawam, K. A., & Alsaadi, M. M. (2021). Application of aerosol therapy in respiratory diseases in children: A Saudi expert consensus. *Annals of Thoracic Medicine*, *16*(2), 188–218. https://doi.org/10.4103/atm.atm_74_21

AL-Jahdali, H., Ahmed, A., AL-Harbi, A., Khan, M., Baharoon, S., Bin Salih, S., Halwani, R., & Al-Muhsen, S. (2013). Improper inhaler technique is associated with poor asthma control and frequent emergency department visits. *Allergy, Asthma & Clinical Immunology*, *9*(1), 8. https://doi.org/10.1186/1710-1492-9-8

Alowais, S. A., Alghamdi, S. S., Alsuhebany, N., Alqahtani, T., Alshaya, A. I., Almohareb, S. N., Aldairem, A., Alrashed, M., Bin Saleh, K., Badreldin, H. A., Al Yami, M. S., Al Harbi, S., & Albekairy, A. M. (2023). Revolutionizing healthcare: The role of artificial intelligence in clinical practice. *BMC Medical Education*, *23*(1), 689. https://doi.org/10.1186/s12909-023-04698-z

Alsayed, A. R., Abu-Samak, M. S., & Alkhatib, M. (2023). Asthma-COPD overlap in clinical practice (ACO_CP 2023): Toward precision medicine. *Journal of Personalized Medicine*, *13*(4), 677. https://doi.org/10.3390/jpm13040677

Andeobu, L., Wibowo, S., & Grandhi, S. (2021). A systematic review of E-waste generation and environmental management of asia pacific countries. *International Journal of Environmental Research and Public Health*, *18*(17), 9051. https://doi.org/10.3390/ijerph18179051

Anderson, C. F., Grimmett, M. E., Domalewski, C. J., & Cui, H. (2020). Inhalable nanotherapeutics to improve treatment efficacy for common lung diseases. *Wiley Interdisciplinary Reviews. Nanomedicine and Nanobiotechnology*, *12*(1), e1586. https://doi.org/10.1002/wnan.1586

Anderson, P. (2005). History of aerosol therapy: Liquid nebulization to MDIs to DPIs. *Respiratory Care*, *50*, 1139–1150.

Anderson, P. (2006). Use of Respimat® Soft Mist™ Inhaler in COPD patients. *International Journal of Chronic Obstructive Pulmonary Disease*, *1*(3), 251–259.

Arafa, A., Sheerah, H. A., & Alsalamah, S. (2023). Emerging digital technologies in healthcare with a spotlight on cybersecurity: A narrative review. *Information*, *14*(12), 640. https://doi.org/10.3390/info14120640

Aremu, T. O., Oluwole, O. E., Adeyinka, K. O., & Schommer, J. C. (2022). Medication adherence and compliance: Recipe for improving patient outcomes. *Pharmacy: Journal of Pharmacy Education and Practice*, *10*(5), 106. https://doi.org/10.3390/pharmacy10050106

Ari, A. (2014). Jet, ultrasonic, and mesh nebulizers: An evaluation of nebulizers for better clinical outcomes. *Euras J Pulm*, *16*, 1–7. https://doi.org/10.5152/ejp.2014.00087

Baloira, A., Abad, A., Fuster, A., García Rivero, J. L., García-Sidro, P., Márquez-Martín, E., Palop, M., Soler, N., Velasco, J. L., & González-Torralba, F. (2021). Lung deposition and inspiratory flow rate in patients with chronic obstructive pulmonary disease using different inhalation devices: A systematic literature review and expert opinion. *International Journal of Chronic Obstructive Pulmonary Disease*, *16*, 1021–1033. https://doi.org/10.2147/COPD.S297980

Bao, L. K., Khoa, N. D., Chi, L. T. K., & Anh, N. T. (2023). Prevalence and factors affecting appropriate inhaler use in elderly patients with chronic obstructive pulmonary disease: A prospective study. *Journal of Clinical Medicine*, *12*(13), 4420. https://doi.org/10.3390/jcm12134420

Barello, S., Triberti, S., Graffigna, G., Libreri, C., Serino, S., Hibbard, J., & Riva, G. (2016). eHealth for patient engagement: A systematic review. *Frontiers in Psychology*, *6*, 2013. https://doi.org/10.3389/fpsyg.2015.02013

Baryakova, T. H., Pogostin, B. H., Langer, R., & McHugh, K. J. (2023). Overcoming barriers to patient adherence: The case for developing innovative drug delivery systems. *Nature Reviews Drug Discovery*, *22*(5), 387–409. https://doi.org/10.1038/s41573-023-00670-0

Berkenfeld, K., Lamprecht, A., & McConville, J. T. (2015). Devices for dry powder drug delivery to the lung. *AAPS PharmSciTech*, *16*(3), 479–490. https://doi.org/10.1208/s12249-015-0317-x

Biddell, C. B., Waters, A. R., Angove, R. S. M., Gallagher, K. D., Rosenstein, D. L., Spees, L. P., Kent, E. E., Planey, A. M., & Wheeler, S. B. (2023). Facing financial barriers to healthcare: Patient-informed adaptation of a conceptual framework for adults with a history of cancer. *Frontiers in Psychology, 14*, 1178517. https://doi.org/10.3389/fpsyg.2023.1178517

Biddiscombe, M. F., & Usmani, O. S. (2018). Is there room for further innovation in inhaled therapy for airways disease? *Breathe, 14*(3), 216–224. https://doi.org/10.1183/20734735.020318

Bittner, B., Schmit Chiesi, C., Kharawala, S., Kaur, G., & Schmidt, J. (2019). Connected drug delivery devices to complement drug treatments: Potential to facilitate disease management in home setting. *Medical Devices (Auckland, N.Z.), 12*, 101–127. https://doi.org/10.2147/MDER.S198943

Blakey, J. D., Bender, B. G., Dima, A. L., Weinman, J., Safioti, G., & Costello, R. W. (2018). Digital technologies and adherence in respiratory diseases: The road ahead. *The European Respiratory Journal, 52*(5), 1801147. https://doi.org/10.1183/13993003.01147-2018

Borghardt, J. M., Kloft, C., & Sharma, AJ. (2018). Inhaled therapy in respiratory disease: The complex interplay of pulmonary kinetic processes. *Canadian Respiratory Journal, 2018*, e2732017. https://doi.org/10.1155/2018/2732017

Bosnic-Anticevich, S., Bakerly, N. D., Chrystyn, H., Hew, M., & Palen, J. van der. (2023a). Advancing digital solutions to overcome longstanding barriers in asthma and COPD management. *Patient Preference and Adherence, 17*, 259–272. https://doi.org/10.2147/PPA.S385857

Bosnic-Anticevich, S., Bender, B. G., Shuler, M. T., Hess, M., & Kocks, J. W. H. (2023b). Recognizing and tackling inhaler technique decay in asthma and chronic obstructive pulmonary disease (COPD) clinical practice. *The Journal of Allergy and Clinical Immunology: In Practice, 11*(8), 2355–2364.e5. https://doi.org/10.1016/j.jaip.2023.04.031

Brennan, V., Mulvey, C., Greene, G., Hale, E. M., & Costello, R. W. (2021). A clinical perspective on the role of electronic devices in monitoring and promoting adherence in airways disease. *Frontiers in Medical Technology, 3*, 604475. https://doi.org/10.3389/fmedt.2021.604475

Burkhardt, R., & Pankow, W. (2014). The diagnosis of chronic obstructive pulmonary disease. *Deutsches Ärzteblatt International, 111*(49), 834–846. https://doi.org/10.3238/arztebl.2014.0834

Burnier, M., & Egan, B. M. (2019). Adherence in hypertension. *Circulation Research, 124*(7), 1124–1140. https://doi.org/10.1161/CIRCRESAHA.118.313220

Buttini, F., Glieca, S., Sonvico, F., & Lewis, D. A. (2023). Metered dose inhalers in the transition to low GWP propellants: What we know and what is missing to make it happen. *Expert Opinion on Drug Delivery, 20*(8), 1131–1143. https://doi.org/10.1080/17425247.2023.2264184

Cachim, A., Pereira, A. M., Almeida, R., Amaral, R., Alves-Correia, M., Vieira-Marques, P., Chaves-Loureiro, C., Ribeiro, C., Cardia, F., Gomes, J., Vidal, C., Silva, E., Rocha, S., Rocha, D., Marques, M. L., Páscoa, R., Morais, D., Cruz, A. M., Santalha, M., … Jácome, C. (2023). Measuring adherence to inhaled control medication in patients with asthma: Comparison among an asthma app, patient self-report and physician assessment. *Clinical and Translational Allergy, 13*(2), e12210. https://doi.org/10.1002/clt2.12210

Cavaillès, A., Brinchault-Rabin, G., Dixmier, A., Goupil, F., Gut-Gobert, C., Marchand-Adam, S., Meurice, J.-C., Morel, H., Person-Tacnet, C., Leroyer, C., & Diot, P. (2013). Comorbidities of COPD. *European Respiratory Review, 22*(130), 454–475. https://doi.org/10.1183/09059180.00008612

Chan, A., Pleasants, R., Dhand, R., Tilley, S., Schworer, S., Costello, R., & Merchant, R. (2021). Digital inhalers for asthma or chronic obstructive pulmonary disease: A scientific perspective. *Pulmonary Therapy, 7*. https://doi.org/10.1007/s41030-021-00167-4

Chan, J. G. Y., Wong, J., Zhou, Q. T., Leung, S. S. Y., & Chan, H.-K. (2014). Advances in device and formulation technologies for pulmonary drug delivery. *AAPS PharmSciTech, 15*(4), 882–897. https://doi.org/10.1208/s12249-014-0114-y

Charles, T., Quinn, D., Weatherall, M., Aldington, S., Beasley, R., & Holt, S. (2007). An audiovisual reminder function improves adherence with inhaled corticosteroid therapy in asthma. *Journal of Allergy and Clinical Immunology, 119*(4), 811–816. https://doi.org/10.1016/j.jaci.2006.11.700

Chow, M. Y. T., Pan, H. W., & Lam, J. K. W. (2023). Delivery Technology of Inhaled Therapy for Asthma and COPD. In W. S. F. Wong (Ed.), *Advances in Pharmacology* (Vol. 98, pp. 273–311). Academic Press. https://doi.org/10.1016/bs.apha.2023.03.001

Chrystyn, H. (2007). The Diskus™: A review of its position among dry powder inhaler devices. *International Journal of Clinical Practice, 61*(6), 1022–1036. https://doi.org/10.1111/j.1742-1241.2007.01382.x

Chrystyn, H., Audibert, R., Keller, M., Quaglia, B., Vecellio, L., & Roche, N. (2019). Real-life inhaler adherence and technique: Time to get smarter! *Respiratory Medicine, 158*, 24–32. https://doi.org/10.1016/j.rmed.2019.09.008

Coombs, N. C., Campbell, D. G., & Caringi, J. (2022). A qualitative study of rural healthcare providers' views of social, cultural, and programmatic barriers to healthcare access. *BMC Health Services Research*, *22*(1), 438. https://doi.org/10.1186/s12913-022-07829-2

Craven, V. E., Morton, R. W., Spencer, S., Devadason, S. G., & Everard, M. L. (2018). Electronic monitoring and reminding devices for improving adherence to inhaled therapy in patients with asthma. The *Cochrane Database of Systematic Reviews*, *2018*(4), CD011554. https://doi.org/10.1002/14651858.CD011554.pub2

Cutler, R. L., Fernandez-Llimos, F., Frommer, M., Benrimoj, C., & Garcia-Cardenas, V. (2018). Economic impact of medication non-adherence by disease groups: A systematic review. *BMJ Open*, *8*(1), e016982. https://doi.org/10.1136/bmjopen-2017-016982

Czechtizky, W., Su, W., Ripa, L., Schiesser, S., Höijer, A., & Cox, R. J. (2022). Advances in the Design of New Types of Inhaled Medicines. In D. R. Witty & B. Cox (Eds.), *Progress in Medicinal Chemistry* (Vol. 61, pp. 93–162). Elsevier. https://doi.org/10.1016/bs.pmch.2022.04.001

D'Arcy, S., MacHale, E., Seheult, J., Holmes, M. S., Hughes, C., Sulaiman, I., Hyland, D., O'Reilly, C., Glynn, S., Al-Zaabi, T., McCourt, J., Taylor, T., Keane, F., Killane, I., Reilly, R. B., & Costello, R. W. (2014). A method to assess adherence in inhaler use through analysis of acoustic recordings of inhaler events. *PLoS ONE*, *9*(6), e98701. https://doi.org/10.1371/journal.pone.0098701

Dalby, R. N., Eicher, J., & Zierenberg, B. (2011). Development of Respimat® Soft Mist™ Inhaler and its clinical utility in respiratory disorders. *Medical Devices (Auckland, N.Z.)*, *4*, 145. https://doi.org/10.2147/MDER.S7409

Dampage, U., Ariyasinghe, M., & Pulleperuma, S. (2022). An automated jet nebulizer with dynamic flow regulation. *Journal of Pharmaceutical Innovation*, *17*(3), 725–735. https://doi.org/10.1007/s12247-021-09557-2

Damschroder, L. J., Reardon, C. M., Widerquist, M. A. O., & Lowery, J. (2022). The updated consolidated framework for implementation research based on user feedback. *Implementation Science*, *17*(1), 75. https://doi.org/10.1186/s13012-022-01245-0

de Boer, A. H., Hagedoorn, P., & Grasmeijer, F. (2022). Dry powder inhalation, part 2: The present and future. *Expert Opinion on Drug Delivery*, *19*(9), 1045–1059. https://doi.org/10.1080/17425247.2022.2112570

Dey, A., Haque, K., Nayan, A.-A., & Kibria, M. (2020). *IoT Based Smart Inhaler For Context-Aware Service Provisioning* (p. 415). https://doi.org/10.1109/ICAICT51780.2020.9333427

Dhadge, N., Shevade, M., Kale, N., Narke, G., Pathak, D., Barne, M., Madas, S., & Salvi, S. (2020). Monitoring of inhaler use at home with a smartphone video application in a pilot study. *NPJ Primary Care Respiratory Medicine*, *30*(1), 46. https://doi.org/10.1038/s41533-020-00203-x

Dhanani, J., Fraser, J. F., Chan, H.-K., Rello, J., Cohen, J., & Roberts, J. A. (2016). Fundamentals of aerosol therapy in critical care. *Critical Care*, *20*(1), 269. https://doi.org/10.1186/s13054-016-1448-5

Dhand, R. (2017). How should aerosols be delivered during invasive mechanical ventilation? *Respiratory Care*, *62*(10), 1343–1367. https://doi.org/10.4187/respcare.05803

Dhruve, H., & Jackson, D. J. (2022). Assessing adherence to inhaled therapies in asthma and the emergence of electronic monitoring devices. *European Respiratory Review*, *31*(164). https://doi.org/10.1183/16000617.0271-2021

DiBlasi, R. M. (2015). Clinical controversies in aerosol therapy for infants and children. *Respiratory Care*, *60*(6), 894–916. https://doi.org/10.4187/respcare.04137

Eikholt, A. A., Wiertz, M. B. R., Hew, M., Chan, A. H. Y., & van Boven, J. F. M. (2023). Electronic monitoring devices to support inhalation technique in patients with asthma: A narrative review. *Current Treatment Options in Allergy*, *10*(1), 28–52. https://doi.org/10.1007/s40521-023-00328-7

Eriksson, J., Thörn, H., Lennernäs, H., & Sjögren, E. (2020). Pulmonary drug absorption and systemic exposure in human: Predictions using physiologically based biopharmaceutics modeling. *European Journal of Pharmaceutics and Biopharmaceutics*, *156*, 191–202. https://doi.org/10.1016/j.ejpb.2020.09.004

Foster, J. M., Usherwood, T., Smith, L., Sawyer, S. M., Xuan, W., Rand, C. S., & Reddel, H. K. (2014). Inhaler reminders improve adherence with controller treatment in primary care patients with asthma. *The Journal of Allergy and Clinical Immunology*, *134*(6), 1260–1268.e3. https://doi.org/10.1016/j.jaci.2014.05.041

Garin, N., Zarate-Tamames, B., Gras-Martin, L., Milà, R., Crespo-Lessmann, A., Curto, E., Hernandez, M., Mestres, C., & Plaza, V. (2023). Clinical impact of electronic monitoring devices of inhalers in adults with asthma or COPD: A systematic review and meta-analysis. *Pharmaceuticals*, *16*(3), 414. https://doi.org/10.3390/ph16030414

Ge, Y., Tong, Z., Li, R., Huang, F., & Yu, J. (2021). Numerical and experimental investigation on key parameters of the Respimat® spray inhaler. *Processes*, *9*(1), 44. https://doi.org/10.3390/pr9010044

Genes, N., Violante, S., Cetrangol, C., Rogers, L., Schadt, E. E., & Chan, Y.-F. Y. (2018). From smartphone to EHR: A case report on integrating patient-generated health data. *NPJ Digital Medicine*, *1*(1), 1–6. https://doi.org/10.1038/s41746-018-0030-8

Georgantzinos, S. K., Giannopoulos, G. I., & Bakalis, P. A. (2021). Additive manufacturing for effective smart structures: The idea of 6D printing. *Journal of Composites Science*, *5*(5), 119. https://doi.org/10.3390/jcs5050119

George, M. (2018). Adherence in asthma and COPD: New strategies for an old problem. *Respiratory Care*, *63*, 818–831. https://doi.org/10.4187/respcare.05905

Goyal, A., Cusick, A. S., & Bhutta, B. S. (2024). Peripheral Edema. In *StatPearls*. StatPearls Publishing. http://www.ncbi.nlm.nih.gov/books/NBK554452/

Graetz, I., Huang, J., Brand, R., Hsu, J., Yamin, C. K., & Reed, M. E. (2018). Bridging the digital divide: Mobile Access to personal health records among patients with diabetes. *The American Journal of Managed Care*, *24*(1), 43–48.

Greene, G., & Costello, R. (2019). Personalizing medicine—Could the smart inhaler revolutionize treatment for COPD and asthma patients? *Expert Opinion on Drug Delivery*, *16*. https://doi.org/10.1080/17425247.2019.1628017

Guo, Y., Bera, H., Shi, C., Zhang, L., Cun, D., & Yang, M. (2021). Pharmaceutical strategies to extend pulmonary exposure of inhaled medicines. *Acta Pharmaceutica Sinica B*, *11*(8), 2565–2584. https://doi.org/10.1016/j.apsb.2021.05.015

Hagmeyer, L., van Koningsbruggen-Rietschel, S., Matthes, S., Rietschel, E., & Randerath, W. (2023). From the infant to the geriatric patient—Strategies for inhalation therapy in asthma and chronic obstructive pulmonary disease. *The Clinical Respiratory Journal*, *17*(6), 487–498. https://doi.org/10.1111/crj.13610

Harsha, N., Papp, M., Kőrösi, L., Czifra, Á., Ádány, R., & Sándor, J. (2019). Enhancing primary adherence to prescribed medications through an organized health status assessment-based extension of primary healthcare services. *International Journal of Environmental Research and Public Health*, *16*(20), 3797. https://doi.org/10.3390/ijerph16203797

He, S., Gui, J., Xiong, K., Chen, M., Gao, H., & Fu, Y. (2022). A roadmap to pulmonary delivery strategies for the treatment of infectious lung diseases. *Journal of Nanobiotechnology*, *20*(1), 101. https://doi.org/10.1186/s12951-022-01307-x

Hei, S., Poot, C., van den Berg, L., Meijer, E., Van Boven, J., Blok, B., Postma, M., Chavannes, N., & Kocks, J. (2022). Effectiveness, usability and acceptability of a smart inhaler programme in patients with asthma: Protocol of the multicentre, pragmatic, open-label, cluster randomised controlled ACCEPTANCE trial. *BMJ Open Respiratory Research*, *9*, e001400. https://doi.org/10.1136/bmjresp-2022-001400

Hein, A.-E., Vrijens, B., & Hiligsmann, M. (2020). A digital innovation for the personalized management of adherence: Analysis of strengths, weaknesses, opportunities, and threats. *Frontiers in Medical Technology*, *2*. https://doi.org/10.3389/fmedt.2020.604183

Hesso, I., Nabhani Gebara, S., Greene, G., Co stello, R. W., & Kayyali, R. (2020). A quantitative evaluation of adherence and inhalation technique among respiratory patients: An observational study using an electronic inhaler assessment device. *International Journal of Clinical Practice*, *74*(2), e13437. https://doi.org/10.1111/ijcp.13437

Himes, B. E., Leszinsky, L., Walsh, R., Hepner, H., & Wu, A. C. (2019). Mobile Health and inhaler-based monitoring devices for asthma management. *The Journal of Allergy and Clinical Immunology. In Practice*, *7*(8), 2535–2543. https://doi.org/10.1016/j.jaip.2019.08.034

Hoffmann, K., Pelz, A., Karg, E., Gottschalk, A., Zerjatke, T., Schuster, S., Böhme, H., Glauche, I., & Roeder, I. (2023). Data integration between clinical research and patient care: A framework for context-depending data sharing and in silico predictions. *PLOS Digital Health*, *2*(5), e0000140. https://doi.org/10.1371/journal.pdig.0000140

Holland, A., Smith, F., Penny, K., McCrossan, G., Veitch, L., & Nicholson, C. (2013). Metered dose inhalers versus nebulizers for aerosol bronchodilator delivery for adult patients receiving mechanical ventilation in critical care units. *The Cochrane Database of Systematic Reviews*, *2013*(6), CD008863. https://doi.org/10.1002/14651858.CD008863.pub2

Homętowska, H., Świątoniowska-Lonc, N., Klekowski, J., Chabowski, M., & Jankowska-Polańska, B. (2022). Treatment adherence in patients with obstructive pulmonary diseases. *International Journal of Environmental Research and Public Health*, *19*(18), 11573. https://doi.org/10.3390/ijerph191811573

Honkoop, P., Usmani, O., & Bonini, M. (2022). The current and future role of technology in respiratory care. *Pulmonary Therapy*, *8*(2), 167–179. https://doi.org/10.1007/s41030-022-00191-y

Hou, S., Wu, J., Li, X., & Shu, H. (2015). Practical, regulatory and clinical considerations for development of inhalation drug products. *Asian Journal of Pharmaceutical Sciences*, *10*(6), 490–500. https://doi.org/10.1016/j.ajps.2015.08.008

Hoyte, F. C. L., Mosnaim, G. S., Rogers, L., Safioti, G., Brown, R., Li, T., DePietro, M., Reich, M., Hill, T. D., & Wechsler, M. E. (2022). Effectiveness of a digital inhaler system for patients with asthma: A 12-week, open-label, randomized study (CONNECT1). *The Journal of Allergy and Clinical Immunology: In Practice*, *10*(10), 2579–2587. https://doi.org/10.1016/j.jaip.2022.08.023

Hugtenburg, J. G., Timmers, L., Elders, P. J., Vervloet, M., & van Dijk, L. (2013). Definitions, variants, and causes of nonadherence with medication: A challenge for tailored interventions. *Patient Preference and Adherence*, *7*, 675–682. https://doi.org/10.2147/PPA.S29549

Hui, C. Y., McKinstry, B., Mclean, S., Buchner, M., & Pinnock, H. (2022). Assessing the technical feasibility of a flexible, integrated Internet-of-things connected for asthma (C4A) system to support self-management: A mixed method study exploring patients and healthcare professionals perspectives. *JAMIA Open*, *5*(4), ooac110. https://doi.org/10.1093/jamiaopen/ooac110

Humenberger, M., Horner, A., Labek, A., Kaiser, B., Frechinger, R., Brock, C., Lichtenberger, P., & Lamprecht, B. (2018). Adherence to inhaled therapy and its impact on chronic obstructive pulmonary disease (COPD). *BMC Pulmonary Medicine*, *18*, 163. https://doi.org/10.1186/s12890-018-0724-3

Ibrahim, M., Verma, R., & Garcia-Contreras, L. (2015). Inhalation drug delivery devices: Technology update. *Medical Devices (Auckland, N.Z.)*, *8*, 131–139. https://doi.org/10.2147/MDER.S48888

Irish Thoracic Society. (2023). Irish Thoracic Society Annual Scientific Meeting 2023. *Irish Journal of Medical Science (1971–)*, *192*(2), 41–120. https://doi.org/10.1007/s11845-023-03534-z

Iuga, A. O., & McGuire, M. J. (2014). Adherence and health care costs. *Risk Management and Healthcare Policy*, *7*, 35–44. https://doi.org/10.2147/RMHP.S19801

Iwanaga, T., Tohda, Y., Nakamura, S., & Suga, Y. (2019). The Respimat® soft mist inhaler: Implications of drug delivery characteristics for patients. *Clinical Drug Investigation*, *39*(11), 1021–1030. https://doi.org/10.1007/s40261-019-00835-z

Iyengar, V., Wolf, A., Brown, A., & Close, K. (2016). Challenges in diabetes care: Can digital health help address them? *Clinical Diabetes: A Publication of the American Diabetes Association*, *34*(3), 133–141. https://doi.org/10.2337/diaclin.34.3.133

Jácome, C., Almeida, R., Pereira, A. M., Amaral, R., Vieira-Marques, P., Mendes, S., Alves-Correia, M., Ferreira, J. A., Lopes, I., Gomes, J., Araújo, L., Couto, M., Chaves Loureiro, C., Santos, L. M., Arrobas, A., Valério, M., Todo Bom, A., Azevedo, J., Teixeira, M. F., … Fonseca, J. A. (2021). Monitoring adherence to asthma inhalers using the InspirerMundi app: Analysis of real-world, medium-term feasibility studies. *Frontiers in Medical Technology*, *3*, 649506. https://doi.org/10.3389/fmedt.2021.649506

Janjua, S., Carter, D., Threapleton, C. J., Prigmore, S., & Disler, R. T. (2021). Telehealth interventions: Remote monitoring and consultations for people with chronic obstructive pulmonary disease (COPD). *The Cochrane Database of Systematic Reviews*, *2021*(7), CD013196. https://doi.org/10.1002/14651858.CD013196.pub2

Jardim, J. R., & Nascimento, O. A. (2019). The importance of inhaler adherence to prevent COPD exacerbations. *Medical Sciences*, *7*(4), 54. https://doi.org/10.3390/medsci7040054

Javadzadeh, Y., & Yaqoubi, S. (2017). Therapeutic Nanostructures for Pulmonary Drug Delivery. In E. Andronescu & A. M. Grumezescu (Eds.), *Nanostructures for Drug Delivery* (pp. 619–638). Elsevier. https://doi.org/10.1016/B978-0-323-46143-6.00020-8

Javaid, M., Haleem, A., Singh, R. P., Suman, R., & Gonzalez, E. S. (2022). Understanding the adoption of industry 4.0 technologies in improving environmental sustainability. *Sustainable Operations and Computers*, *3*, 203–217. https://doi.org/10.1016/j.susoc.2022.01.008

Jha, S., & Chandi, D. (n.d). Recent advances in the devices for the treatment of chronic obstructive pulmonary disease: A review. *Cureus*, *15*(11), e49371. https://doi.org/10.7759/cureus.49371

Jimmy, B., & Jose, J. (2011). Patient medication adherence: Measures in daily practice. *Oman Medical Journal*, *26*(3), 155–159. https://doi.org/10.5001/omj.2011.38

Junaid, S. B., Imam, A. A., Balogun, A. O., Silva, L. C. D., Surakat, Y. A., Kumar, G., Abdulkarim, M., Shuaibu, A. N., Garba, A., Sahalu, Y., Mohammed, A., Mohammed, T. Y., Abdulkadir, B. A., Abba, A. A., Kakumi, N. A. I., & Mahamad, S. (2022). Recent advancements in emerging technologies for healthcare management systems: A survey. *Healthcare*, *10*(10). https://doi.org/10.3390/healthcare10101940

Kanyongo, W., & Ezugwu, A. E. (2023). Machine learning approaches to medication adherence amongst NCD patients: A systematic literature review. *Informatics in Medicine Unlocked*, *38*, 101210. https://doi.org/10.1016/j.imu.2023.101210

Kaplan, A., Boivin, M., Bouchard, J., Kim, J., Hayes, S., & Licskai, C. (2023). The emerging role of digital health in the management of asthma. *Therapeutic Advances in Chronic Disease, 14*. https://doi.org/10.1177/20406223231209329

Kaplan, R. M., & Milstein, A. (2019). Contributions of health care to longevity: A review of 4 estimation methods. *Annals of Family Medicine, 17*(3), 267–272. https://doi.org/10.1370/afm.2362

Kasoju, N., Remya, N. S., Sasi, R., Sujesh, S., Soman, B., Kesavadas, C., Muraleedharan, C. V., Varma, P. R. H., & Behari, S. (2023). Digital health: Trends, opportunities and challenges in medical devices, pharma and bio-technology. *CSI Transactions on ICT, 11*(1), 11–30. https://doi.org/10.1007/s40012-023-00380-3

Khalid, K., Padda, J., Komissarov, A., Colaco, L. B., Padda, S., Khan, A. S., Campos, V. M., & Jean-Charles, G. (n.d). The coexistence of chronic obstructive pulmonary disease and heart failure. *Cureus, 13*(8), e17387. https://doi.org/10.7759/cureus.17387

Khatiwada, P., Yang, B., Lin, J.-C., & Blobel, B. (2024). Patient-generated health data (PGHD): Understanding, requirements, challenges, and existing techniques for data security and privacy. *Journal of Personalized Medicine, 14*(3), 282. https://doi.org/10.3390/jpm14030282

Kim, M. O., Coiera, E., & Magrabi, F. (2017). Problems with health information technology and their effects on care delivery and patient outcomes: A systematic review. *Journal of the American Medical Informatics Association: JAMIA, 24*(2), 246–250. https://doi.org/10.1093/jamia/ocw154

Kladovasilakis, N., Pechlivani, E. M., Kanlis, A., Tsita, A., Mitsopoulou, E. E., Votis, K., & Tzovaras, D. (2023). Smart inhaler mask for portable medical gas cylinders utilizing IoT technologies. *Journal of Engineering Research*. https://doi.org/10.1016/j.jer.2023.10.037

Klijn, S. L., Hiligsmann, M., Evers, S. M. A. A., Román-Rodríguez, M., van der Molen, T., & van Boven, J. F. M. (2017). Effectiveness and success factors of educational inhaler technique interventions in asthma & COPD patients: A systematic review. *NPJ Primary Care Respiratory Medicine, 27*, 24. https://doi.org/10.1038/s41533-017-0022-1

Kochanek, K. D. (2017). *Mortality in the United States, 2016. 293.*

Kocks, J., Bosnic-Anticevich, S., van Cooten, J., Correia de Sousa, J., Cvetkovski, B., Dekhuijzen, R., Dijk, L., Garcia Pardo, M., Gardev, A., Gawlik, R., van der Ham, I., Janse, Y., Lavorini, F., Maricoto, T., Meijer, J., Metz, B., Price, D., Roman Rodriguez, M., Schuttel, K., … Leving, M. T. (2023). Identifying critical inhalation technique errors in Dry Powder Inhaler use in patients with COPD based on the association with health status and exacerbations: Findings from the multi-country cross-sectional observational PIFotal study. *BMC Pulmonary Medicine, 23*, 302. https://doi.org/10.1186/s12890-023-02566-6

Kraus, S., Schiavone, F., Pluzhnikova, A., & Invernizzi, A. C. (2021). Digital transformation in healthcare: Analyzing the current state-of-research. *Journal of Business Research, 123*, 557–567. https://doi.org/10.1016/j.jbusres.2020.10.030

Kumar, R., Mehta, P., Shankar, K. R., Rajora, M. A. K., Mishra, Y. K., Mostafavi, E., & Kaushik, A. (2022). Nanotechnology-assisted metered-dose inhalers (MDIs) for high-performance pulmonary drug delivery applications. *Pharmaceutical Research, 39*(11), 2831–2855. https://doi.org/10.1007/s11095-022-03286-y

Labiris, N. R., & Dolovich, M. B. (2003). Pulmonary drug delivery. Part I: Physiological factors affecting therapeutic effectiveness of aerosolized medications. *British Journal of Clinical Pharmacology, 56*(6), 588–599. https://doi.org/10.1046/j.1365-2125.2003.01892.x

Lange, C. F., & Finlay, W. H. (2006). Liquid atomizing: Nebulizing and other methods of producing aerosols. *Journal of Aerosol Medicine: The Official Journal of the International Society for Aerosols in Medicine, 19*(1), 28–35. https://doi.org/10.1089/jam.2006.19.28

Laube, B. L., Janssens, H. M., de Jongh, F. H. C., Devadason, S. G., Dhand, R., Diot, P., Everard, M. L., Horvath, I., Navalesi, P., Voshaar, T., & Chrystyn, H. (2011). What the pulmonary specialist should know about the new inhalation therapies. *European Respiratory Journal, 37*(6), 1308–1417. https://doi.org/10.1183/09031936.00166410

Lavorini, F. (2013). The challenge of delivering therapeutic aerosols to asthma patients. *ISRN Allergy, 2013*, 102418. https://doi.org/10.1155/2013/102418

Levy, M. L., Carroll, W., Izquierdo Alonso, J. L., Keller, C., Lavorini, F., & Lehtimäki, L. (2019). Understanding dry powder inhalers: Key technical and patient preference attributes. *Advances in Therapy, 36*(10), 2547–2557. https://doi.org/10.1007/s12325-019-01066-6

Li, C., Wang, J., Wang, S., & Zhang, Y. (2024). A review of IoT applications in healthcare. *Neurocomputing, 565*, 127017. https://doi.org/10.1016/j.neucom.2023.127017

Longuespée, R., Theile, D., Fresnais, M., Burhenne, J., Weiss, J., & Haefeli, W. E. (2021). Approaching sites of action of drugs in clinical pharmacology: New analytical options and their challenges. *British Journal of Clinical Pharmacology, 87*(3), 858–874. https://doi.org/10.1111/bcp.14543

López-Campos, J. L., Quintana Gallego, E., & Carrasco Hernández, L. (2019). Status of and strategies for improving adherence to COPD treatment. *International Journal of Chronic Obstructive Pulmonary Disease*, *14*, 1503–1515. https://doi.org/10.2147/COPD.S170848

Louie, S., Zeki, A. A., Schivo, M., Chan, A. L., Yoneda, K. Y., Avdalovic, M., Morrissey, B. M., & Albertson, T. E. (2013). The asthma–chronic obstructive pulmonary disease overlap syndrome: Pharmacotherapeutic considerations. *Expert Review of Clinical Pharmacology*, *6*(2), 197–219. https://doi.org/10.1586/ecp.13.2

Lu, Z., Coll, P., Maitre, B., Epaud, R., & Lanone, S. (2022). Air pollution as an early determinant of COPD. *European Respiratory Review*, *31*(165). https://doi.org/10.1183/16000617.0059-2022

Lythreatis, S., Singh, S. K., & El-Kassar, A.-N. (2022). The digital divide: A review and future research agenda. *Technological Forecasting and Social Change*, *175*, 121359. https://doi.org/10.1016/j.techfore.2021.121359

Mahler, D. A. (2020). The role of inspiratory flow in selection and use of inhaled therapy for patients with chronic obstructive pulmonary disease. *Respiratory Medicine*, *161*, 105857. https://doi.org/10.1016/j.rmed.2019.105857

Man, F., Tang, J., Swedrowska, M., Forbes, B., & de Rosales, R. T. M. (2023). Imaging drug delivery to the lungs: Methods and applications in oncology. *Advanced Drug Delivery Reviews*, *192*, 114641. https://doi.org/10.1016/j.addr.2022.114641

Mason, M., Cho, Y., Rayo, J., Gong, Y., Harris, M., & Jiang, Y. (2022). Technologies for medication adherence monitoring and technology assessment criteria: Narrative review. *JMIR mHealth and uHealth*, *10*(3), e35157. https://doi.org/10.2196/35157

McCarthy, S. D., González, H. E., & Higgins, B. D. (2020). Future trends in nebulized therapies for pulmonary disease. *Journal of Personalized Medicine*, *10*(2). https://doi.org/10.3390/jpm10020037

Mehta, P. (2020). Dry powder inhalers: A concise summary of the electronic monitoring devices. *Therapeutic Delivery*, *12*. https://doi.org/10.4155/tde-2020-0091

Mishra, J., Acharya, S., Taksande, A. B., Prasad, R., Munjewar, P. K., & Wanjari, M. B. (n.d). Occupational risks and chronic obstructive pulmonary disease in the Indian subcontinent: A critical review. *Cureus*, *15*(6), e41149. https://doi.org/10.7759/cureus.41149

Mohan, A. R., Wang, Q., Dhapare, S., Bielski, E., Kaviratna, A., Han, L., Boc, S., & Newman, B. (2022). Advancements in the design and development of dry powder inhalers and potential implications for generic development. *Pharmaceutics*, *14*(11), 2495. https://doi.org/10.3390/pharmaceutics14112495

Morag, I., Kedmi-Shahar, E., & Arad, D. (2023). Remote communications between patients and general practitioners: Do patients choose the most effective communication routes? *International Journal of Environmental Research and Public Health*, *20*(24), 7188. https://doi.org/10.3390/ijerph20247188

Morton, R. W., Elphick, H. E., Craven, V., Shields, M. D., & Kennedy, L. (2020). Aerosol therapy in Asthma–Why we are failing our patients and how we can do better. *Frontiers in Pediatrics*, *8*, 305. https://doi.org/10.3389/fped.2020.00305

Mosnaim, G., Safioti, G., Brown, R., DePietro, M., Szefler, S. J., Lang, D. M., Portnoy, J. M., Bukstein, D. A., Bacharier, L. B., & Merchant, R. K. (2021). Digital health technology in asthma: A comprehensive scoping review. *The Journal of Allergy and Clinical Immunology: In Practice*, *9*(6), 2377–2398. https://doi.org/10.1016/j.jaip.2021.02.028

Moudatsou, M., Stavropoulou, A., Philalithis, A., & Koukouli, S. (2020). The role of empathy in health and social care professionals. *Healthcare*, *8*(1), 26. https://doi.org/10.3390/healthcare8010026

Myrdal, P. B., Sheth, P., & Stein, S. W. (2014). Advances in metered dose inhaler technology: Formulation development. *AAPS PharmSciTech*, *15*(2), 434–455. https://doi.org/10.1208/s12249-013-0063-x

Nahum-Shani, I., Shaw, S. D., Carpenter, S. M., Murphy, S. A., & Yoon, C. (2022). Engagement in digital interventions. *The American Psychologist*, *77*(7), 836–852. https://doi.org/10.1037/amp0000983

Newman, S. (2014). Improving inhaler technique, adherence to therapy and the precision of dosing: Major challenges for pulmonary drug delivery. *Expert Opinion on Drug Delivery*, *11*(3), 365–378. https://doi.org/10.1517/17425247.2014.873402

Newman, S. P. (2017). Drug delivery to the lungs: Challenges and opportunities. *Therapeutic Delivery*, *8*(8), 647–661. https://doi.org/10.4155/tde-2017-0037

Nieuwlaat, R., Wilczynski, N., Navarro, T., Hobson, N., Jeffery, R., Keepanasseril, A., Agoritsas, T., Mistry, N., Iorio, A., Jack, S., Sivaramalingam, B., Iserman, E., Mustafa, R. A., Jedraszewski, D., Cotoi, C., & Haynes, R. B. (2014). Interventions for enhancing medication adherence. *The Cochrane Database of Systematic Reviews*, *2014*(11), CD000011. https://doi.org/10.1002/14651858.CD000011.pub4

North, M., Bourne, S., Green, B., Chauhan, A. J., Brown, T., Winter, J., Jones, T., Neville, D., Blythin, A., Watson, A., Johnson, M., Culliford, D., Elkes, J., Cornelius, V., & Wilkinson, T. M. A. (2020). A randomised controlled feasibility trial of e-health application supported care vs usual care after exacerbation of COPD: The RESCUE trial. *NPJ Digital Medicine*, *3*(1), 1–8. https://doi.org/10.1038/s41746-020-00347-7

Ohar, J. A., Ferguson, G. T., Mahler, D. A., Drummond, M. B., Dhand, R., Pleasants, R. A., Anzueto, A., Halpin, D. M. G., Price, D. B., Drescher, G. S., Hoy, H. M., Haughney, J., Hess, M. W., & Usmani, O. S. (2022). Measuring peak inspiratory flow in patients with chronic obstructive pulmonary disease. *International Journal of Chronic Obstructive Pulmonary Disease*, *17*, 79–92. https://doi.org/10.2147/COPD.S319511

Pahal, P., Avula, A., & Sharma, S. (2024). Emphysema. In *StatPearls*. StatPearls Publishing. http://www.ncbi.nlm.nih.gov/books/NBK482217/

Pannunzio, V., Kleinsmann, M., Snelders, D., & Raijmakers, J. (2023). From digital health to learning health systems: Four approaches to using data for digital health design. *Health Systems*, *12*(4), 481–494. https://doi.org/10.1080/20476965.2023.2284712

Parray, H. A., Shukla, S., Perween, R., Khatri, R., Shrivastava, T., Singh, V., Murugavelu, P., Ahmed, S., Samal, S., Sharma, C., Sinha, S., Luthra, K., & Kumar, R. (2021). Inhalation monoclonal antibody therapy: A new way to treat and manage respiratory infections. *Applied Microbiology and Biotechnology*, *105*(16–17), 6315–6332. https://doi.org/10.1007/s00253-021-11488-4

Patel, M., Pilcher, J., Travers, J., Perrin, K., Shaw, D., Black, P., Weatherall, M., & Beasley, R. (2013). Use of metered-dose inhaler electronic monitoring in a real-world asthma randomized controlled trial. *The Journal of Allergy and Clinical Immunology. In Practice*, *1*(1), 83–91. https://doi.org/10.1016/j.jaip.2012.08.004

Paukkonen, L., Oikarinen, A., Kähkönen, O., & Kaakinen, P. (2022). Patient activation for self-management among adult patients with multimorbidity in primary healthcare settings. *Health Science Reports*, *5*(4), e735. https://doi.org/10.1002/hsr2.735

Pérez-Jover, V., Sala-González, M., Guilabert, M., & Mira, J. J. (2019). Mobile Apps for increasing treatment adherence: Systematic review. *Journal of Medical Internet Research*, *21*(6), e12505. https://doi.org/10.2196/12505

Perumal, R., Leite, M., & van Zyl-Smit, R. N. (2020). The relationship between clinical trial participation and inhaler technique errors in asthma and COPD patients. *International Journal of Chronic Obstructive Pulmonary Disease*, *15*, 1217–1224. https://doi.org/10.2147/COPD.S249620

Pinnock, H., Hui, C. Y., & van Boven, J. F. M. (2023). Implementation of digital home monitoring and management of respiratory disease. *Current Opinion in Pulmonary Medicine*, *29*(4), 302. https://doi.org/10.1097/MCP.0000000000000965

Pleasants, R., Chan, A., Mosnaim, G., Costello, R., Dhand, R., Schworer, S., Merchant, R., & Tilley, S. (2022). Integrating digital inhalers into clinical care of patients with asthma and chronic obstructive pulmonary disease. *Respiratory Medicine*, *205*, 107038. https://doi.org/10.1016/j.rmed.2022.107038

Pleasants, R. A., & Hess, D. R. (2018). Aerosol delivery devices for obstructive lung diseases. *Respiratory Care*, *63*(6), 708–733. https://doi.org/10.4187/respcare.06290

Plekhanov, D., Franke, H., & Netland, T. H. (2023). Digital transformation: A review and research agenda. *European Management Journal*, *41*(6), 821–844. https://doi.org/10.1016/j.emj.2022.09.007

Pollack, A. H., Backonja, U., Miller, A. D., Mishra, S. R., Khelifi, M., Kendall, L., & Pratt, W. (2016). Closing the gap: Supporting Patients' transition to self-management after hospitalization. *Proceedings of the SIGCHI Conference on Human Factors in Computing Systems CHI Conference*, *2016*, 5324–5336. https://doi.org/10.1145/2858036.2858240

Poot, C. C., de Boer, J., Goto, L., van de Hei, S. J., Chavannes, N. H., Visch, V. T., & Meijer, E. (2023). The design of a persuasive game to motivate people with asthma in adherence to their maintenance medication. *Patient Preference and Adherence*, *17*, 2719–2736. https://doi.org/10.2147/PPA.S423161

Price, D., Jones, R., Pfister, P., Cao, H., Carter, V., Kemppinen, A., Holzhauer, B., Kaplan, A., Clark, A., Halpin, D. M. G., Pinnock, H., Chalmers, J. D., van Boven, J. F. M., Beeh, K. M., Kostikas, K., Roche, N., Usmani, O., & Mastoridis, P. (2021). Maximizing adherence and gaining new information for your chronic obstructive pulmonary disease (MAGNIFY COPD): Study protocol for the pragmatic, cluster randomized trial evaluating the impact of dual bronchodilator with add-on sensor and electronic monitoring on clinical outcomes. *Pragmatic and Observational Research*, *12*, 25–35. https://doi.org/10.2147/POR.S302809

Rafi, Md. A., Tahmin, C. I., Tashrik, S., Bonna, A. S., Jannat, F., Mily, S. J., Shrestha, A. B., Seemanta, S., Rashid, A., Mahjabeen, M., Nura, N., Shahriar, T., Mahadi, A. R., Ahmed, K., Hasan, M. J., Haque, Md. A., & Hossain, Md. G. (2022). Adherence to inhalers and associated factors among adult asthma patients: An outpatient-based study in a tertiary hospital of Rajshahi, Bangladesh. *Asthma Research and Practice*, *8*, 1. https://doi.org/10.1186/s40733-022-00083-7

Raghupathi, W., & Raghupathi, V. (2014). Big data analytics in healthcare: Promise and potential. *Health Information Science and Systems*, *2*, 3. https://doi.org/10.1186/2047-2501-2-3

Raikar, A. S., Kumar, P., Raikar, G. (Vedant) S., & Somnache, S. N. (2023). Advances and challenges in IoT-based smart drug delivery systems: A comprehensive review. *Applied System Innovation*, *6*(4), 62. https://doi.org/10.3390/asi6040062

Ramadan, W. H., & Sarkis, A. T. (2017). Patterns of use of dry powder inhalers versus pressurized metered-dose inhalers devices in adult patients with chronic obstructive pulmonary disease or asthma: An observational comparative study. *Chronic Respiratory Disease*, *14*(3), 309–320. https://doi.org/10.1177/1479972316687209

Rana, J. S., Khan, S. S., Lloyd-Jones, D. M., & Sidney, S. (2021). Changes in mortality in top 10 causes of death from 2011 to 2018. *Journal of General Internal Medicine*, *36*(8), 2517–2518. https://doi.org/10.1007/s11606-020-06070-z

Rinderknecht, C. H., Ning, M., Wu, C., Wilson, M. S., & Gampe, C. (2024). Designing inhaled small molecule drugs for severe respiratory diseases: An overview of the challenges and opportunities. *Expert Opinion on Drug Discovery*, *19*(4), 493–506. https://doi.org/10.1080/17460441.2024.2319049

Roche, N., & Dekhuijzen, P. N. R. (2016). The evolution of pressurized metered-dose inhalers from early to modern devices. *Journal of Aerosol Medicine and Pulmonary Drug Delivery*, *29*(4), 311–327. https://doi.org/10.1089/jamp.2015.1232

Rogliani, P., Calzetta, L., Coppola, A., Cavalli, F., Ora, J., Puxeddu, E., Matera, M. G., & Cazzola, M. (2017). Optimizing drug delivery in COPD: The role of inhaler devices. *Respiratory Medicine*, *124*, 6–14. https://doi.org/10.1016/j.rmed.2017.01.006

Ryu, M. H., Murphy, S., Hinkley, M., & Carlsten, C. (2024). COPD exposed to air pollution: A path to understand and protect a susceptible population. *CHEST*, *165*(4), 836–846. https://doi.org/10.1016/j.chest.2023.11.012

Samal, L., Fu, H. N., Camara, D. S., Wang, J., Bierman, A. S., & Dorr, D. A. (2021). Health information technology to improve care for people with multiple chronic conditions. *Health Services Research*, *56*(Suppl 1), 1006. https://doi.org/10.1111/1475-6773.13860

Scichilone, N., Spatafora, M., Battaglia, S., Arrigo, R., Benfante, A., & Bellia, V. (2013). Lung penetration and patient adherence considerations in the management of asthma: Role of extra-fine formulations. *Journal of Asthma and Allergy*, *6*, 11–21. https://doi.org/10.2147/JAA.S14743

Shojaei, P., Vlahu-Gjorgievska, E., & Chow, Y.-W. (2024). Security and privacy of technologies in health information systems: A systematic literature review. *Computers*, *13*(2), 41. https://doi.org/10.3390/computers13020041

Siminerio, L. M. (2010). The role of technology and the chronic care model. *Journal of Diabetes Science and Technology*, *4*(2), 470. https://doi.org/10.1177/193229681000400229

Sivarajah, U., Kamal, M. M., Irani, Z., & Weerakkody, V. (2017). Critical analysis of big data challenges and analytical methods. *Journal of Business Research*, *70*, 263–286. https://doi.org/10.1016/j.jbusres.2016.08.001

Slevin, P., Kessie, T., Cullen, J., Butler, M. W., Donnelly, S. C., & Caulfield, B. (2019). Exploring the potential benefits of digital health technology for the management of COPD: A qualitative study of patient perceptions. *ERJ Open Research*, *5*(2). https://doi.org/10.1183/23120541.00239-2018

Soni, P., Siddiqui, A. A., Dwivedi, J., & Soni, V. (2012). Pharmacological properties of *Datura stramonium* L. as a potential medicinal tree: An overview. *Asian Pacific Journal of Tropical Biomedicine*, *2*(12), 1002–1008. https://doi.org/10.1016/S2221-1691(13)60014-3

Sosnowski, T. R. (2024). Towards more precise targeting of inhaled aerosols to different areas of the respiratory system. *Pharmaceutics*, *16*(1), 97. https://doi.org/10.3390/pharmaceutics16010097

Stein, S. W., & Thiel, C. G. (2017). The history of therapeutic aerosols: A chronological review. *Journal of Aerosol Medicine and Pulmonary Drug Delivery*, *30*(1), 20–41. https://doi.org/10.1089/jamp.2016.1297

Stewart, S.-J. F., Moon, Z., & Horne, R. (2023). Medication nonadherence: Health impact, prevalence, correlates and interventions. *Psychology & Health*, *38*(6), 726–765. https://doi.org/10.1080/08870446.2022.2144923

Tan, W. C., & Ng, T. P. (2008). COPD in Asia. *Chest*, *133*(2), 517–527. https://doi.org/10.1378/chest.07-1131

Taylor, T. E., Zigel, Y., De Looze, C., Sulaiman, I., Costello, R. W., & Reilly, R. B. (2018). Advances in audio-based systems to monitor patient adherence and inhaler drug delivery. *Chest*, *153*(3), 710–722. https://doi.org/10.1016/j.chest.2017.08.1162

Terry, P. D., & Dhand, R. (2020). Maintenance therapy with nebulizers in patients with stable COPD: Need for reevaluation. *Pulmonary Therapy*, *6*(2), 177–192. https://doi.org/10.1007/s41030-020-00120-x

Teva (2020). *Teva Launches Two Digital Inhalers in the U.S., AirDuo® Digihaler® (fluticasone propionate and salmeterol) Inhalation Powder and ArmonAir® Digihaler® (fluticasone propionate) Inhalation Powder* (September 21). https://www.tevapharm.com/news-and-media/latest-news/teva-launches-two-digital-inhalers-in-the-u.s.-airduo-digihaler-fluticasone-propionate-and-salmeterol/

Tsiligianni, I., & Kocks, J. W. H. (2020). Daytime symptoms of chronic obstructive pulmonary disease: A systematic review. *NPJ Primary Care Respiratory Medicine*, *30*, 6. https://doi.org/10.1038/s41533-020-0163-5

Turégano-Yedro, M., Trillo-Calvo, E., Navarro i Ros, F., Maya-Viejo, J. D., González Villaescusa, C., Echave Sustaeta, J. M., Doña, E., & Alcázar Navarrete, B. (2023). Inhaler adherence in COPD: A crucial step towards the correct treatment. *International Journal of Chronic Obstructive Pulmonary Disease*, *18*, 2887–2893. https://doi.org/10.2147/COPD.S431829

Usmani, O. S. (2014). Small airways dysfunction in asthma: Evaluation and management to improve asthma control. *Allergy, Asthma & Immunology Research*, *6*(5), 376. https://doi.org/10.4168/aair.2014.6.5.376

Usmani, O. S., Lavorini, F., Marshall, J., Dunlop, W. C. N., Heron, L., Farrington, E., & Dekhuijzen, R. (2018). Critical inhaler errors in asthma and COPD: A systematic review of impact on health outcomes. *Respiratory Research*, *19*(1), 10. https://doi.org/10.1186/s12931-017-0710-y

Vachon, B., Giasson, G., Gaboury, I., Gaid, D., Noël De Tilly, V., Houle, L., Bourbeau, J., & Pomey, M.-P. (2022). Challenges and strategies for improving COPD primary care services in Quebec: Results of the experience of the COMPAS+ quality improvement collaborative. *International Journal of Chronic Obstructive Pulmonary Disease*, *17*, 259–272. https://doi.org/10.2147/COPD.S341905

van de Hei, S. J., Kim, C. H., Honkoop, P. J., Sont, J. K., Schermer, T. R. J., MacHale, E., Costello, R. W., Kocks, J. W. H., Postma, M. J., & van Boven, J. F. M. (2023). Long-term cost-effectiveness of digital inhaler adherence technologies in difficult-to-treat asthma. *The Journal of Allergy and Clinical Immunology: In Practice*, *11*(10), 3064–3073.e15. https://doi.org/10.1016/j.jaip.2023.06.051

Vincken, W., Levy, M. L., Scullion, J., Usmani, O. S., Dekhuijzen, P. N. R., & Corrigan, C. J. (2018). Spacer devices for inhaled therapy: Why use them, and how? *ERJ Open Research*, *4*(2), 00065–02018. https://doi.org/10.1183/23120541.00065-2018

Virchow, J. C., Crompton, G. K., Dal Negro, R., Pedersen, S., Magnan, A., Seidenberg, J., & Barnes, P. J. (2008). Importance of inhaler devices in the management of airway disease. *Respiratory Medicine*, *102*(1), 10–19. https://doi.org/10.1016/j.rmed.2007.07.031

Volerman, A., Kan, K., Carpenter, D., & Press, V. G. (2021). Strategies for improving inhalation technique in children: A narrative review. *Patient Preference and Adherence*, *15*, 665–675. https://doi.org/10.2147/PPA.S267053

Vukotich, G. (2023). Healthcare and cybersecurity: Taking a zero trust approach. *Health Services Insights*, *16*. https://doi.org/10.1177/11786329231187826

Wachtel, H., Emerson-Stadler, R., Langguth, P., Hohlfeld, J. M., & Ohar, J. (2024). Aerosol plumes of inhalers used in COPD. *Pulmonary Therapy*, *10*(1), 109–122. https://doi.org/10.1007/s41030-023-00249-5

Wachtel, H., Ertunç, Ö., Koksoy, C., & Delgado, A. (2008). Aerodynamic optimization of handihaler and respimat: The roles of computational fluid dynamics and flow visualization. *Respiratory Drug Delivery*, *1*, 165–174.

Walquist, M. J., Svendsen, K., Garcia, B. H., Bergmo, T. S., Eggen, A. E., Halvorsen, K. H., Småbrekke, L., Ringberg, U., Lehnbom, E. C., & Waaseth, M. (2022). Self-reported medication information needs among medication users in a general population aged 40 years and above – the Tromsø study. *BMC Public Health*, *22*. https://doi.org/10.1186/s12889-022-14573-z

Watase, M., Miyata, J., Terai, H., Sunata, K., Matsuyama, E., Asakura, T., Namkoong, H., Masaki, K., Yagi, K., Ohgino, K., Chubachi, S., Kawada, I., Mochimaru, T., Satomi, R., Oyamada, Y., Kobayashi, K., Hirano, T., Inoue, T., Lee, H., … Fukunaga, K. (2023). Cough and sputum in long COVID are associated with severe acute COVID-19: A Japanese cohort study. *Respiratory Research*, *24*(1), 283. https://doi.org/10.1186/s12931-023-02591-3

Watson, A., & Wilkinson, T. (2022). Digital healthcare in COPD management: A narrative review on the advantages, pitfalls and need for further research. *Therapeutic Advances in Respiratory Disease*. https://doi.org/10.1177/17534666221075493

Weers, J. (2022). Suboptimal inspiratory flow rates with passive dry powder inhalers: Big issue or overstated problem? *Frontiers in Drug Delivery*, *2*. https://doi.org/10.3389/fddev.2022.855234

Xiroudaki, S., Schoubben, A., Giovagnoli, S., & Rekkas, D. M. (2021). Dry powder inhalers in the digitalization era: Current status and future perspectives. *Pharmaceutics*, *13*(9), 1455. https://doi.org/10.3390/pharmaceutics13091455

Yang, D., Li, K., Mingwei Chua, D., Song, Y., Bai, C., & Powell, C. A. (2022). Application of internet of things in chronic respiratory disease prevention, diagnosis, treatment and management. *Clinical eHealth*, *5*, 10–16. https://doi.org/10.1016/j.ceh.2021.08.001

Ye, Y., Ma, Y., & Zhu, J. (2022). The future of dry powder inhaled therapy: Promising or discouraging for systemic disorders? *International Journal of Pharmaceutics*, *614*, 121457. https://doi.org/10.1016/j.ijpharm.2022.121457

Yuan, P., Li, J., Liu, J., Liu, H., Gong, S., Wang, L., Liu, J., Li, C., Wang, Y., Xia, X., & Zheng, J. C. (n.d). Cognitive dysfunction in patients with pulmonary hypertension. *American Journal of Respiratory and Critical Care Medicine*, *206*(10), 1289–1293. https://doi.org/10.1164/rccm.202204-0726LE

Zabczyk, C., & Blakey, J. D. (2021). The effect of connected "Smart" inhalers on medication adherence. *Frontiers in Medical Technology*, *3*, 657321. https://doi.org/10.3389/fmedt.2021.657321

Zhang, D., Wang, F., Li, P., & Gao, Y. (2022). Mitochondrial Ca^{2+} homeostasis: Emerging roles and clinical significance in cardiac remodeling. *International Journal of Molecular Sciences*, *23*(6), 3025. https://doi.org/10.3390/ijms23063025

Zhang, P., & Kamel Boulos, M. N. (2023). Generative AI in medicine and healthcare: Promises, opportunities and challenges. *Future Internet*, *15*(9), 286. https://doi.org/10.3390/fi15090286

6 Nanotechnology in Pulmonary Drug Delivery Systems

Prachi Saini, Kanika, Girish Kumar, and Tarun Virmani
School of Pharmaceutical Sciences, MVN University
Palwal, Haryana, India

Kamla Pathak
Faculty of Pharmacy, Uttar Pradesh, University of Medical Sciences
Etawah, Uttar Pradesh, India

6.1 OVERVIEW OF PULMONARY DRUG DELIVERY

The respiratory system is a biological system found in both plants and animals that is made of organs and structures used for gas exchange. The organism's size, its living habitat, and its evolutionary background all have a significant impact on the anatomy and physiology that enable it. The respiratory surface is internalized as the linings of the lungs in land animals. The number of cases of respiratory disorders is rising due to rising air pollution levels, quick lifestyle changes, and frequent outbreaks of microbial infections (Virmani *et al.*, 2022). This is especially true for elderly people and children with compromised immune systems. Every year, respiratory illnesses claim the lives of more than 4 million individuals globally (Ferkol and Schraufnagel, 2014). Since the lungs can absorb medications for systemic or local distribution, pulmonary drug delivery has gained significant scientific and biomedical attention and has become a desirable focus in the field of healthcare research. The creation of the fluid that lines the airways and the control of airway tone are largely dependent on the respiratory epithelial cells. Due to the lungs' high permeability and sizable absorptive surface area, increasing consideration has been given to the possibility of using the pulmonary route as a non-invasive administration method for systemic and local delivery of medicinal drugs and a healthy blood flow (Groneberg *et al.*, 2003). The lung's alveolar epithelium has been shown to be an absorption site for the majority of medications and other macromolecules (Tuncer Degim and Celebi, 2007).

In addition, advantages over oral administration include the drug's quick absorption, comparatively low enzymatic activity, and ability to bypass first-pass metabolism. It has previously been documented that administering medication via the pulmonary route is an effective way to treat certain systemic diseases as well as local respiratory ailments. This covers the systemic use of insulin, human growth hormones, and oxytocin, as well as the topical therapy of asthma, pulmonary hypertension, and local infectious illnesses (Flume and Klepser, 2002). Currently, this is the case for a number of biotherapeutics that are administered intravenously, including insulin, glucagon, and growth hormones. However, inhalation treatment may be a more effective way to deliver these drugs to humans. Inhalation treatment relies heavily on our ability to comprehend the movement and accumulation of inhaled particles (Ahalwat *et al.*, 2023). The technological, physiological, and efficacious elements of the pulmonary medication delivery system are discussed here. The mechanisms of pulmonary administration and transepithelial transport are also the subject of this review. Furthermore, a compilation of polymer choices for dose and delivery device kinds has been made. Active pharmaceutical ingredients (APIs) are interesting targets for pulmonary administration using a number of drug delivery methods (Azarmi *et al.*, 2008). In addition, this method overcomes

DOI: 10.1201/9781003498926-6

the first-pass effect and has many advantages over traditional oral administration, including a high surface area with quick absorption due to strong vascularization (Sung *et al.*, 2007). Many drug delivery strategies have been created in the past few decades in an effort to get the best products across physicochemical or biological barriers (R. Virmani *et al.*, 2023). Because of its selectivity, the medicine may be delivered specifically, which lessens its negative effects (Beck-Broichsitter *et al.*, 2009).

Colloidal drug delivery systems have been thoroughly studied as drug carrier systems for administering various medications through various routes of administration (Dhir *et al.*, 2023; Kumar *et al.*, 2022a, 2024). One of the most interesting colloidal systems that has been studied for more than ten years is solid lipid nanoparticles (SLN) (Müller *et al.*, 2000). Water-based nanoscale suspensions known as SLN are mostly made of physiologically tolerated triglycerides and phospholipids (Nassimi *et al.*, 2010). Because of the prolonged release of APIs, biodegradable polymeric nanoparticles—along with SLNs—are becoming increasingly significant (Sharma *et al.*, 2023). Lipophilic medicines, which are poorly soluble in aqueous environments, are well suited for these systems. Consequently, improved solubility in a lipophilic matrix improves therapeutic efficacy and pharmacokinetics. By changing the physical characteristics of the particles—such as enhancing drug solubility, encapsulation effectiveness, and surface changes to improve drug release profiles and achieve a maximal effect—nanoparticle-mediated drug delivery systems bring up new possibilities (Menon *et al.*, 2014).

6.2 CONVENTIONAL TREATMENT OF PULMONARY DRUG DELIVERY

It was recently claimed that a variety of conventional methods may produce dry powder inhaler (DPI) formulations. Particle size, size distribution, form, and the inability to effectively manage powder crystallinity are only a few of the approach's difficulties. Specific cutting processes can be used to solve these issues. The most effective technique for producing nanoparticles intended for pulmonary medication delivery is the jet-milling of the drug under nitrogen gas using a novel nano jet-milling apparatus. A discussion of a few key strategies is provided next.

6.2.1 Supercritical Fluid Technology

The primary characteristic of this procedure is the regulated solidification of medications from their dispersion in supercritical carbon dioxide fluids. The pharmaceutical industry has produced liposomes, inclusion complexes, nanoparticles, and microparticles using this technique (Rehman *et al.*, 2004). This technique is employed to enhance the formulation characteristics of some medication options as well as to produce particulate pulmonary drug delivery systems with proteins and peptides (Chattopadhyay *et al.*, 2007).

6.2.2 Spray Freeze Drying Method

The first decade of the 1990s saw the exploration of this technique for use in pharmaceuticals. It is a sophisticated particle engineering technique that combines the processing stages of freeze-drying and spray-drying. It entails lyophilizing the medication solution after it has been sprayed into liquid nitrogen to act as a freezing medium (Rogers *et al.*, 2001). At sub-ambient temperatures, this approach yields nearly 100% of light, porous particles with a high fine particle percentage and increased aerosol performance (Yu *et al.*, 2004). It is also possible to formulate peptide and protein molecules that are thermostable, such insulin and plasmid DNA, into DPI products (Kuo and Hwang, 2004). But this is a costly procedure limited to expensive medications only.

6.2.3 Particles Replicates in Non-wetting Templates

The top-down method for creating particles, known as particle replication in non-wetting templates (PRINT), was created by Dr. Joseph DeSimone and his team. This method helps load small organic medicines, proteins, peptides, oligonucleotides, siRNA contrast agents, radiotracers, and fluorophores. It can also manufacture uniform-sized organic micro- and nanoparticles with total control over size, shape, and surface functionality (Gratton *et al.*, 2007).

6.2.4 Spray Drying Technique

Spray drying is a state-of-the-art method for producing respirable colloidal particles in a solid form in pharmaceutical manufacture (Mosén *et al.*, 2005). In the 1980s, spray drying was investigated as a potential substitute method for creating fine particles for pulmonary administration. The feed solution is delivered to the nozzle in this procedure at room temperature, where the nozzle gas atomizes it. After the solution has been atomized, it is dried in a specific chamber using heated drying gas to eliminate any remaining moisture to produce dry particles (Gilani *et al.*, 2005). When it comes to creating particles larger than 2 μm, this approach shows greater promise. According to reports, this approach can be easily adapted for large-scale production because it has superior control over particle formation. Because this approach avoids mechanical high-energy input, it is also appropriate for thermolabile materials like proteins and peptides (White *et al.*, 2005).

6.3 PROBLEMS ASSOCIATED WITH CONVENTIONAL TREATMENT

In pulmonary drug delivery, the lungs' five lobes and their 300 million or more alveoli constitute 280 billion networks of capillaries, which serve as the primary pathway for the particles to be transported once they are administered to the system (Valiulin *et al.*, 2021). Particle size will affect where they settle in the lungs via diffusion, impaction, and sedimentation. Particles larger than 5 μm are inhaled and travel at a rapid speed through the upper airways and oropharynx. The centrifugal forces then strike the respiratory wall and settle in the oropharynx areas (Kumar *et al.*, 2023). In the case of the sedimentation mechanism, particles with sufficient mass and diameters between 1 μm and 5 μm are deposited in the bronchioles and smaller airways due to gravity forces.

Meanwhile, the random movement of the surrounding molecules causes the APIs to release and eventually dissolve in the alveolar fluid due to the Brownian motion of the particles. One of these polymeric system's main drawbacks is that improper use might lead to uneven dosing. Breath holding, deep inhalation, and a fairly low inspiratory flow are necessary for the effective deposition of aerosols from pressurized metered-dose inhalers (pMDIs) (Lim *et al.*, 2016).

6.4 ADVANTAGES OF NANOTECHNOLOGY OVER CONVENTIONAL TREATMENT

Many electronic devices, processes, and applications have the potential to be completely transformed by nanotechnology. When it comes to electronic devices, fields such as nanotransistors, nanodiodes, organic light-emitting diode (OLED), plasma displays, quantum computers, and many more stand to gain from the ongoing advancement of nanotechnology (Gramling *et al.*, 2017).

The power industry can gain from nanotechnology as well. This technology can lead to the development of more efficient energy-producing, energy-absorbing, and energy-storing products in smaller, more efficient systems. With this technology, products like batteries, fuel cells, and solar cells can be constructed more compactly while yet being more efficient (Mahmud, 2023).

The manufacturing sector, which requires materials like nanotubes, aerogels, nanoparticles, and other comparable goods with which to make their products, is another business that stands to benefit from nanotechnology. Compared to materials that are not created using nanotechnology, these materials are frequently stronger, more resilient, and lighter (Jasmani *et al.*, 2020). Nanotechnology is also seen favourably in the medical field since it can be used to create smart medications. These aid in patient recovery more quickly and without the negative effects of other conventional medications. Additionally, medical nanotechnology research is currently concentrated on issues like bone healing, tissue regeneration, immunology, and even the treatment of serious illnesses like cancer and diabetes (Malik *et al.*, 2023)

6.5 NANOTECHNOLOGY FOR PULMONARY DRUG DELIVERY

In medication delivery systems intended for parenteral administration, polymeric nanoparticles are extensively researched (Plumley *et al.*, 2009). However, there is also general acceptance of their relevance to the pulmonary routes. Polymeric nanoparticles play three primary functions in drug delivery systems: they transport the drug molecules, shield them from deterioration, and regulate the release of the pharmaceuticals (Müller *et al.*, 2000). Biodegradable or biocompatible materials, such as poly (ε-caprolactone) (PCL), poly (lactic acid) (PLA), poly (lactic-co-glycolic acid) (PLGA), alginic acid, gelatine, and chitosan, are the building blocks of therapeutically employed polymeric nanoparticles. It displays several polymers for pulmonary medication delivery using nanocarrier systems (T. Virmani *et al.*, 2023b). The multidisciplinary field of nanotechnology includes elements of engineering, physics, chemistry, and biology (Byrne *et al.*, 2015). The environment contains dust, smoke, combustion products, and ultrafine particles that act as air pollutants. These are ingested and can enter the respiratory system of the body, and many of these have negative impacts on human health. Nanoparticles developed and researched in the field of nanotechnology are more appropriate for our uses. Nanotechnology has several potential uses in the biomedical field, including targeted therapy, medication administration, imaging, diagnostics, and even surgery (Kumar *et al.*, 2023). The field that most likely draws interest is the delivery of drugs or genes using nanoparticles and nanostructures. Because of their various advantages, nanoparticles are especially well suited for this application. Their ability to effectively cross biological barriers such as cell membranes has led to their application as gene delivery vectors (Dubey *et al.*, 2024). It is generally accepted that particles need to be approximately 200 nm to be absorbed by a cell. While manufacturing a nanoparticle for the delivery of an inhalation treatment, the respiratory tract's subsequent particle deposition may occur (Alton, 2004). The results of an *in vivo* study, size and surface characteristics were the most important factors in determining a nanoparticle's ability to enter the systemic circulation when it was injected into the lung. The dispersion into the bloodstream did not seem to be affected by the type of material applied. The research, which used a mouse model as a test subject, revealed that particles in the low nanometre range are quickly transported to the lymph nodes nearby. Smaller particles were discovered to be released in the urine in as little as 30 minutes due to their rapid elimination. Larger particles had very little systemic absorption (Joseph *et al.*, 2023). Various nanotechnology-based drug delivery systems for the pulmonary route were discussed in the previous section. Table 6.1 highlights the application of nanotechnology in pulmonary drug delivery. Various drugs are encapsulated or formulated with diverse polymers to achieve nanoscale sizes, ranging from 77 nm to 500 nm. These formulations include therapeutic agents such as insulin, plasmid DNA, rifampicin, doxorubicin, and vasoactive intestinal peptides, showcasing the versatility of nanocarriers like chitosan, sodium alginate, and PEGylated polymers in optimizing drug delivery efficiency and targeting. Table 6.2 highlights various nanoparticle (NP) systems utilized for drug delivery, showcasing different types of particles, their associated active pharmaceutical ingredients (APIs), and relevant references. This comparison emphasizes the versatility of NP-based systems, including polymeric nanoparticles, solid lipid nanoparticles (SLN), and lipid nanocapsules (NC), in enhancing the delivery of therapeutic agents such as budesonide, indomethacin, pirfenidone, beclomethasone, and methotrexate.

TABLE 6.1
Nanotechnology for Pulmonary Drug Delivery

Drug	Polymer	Size	References
Insulin	Chitosan/tripolyphosphate	300–388 nm	(Lee and Mohapatra, 2008)
Plasmid DNA	Trisaccharide-substituted chitosan	77–90 nm	(Howard *et al.*, 2006)
Rifampicin, isoniazid, pyrazinamide	Sodium alginate	235.5 nm	(Kimura *et al.*, 2009)
Doxorubicin	Polybutylcyanoacrylate	173 ± 43 nm	(Sham *et al.*, 2004)
Plasmid DNA	PEGylated gelatin	100–500 nm	(Brzoska *et al.*, 2004)
Fluoresceinamine	Gelatin type A	277.8 nm	(Grenha *et al.*, 2005)
Vasoactive intestinal peptide	Protamine-oligonucleotide	177–318 nm	(Park *et al.*, 2008)
Plasmid DNA	PEGylated poly-L-lysine	211 ± 29 nm	(Nguyen *et al.*, 2008)

TABLE 6.2
Various Nanoparticle Systems with Nanoparticles (NPs)

Types of Particles	API Category	Reference
Polymeric NP, SLN	Budesonide	(Elhissi *et al.*, 2010)
Lipid NP	Indomethacin	(Castelli *et al.*, 2005)
Polymeric NP	Pirfenidone	(Trivedi *et al.*, 2012)
Lipid NC	Beclomethasone	(Jaafar-Maalej *et al.*, 2011)
Polymeric NP	Methotrexate	(Li *et al.*, 2013)
SLN	Silibinin	(Xu *et al.*, 2013)
SLM, liposomes	Antioxidants multiple types	(Silva *et al.*, 2013)
Liposomes	Ciprofloxacin	(Liu *et al.*, 2015)
Polymeric NP	Voriconazole	(Sinha *et al.*, 2013)
SLN	Insulin	(Liu *et al.*, 2008)

Abbreviation: NP = Nanoparticle; SLN = Solid Lipid Nanoparticle; API = Active Pharmaceutical Ingredient; NC = Nanocarrier; SLM = Solid Lipid Microparticle

6.5.1 Solid Lipid Nanoparticle

For many years, SLNs have been the focus of much research about their potential for the delivery of medications to the lungs. SLNs are aqueous suspensions at the nanoscale made of physiological lipids, mostly phospholipids and triglycerides. Because the formulations are based on the use of physiological components, they are more acceptable for pulmonary drug delivery because they are less toxic. The deep lung regions are overflowing with phospholipids, which are vital for the respiratory system to operate that (Kumar *et al.*, 2022b). For the lung tissue to stay at its ideal surface tension and experience less friction, phospholipid-based surfactant proteins must be present at the alveolar surface (Beck-Broichsitter *et al.*, 2011). It established the application of triglyceride- and phospholipid-based nanoparticles for possible pulmonary uses. In addition to measuring cytokine activation, they assessed the toxicity profile of these SLNs in *in vitro*, *ex vivo*, and *in vivo* models. When the SLN was nebulized into mice, no activation of pro-inflammatory cytokines (Nassimi *et al.*, 2010) was seen. Because the lipid matrix base used in the research is comparable, these SLNs may be appropriate for the pulmonary delivery of API to treat pulmonary conditions (Nassimi *et al.*, 2009). A further recent investigation examined the physicochemical characteristics of quercetin-loaded solid lipid microparticles (SLMs) and examined their potential as an asthma treatment due

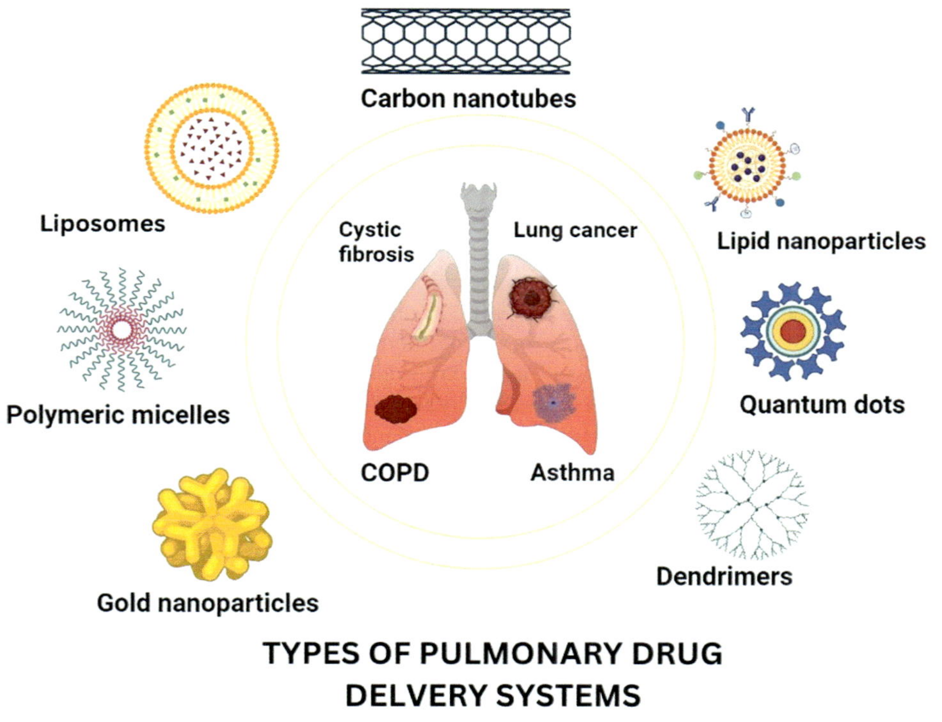

FIGURE 6.1 Nanoparticles in pulmonary drug delivery systems.

to the flavonoid's antioxidant and anti-inflammatory qualities (Silva *et al.*, 2013). Different types of pulmonary drug delivery systems are mentioned in Figure 6.1.

6.5.2 Liposomes

Phospholipids, which are naturally occurring in the lungs, are the primary ingredient in liposome preparation, providing an attractive drug delivery mechanism, particularly for pulmonary applications (Virmani and Gupta, 2017). They are made with cholesterol, phospholipids, lung surfactants, etc. With their continuous release characteristics, liposomes allow for a prolonged duration of the greatest therapeutic effect. Purified buffalo surfactant was the first liposomal product to be released in the 1990s for the treatment of acute respiratory distress syndrome (ARDS) in babies through pulmonary instillation. Subsequently, parenteral rather than pulmonary administration of amphotericin B-loaded liposomes was introduced (Adler-Moore and Proffitt, 2002). Liposomes inhaled constitute a difficulty. Keeping liposomes' essential physical characteristics after nebulization is key to a good liposomal product (Gaspar *et al.*, 2008). A double-blind, controlled experiment with a temperature of individuals with *Pseudomonas aeruginosa* infection was conducted as part of a recent clinical investigation (Clancy *et al.*, 2013). Over 28 days, the patients were split into two groups: one receiving ampicillin aerosol once daily and the other receiving a control group. When comparing the sputum of the amikacin-treated group to the saline group, the investigators noticed a lower density of *P. aeruginosa*. In addition, groups treated with amikacin showed improvements in lung functioning compared to the placebo. Based on the findings, they discovered that amikacin was safe and tolerable, lung functions were enhanced, and the density of *P. aeruginosa* in the sputum of cystic fibrosis patients had decreased. Ciprofloxacin liposomes were examined in a similar manner for lung infections (Liu *et al.*, 2015).

6.5.3 Polymeric Nanoparticles

In the delivery of drugs to the lungs, polymers are becoming progressively more important. Numerous polymers have been researched for use in the lungs. Polymers offer an assortment of benefits, including altered surface characteristics, increased drug encapsulation and defence against degradation, extended drug delivery, and an extended shelf life. The most often utilized polymers for medicinal purposes include alginate, chitosan, gelatin base, PLA, PLGA, and PCL (Menon *et al.*, 2014). To make these biodegradable, their surface and chemical characteristics are changed (Beck-Broichsitter *et al.*, 2012). A new investigation by Beck-Broichsitter and links evaluated the impact of polymeric nanoparticles on the surface characteristics of a lung surfactant. The impact of synthetic and biodegradable polymeric nanoparticles on the lung surfactant was also contrasted by the scientists. They discovered variations in the pulmonary surfactant's surface tension that were dose dependent. Protein and gene delivery using chitosan-based nanoparticles has been the subject of numerous investigations. In studies conducted by low molecular-weight heparin was delivered to the lungs using phospholipid and chitosan-containing nanoparticles that were made using the ionic gelation process under various pH conditions. From *in vivo* experiments on mice, they found that nanoparticles prepared in a neutral setting increased the coagulation time, those made in an acidic environment conducted alternatively (Trapani *et al.*, 2013).

Generally, with the possibility of biodegradability, polymeric particles need their rate of disintegration tested. Furthermore, toxicity profiles in different *in vitro*, *ex vivo*, and *in vivo* models need to be thoroughly examined. Polymeric nanoparticle-encapsulated medications are emerging for pulmonary delivery to treat various respiratory disorders.

6.6 NANOPARTICLES IN DIFFERENT TYPES OF PULMONARY DRUG DELIVERY SYSTEMS

6.6.1 Asthma

Asthma is a chronic, non-communicable condition that affects more than 300 million people globally. It is affected by a variety of hereditary and environmental variables, such as dusting, illnesses, smoking, temperature fluctuations, animals, perfume odours, home dust mites, hormone changes, rain, exercise, and virus infections. Symptoms of the condition include bronchoconstriction, increased mucus output, and persistent inflammation. Because of its complicated aetiology, asthma has no particular medication. However, long-term systematic therapies can reduce symptoms and improve prognosis. The first-line strategy to asthma care is to use a combination of steroids and bronchodilators (Grover *et al.*, 2021). However, steroid-resistant asthma accounts for more than 60% of asthma-related medical expenses. Other therapies for moderate to severe refractory asthma include human monoclonal antibodies, cytokine/chemokine antagonists, and nebulized glucocorticoids. The nanoparticles used in asthma treatment are mentioned in Table 6.3.

6.6.2 COPD

Chronic obstructive pulmonary disease (COPD) is a public health concern and the third leading cause of mortality worldwide. It is prominent in low-income and underdeveloped countries, resulting in poor health and a worse quality of life. Tobacco smoke, biomass, pollution, and previous tuberculosis infection are all major risk factors. The primary cause of COPD-mediated airway damage is excessive free radical production and chronic inflammation in the lungs, which causes airway blockage and difficulty breathing. Currently, therapy options include bronchodilators, smoking cessation, antibiotic medication, and corticosteroids. However, these medications do not address inflammation, they are only useful in the therapeutic therapy of COPD (Singh *et al.*, 2018). Roflumilast, an anti-inflammatory medication, has demonstrated encouraging effects in increasing

TABLE 6.3
Nanoparticles for the Treatment of Asthma

Drug	Types of Nanoparticles	Developmental Stage	Findings	References
Doxofylline	Liposomal	*In vitro*	Liposomes are better retained than controlled-release formulations.	(Nsairat *et al.*, 2022)
Salbutamol sulphate	Liposomal	Phase I	Entrapping drugs in niosomes, which may be packaged into aerosols that follow United State Pharmacopoeia (USP) standards, is a viable alternative for constructing a regulated drug delivery system.	(Arafa and Ayoub, 2017)
R-terbutaline hydrochloride	Liposomal	Preclinical studies	The R-terbutaline hydrochloride liposome group had a longer anti-asthmatic effect than the R-terbutaline hydrochloride solution group.	(Li *et al.*, 2018)
Budesonide	Liposomal	Preclinical trials	Reduces toxicity and boosts compliance in asthma patients.	(Konduri *et al.*, 2003)
Methylprednisolone	Dendrimer	Preclinical studies	Dendrimer-conjugated medicines may help with medication retention in the lungs for treating inflammatory disorders like asthma.	(Inapagolla *et al.*, 2010)
Beclomethasone dipropionate	Dendrimer	Preclinical studies	Dexamethasone nanoparticles are more effective than free dexamethasone in decreasing airway hyperresponsiveness.	(Nasr *et al.*, 2014)
Thymulin	Polymeric	Preclinical studies	Nanoparticle-based gene therapies provide therapeutic genes for asthma in a safe and effective manner.	(da Silva *et al.*, 2014)
Quercetin	Polymeric	*In vitro*	Effective in lowering the synthesis of major pro-inflammatory cytokines linked to the pathogenesis of asthma.	(Cote *et al.*, 2015)
Baicalein	Polymeric	Preclinical studies	Encapsulated and loaded Baicalein nanoparticles decreased airway hyperresponsiveness and inflammation, making them promising anti-asthmatic medicines.	(Wang *et al.*, 2021)
Andrographolide	Polymeric	Preclinical studies	Nanoparticle formulations provide superior bioavailability, effectiveness, and efficiency compared to other formulations.	(Chakraborty *et al.*, 2019)
Isoliquiritigenin	Nano suspension	Preclinical studies	Isoliquiritigenin is a self-nano emulsifying drug delivery system was proven to have much greater bioavailability and anti-asthma impact than isoliquiritigenin suspensions.	(Cao *et al.*, 2020)
Chafuroside A	Nanomicelles	Preclinical studies	Chafuroside A, a self-assembled micellar formulation, can be used to treat asthma.	(Onoue *et al.*, 2012)
Curcumin	Nanomicelles	Preclinical studies	Micellarcurcumin has a long-term anti-asthmatic action.	(Chawla *et al.*, 2022)
Chil3 and chil4 siRNA	Nanomicelles	*In vitro*	Asthma symptoms such as mucus production were decreased.	(Choi *et al.*, 2020)

TABLE 6.4
Nanoparticles for the Treatment of COPD

Drug	Nano Carrier Composition	Types of Nanoparticle	Route of Administration	Mode of Action	References
Au	Gold NPs	Inorganic nanoparticles	Inhalation	AuNP acts as a nanocarrier (rapid binding to the alveolar epithelium).	(Geiser *et al.*, 2013)
SOD and CAT	Cerium oxide (IV) NPs	Inorganic nanoparticles	SOD enzymatic assay	Reduce ROS	(Gil *et al.*, 2017)
Antibody conjugates	Ferrous and ferric chlorides	Inorganic nanoparticles	Intravenous	Enable endothelial distribution of active substances while protecting against proteolysis CAT and SOD.	(Hood *et al.*, 2014)
Carvacrol	Lipid	Solid lipid NPs	H441 cells	Antagonisis ROS	(Carvalho *et al.*, 2020)
Budesonide	PEG-DSPE	Polymer NPs	Inhalation	–	(Jacobs and Müller, 2002)
Curcumin	PVP, PVA	Polymer NPs	Inhalation	Reduce LPS-induced inflammation in alveolar macrophages in a time-dependent manner.	(Lee *et al.*, 2016)
microRNAs	Cationic lipid DOTAP	Polymer NPs	Human alveolar adenocarcinoma A549 cells	Reduce IRAK1 expression and lower IL-8 promoter reporter output.	(Mohamed *et al.*, 2019)
siRNA	PAMAM dendrimers	Dendrimers	Inhalation	Target and mute genes in lung alveolar epithelial cells (A549).	(Conti *et al.*, 2014)
TNF-α siRNA	PAMAM dendrimer	Dendrimers	Intranasal	Gene silencing that targets TNF-α	(Bohr *et al.*, 2020)
Budesonide	Lecithin	Nanocrystals	–	Facilitate simpler industrial usage of nanocrystals.	(Raula *et al.*, 2013)

Abbreviation: ROS = Reactive oxygen species; IV = Intravenous; PEG-DSPE = Poly(ethylene glycol)–distearoylphosphatidylethanolamine; PVA = Polyvinyl alcohol; IL-8 = Intraleukins-8; DOTAP = 1,2-dioleoyl-3-trimethylammonium propane; PAMAM = Poly (amidoamine); TNF-α = Tumour Nacrosis Factor-α

lung function and avoiding COPD exacerbations. There is a significant unmet therapeutic need to provide a safe and effective medication for the clinical treatment of COPD (Gershon *et al.*, 2011).

Smart drug delivery systems have developed an alternative therapy strategy for a variety of disorders. Polymeric nanoparticles (PLGA), a synthetic polymer, are attractive instruments for drug delivery systems against a variety of illnesses since they significantly increase the efficacy of many current medications. PLGA has several advantages, including excellent biocompatibility, biodegradability, adaptability, and adjustable mechanical characteristics (Saxena *et al.*, 2022). Some polymeric PLGA are discussed in Table 6.4.

6.6.3 Cystic Fibrosis

Cystic fibrosis (CF) is a genetic condition that affects about 70,000 people globally. It is mainly caused by an excess of thick mucus in the lungs, which results in airway blockage and bacterial infections. This leads to inflammation that causes bronchiectasis, respiratory failure, and mortality. Despite the obstacles, symptomatic therapy and gene therapy have shown good outcomes in the treatment of cystic fibrosis (Virmani *et al.*, 2023a). However, the absence of efficient delivery techniques,

TABLE 6.5
Nanoformulations for the Treatment of Cystic Fibrosis

Drug	Composition	Nanoparticles	Mode of Action	References
Curcumin	PLGA	Polymeric nanoparticles	Improve bioavailability and effectiveness over free drug.	(Cartiera *et al.*, 2010)
Ciprofloxacin	PLGA	Polymeric nanoparticles	The antibacterial activity of drug and enhanced mucus penetration.	(Günday Türeli *et al.*, 2017)
Tobramycin	PEG/PLGA	Polymeric nanoparticles	Improve penetration of mucus, increased antibacterial activity.	(Ernst *et al.*, 2018)
Plasmid DNA	PLGA	Polymeric nanoparticles	PLGA-mediated gene transfer can extend gene expression, however efficiency has to be increased.	(Stass *et al.*, 2013)
Cysteamine	PAMAM-DEN^{CYS}	Dendrimers	PA infection and growth decreased, and rescue of the CFTR protein.	(Brockman *et al.*, 2017)
siRNA	PAMAM G4	Dendrimers	Gene silencing and cellular uptake.	(Conti *et al.*, 2014)
Gentamicin	PC/Chol/DOTAP	Liposomes	Increased activity of antibiotics at lower concentrations and improved bacterial penetration.	(Drulis-Kawa *et al.*, 2006a)
Amikacin	DPPC/Chol	Liposomes	Improved penetration within PA biofilm.	(Mugabe et al., 2005)
Polymyxin B	DPPC/Chol	Liposomes	Enhanced absorption and antibacterial action against PA.	(Alipour *et al.*, 2008)
Meropenem	PC/DOPE/SA	Liposomes	The encapsulation drugs in cationic liposomes enhances their effectiveness	(Drulis-Kawa *et al.*, 2006 b)
Myriocin	SA/PA	Solid lipid NPs (SLNs)	Reduces lung inflammation.	(Caretti *et al.*, 2014)
Ciprofloxacin	Albumin gelatin	SLNs	SR and CR.	(Jain and Banerjee, 2008)
Budesonide	GMS	SLNs	Better drug dispersion inside the lipid matrix.	(Zhang *et al.*, 2011)

Abbreviation: PEG = Polyethylene glycol; PAMAM-DEN^{Cys} = Poly (amidoamine)-Cysteamine analogue; PA = Pseudomonas aeruginosa; PC/Chol/DOTAP = Phosphatidylcholine/Cholesterol/1,2-dioleoyloxy-3-trimethyl ammonium-propane; DPPC/Chol = Dipalmitoylphosphatidylcholine/Cholesterol; DOPE = 1,2-dioleoyl-sn-glycero-3-phosphoethanolamine; DOTAP = 1,2-Dioleoyloxy-3-trimethylammonium-propane; GMS = Glycerol Monostearate; siRNA = Small Interfering RNA

particularly those targeting lung tissues and cells, restricts therapy efficacy. Nanomedicine offers the potential to enhance current medicines and provide novel treatments for CF. The implementation of nanotechnology to improve medicine delivery or gene treatments is a viable approach to addressing this condition (Velino *et al.*, 2019). Some nanoparticles are listed in Table 6.5.

6.6.4 Lung Cancer

Targeted drug delivery systems based on nanoparticles make use of the different site-specific elements in the intended tumour site, such as the vasculature, pH, enzyme concentrations, genetic mutation products, and other elements in the tumour microenvironment. The anticancer medications are released by the stimuli-responsive drug delivery systems when certain stimulants are present (Doroudian *et al.*, 2021). The stimuli can be categorized as "exogenous," such as light, ultrasonography, and magnetic field, or "endogenous," which includes pH, redox agents, and enzymes. Strengthening these variables increases the targeted absorption of nanoparticles into neoplastic tissue (Damodharan, 2021). Currently, the U.S. Food and Drug Administration (FDA) has approved or is in the process of approving more than 20 commercial nanomaterial-based cancer treatments, of which roughly 85% are based on liposomes and polymer micelles (Chen *et al.*, 2014). To improve the therapeutic efficacy of the medications, attention is also being paid to putting biomolecules like proteins, peptides, DNA, RNA, and other components into nanoparticles in addition to liposomes and polymers (Yetisgin *et al.*, 2020). Different nanoformulations are discussed in Table 6.6.

TABLE 6.6
Nanoformulations in Lung Cancer

Nanoparticle Types	Drug	Phase	Types of Lung Cancer	Uses	References
Polymeric	Docetaxel	I	Lung cancer is one type of advanced or metastatic cancer.	It treats lung cancer, prostate cancer and stomach cancers.	(Chiang *et al.*, 2021)
Nanoparticle	Hafnium oxide	I	NSCLC	To create control rods for use in nuclear submarines, etc.	(Doroudian *et al.*, 2021)
Nanoparticles albumin based	Paclitaxel	III	Non-squamous NSCLC	To treat Kaposi sarcoma, NSCLC, advanced ovarian cancer, and breast cancer.	(Bhardwaj *et al.*, 2019)
Polymeric	Camptothecin	I/II	Small cell lung cancer, lung neoplasm	It is used to treat leukaemia as a chemotherapeutic agent.	(Tran *et al.*, 2017)
Polymeric	Irinotecan	II	Breast cancer and lung	To treat a variety of solid tumours.	(Doroudian *et al.*, 2021)
Micellar	Cisplatin	I/II	Solid tumours	To treat cervical, head and neck, testicular, ovarian, bladder, and lung cancer.	(Gold and Raja, 2024)
Liposomal	Topotecan	II/III	Small cell lung cancer	To treat individuals with ovarian cancer that have spread when all other treatments have failed.	(Ardizzoni *et al.*, 1997)
Polymeric micelle	SN-38	II	Small cell lung cancer	Research on the treatment of cancer, advanced solid tumours, and triple negative breast cancer has involved the use of SN 38.	(Tseng *et al.*, 2020)
Liposome	Doxorubicin	II	Lung cancer	To treat malignancies of the breast, ovary, bladder, and thyroid as well as soft tissue and bone sarcomas.	(Tap *et al.*, 2016)
Polymer	Leuprolide acetate	III	Prostate cancer	Treatment for *in vitro* fertilization, endometriosis, uterine fibroids, central precocious puberty, and prostate cancer.	(Wilson *et al.*, 2007)

Abbreviation: NSCLC: non-small cell lung cancer.

TABLE 6.7
Nanoformulations under Clinical Applications

API	Nanoformulations	Disease	NCT No.	Status
ILL-4Ra antagonist (PRS-060)	Lipocalin-1	Asthma	NCT03574805	Phase I
MRT5005 (CFTR mRNA)	mRNA-refd LNP	Cystic fibrosis	NCT03375047	Phase I/II
Remdesivir (GS-5734) and NA-831 (NEUROSIVIR)	Inhaled nanoparticle	Covid-19, SARS, etc.	NCT04480333	Phase I
Amikacin	Liposome	*Mycobacterium avium* complex (MAC) lungs disease	ARIKACE® KIT	Approved by FDA in 2018
Amikacin	Liposome	Cystic fibrosis	NCT01316276	Phase III
Amikacin	Liposome	NTM lung infection due to MAC	NCT02344004	Phase III
Amikacin	Liposome	Bronchiectasis	NCT00775138	Phase II
Ciprofloxacin	Liposome	Cystic fibrosis	NCT00645788	Phase II
PGM 169/GL67A (Plasmid DNA)	Gene product/lipid vector	Cystic fibrosis	NCT01621867	Phase II
Salbutamol sulphate	Nanovesicles/niosomes	Pulmonary disease	NCT03059017	Phase I

Abbreviation: LNP = Lipid Nanoparticle; NTM = Non-tuberculous mycobacterium

6.7 CLINICAL TRIALS AND PATENTS

Inhalable nanoformulations have been used in recent years, although translational research on them is still in its early stages. Currently, there is just one commercialized product, with a few inhalable nanoformulations entering clinical studies. In 2018, the sole commercial inhalable nanoformulation was the FDA-approved ARIKAYCE® (an amikacin liposome suspension) inhalation suspension. It is logical for medication formulation research and development to begin with a commercial product (Peng *et al.*, 2024). Some nanoformulaions are listed in Table 6.7.

6.8 CONCLUSION

Nanotechnology-based pulmonary drug delivery shows great promise to treat various respiratory disorders such as COPD, cystic fibrosis, and lung cancer. These delivery systems offer various benefits of improved bioavailability, increased efficiency, decreased biodistribution, reduced adverse effects, targetability, and increased patient compliance. The nano-sized results in increased surface area cause improved drug solubility and uniform drug distribution within the alveoli. In addition, the concentration of a drug at a target site is improved, resulting in a reduced dose. The loading of drugs in nanocarriers enabled the sustained release of the drugs in the lungs, leading to decreased dose frequency and increased patient compliance. Various nanotechnology-based pulmonary drug delivery systems have granted patents, of which some are under clinical trials. As these nanoformulations are intended for pulmonary delivery, the risk of potential toxicity of excipients used for their fabrication is a major issue that must be addressed in the future.

REFERENCES

Adler-Moore, J. and Proffitt, R.T. (2002) 'AmBisome: Liposomal formulation, structure, mechanism of action and pre-clinical experience', *Journal of Antimicrobial Chemotherapy*, 49(suppl_1), pp. 21–30. https://doi.org/10.1093/jac/49.suppl_1.21.

Ahalwat, S. *et al.* (2023) 'Mannose-functionalized isoniazid-loaded nanostructured lipid carriers for pulmonary delivery: In vitro prospects and in vivo therapeutic efficacy assessment', *Pharmaceuticals*, 16(8), p. 1108. https://doi.org/10.3390/ph16081108.

Alipour, M. *et al.* (2008) 'Antimicrobial effectiveness of liposomal polymyxin B against resistant gram-negative bacterial strains', *International Journal of Pharmaceutics*, 355(1–2), pp. 293–298. https://doi.org/10.1016/j.ijpharm.2007.11.035.

Alton, E.W.F.W. (2004) 'Use of nonviral vectors for cystic fibrosis gene therapy', *Proceedings of the American Thoracic Society*, 1(4), pp. 296–301. https://doi.org/10.1513/pats.200404-031MS.

Arafa, M.G. and Ayoub, B.M. (2017) 'nano-vesicles of salbutamol sulphate in metered dose inhalers: Formulation, characterization and in vitro evaluation', *International Journal of Applied Pharmaceutics*, 9(6), p. 100. https://doi.org/10.22159/ijap.2017v9i6.22448.

Ardizzoni, A. *et al.* (1997) 'Topotecan, a new active drug in the second-line treatment of small-cell lung cancer: A phase II study in patients with refractory and sensitive disease. The European organization for research and treatment of cancer early clinical studies group and new drug development office, and the lung cancer cooperative group', *Journal of Clinical Oncology: Official Journal of the American Society of Clinical Oncology*, 15(5), pp. 2090–2096. https://doi.org/10.1200/JCO.1997.15.5.2090.

Azarmi, S., Roa, W.H. and Löbenberg, R. (2008) 'Targeted delivery of nanoparticles for the treatment of lung diseases', *Advanced Drug Delivery Reviews*, 60(8), pp. 863–875. https://doi.org/10.1016/j.addr.2007.11.006.

Beck-Broichsitter, M. *et al.* (2009) 'Pulmonary drug delivery with aerosolizable nanoparticles in an *ex vivo* lung model', *International Journal of Pharmaceutics*, 367(1), pp. 169–178. https://doi.org/10.1016/j.ijpharm.2008.09.017.

Beck-Broichsitter, M. *et al.* (2011) 'Biophysical investigation of pulmonary surfactant surface properties upon contact with polymeric nanoparticles in vitro', *Nanomedicine: Nanotechnology, Biology and Medicine*, 7(3), pp. 341–350. https://doi.org/10.1016/j.nano.2010.10.007.

Beck-Broichsitter, M. *et al.* (2012) 'Development of a biodegradable nanoparticle platform for sildenafil: Formulation optimization by factorial design analysis combined with application of charge-modified branched polyesters', *Journal of Controlled Release*, 157(3), pp. 469–477. https://doi.org/10.1016/j.jconrel.2011.09.058.

Bhardwaj, V. *et al.* (2019) 'Recalcitrant issues and new frontiers in nano-pharmacology', *Frontiers in Pharmacology*, 10. https://doi.org/10.3389/fphar.2019.01369.

Bharti Sharma, J. *et al.* (2023) 'Statistical optimization of tetrahydrocurcumin loaded solid lipid nanoparticles using box behnken design in the management of streptozotocin-induced diabetes mellitus', *Saudi Pharmaceutical Journal*, p. 101727. https://doi.org/10.1016/j.jsps.2023.101727.

Bohr, A. *et al.* (2020) 'Treatment of acute lung inflammation by pulmonary delivery of anti-TNF-α siRNA with PAMAM dendrimers in a murine model', *European Journal of Pharmaceutics and Biopharmaceutics*, 156, pp. 114–120. https://doi.org/10.1016/j.ejpb.2020.08.009.

Brockman, S.M. *et al.* (2017) 'Dendrimer-based selective autophagy-induction rescues ΔF508-CFTR and inhibits *Pseudomonas aeruginosa* infection in cystic fibrosis', *PloS One*, 12(9), p. e0184793. https://doi.org/10.1371/journal.pone.0184793.

Brzoska, M. *et al.* (2004) 'Incorporation of biodegradable nanoparticles into human airway epithelium cells—in vitro study of the suitability as a vehicle for drug or gene delivery in pulmonary diseases', *Biochemical and Biophysical Research Communications*, 318(2), pp. 562–570. https://doi.org/10.1016/j.bbrc.2004.04.067.

Byrne, M. *et al.* (2015) 'Star-shaped polypeptides: Synthesis and opportunities for delivery of therapeutics', *Macromolecular Rapid Communications*, 36(21), pp. 1862–1876. https://doi.org/10.1002/marc.201500300.

Cao, M. *et al.* (2020) 'Development of an orally bioavailable isoliquiritigenin self-nanoemulsifying drug delivery system to effectively treat ovalbumin-induced asthma', *International Journal of Nanomedicine*, 15, pp. 8945–8961. https://doi.org/10.2147/IJN.S269982.

Caretti, A. *et al.* (2014) 'Anti-inflammatory action of lipid nanocarrier-delivered myriocin: Therapeutic potential in cystic fibrosis', *Biochimica Et Biophysica Acta*, 1840(1), pp. 586–594. https://doi.org/10.1016/j.bbagen.2013.10.018.

Cartiera, M.S. *et al.* (2010) 'Partial correction of cystic fibrosis defects with PLGA nanoparticles encapsulating curcumin', *Molecular Pharmaceutics*, 7(1), p. 86. https://doi.org/10.1021/mp900138a.

Carvalho, F.O. *et al.* (2020) 'Effects of the solid lipid nanoparticle of carvacrol on rodents with lung injury from smoke inhalation', *Naunyn-Schmiedeberg's Archives of Pharmacology*, 393(3), pp. 445–455. https://doi.org/10.1007/s00210-019-01731-1.

Castelli, F. *et al.* (2005) 'Characterization of indomethacin-loaded lipid nanoparticles by differential scanning calorimetry', *International Journal of Pharmaceutics*, 304(1), pp. 231–238. https://doi.org/10.1016/j.ijpharm.2005.08.011.

Chakraborty, S. *et al.* (2019) 'Therapeutic potential of andrographolide-loaded nanoparticles on a murine asthma model', *Nanomedicine: Nanotechnology, Biology and Medicine*, 20, p. 102006. https://doi.org/10.1016/j.nano.2019.04.009.

Chattopadhyay, P. *et al.* (2007) 'Production of solid lipid nanoparticle suspensions using supercritical fluid extraction of emulsions (SFEE) for pulmonary delivery using the AERx system', *Advanced Drug Delivery Reviews*, 59(6), pp. 444–453. https://doi.org/10.1016/j.addr.2007.04.010.

Chawla, R. *et al.* (2022) 'Intranasal micellar curcumin for the treatment of chronic asthma', *Journal of Drug Delivery Science and Technology*, 67, p. 102922. https://doi.org/10.1016/j.jddst.2021.102922.

Chen, Z. *et al.* (2014) 'Non-small-cell lung cancers: A heterogeneous set of diseases', *Nature Reviews Cancer*, 14(8), pp. 535–546. https://doi.org/10.1038/nrc3775.

Chiang, C.-L., Cheng, M.-H. and Lin, C.-H. (2021) 'From nanoparticles to cancer nanomedicine: Old problems with new solutions', *Nanomaterials*, 11(7), p. 1727. https://doi.org/10.3390/nano11071727.

Choi, M. *et al.* (2020) 'Targeted delivery of Chil3/Chil4 siRNA to alveolar macrophages using ternary complexes composed of HMG and oligoarginine micelles', *Nanoscale*, 12(2), pp. 933–943. https://doi.org/10.1039/C9NR06382J.

Clancy, J.P. *et al.* (2013) 'Phase II studies of nebulised Arikace in CF patients with *Pseudomonas aeruginosa* infection', *Thorax*, 68(9), pp. 818–825. https://doi.org/10.1136/thoraxjnl-2012-202230.

Conti, D.S. *et al.* (2014) 'Poly(amidoamine) dendrimer nanocarriers and their aerosol formulations for siRNA delivery to the lung epithelium', *Molecular Pharmaceutics*, 11(6), pp. 1808–1822. https://doi.org/10.1021/mp4006358.

Cote, B. *et al.* (2015) 'Combinatorial resveratrol and quercetin polymeric micelles mitigate doxorubicin induced cardiotoxicity in vitro and in vivo', *Journal of Controlled Release*, 213, pp. 128–133. https://doi.org/10.1016/j.jconrel.2015.06.040.

da Silva, A.L. *et al.* (2014) 'DNA nanoparticle-mediated thymulin gene therapy prevents airway remodeling in experimental allergic asthma', *Journal of Controlled Release*, 180, pp. 125–133. https://doi.org/10.1016/j.jconrel.2014.02.010.

Damodharan, J. (2021) 'Nanomaterials in medicine – An overview', *Materials Today: Proceedings*, 37, pp. 383–385. https://doi.org/10.1016/j.matpr.2020.05.380.

Dhir, S. *et al.* (2023) 'Amomum subulatum fruit extract mediated green synthesis of silver and copper oxide nanoparticles: Synthesis, characterization, antibacterial and anticancer Activities', *Processes*, 11(9), p. 2698. https://doi.org/10.3390/pr11092698.

Doroudian, M. *et al.* (2021) 'Smart nanotherapeutics and lung cancer', *Pharmaceutics*, 13(11), p. 1972. https://doi.org/10.3390/pharmaceutics13111972.

Drulis-Kawa, Z. *et al.* (2006a) 'A comparison of the in vitro antimicrobial activity of liposomes containing meropenem and gentamicin', *Cellular & Molecular Biology Letters*, 11(3), pp. 360–375. https://doi.org/10.2478/s11658-006-0030-6.

Drulis-Kawa, Z. *et al.* (2006b) 'In vitro antimicrobial activity of liposomal meropenem against *Pseudomonas aeruginosa* strains', *International Journal of Pharmaceutics*, 315(1–2), pp. 59–66. https://doi.org/10.1016/j.ijpharm.2006.02.017.

Dubey, S. *et al.* (2024) 'Breaking barriers in eco-Friendly synthesis of plant-mediated metal/metal oxide/bimetallic nanoparticles: Antibacterial, anticancer, mechanism elucidation, and versatile utilizations', *Journal of Nanomaterials*, 2024, p. e9914079. https://doi.org/10.1155/2024/9914079.

Elhissi, A.M.A. *et al.* (2010) 'Development and characterisation of freeze-dried liposomes containing two anti-asthma drugs', *Micro & Nano Letters*, 5(3), pp. 184–188. https://doi.org/10.1049/mnl.2010.0032.

Ernst, J. *et al.* (2018) 'Polyester-based particles to overcome the obstacles of mucus and biofilms in the lung for tobramycin application under static and dynamic fluidic conditions', *European Journal of Pharmaceutics and Biopharmaceutics: Official Journal of Arbeitsgemeinschaft Fur Pharmazeutische Verfahrenstechnik e.V*, 131, pp. 120–129. https://doi.org/10.1016/j.ejpb.2018.07.025.

Ferkol, T. and Schraufnagel, D. (2014) 'The global burden of respiratory disease', *Annals of the American Thoracic Society*, 11(3), pp. 404–406. https://doi.org/10.1513/AnnalsATS.201311-405PS.

Flume, P. and Klepser, M.E. (2002) 'The rationale for aerosolized Antibiotics', *Pharmacotherapy: The Journal of Human Pharmacology and Drug Therapy*, 22(3P2), pp. 71S–79S. https://doi.org/10.1592/phco.22.6.71S.33909.

Gaspar, M.M., Bakowsky, U. and Ehrhardt, C. (2008) 'Inhaled liposomes–current strategies and future challenges', *Journal of Biomedical Nanotechnology*, 4(3), pp. 245–257. https://doi.org/10.1166/jbn.2008.334.

Geiser, M. *et al.* (2013) 'Cellular uptake and localization of inhaled gold nanoparticles in lungs of mice with chronic obstructive pulmonary disease', *Particle and Fibre Toxicology*, 10, p. 19. https://doi.org/10.1186/1743-8977-10-19.

Gershon, A.S. *et al.* (2011) 'Lifetime risk of developing chronic obstructive pulmonary disease: A longitudinal population study', *The Lancet*, 378(9795), pp. 991–996. https://doi.org/10.1016/S0140-6736(11)60990-2.

Gil, D. *et al.* (2017) 'Antioxidant activity of SOD and catalase conjugated with nanocrystalline ceria', *Bioengineering*, 4(1), p. 18. https://doi.org/10.3390/bioengineering4010018.

Gilani, K. *et al.* (2005) 'The effect of water to ethanol feed ratio on physical properties and aerosolization behavior of spray dried cromolyn sodium particles', *Journal of Pharmaceutical Sciences*, 94(5), pp. 1048–1059. https://doi.org/10.1002/jps.20315.

Gold, J.M. and Raja, A. (2024) 'Cisplatin', in *StatPearls*. Treasure Island (FL): StatPearls Publishing. Available at: http://www.ncbi.nlm.nih.gov/books/NBK547695/ (Accessed: 30 April 2024).

Gramling, H., Kiziroglou, M. and Yeatman, E. (2017) 'Nanotechnology for Consumer Electronics: Materials, Devices, Applications', in *Nanoelectronics: Materials, Devices, Applications*, pp. 501–526. https://doi.org/10.1002/9783527800728.ch22.

Gratton, S.E.A. *et al.* (2007) 'Nanofabricated particles for engineered drug therapies: A preliminary biodistribution study of PRINT™ nanoparticles', *Journal of Controlled Release*, 121(1), pp. 10–18. https://doi.org/10.1016/j.jconrel.2007.05.027.

Grenha, A., Seijo, B. and Remuñán-López, C. (2005) 'Microencapsulated chitosan nanoparticles for lung protein delivery', *European Journal of Pharmaceutical Sciences*, 25(4), pp. 427–437. https://doi.org/10.1016/j.ejps.2005.04.009.

Groneberg, D.A. *et al.* (2003) 'Fundamentals of pulmonary drug delivery', *Respiratory Medicine*, 97(4), pp. 382–387. https://doi.org/10.1053/rmed.2002.1457.

Grover, M. *et al.* (2021) 'In vitro phytochemical screening, cytotoxicity studies of curcuma Longa extracts with isolation and characterisation of their isolated compounds', *Molecules*, 26(24), p. 7509. https://doi.org/10.3390/molecules26247509.

Günday Türeli, N. *et al.* (2017) 'Ciprofloxacin-loaded PLGA nanoparticles against cystic fibrosis *P. aeruginosa* lung infections', *European Journal of Pharmaceutics and Biopharmaceutics: Official Journal of Arbeitsgemeinschaft Fur Pharmazeutische Verfahrenstechnik e.V*, 117, pp. 363–371. https://doi.org/10.1016/j.ejpb.2017.04.032.

Hood, E.D. *et al.* (2014) 'Endothelial targeting of nanocarriers loaded with antioxidant enzymes for protection against vascular oxidative stress and inflammation', *Biomaterials*, 35(11), pp. 3708–3715. https://doi.org/10.1016/j.biomaterials.2014.01.023.

Howard, K.A. *et al.* (2006) 'RNA interference in vitro and in vivo using a novel chitosan/siRNA nanoparticle system', *Molecular Therapy*, 14(4), pp. 476–484. https://doi.org/10.1016/j.ymthe.2006.04.010.

Inapagolla, R. *et al.* (2010) '*In vivo* efficacy of dendrimer–methylprednisolone conjugate formulation for the treatment of lung inflammation', *International Journal of Pharmaceutics*, 399(1), pp. 140–147. https://doi.org/10.1016/j.ijpharm.2010.07.030.

Jaafar-Maalej, C. *et al.* (2011) 'Beclomethasone-loaded lipidic nanocarriers for pulmonary drug delivery: Preparation, characterization and in vitro drug release', *Journal of Nanoscience and Nanotechnology*, 11(3), pp. 1841–1851. https://doi.org/10.1166/jnn.2011.3119.

Jacobs, C. and Müller, R.H. (2002) 'Production and characterization of a budesonide nanosuspension for pulmonary administration', *Pharmaceutical Research*, 19(2), pp. 189–194. https://doi.org/10.1023/a:1014276917363.

Jain, D. and Banerjee, R. (2008) 'Comparison of ciprofloxacin hydrochloride-loaded protein, lipid, and chitosan nanoparticles for drug delivery', *Journal of Biomedical Materials Research. Part B, Applied Biomaterials*, 86(1), pp. 105–112. https://doi.org/10.1002/jbm.b.30994.

Jasmani, L. *et al.* (2020) 'Application of nanotechnology in wood-based products industry: A review', *Nanoscale Research Letters*, 15(1), p. 207. https://doi.org/10.1186/s11671-020-03438-2.

Joseph, T.M. *et al.* (2023) 'Nanoparticles: Taking a unique position in medicine', *Nanomaterials*, 13(3), p. 574. https://doi.org/10.3390/nano13030574.

Kimura, S. *et al.* (2009) 'Nanoparticle-mediated delivery of nuclear factor κB decoy into lungs ameliorates monocrotaline-induced pulmonary arterial hypertension', *Hypertension*, 53(5), pp. 877–883. https://doi.org/10.1161/HYPERTENSIONAHA.108.121418.

Konduri, K.S. *et al.* (2003) 'Efficacy of liposomal budesonide in experimental asthma', *Journal of Allergy and Clinical Immunology*, 111(2), pp. 321–327. https://doi.org/10.1067/mai.2003.104.

Kumar, G. *et al.* (2023) 'Codelivery of phytochemicals with conventional anticancer drugs in form of nanocarriers', *Pharmaceutics*, 15(3), p. 889. https://doi.org/10.3390/pharmaceutics15030889.

Kumar, G. *et al.* (2024) 'Nanoparticulate Drug Delivery Systems for Colonic Disorders', in H. Dureja et al. (eds) *Advanced Drug Delivery Systems for Colonic Disorders*. Academic Press, pp. 317–344. https://doi.org/10.1016/B978-0-443-14044-0.00009-0.

Kumar, G. *et al.* (2022a) 'Central composite design implemented azilsartan medoxomil loaded nanoemulsion to improve its aqueous solubility and intestinal permeability: In vitro and ex vivo evaluation', *Pharmaceuticals*, 15(11), p. 1343. https://doi.org/10.3390/ph15111343.

Kumar, G., Virmani, T., Pathak, K. and Alhalmi, A. (2022b) 'A revolutionary blueprint for mitigation of hypertension via nanoemulsion', *BioMed Research International*, 2022, p. e4109874. https://doi.org/10.1155/2022/4109874.

Kumar, Mahesh *et al.* (2023a) 'Nanocarriers in tuberculosis treatment: Challenges and delivery strategies', *Pharmaceuticals*, 16(10), p. 1360. https://doi.org/10.3390/ph16101360.

Kumar, Mohit *et al.* (2023b) 'Micro and nano-carriers-based pulmonary drug delivery system: Their current updates, challenges, and limitations – A review', *JCIS Open*, 12, p. 100095. https://doi.org/10.1016/j.jciso.2023.100095.

Kuo, J.S. and Hwang, R. (2004) 'Preparation of DNA dry powder for non-viral gene delivery by spray-freeze drying: Effect of protective agents (polyethyleneimine and sugars) on the stability of DNA', *Journal of Pharmacy and Pharmacology*, 56(1), pp. 27–33. https://doi.org/10.1211/0022357022494.

Lee, D. and Mohapatra, S.S. (2008) 'Chitosan Nanoparticle-Mediated Gene Transfer', in J.M.L. Doux (ed.) *Gene Therapy Protocols: Production and In Vivo Applications of Gene Transfer Vectors*. Totowa, NJ: Humana Press, pp. 127–140. https://doi.org/10.1007/978-1-59745-237-3_8.

Lee, W.-H. *et al.* (2016) 'Curcumin nanoparticles attenuate production of pro-inflammatory markers in lipopolysaccharide-induced macrophages', *Pharmaceutical Research*, 33(2), pp. 315–327. https://doi.org/10.1007/s11095-015-1789-9.

Li, M. *et al.* (2013) 'Methotrexate-conjugated and hyperbranched polyglycerol-grafted Fe3O4 magnetic nanoparticles for targeted anticancer effects', *European Journal of Pharmaceutical Sciences*, 48(1), pp. 111–120. https://doi.org/10.1016/j.ejps.2012.10.008.

Li, Q. *et al.* (2018) 'Preparation of a sustained-release nebulized aerosol of r-terbutaline hydrochloride liposome and evaluation of its anti-asthmatic effects via pulmonary delivery in Guinea Pigs', *AAPS PharmSciTech*, 19(1), pp. 232–241. https://doi.org/10.1208/s12249-017-0816-z.

Lim, Y.H. *et al.* (2016) 'Polymeric nanoparticles in development for treatment of pulmonary infectious diseases', *WIREs Nanomedicine and Nanobiotechnology*, 8(6), pp. 842–871. https://doi.org/10.1002/wnan.1401.

Liu, C. *et al.* (2015) 'In-vitro and in-vivo evaluation of ciprofloxacin liposomes for pulmonary administration', *Drug Development and Industrial Pharmacy*, 41(2), pp. 272–278. https://doi.org/10.3109/03639045.2013.858740.

Liu, J. *et al.* (2008) 'Solid lipid nanoparticles for pulmonary delivery of insulin', *International Journal of Pharmaceutics*, 356(1), pp. 333–344. https://doi.org/10.1016/j.ijpharm.2008.01.008.

Mahmud, M.Z.A. (2023) 'A concise review of nanoparticles utilized energy storage and conservation', *Journal of Nanomaterials*, 2023, p. e5432099. https://doi.org/10.1155/2023/5432099.

Malik, S., Muhammad, K. and Waheed, Y. (2023) 'Emerging applications of nanotechnology in healthcare and medicine', *Molecules*, 28(18), p. 6624. https://doi.org/10.3390/molecules28186624.

Matschiner, G. *et al.* (2023) 'Elarekibep (PRS-060/AZD1402), a new class of inhaled Anticalin medicine targeting IL-4Ra for type 2 endotype asthma', *Journal of Allergy and Clinical Immunology*, 151(4), pp. 966–975. https://doi.org/10.1016/j.jaci.2022.12.815.

Menon, J.U. *et al.* (2014) 'Polymeric nanoparticles for pulmonary protein and DNA delivery', *Acta Biomaterialia*, 10(6), pp. 2643–2652. https://doi.org/10.1016/j.actbio.2014.01.033.

Mohamed, A. *et al.* (2019) 'Pulmonary delivery of Nanocomposite Microparticles (NCMPs) incorporating miR-146a for treatment of COPD', *International Journal of Pharmaceutics*, 569, p. 118524. https://doi.org/10.1016/j.ijpharm.2019.118524.

Mosén, K. *et al.* (2005) 'Particle formation and capture during spray drying of inhalable particles', *Pharmaceutical Development and Technology*, 9(4), pp. 409–417. https://doi.org/10.1081/PDT-200035795.

Mugabe, C., Azghani, A.O. and Omri, A. (2005) 'Liposome-mediated gentamicin delivery: Development and activity against resistant strains of *Pseudomonas aeruginosa* isolated from cystic fibrosis patients', *The Journal of Antimicrobial Chemotherapy*, 55(2), pp. 269–271. https://doi.org/10.1093/jac/dkh518.

Müller, R.H., Mäder, K. and Gohla, S. (2000) 'Solid lipid nanoparticles (SLN) for controlled drug delivery – A review of the state of the art', *European Journal of Pharmaceutics and Biopharmaceutics*, 50(1), pp. 161–177. https://doi.org/10.1016/S0939-6411(00)00087-4.

Nasr, M. *et al.* (2014) 'PAMAM dendrimers as aerosol drug nanocarriers for pulmonary delivery via nebulization', *International Journal of Pharmaceutics*, 461(1), pp. 242–250. https://doi.org/10.1016/j.ijpharm.2013.11.023.

Nassimi, M. *et al.* (2009) 'Low cytotoxicity of solid lipid nanoparticles in in vitro and ex vivo lung models', *Inhalation Toxicology*, 21(supl), pp. 104–109. https://doi.org/10.1080/08958370903005769.

Nassimi, M. *et al.* (2010) 'A toxicological evaluation of inhaled solid lipid nanoparticles used as a potential drug delivery system for the lung', *European Journal of Pharmaceutics and Biopharmaceutics*, 75(2), pp. 107–116. https://doi.org/10.1016/j.ejpb.2010.02.014.

Nguyen, J. *et al.* (2008) 'Effects of cell-penetrating peptides and pegylation on transfection efficiency of polyethylenimine in mouse lungs', *The Journal of Gene Medicine*, 10(11), pp. 1236–1246. https://doi.org/10.1002/jgm.1255.

Nsairat, H. *et al.* (2022) 'Liposomes: Structure, composition, types, and clinical applications', *Heliyon*, 8(5), p. e09394. https://doi.org/10.1016/j.heliyon.2022.e09394.

Onoue, S. *et al.* (2012) 'Self-assembled micellar formulation of chafuroside a with improved anti-inflammatory effects in experimental asthma/COPD-model rats', *European Journal of Pharmaceutical Sciences*, 45(1), pp. 184–189. https://doi.org/10.1016/j.ejps.2011.11.003.

Park, J.H. *et al.* (2008) 'Polymeric nanomedicine for cancer therapy', *Progress in Polymer Science*, 33(1), pp. 113–137. https://doi.org/10.1016/j.progpolymsci.2007.09.003.

Peng, S. *et al.* (2024) 'Nano-formulations for pulmonary delivery: Past, present, and future perspectives', *Pharmaceutics*, 16(2), p. 161. https://doi.org/10.3390/pharmaceutics16020161.

Plumley, C. *et al.* (2009) 'Nifedipine nanoparticle agglomeration as a dry powder aerosol formulation strategy', *International Journal of Pharmaceutics*, 369(1), pp. 136–143. https://doi.org/10.1016/j.ijpharm.2008.10.016.

Raula, J. *et al.* (2013) 'Coated particle assemblies for the concomitant pulmonary administration of budesonide and salbutamol sulphate', *International Journal of Pharmaceutics*, 441(1–2), pp. 248–254. https://doi.org/10.1016/j.ijpharm.2012.11.036.

Rehman, M. *et al.* (2004) 'Optimisation of powders for pulmonary delivery using supercritical fluid technology', *European Journal of Pharmaceutical Sciences*, 22(1), pp. 1–17. https://doi.org/10.1016/j.ejps.2004.02.001.

Rogers, T.L., Johnston, K.P. and Williams, R.O. III. (2001) 'Solution-based particle formation of pharmaceutical powders by supercritical or compressed fluid Co_2 and cryogenic spray-freezing technologies', *Drug Development and Industrial Pharmacy*, 27(10), pp. 1003–1015. https://doi.org/10.1081/DDC-100108363.

Saxena, J. *et al.* (2022) 'Targeting COPD with PLGA-based nanoparticles: Current status and prospects', *BioMed Research International*, 2022, p. e5058121. https://doi.org/10.1155/2022/5058121.

Sham, J.O.-H. *et al.* (2004) 'Formulation and characterization of spray-dried powders containing nanoparticles for aerosol delivery to the lung', *International Journal of Pharmaceutics*, 269(2), pp. 457–467. https://doi.org/10.1016/j.ijpharm.2003.09.041.

Silva, L.F.C. *et al.* (2013) 'Preparation and characterization of quercetin-loaded solid lipid microparticles for pulmonary delivery', *Powder Technology*, 239, pp. 183–192. https://doi.org/10.1016/j.powtec.2013.01.037.

Singh, S. *et al.* (2018) 'Fast dissolving drug delivery systems: Formulation, preparation techniques and evaluation', *Universal Journal of Pharmaceutical Research*, 3(4), pp. 60–69.

Sinha, B., Mukherjee, B. and Pattnaik, G. (2013) 'Poly-lactide-*co*-glycolide nanoparticles containing voriconazole for pulmonary delivery: In vitro and in vivo study', *Nanomedicine: Nanotechnology, Biology and Medicine*, 9(1), pp. 94–104. https://doi.org/10.1016/j.nano.2012.04.005.

Stass, H. *et al.* (2013) 'Tolerability and pharmacokinetic properties of ciprofloxacin dry powder for inhalation in patients with cystic fibrosis: A phase I, randomized, dose-escalation study', *Clinical Therapeutics*, 35(10), pp. 1571–1581. https://doi.org/10.1016/j.clinthera.2013.08.003.

Sung, J.C., Pulliam, B.L. and Edwards, D.A. (2007) 'Nanoparticles for drug delivery to the lungs', *Trends in Biotechnology*, 25(12), pp. 563–570. https://doi.org/10.1016/j.tibtech.2007.09.005.

Tap, W.D. *et al.* (2016) 'Olaratumab and doxorubicin versus doxorubicin alone for treatment of soft-tissue sarcoma: An open-label phase 1b and randomised phase 2 trial', *Lancet (London, England)*, 388(10043), pp. 488–497. https://doi.org/10.1016/S0140-6736(16)30587-6.

Tran, S. *et al.* (2017) 'Cancer nanomedicine: A review of recent success in drug delivery', *Clinical and Translational Medicine*, 6(1), p. 44. https://doi.org/10.1186/s40169-017-0175-0.

Trapani, A. *et al.* (2013) 'Systemic heparin delivery by the pulmonary route using chitosan and glycol chitosan nanoparticles', *International Journal of Pharmaceutics*, 447(1), pp. 115–123. https://doi.org/10.1016/j.ijpharm.2013.02.035.

Trivedi, R. *et al.* (2012) 'Local delivery of biodegradable pirfenidone nanoparticles ameliorates bleomycin-induced pulmonary fibrosis in mice', *Nanotechnology*, 23(50), p. 505101. https://doi.org/10.1088/0957-4484/23/50/505101.

Tseng, Y.-Y. *et al.* (2020) 'Injectable SN-38-embedded polymeric microparticles promote antitumor efficacy against malignant glioma in an animal Model', *Pharmaceutics*, 12(5), p. 479. https://doi.org/10.3390/pharmaceutics12050479.

Tuncer Degim, I. and Celebi, N. (2007) 'Controlled delivery of peptides and Proteins', *Current Pharmaceutical Design*, 13(1), pp. 99–117. https://doi.org/10.2174/138161207779313795.

Valiulin, S.V. *et al.* (2021) 'Aerosol inhalation delivery of triazavirin in mice: Outlooks for advanced therapy against novel viral infections', *Journal of Pharmaceutical Sciences*, 110(3), pp. 1316–1322. https://doi.org/10.1016/j.xphs.2020.11.016.

Velino, C. *et al.* (2019) 'Nanomedicine approaches for the pulmonary treatment of cystic fibrosis', *Frontiers in Bioengineering and Biotechnology*, 7, p. 406. https://doi.org/10.3389/fbioe.2019.00406.

Virmani, R. *et al.* (2023) 'Nanotechnology in Pulmonary Tissue Engineering', in A. Mondal, A.K. Nayak, and P. Chakraborty (eds) *Nanostructured Materials for Tissue Engineering*. Elsevier (Nanotechnology in Biomedicine), pp. 537–556. https://doi.org/10.1016/B978-0-323-95134-0.00017-1.

Virmani, T. *et al.* (2022) 'Nanocarrier-based approaches to combat chronic obstructive pulmonary disease', *Nanomedicine (London, England)*, 17(24), pp. 1833–1854. https://doi.org/10.2217/nnm-2021-0403.

Virmani, T. *et al.* (2023) 'Amelioration of cancer employing chitosan, its derivatives, and chitosan-based nanoparticles: Recent updates', *Polymers*, 15(13), p. 2928.

Virmani, T. *et al.* (2023) 'Xanthan Gum-Based Drug Delivery Systems for Respiratory Diseases', in H. Dureja et al. (eds) *Natural Polymeric Materials Based Drug Delivery Systems in Lung Diseases*. Singapore: Springer Nature, pp. 279–295. https://doi.org/10.1007/978-981-19-7656-8_16.

Virmani, T. and Gupta, J. (2017) 'Pharmaceutical application of microspheres: An approach for the treatment of various diseases', *International Journal of Pharmaceutical Sciences and Research*, 8(8), pp. 3253–3260.

Wang, D., Mehrabi Nasab, E. and Athari, S.S. (2021) 'Study effect of Baicalein encapsulated/loaded Chitosan-nanoparticle on allergic Asthma pathology in mouse model', *Saudi Journal of Biological Sciences*, 28(8), pp. 4311–4317. https://doi.org/10.1016/j.sjbs.2021.04.009.

White, S. *et al.* (2005) 'EXUBERA®: Pharmaceutical development of a novel product for pulmonary delivery of insulin', *Diabetes Technology & Therapeutics*, 7(6), pp. 896–906. https://doi.org/10.1089/dia.2005.7.896.

Wilson, A.C. *et al.* (2007) 'Leuprolide acetate: A drug of diverse clinical applications', *Expert Opinion on Investigational Drugs*, 16(11), pp. 1851–1863. https://doi.org/10.1517/13543784.16.11.1851.

Xu, P. *et al.* (2013) 'Synergistic inhibition of breast cancer metastasis by silibinin-loaded lipid nanoparticles containing TPGS', *International Journal of Pharmaceutics*, 454(1), pp. 21–30. https://doi.org/10.1016/j.ijpharm.2013.06.053.

Yetisgin, A.A. *et al.* (2020) 'Therapeutic nanoparticles and their targeted delivery applications', *Molecules*, 25(9), p. 2193. https://doi.org/10.3390/molecules25092193.

Yu, Z. *et al.* (2004) 'Spray freezing into liquid nitrogen for highly stable protein nanostructured microparticles', *European Journal of Pharmaceutics and Biopharmaceutics*, 58(3), pp. 529–537. https://doi.org/10.1016/j.ejpb.2004.04.018.

Zhang, P.-R. *et al.* (2011) 'Preparation and characterization of budesonide-loaded solid lipid nanoparticles for pulmonary delivery', *Journal of Chinese Pharmaceutical Sciences*, 20. https://doi.org/10.5246/jcps.2011.04.049.

7 Liposomal Drug Delivery for the Management of Chronic Obstructive Pulmonary Disease

Sarath Chandran C, Rahul K, Hiba K, Abhijith. K. Shaji, Reshma K, and Gopika Nath C
College of Pharmaceutical Sciences, Govt. Medical College Kannur
Kerala, India

Shijina Kappally
College of Pharmacy, Sharjah University
UAE

7.1 INTRODUCTION

Liposomes are among the most extensively studied lipid-based drug delivery systems and offer significant advantages for managing chronic obstructive pulmonary disease (COPD). Over the past six decades, liposomes have garnered significant attention as drug reservoirs for various applications in the diagnosis, treatment, and imaging of various diseases (1). Liposomes, also referred to as phospholipid vesicles, were initially observed during the investigation of model biological membranes in the 1960s (2). Liposomes exhibit self-assembled structures comprising amphiphilic phospholipids featuring one or more lipid bilayers encapsulating an aqueous core. The ability of liposomes to entrap both hydrophobic and hydrophilic agents is attributed to their inherent amphiphilic nature. The size of liposomes can vary greatly, typically falling between 30 nanometers and several micrometers, and is influenced by factors such as temperature, environmental conditions, and the methods used for their preparation (3).

Liposomes are composed of phospholipids and cholesterol. Phospholipids are integral components of cell membranes and are typically composed of diglycerides, a phosphate group (phosphoric acid molecule), and an organic molecule (choline). Diglycerides consist of two fatty acid chains covalently bound to a single glycerol molecule via ester linkage. The presence of three hydroxyl groups (–OH) in glycerol, the central scaffold of a phospholipid molecule, contributes to its hydrophilicity. This polyol backbone serves as the attachment point for both fatty acid chains and the phosphate group (4). The fatty acid tails, either saturated or unsaturated, contribute to the hydrophobic character of the phospholipid molecule. This contrasts with the hydrophilic head group formed by the phosphate and glycerol moieties. These opposing properties drive the self-assembly of phospholipids into bilayer structures in aqueous environments. In these bilayers, the hydrophobic tails orient toward each other, minimizing contact with water, while the hydrophilic head groups face the surrounding water, ensuring stability. Phospholipids can occur naturally (e.g., soya phosphatidylcholine, egg phosphatidylcholine) or synthetic (e.g., dipalmitoyl phosphatidylcholine, dimyrestoyl phosphatidylcholine). Cholesterol is incorporated into liposome preparations to enhance bilayer properties, including modification of fluidity and reduction of molecular permeability across the

 DOI: 10.1201/9781003498926-7

bilayer membranes (5). Liposomal surface engineering with hydrophilic polymers, like polyethylene glycol (PEG), is a prevalent strategy to generate long-circulating or "stealth" liposomes. PEGylation extends circulation time in the bloodstream by evading recognition by the phagocytic system. This innovation has facilitated the development of numerous efficacious liposomal formulations that encapsulate active molecules with enhanced targeting and activity. The composition of liposomal membranes can be manipulated to create neutral, anionic, or cationic variants. Liposomes are classified into multilamellar vesicles (MLVs), oligolamellar vesicles (OLVs), large unilamellar vesicles (LUVs), small unilamellar vesicles (SUVs), and multivesicular liposomes (MVLs) (2). MLVs contain multiple phospholipid bilayers and are typically prepared using a thin-film hydration method. OLVs are smaller than MLVs and are often generated through extrusion of giant MLVs using polycarbonate membrane filters. LUVs consist of a single phospholipid bilayer and are prepared using techniques such as reverse-phase evaporation. SUVs are characterized by their small dimensions and high stability, and are prepared using the solvent injection method. MVLs contain numerous liposomes enclosed within larger vesicles and are prepared through dual emulsification (2, 5). Liposomes offer a versatile platform for aerosol delivery of several active agents, including drugs, nucleic acids (DNA/RNA), and vitamins. Notably, negatively charged and neutral liposomes exhibit superior biocompatibility compared with their cationic counterparts, making them preferable for applications involving negatively charged nucleic acids to minimize cellular toxicity. Furthermore, the inherent uptake of anionic liposomes by macrophages renders them particularly effective for targeted delivery to alveolar macrophages in diseases like COPD (6).

7.2 LIPOSOMES IN COPD MANAGEMENT

Liposomes represent a well-established and promising platform for drug delivery in COPD owing to their inherent advantages of versatility, biocompatibility, biodegradability, negligible immunogenicity, and ability to encapsulate a broad range of therapeutic agents, both hydrophilic and hydrophobic (Table 7.1) (4, 7). These characteristics, stemming from their resemblance to mammalian cell membranes, allow liposomes to penetrate biofilms and access intracellular compartments including those within macrophages (8). Notably, systemic administration of liposomes leads to their preferential accumulation in the reticuloendothelial system, with organs such as the lungs, spleen, liver, kidneys, bone marrow, and lymph nodes exhibiting high uptake (9). This inherent lung accumulation within the reticuloendothelial system suggests that direct pulmonary delivery of liposomal formulations could be a strategic approach for treating respiratory illnesses, particularly those involving intracellular pathogens (9).

TABLE 7.1
Liposomal-based Products for Drug Administration to the Pulmonary Region

Active Molecule	Process	Indication	Type of Device	Reference
Salbutamol sulphate	Freeze drying	Asthma	Spinhaler®	(10)
Docetaxel	Spray drying	Lung cancer	Reusable device	(11)
N-acetyl cysteine	Spray drying	Manage the inflammation associated with lung disease	HandiHaler®	(12)
Ciprofloxacin	Spray drying	Pneumonia	Osmohaler®	(13)
Licorice extract	Freeze drying	Asthma	Lupihaler®	(14)
Andrographolide	Freeze drying	COPD	Twister®	(15)
Salbutamol sulphate	Spray drying	Asthma	Rotahaler®	(16)
Salbutamol sulphate	Freeze drying	Asthma	Cyclohaler®	(17)
Dapsone	Spray drying	Pneumocystis pneumonia	–	(18)
Clarithromycin	Freeze drying	Respiratory tract infection	–	(19)

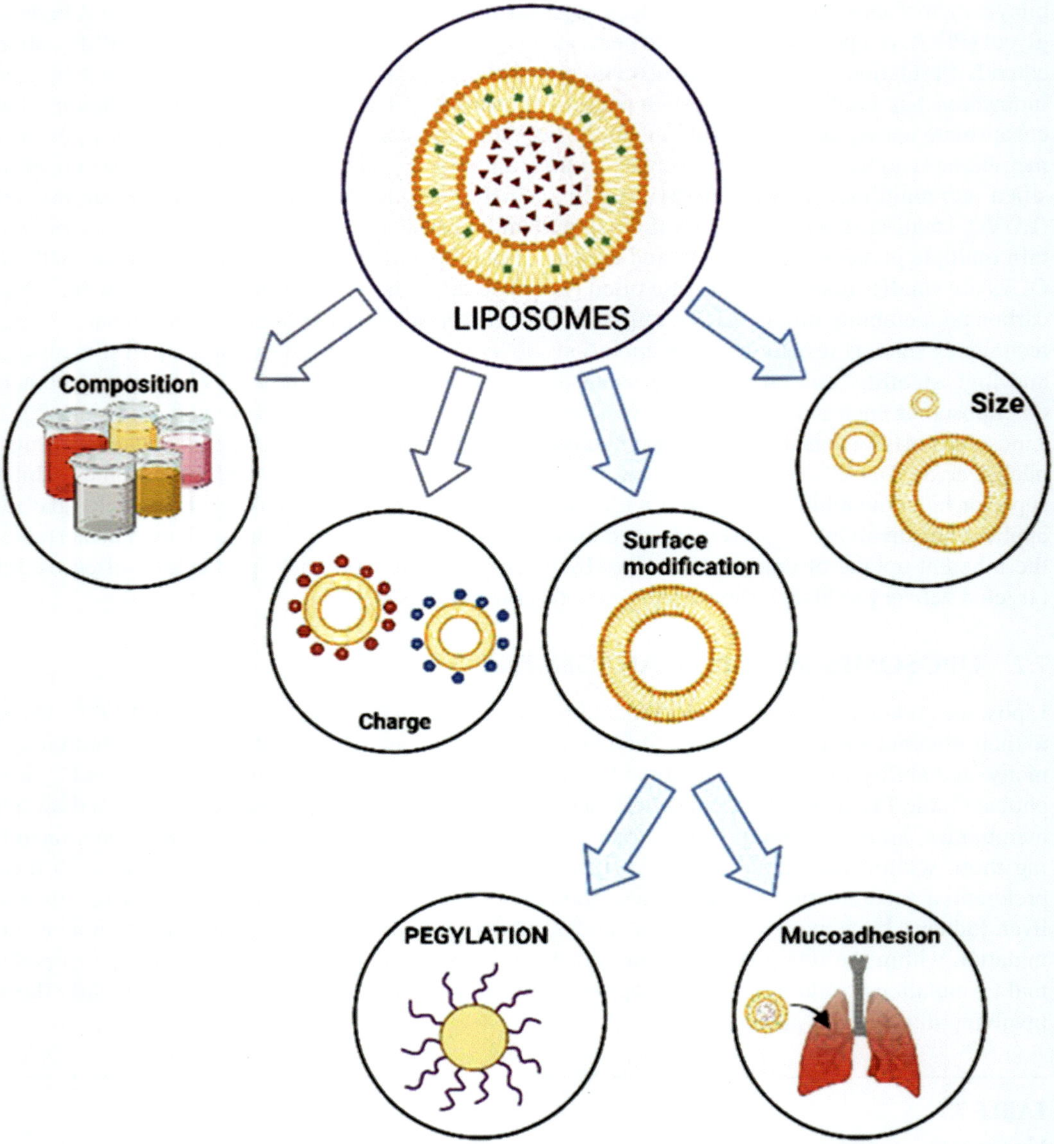

FIGURE 7.1 Mechanisms behind the application of liposomes in the management of chronic obstructive pulmonary disease.

Various strategies can be developed to optimize liposomal functionality and overcome these limitations, thereby enhancing their potential as drug delivery carriers (Figure 7.1). One approach involves ligand conjugation, in which specific ligands or antibodies are attached to the liposomal surface to enable targeted delivery to designated sites. Additionally, PEG modification improves liposomal stability, circulation time, permeability, and site-specific localization. Liposomes conjugated with mucoadhesive polymers, such as chitosan and carbopol, exhibit enhanced adherence to mucous membranes, which is a crucial feature for applications in COPD (20, 21). Furthermore, the electrostatic properties of liposomes influence their interactions with biological systems. While cationic liposomes can bind to negatively charged membranes for targeted delivery, they can also

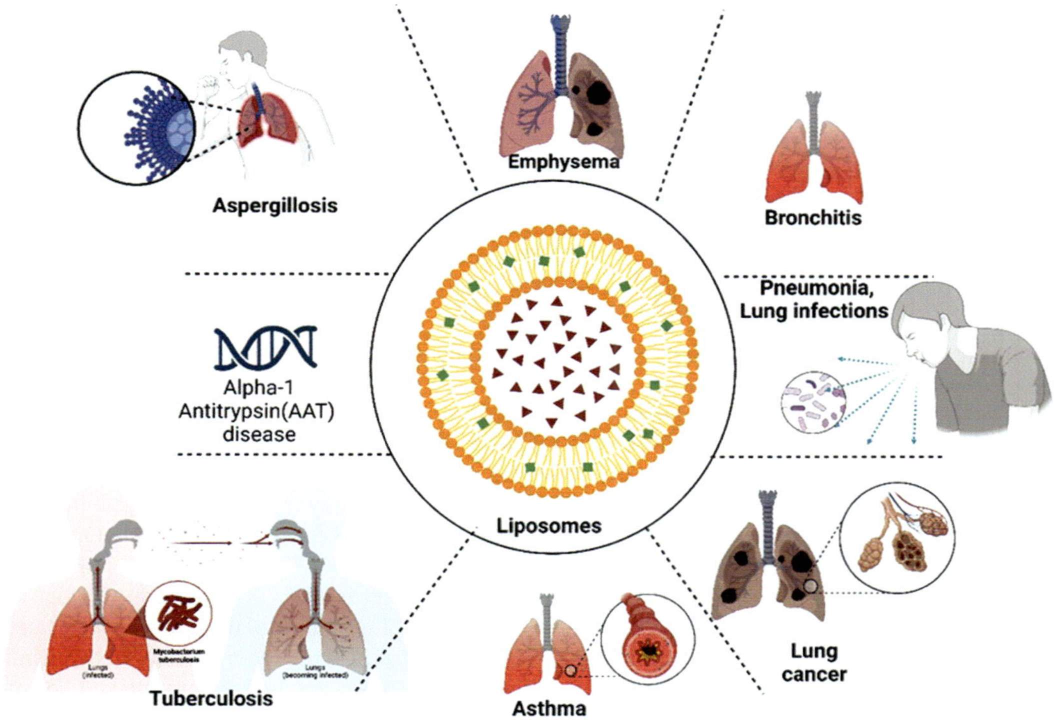

FIGURE 7.2 Liposomes for the management of chronic respiratory conditions.

trigger leukocyte activation and release of inflammatory mediators. In such scenarios, negatively charged liposomes offer the advantage of suppressing the macrophage effects on alveolar cells during COPD treatment. The electrostatic characteristics of liposomes may also influence their interactions in biological systems. In some cases, cationic liposomes can bind to negatively charged membranes at targeted sites, but this can also lead to leukocyte activation and the release of inflammatory molecules. In such situations, the application of negatively charged liposomes in COPD treatment has shown to potentially modulating the interaction between macrophages and alveolar cells (22–24).

Over the past several decades, liposomes have been important in managing various COPD manifestations, including asthma, emphysema, bronchitis, and Alpha-1antitrypsindeficiency (AATD). Additionally, liposomes have the potential to treat comorbidities associated with COPD such as pneumonia, cancer, and tuberculosis (TB) (Figure 7.2). Liposome-based drug carriers have been reported to have enhanced drug delivery profiles, especially by targeting the pulmonary region for better therapeutic outcomes.

7.3 LIPOSOMES IN ASTHMA MANAGEMENT

Asthma is a persistent inflammatory disorder that typically presents as recurring or intermittent episodes of shortness of breath and wheezing, frequently occurring at night or early in the morning. It is often caused by airflow obstruction, bronchospasm, and inflammation. The key aspect of the pathophysiology of asthma involves airway obstruction, which can be influenced by a variety of factors such as genetics, environment, sex, and respiratory infections that can trigger asthma attacks (25). Exposure to allergens or irritants triggers mast-cell degranulation in individuals susceptible to acute asthma. This process releases immunoglobulin E (IgE)-dependent mediators including histamine, tryptase, leukotrienes, and prostaglandins. Collectively, these mediators induce bronchial smooth muscle contractions, leading to restricted airflow. Furthermore, the ensuing inflammatory response characterized by mucous hypersecretion, edema formation, and structural alterations within smooth muscle contributes to a progressive decline in airway patency (26). Salbutamol sulfate is one of the most commonly prescribed drugs and is often administered in various formulations including aerosols, injections, tablets, and combinations. However, frequent administration of large doses of salbutamol sulfate, either intravenously or orally, can lead to undesirable consequences such as fetal and maternal tachycardia, maternal hyperglycemia, dyspnea, palpitations, anxiety, and oliguria when administered during pregnancy (26, 27). Although aerosol sprays are the most common route used to treat asthma, the low kinetic profile of salbutamol sulfate requires frequent dosing to effectively manage the condition. To address this challenge, salbutamol sulfate-loaded liposomes were developed and investigated, which release the drug at predetermined intervals to provide a longer-lasting effect (7, 28).

The therapeutic efficacy of liposome-encapsulated salbutamol sulfate was compared against free salbutamol sulfate in various animal models by Yang et al. The study demonstrated the potential of liposome as a drug carrier, particularly for hydrophilic salbutamol sulfate that produced a longer lasting effect than free salbutamol sulfate. Asian toad pulmonary membrane permeability was investigated in an *in vitro* model, and the results indicated a slower transfer rate of salbutamol sulfate from liposomes than that of the free salbutamol sulfate solution. A rat model using pulmonary delivery of liposomes demonstrated that the liposomes were efficiently dispersed throughout the lungs and respiratory system, and that the release of salbutamol sulfate from the liposomes persisted for a minimum of 48 hours. Pharmacodynamic research on a guinea pig reported the extended anti-asthmatic effect of 18 hours from salbutamol sulfate-loaded liposomes, which was almost two-fold higher than free salbutamol sulfate solution. Overall, these liposomes increased the concentration and retention time of salbutamol sulfate in the lungs, thereby maintaining a longer therapeutic concentration of the drug and prolonging its therapeutic effect (29).

Current asthma management relies heavily on inhaled corticosteroids such as budesonide for prophylaxis. While offering a superior safety profile compared to oral corticosteroids, concerns regarding the frequent dosing and potential side effects of budesonide persist. In this context, Parmar et al., identified a significant advantage of liposomes: their composition aligns well with lung surfactant components, making them particularly suitable for prolonged pulmonary drug delivery (30). Liposomes offer dual benefits: they can encapsulate poorly soluble drugs such as budesonide and act as a sustained-release reservoir within the lungs. Additionally, studies have shown that freeze-dried liposomes significantly reduce drug leakage compared to their aqueous counterparts, leading to improved formulation stability (30). Notably, this reduction in leakage appears to correlate with an increase in the particle size (30). Li et al., explored the potential benefits of liposomal encapsulation of the bronchodilator R-terbutaline hydrochloride for treating asthma (31). They investigated the efficacy of nebulized liposomal R-terbutaline hydrochloride, in controlling asthma. Compared to the control group receiving a free R-terbutaline hydrochloride solution aerosol, the animal model treated with the liposomal formulation produced extended control over asthma attacks. Notably, the therapeutic action of the liposomal formulation persisted 4 hours after the initial nebulization, whereas the free R-terbutaline hydrochloride solution exhibited no effect at this time point. The authors attributed the sustained efficacy of liposomal R-terbutaline hydrochloride to its protection against rapid drug metabolism within the lungs. This protection allows for controlled release of the therapeutic dose over a longer period, potentially explaining the observed extended anti-asthmatic activity.

In an investigation, Honmane et al. developed a dry powder inhalation containing nanoliposomal salbutamol sulfate that could address the limitations associated with oral salbutamol sulfate therapy for asthma (16). The salbutamol sulfate from nanoliposomal formulation was delivered directly to the lungs, bypassing the limitations of oral administration, which included systemic toxicity at high dose, short half-life and first-pass metabolism, and low bioavailability. The *in vitro* release of the liposomal aerosol product exhibited 90% release over 14 hours. This release profile followed the Higuchi model for controlled release, suggesting gradual and prolonged delivery of the drug. Honmane et al., proposed this novel liposomal formulation as a promising alternative to conventional therapies for asthma and other respiratory illnesses (16). The liposomes may be an excellent carrier for salbutamol sulfate to deliver at lungs, potentially improving treatment efficacy and reducing systemic side effects. In a study of liposomal polyinosinic-polycytidylic acid for asthma. The potential of liposomes in targeted immunomodulation during asthma therapy was investigated by Dauletbaev (32). Their study focused on liposomal polyinosinic-polycytidylic acid and its ability to stimulate interferon-beta production, leading to reduced inflammatory responses. *In vitro* experiments revealed distinct activation pathways for the liposomal and non-liposomal forms of polyinosinic-polycytidylic acid. The liposomal formulation triggered the stimulation of interferon-beta through the RIG-I/MAVS pathway. During the process, there is minimal co-stimulation of interleukin-8. When compared to non-liposomal formulation where stimulation of interferon-beta takes place via the TLR3/TRIF pathway and it required significant co-stimulation by interleukin-8. The study indicated a better targeted approach for immunomodulation in asthma using liposomal polyinosinic-polycytidylic acid. Liposomes appear to direct the activation pathway toward a less inflammatory response, potentially leading to improved therapeutic outcomes. This study provides stronger evidence to prove the ability of liposomes as a drug carrier in asthma treatment.

7.4 LIPOSOMES IN ALPHA-1 ANTITRYPSIN DEFICIENCY DISEASE

AATD, a genetic disorder, remains an under-diagnosed cause of respiratory and liver diseases, making it a global health concern (33, 34). Alpha-1 antitrypsin (AAT), being the abundant serine protease inhibitor in human serum, has been predominantly produced by the liver (hepatocytes). AAT has a critical role in protecting lungs from destructive enzymes and regulating inflammatory responses (24). Laurell and Eriksson first identified AATD in 1963 using serum protein electrophoresis (24). This condition affects an estimated 3.4 million individuals globally, with variations classified based on specific protein phenotypes (ZZ, SZ, or SS) (25). Despite significant research advancements in the understanding of AATD genotypes, phenotypes, and diagnostic methods, several key questions remain unanswered (24, 25).

In Canada, primary care physicians oversee the management of 80–90% of asthma and COPD cases, including those genetically predisposed to rapid lung function decline due to AATD. Despite its prevalence in North America, akin to cystic fibrosis, diagnosis remains low, with a delay of 5–10 years even when recognized. This review integrates evidence from the research literature and guidelines to explore the pathogenesis, clinical features, diagnosis, and treatment of AATD, drawing on COPD management strategies due to the absence of specific treatments. AATD is associated with pulmonary emphysema, COPD, and asthma-like symptoms. Heterozygotes also exhibit varying risks for emphysema and bronchiectasis. Additionally, bronchiectasis has been linked to AATD, although the exact mechanism remains unclear. A significant proportion of patients with severe AATD also have bronchiectasis, suggesting a need for screening in this population (35).

Despite a lack of objective bronchodilator responsiveness, most AATD patients with obstructive lung disease experience symptomatic improvement with bronchodilators. For patients with confirmed bronchial hyperreactivity, inhaled corticosteroids may be prescribed. The rationale behind this approach is that reduced bronchial inflammation could potentially slow the reduction in forced expiratory volume in one second (FEV1) over time. While one study suggested the potential benefits of inhaled steroids in a subset of AATD-related lung disease patients, the specific patient population that might benefit remains unclear. Further research is necessary to elucidate which patients respond favorably to inhaled corticosteroids in this context (33).

Liposomal mRNA therapy for AATD shows a promising potential for disease management. A recent study explored a promising new avenue for liposomal mRNA therapy of AATD(27, 28).

Liposomes act as microscopic carriers, protecting mRNA from degradation and facilitating its cellular uptake. The study investigated liposomes containing AAT-encoding mRNA and the results showed high levels of human AAT protein expression within cells, lasting up to seven days in both *in vitro* and *in vivo* experiments upon liposomal delivery (27, 28). While these findings are positive, further research is necessary to optimize this approach, and the optimal dose and frequency of liposomal AAT therapy needs to be established through studies using appropriate animal models (27, 28). The development of liposomal mRNA therapy for AATD has the potential to provide several benefits. One of these benefits is the ability of liposomes to specifically target cells for AAT production, which could maximize their therapeutic effect. Studies have shown extended protein expression, indicating that sustained AAT delivery within the body is possible. Compared to the current treatment options, liposomal mRNA therapy may require less frequent administration. The development of liposomal mRNA therapy represents a promising future direction for the management of AATD. Further research is necessary, but this approach has the potential to greatly improve disease management (33, 36, 37).

7.5 LIPOSOME IN EMPHYSEMA MANAGEMENT

Pulmonary emphysema arises from the deterioration of lung air sacs, marked by gradual enlargement and deterioration of alveoli, contributing to respiratory difficulties. While the exact cause for emphysema is not completely understood, the inflammation and oxidative stress of lung tissue results in its progression. Exposure to external toxins triggers the recruitment of inflammatory cells such as neutrophils and macrophages into the alveoli, exacerbating the release of inflammatory molecules. Moreover, the disruption of the balance between oxidation and antioxidation leads to oxidative stress and inflammatory reactions. Consequently, inflammation and oxidative stress induce cell death and harm the extracellular matrix of the lung, culminating in alveolar wall destruction and lung tissue damage. Existing clinical treatments, notably corticosteroids, are limited by steroid resistance, oral candidiasis, and cataracts, underscoring the urgency for novel therapeutic approaches to mitigate emphysema (38–43).

Liposomes have emerged as crucial tools for drug delivery in emphysema treatment. A notable case describes a promising yet limited exploration of liposomal all-trans retinoic acid for the treatment of emphysema. A 45-year-old non-smoking Caucasian male with no family history of lung disease and extensive emphysema received treatment with liposomal all-trans retinoic acid through controlled inhalation (44). The dosage was gradually increased from 0.1 mg to 3 mg per day, and then maintained at 1 mg for four days a week over six months, with subsequent observation. Over six months, the results showed increase in FEV1 increased from 16% to 21%, walking distance in the six-minute walk test improved (400 m to 615 m) improving exercise capacity, and biochemical markers indicated reduced inflammation and lung tissue destruction, despite no visible changes in lung structure on CT scans. The observed improvements in lung function, exercise capacity, and biochemical markers suggest a potential benefit of liposomal all-trans retinoic acid in some emphysema cases (44).

A recent study explored the therapeutic potential of carboxymethyl chitosan-modified oxymatrine liposomes in the treatment of emphysema via pulmonary administration. Oxymatrine, previously investigated for its efficacy in hepatitis management, was encapsulated within liposomes and assessed for its ability to alleviate emphysema in a murine model. The anti-inflammatory and antioxidant properties of oxymatrine have been hypothesized to contribute to its therapeutic effects. Pulmonary delivery of the liposomal formulation was achieved via intratracheal nebulization using a handheld aerosolizer. These findings revealed that oxymatrine administration significantly reduced emphysema induced by porcine pancreatic elastase. This protective effect was attributed to the modulation of the Nrf2/HO-1 and NF-κB p65 signaling pathways, confirming the anti-inflammatory and antioxidant properties of oxymatrine. Notably, the carboxymethyl chitosan modification of liposomes further enhanced the therapeutic efficacy of oxymatrine in this emphysema model. These results highlight the promise of surface-modified liposomes, particularly those incorporating carboxymethyl chitosan, as a proficient pulmonary drug delivery system for oxymatrine to treat emphysema (45).

In a comparable investigation, the protective potential of 1,8-cineole was examined in a porcine pancreatic elastase-induced emphysema mouse model using precise pulmonary delivery (41). To address the challenges associated with its volatility, instability, and poor water solubility, researchers have developed hyaluronic acid-coated liposomes to encapsulate drugs. Drug administration in emphysema mouse models was facilitated using microsprayers. The results were comparable with previous investigations, demonstrating 1,8-cineole notably reduces inflammation and oxidative stress levels, consequently reducing lung cell apoptosis. Notably, hyaluronic acid-coated liposomes encapsulating 1,8-cineole exhibited superior protective effects compared to free drugs and conventional liposome-encapsulated 1,8-cineole, effectively preventing porcine pancreatic elastase-induced emphysema in mice. Additionally, therapeutic mechanisms were found to be primarily mediated by the Nrf2/NQO1 and NF-κB/IκBα pathways. Consequently, researchers concluded that the quantitative pulmonary administration of 1,8-cineole could mitigate emphysema progression, with further enhancement of therapeutic effects achievable through the utilization of hyaluronic acid-coated liposomal carriers (41).

A separate investigation focused on developing a novel therapeutic approach for pulmonary fibrosis management. Researchers have designed an aerosolized naringin delivery system utilizing an endogenous surfactant-based liposomal formulation to target pulmonary mechanics and counteract disease progression (43). Phosphatidylcholine-based liposomes exhibited a high encapsulation efficiency for naringin >90% with a desirable size >180 nm and optimum zeta potential indicating good stability. Electron microscopy analysis verified the liposomes' unilamellar structure and spherical shape, which are ideal characteristics for drug delivery(43).The formulation displayed a surface tension (32.6 ± 0.96 mN/m) conducive to deep lung deposition and maintained excellent airway patency (97 ± 2.5%) over a 120-second test, ensuring effective delivery to the alveoli. *In vitro* experiments using a twin-stage impinger revealed that a significant proportion (79 ± 1.5%) of the aerosolized formulation reached the lower airways. Subsequently, the efficacy of this approach via inhalation was evaluated *in vivo,* using a lung fibrosis induced rats. In the bronchoalveolar lavage fluid, liposomal naringin therapy dramatically decreased lactate dehydrogenase activity, total protein content, and inflammatory cell infiltration, indicating decreased lung injury and inflammation. Moreover, the formulation reduced oxidative stress, as seen by the antioxidant enzyme levels returning to normal. Masson's trichrome staining was used in the histopathological examination of lung tissue, and the results showed a marked improvement in tissue architecture, with less collagen deposition and little evidence of fibrosis, edema, hemorrhage, or inflammation. All of these results point to the non-invasive inhaled liposomal naringin formulation's promising therapeutic potential for the efficient treatment of pulmonary fibrosis (43).

Oxidative stress is a major factor in the development of COPD, encompassing emphysema, chronic inflammation, and bronchiolar obstruction (46–48). This stress upregulates the expression of matrix metalloproteinase 9, contributing to elastin fiber breakdown, which is a critical mechanism in emphysema development. In addition, oxidative stress exacerbates chronic inflammation, further promoting emphysema. Environmental factors, such as cigarette smoke and air pollutants, along with reactive oxygen species (ROS) and reactive nitrogen species (RNS) released from leukocytes and macrophages in the lungs of COPD patients, contribute to oxidative stress. Lung susceptibility to oxidative stress arises from its oxygen-rich environment, increased blood supply, and exposure to pathogens and toxins (46–51).

In a recent study targeting oxidative stress induced by cigarette smoke, beclomethasone dipropionate was encapsulated into liposomes and hyalurosomes modified with mucin to enhance their ability to counteract smoke-induced airway damage (49, 51). The vesicles were prepared using a solvent-free method involving dispersion and sonication, resulting in small (approximately 117 nm), homogeneous, and negatively charged vesicles. These formulations were stable for five months at room temperature. *In vitro* studies using the next-generation impactor demonstrated the suitability of the formulations for nebulization, effectively reaching deeper airway regions. Biocompatibility assessments with 16HBE bronchial epithelial cells confirmed the safety and high biocompatibility of all the vesicles. Notably, mucin-hyalurosomes were identified as the only formulation capable of effectively mitigating ROS production induced by cigarette smoke extract. This suggests the potential of this formulation to reduce cigarette smoke-induced lung tissue damage, thereby mitigating the pathogenesis of associated diseases, such as COPD, emphysema, and cancer (51).

These findings suggest that liposomes offer significant potential for treating emphysema by delivering therapeutic agents directly to affected lung tissues. Encapsulation within liposomal carriers protects drugs from degradation and systemic side effects, while maximizing their concentration in diseased lung areas. The versatility of liposomal formulations allows the incorporation of various drugs, such as anti-inflammatory agents, antioxidants, and protease inhibitors, offering a comprehensive approach to address the complex pathophysiology of emphysema. Clinical research has validated the effectiveness and safety of liposomal formulations in the management of emphysema, leading to enhanced lung function, decreased inflammation, and alleviation of symptoms for patients. Targeted medication distribution also minimizes side effects and improves patient adherence by enabling lower dosages and less frequent administration. However, further research is necessary to optimize liposomal formulations for emphysema therapy, including refining drug loading and release kinetics, improving lung-specific targeting, and evaluating their long-term efficacy and safety. Overcoming challenges such as pulmonary barrier penetration and clearance mechanisms is critical for translating liposomal-based therapies into clinical practice for emphysema management (43, 45).

7.6 LIPOSOMES FOR VARIOUS CONDITIONS AND COMORBIDITIES ASSOCIATED WITH COPD

COPD is associated with various conditions and comorbidities, which indirectly increase its risk. These include pneumonia, aspergillosis, TB, lung cancer, lung infections, congestive heart failure, gastroesophageal reflux disease, and smoking (52–55) (Figure 7.2). Pneumonia, an infection of the lung parenchyma, can stem from respiratory syndrome caused by bacteria, viruses, or fungi. Its pathophysiology involves an immune response in the lower respiratory tract, leading to inflammation mediated by macrophages and cytokines, such as tumor necrosis factor-alpha, interleukin-8, and interleukin-1. Treatment typically involves antimicrobial agents, but challenges, such as impaired host immunity or microbial resistance, may limit drug efficacy. Targeted drug delivery systems, such as liposomes, offer potential solutions by delivering drugs directly to infection sites, as demonstrated by Schiffelers et al.'s study on gentamicin-loaded liposomes in rats with Klebsiella pneumoniae (56).

Thymoquinone's *in vitro* antibacterial and antibiofilm properties against the nosocomial pathogen *Acinetobacter baumannii* were examined in a recent work (57). Furthermore, the therapeutic potential of the liposomal formulation of thymoquinone was evaluated in a murine model of *A. baumannii* infection. To establish an immunocompromised state, mice were administered cyclophosphamide prior to *A. baumannii* infection. The efficacy of both free thymoquinone and liposomal thymoquinone in combating infection was assessed through analysis of survival rates, bacterial load in lung tissues, systemic inflammatory markers, and histological changes. On day 40 post-infection, mice treated with liposomal thymoquinone at a dose of 10 mg/kg had a considerably greater survival rate (60%) than mice treated with free thymoquinone at the same dose (0% survival). Remarkably, treatment with liposomal thymoquinone also resulted in the lowest bacterial load observed in the lung tissue. Additionally, liposomal thymoquinone reduced infection-induced inflammation, oxidative stress, and histological damage in lung tissues. Collectively, these findings point to liposomal thymoquinone as a potentially effective treatment option for pneumonia linked to *A. baumannii*, especially in individuals with weakened immune systems (57).

Another recent investigation explored the efficacy of a novel pulmonary delivery system for *Staphylococcus aureus*-induced pneumonia, liposomal andrographolide dry powder inhalers (15). Andrographolide, a natural compound derived from traditional Chinese herbs with known anti-inflammatory properties in the respiratory system, has been encapsulated within liposomes for targeted pulmonary delivery. The liposomal formulation was prepared using an injection method, followed by freeze-drying to create liposomal andrographolide dry powder inhalers. Characterization of andrographolide liposomes revealed desirable attributes for inhalation, including small particle size and high stability. Upon rehydration, these liposomal andrographolide dry powder inhalers demonstrated suitable characteristics for pulmonary delivery, with a mass median aerodynamic diameter (MMAD) of 4.87 μm and a fine particle fraction (FPF) of 23.03%, exceeding the performance of andrographolide powder alone (MMAD: 10.14 μm; FPF: 8.37%). While *in vitro* studies did not detect significant differences in anti-*S. aureus* activity between andrographolide powder and liposomal andrographolide dry powder inhalers, *in vivo* experiments in a rat model of *S. aureus* pneumonia yielded surprising results. Liposomal andrographolide dry powder inhalers exhibited superior antibacterial efficacy compared to both a ten-fold higher dose of free andrographolide and the penicillin. Mechanistically, liposomal andrographolide dry powder inhalers suppressed the phosphorylation of $I\kappa B$-α in the nuclear factor-κB pathway and drastically downregulated pro-inflammatory cytokines such tumor necrosis factor-alpha and interleukin-1. This suggests that andrographolide effectively modulates the immune response and maintains its antibacterial properties while simultaneously suppressing inflammation, thereby demonstrating a potent therapeutic effect against bacterial pneumonia. Overall, these results underscore the potential of liposomal andrographolide dry powder inhalers as effective pulmonary delivery systems for combating bacterial pneumonia (15).

TB, caused by *Mycobacterium tuberculosis*, continues to be a significant contributor to mortality worldwide and is among the various comorbidities associated with COPD. El-Ridy et al., explored liposomal formulations of antitubercular drugs, finding enhanced efficacy and reduced toxicity compared to conventional therapy. Similarly, Bhardwaj et al., developed liposomal formulations targeting infected cells, demonstrating sustained drug release and highlighting their potential for combating antibiotic-resistant TB (56). One study highlighted the urgent need for novel and efficacious treatment strategies for TB, particularly considering the rising prevalence of drug-resistant strains and the evolving epidemiological landscape (14). Licorice, a plant with established medicinal properties against *M. tuberculosis*, has been investigated as a potential therapeutic candidate. The study explored the potential for licorice-loaded dry powder inhalers as targeted pulmonary delivery. Thin-film hydration technique, followed by lyophilization were used to obtain dry powders of licorice loaded liposomes. A comprehensive *in vitro* and *in vivo* characterization of the formulation was performed and the results showed that the liposomes exhibited favorable particle size of ≥200 nm and high drug entrapment efficiency of 75%. Spherical morphology of the liposomal vesicles was evaluated using transmission electron microscopy. Furthermore, the formulation demonstrated satisfactory flow characteristics and met established standards for inhalable formulations. Analysis using the Anderson Cascade Impactor revealed a desirable size profile and FPF for deep lung deposition. Studies on *in vivo* lung deposition in mice demonstrated effective lung targeting, with ≥46% of the medication given reaching the lungs and 16% remaining after 24 h. Notably, pharmacodynamic evaluation demonstrated a significant reduction in the bacterial burden in both the lungs and spleen of TB-infected mice following licorice-loaded dry powder inhalers treatment. Collectively, these findings suggest that licorice-loaded dry powder inhalers hold promise as a potential therapeutic strategy for TB, either as a standalone treatment or as an adjunct to existing regimens (14).

Another study focused on developing rifampicin-loaded liposomes for targeted pulmonary delivery to enhance the local concentration of antibiotics against *Mycobacterium abscessus* infections (58). Unilamellar liposomal vesicles were designed, formulated, and characterized using a thin-film hydration method. These vesicles were comprised of a combination of anionic 1,2-Dipalmitoyl-sn-glycero-3-phosphorylglycerol sodium salt and hydrogenated phosphatidylcholine derived from soybeans to optimize rifampicin encapsulation. Characterization of both rifampicin-loaded and unloaded liposomes and subsequently, the biological activity of these liposomal formulations was assessed in *M. abscessus*-infected macrophages. The findings revealed that rifampicin-loaded liposomes caused a notably greater decrease in intracellular *M. abscessus* viability compared to treatment with the free drug. These results indicate that liposomal encapsulation holds promise as a strategy to enhance the biological effectiveness of rifampicin against intracellular mycobacterial infections (58).

Aspergillosis, a severe fungal infection, primarily affects immunocompromised individuals, is also another miscellaneous condition leading to COPD. Liposomal amphotericin B shows promise in mitigating amphotericin B toxicity while retaining antifungal efficacy. Ahmad et al.'s study on liposomal amphotericin B in mice revealed improved therapeutic indices compared to free amphotericin B (56). Building upon prior research, a study investigated the efficacy of Dectin-2-coated liposomal amphotericin B for antifungal therapy (59). The researchers proposed that coating these liposomes with the extracellular oligomannan-binding domain of the C-type lectin receptor Dectin-2 could improve their targeting ability toward the fungal pathogens. *In vitro* experiments confirmed this hypothesis, demonstrating that Dectin-2-coated liposomal amphotericin B significantly reduced the effective dose of amphotericin B required to inhibit and kill *Aspergillus fumigatus* compared with their untargeted counterparts. Furthermore, *in vivo* studies using immunosuppressed mice infected with *A. fumigatus*, the formulation revealed a 30-fold increase in the binding efficiency. Significantly, Dectin-2-coated liposomal amphotericin B demonstrated superior efficacy at lower drug doses. The formulation exhibited a remarkable reduction in fungal burden within the lungs of infected mice and extended their survival compared to both untargeted amphotericin B-loaded liposomes and micellar voriconazole. These findings collectively suggest that Dectin-2-coated liposomal amphotericin B represent a promising therapeutic strategy for antifungal treatment, potentially enabling lower drug doses and improving patient outcomes (59).

Lung cancer is a leading cause of cancer-related death and poses significant challenges for treatment. Liposomal formulations are safer and highly effective treatment options. Zhang et al., utilized curcumin-based liposomes to treat lung cancer, demonstrating enhanced drug delivery and cell uptake. Chen et al. investigated PEGylated liposomal docetaxel and showed potent radiosensitization effects in mouse models of lung cancer (56). A study for the treatment of primary lung cancer, investigated the potential of a curcumin loaded liposomal dry powder inhalation therapy (60). Liposomes encapsulating curcumin were prepared and lyophilized so as to obtain the curcumin loaded liposomal dry powder. The resulting inhaler formulation exhibited characteristics optimal for efficient pulmonary delivery. *In vitro* investigations revealed substantially higher and more rapid cellular absorption of curcumin by human lung cancer A549 cells when administered through liposomes in comparison to free curcumin. This enhanced uptake correlated with high cytotoxicity towards A549 cells, while minimal cytotoxicity was observed against normal human bronchial BEAS-2B epithelial cells, resulting in a favorable selectivity index. The mechanism underlying this selectivity may be partly attributed to the increased induction of apoptosis in cancer cells. To assess *in vivo* efficacy, curcumin powder, curcumin-loaded liposomal dry powder, and gemcitabine were administered directly into the lungs of rats with lung cancer via the trachea. liposomal curcumin dry powder inhaler treatment exhibited superior anticancer effects compared to free curcumin and gemcitabine, as evidenced by pathological analysis and expression levels of various cancer-related biomarkers, including vascular endothelial growth factor, malondialdehyde, B-cell lymphoma 2 tumor necrosis factor-alpha, and caspase-3. Collectively, these findings suggest that curcumin loaded liposomal dry powder inhalers hold promise as a novel therapeutic approach for lung cancer, potentially offering high therapeutic efficacy through targeted pulmonary delivery (60). These studies underscore the potential of liposomal drug delivery systems in addressing various pulmonary diseases by offering targeted therapy with enhanced efficacy and reduced toxicity.

7.7 CONCLUSION

The development of versatile liposomes as potent drug delivery systems in pulmonary medicine has been facilitated by the rise of nanotechnology. The use of liposomal drug delivery for COPD management presents a promising avenue for addressing the urgent needs of COPD patients worldwide. Current research suggests the significant potential of liposomes in treating a wide range of lung-related (intrapulmonary) and systemic (extrapulmonary) diseases. The limitations of existing COPD treatments can be overcome by liposomal drug delivery, which delivers drugs directly to inflamed lung tissue, enhances drug bioavailability, and implements controlled-release mechanisms. By avoiding systemic circulation, liposomes minimize systemic side effects while maximizing local efficacy, thereby optimizing therapeutic outcomes for COPD patients with COPD. Engineered liposomes offer a targeted approach for delivering potent anti-inflammatory agents directly to inflammatory cells like macrophages and neutrophils in the lungs of COPD patients. This strategy has the potential to address the underlying pathology of COPD by specifically targeting the sites of inflammation. However, further research is necessary to design, develop, and rigorously evaluate novel liposomal formulations for optimal efficacy and safety in COPD treatment. By leveraging evidence-based insights and real-world implications, liposomal drug delivery has the potential to significantly improve the outcomes and quality of life of COPD patients worldwide. Formulation scientists must devise strategies to achieve long-term success in clinical translation and navigate regulatory hurdles for U.S. Food and Drug Administration approval to ensure that the current state of liposome research leads to reliable results.

REFERENCES

1. Mehta PP, Ghoshal D, Pawar AP, Kadam SS, Dhapte-Pawar VS. Recent advances in inhalable liposomes for treatment of pulmonary diseases: Concept to clinical stance. Journal of Drug Delivery Science and Technology. 2020;56:101509.
2. Elhissi A. Liposomes for pulmonary drug delivery: The role of formulation and inhalation device design. Current Pharmaceutical Design. 2017;23(3):362–72.
3. Kulkarni MP, Kiran PS, Kumar R. Liposomal Drug Delivery: Therapeutic Applications in Chronic Respiratory Diseases. In Chellappan DK, Pabreja K, Faiyazuddin M (eds.) Advanced Drug Delivery Strategies for Targeting Chronic Inflammatory Lung Diseases 2022 (pp. 471–492). Singapore: Springer.
4. Abbasi H, Kouchak M, Mirveis Z, Hajipour F, Khodarahmi M, Rahbar N, et al. What we need to know about liposomes as drug nanocarriers: An updated review. Advanced Pharmaceutical Bulletin. 2023;13(1):7–23.
5. Khan I, Elhissi A, Shah M, Alhnan MA, Ahmed W. Liposome-based carrier systems and devices used for pulmonary drug delivery. In Davim JP (ed.) Biomaterials and Medical Tribology. 2013 (pp. 395–443). Woodhead Publishing.
6. Passi M, Shahid S, Chockalingam S, Sundar IK, Packirisamy G. Conventional and nanotechnology based approaches to combat chronic obstructive pulmonary disease: Implications for chronic airway diseases. International Journal of Nanomedicine. 2020;15:3803–26.
7. Lila AS, Ishida T. Liposomal delivery systems: Design optimization and current applications. Biological and Pharmaceutical Bulletin. 2017;40(1):1–10.
8. Chalmers JD, Van Ingen J, Van Der Laan R, Herrmann JL. Liposomal drug delivery to manage nontuberculous mycobacterial pulmonary disease and other chronic lung infections. European Respiratory Society. 2021;30(161):210010.
9. De Leo V, Ruscigno S, Trapani A, Di Gioia S, Milano F, Mandracchia D, Comparelli R, Castellani S, Agostiano A, Trapani G, Catucci L. Preparation of drug-loaded small unilamellar liposomes and evaluation of their potential for the treatment of chronic respiratory diseases. International Journal of Pharmaceutics. 2018;545(1-2):378–88.
10. Huang WH, Yang ZJ, Wu H, Wong YF, Zhao ZZ, Liu L. Development of liposomal salbutamol sulfate dry powder inhaler formulation. Biological & Pharmaceutical Bulletin. 2010;33(3):512–17.
11. Zhu X, Kong Y, Liu Q, Lu Y, Xing H, Lu X, et al. Inhalable dry powder prepared from folic acid-conjugated docetaxel liposomes alters pharmacodynamic and harmacokinetic properties relevant to lung cancer chemotherapy. Pulmonary Pharmacology & Therapeutics. 2019;55:50–61.

12. Ourique AF, Chaves Pdos S, Souto GD, Pohlmann AR, Guterres SS, Beck RC. Redispersible liposomal-N-acetylcysteine powder for pulmonary administration: Development, in vitro characterization and antioxidant activity. European Journal of Pharmaceutical Sciences. 2014;65:174–82.
13. Khatib I, Khanal D, Ruan J, Cipolla D, Dayton F, Blanchard JD, Chan HK. Ciprofloxacin nanocrystals liposomal powders for controlled drug release via inhalation. International Journal of Pharmaceutics. 2019;566:641–51.
14. Viswanathan V, Pharande R, Bannalikar A, Gupta P, Gupta U, Mukne A. Inhalable liposomes of Glycyrrhiza glabra extract for use in tuberculosis: Formulation, in vitro characterization, in vivo lung deposition, and in vivo pharmacodynamic studies. Drug Development and Industrial Pharmacy. 2019;45(1):11–20.
15. Li M, Zhang T, Zhu L, Wang R, Jin Y. Liposomal andrographolide dry powder inhalers for treatment of bacterial pneumonia via anti-inflammatory pathway. International Journal of Pharmaceutics. 2017;528(1-2):163–71.
16. Honmane S, Hajare A, More H, Osmani RAM, Salunkhe S. Lung delivery of nanoliposomal salbutamol sulfate dry powder inhalation for facilitated asthma therapy. Journal of Liposome Research. 2019;29(4):332–42.
17. Daman Z, Gilani K, Najafabadi A, Eftekhariand H, Barghi M. Formulation of inhalable lipid-based salbutamol sulfatemicroparticles by spray drying technique. Daru. 2014;22(1):50.
18. Chougule M, Padhi B, Misra A. Development of spray dried liposomal dry powder inhaler of Dapsone. AAPS PharmSciTech. 2008;9(1):47–53.
19. Ye T, Yu J, Luo Q, Wang S, Chan H. Inhalable clarithromycin liposomal dry powders using ultrasonic spray freeze drying. Powder Technology. 2017;305:63–70.
20. Lombardo D, Kiselev MA. Methods of liposomes preparation: Formation and control factors of versatile nanocarriers for biomedical and nanomedicine application. Pharmaceutics. 2022;14(3):543.
21. Pasarin D, Ghizdareanu AI, Enascuta CE, Matei CB, Bilbie C, Paraschiv-Palada L, Veres PA. Coating materials to increase the stability of liposomes. Polymers. 2023;15(3):782.
22. Karn PR, Vanić Z, Pepić I, Skalko-Basnet N. Mucoadhesive liposomal delivery systems: The choice of coating material. Drug Development and Industrial Pharmacy. 2011;37(4):482–8.
23. Kelly C, Jefferies C, Cryan SA. Targeted liposomal drug delivery to monocytes and macrophages. Journal of Drug Delivery. 2011;2011:727241.
24. Robertson JD, Ward JR, Avila-Olias M, Battaglia G, Renshaw SA. Targeting neutrophilic inflammation using polymersome-mediated cellular delivery. The Journal of Immunology. 2017; 198(9):3596–604.
25. Quirt J, Hildebrand KJ, Mazza J, Noya F, Kim H. Asthma. Allergy, Asthma & Clinical Immunology. 2018;14(S2):50.
26. Lemanske RF Jr, Busse WW. Asthma: Clinical expression and molecular mechanisms. Journal of Allergy and Clinical Immunology. 2010;125(2):S95–102.
27. Marques L, Vale N. Salbutamol in the management of asthma: A review. International Journal of Molecular Sciences. 2022;23(22):14207.
28. Borghardt JM, Kloft C, Sharma A. Inhaled therapy in respiratory disease: The complex interplay of pulmonary kinetic processes. Canadian Respiratory Journal. 2018;2018.
29. Yang Z, Chen X, Huang W, Wong BC, Yin L, Wong YF, Xu M, et al. Liposomes prolong the therapeutic effect of anti-asthmatic medication via pulmonary delivery. International Journal of Nanomedicine. 2012;1139.
30. Parmar JJ, Singh DJ, Hegde DD, Lohade AA, Soni PS, Samad A, Menon MD. Development and evaluation of inhalational liposomal system of budesonide for better management of asthma. Indian Journal of Pharmaceutical Sciences. 2010;72(4):442.
31. Li Q, Zhan S, Liu Q, Su H, Dai X, Wang H, Beng H, Tan W. Preparation of a sustained-release nebulized aerosol of r-terbutaline hydrochloride liposome and evaluation of its anti-asthmatic effects via pulmonary delivery in guinea pigs. AAPS PharmSciTech. 2018;19:232–41.
32. Dauletbaev N, Cammisano M, Herscovitch K, Lands LC. Stimulation of the RIG-I/MAVS pathway by polyinosinic: Polycytidylic acid upregulates IFN-β in airway epithelial cells with minimal costimulation of IL-8. The Journal of Immunology. 2015;195(6):2829–41.
33. Kelly E, Greene CM, Carroll TP, McElvaney NG, O'Neill SJ. Alpha-1 antitrypsin deficiency. Respiratory Medicine CME. 2011;4(1):1–8.
34. Cazzola M, Stolz D, Rogliani P, Matera MG. α1-antitrypsin deficiency and chronic respiratory disorders. European Respiratory Review. 2020;29(155):190073.
35. Brode SK, Ling SC, Chapman KR. Alpha-1 antitrypsin deficiency: A commonly overlooked cause of lung disease. CMAJ. 2012;184(12):1365–71.

36. Strange C. Alpha-1 antitrypsin deficiency associated COPD. Clinics in Chest Medicine. 2020;41(3):339–45.
37. Michel T, Luft D, Abraham MK, Reinhardt S, Medina ML, Kurz J, Schaller M, Avci-Adali M, Schlensak C, Peter K, Wendel HP. Cationic nanoliposomes meet mRNA: Efficient delivery of modified mRNA using hemocompatible and stable vectors for therapeutic applications. Molecular Therapy-Nucleic Acids. 2017;8:459–68.
38. Thurlbeck WM, Müller NL. Emphysema: Definition, imaging, and quantification. American Journal of Roentgenology. 1994;163(5):1017–25.
39. Taraseviciene-Stewart L, Voelkel NF. Molecular pathogenesis of emphysema. The Journal of Clinical Investigation. 2008;118(2):394–402.
40. Cerveri I, Dore R, Corsico A, Zoia MC, Pellegrino R, Brusasco V, Pozzi E. Assessment of emphysema in COPD: A functional and radiologic study. Chest. 2004;125(5):1714–8.
41. Cai Z, Wang Q, Xu J, Zhou J, Jiang Z, Pan D, Zhang Y, Tao L, Peng J, Chen Y, Shen X. Enhanced protective activity of 1, 8-cineole on emphysema using hyaluronic acid-coated liposomes via quantitative pulmonary administration in mice. Journal of Drug Delivery Science and Technology. 2022;72:103402.
42. McDonough JE, Yuan R, Suzuki M, Seyednejad N, Elliott WM, Sanchez PG, Wright AC, Gefter WB, Litzky L, Coxson HO, Paré PD. Small-airway obstruction and emphysema in chronic obstructive pulmonary disease. New England Journal of Medicine. 2011;365(17):1567–75.
43. Kotta S, Aldawsari HM, Badr-Eldin SM, Binmahfouz LS, Bakhaidar RB, Sreeharsha N, et al. Aerosol delivery of surfactant liposomes for management of pulmonary fibrosis: An approach supporting pulmonary mechanics. Pharmaceutics. 2021;13(11):1851.
44. Frankenberger M, Heimbeck I, Moller W, Mamidi S, Kassner G, Pukelsheim K, et al. Inhaled all-trans retinoic acid in an individual with severe emphysema. The European Respiratory Journal. 2009;34(6):1487–9.
45. Peng J, Cai Z, Wang Q, Zhou J, Xu J, Pan D, et al. Carboxymethyl chitosan modified oxymatrine liposomes for the alleviation of emphysema in mice via pulmonary administration. Molecules. 2022;27(11):3610.
46. McGuinness AJ, Sapey E. Oxidative stress in COPD: Sources, markers, and potential mechanisms. Journal of Clinical Medicine. 2017;6(2):21.
47. Repine JE, Bast AA, Lankhorst ID, Oxidative Stress Study Group. Oxidative stress in chronic obstructive pulmonary disease. American Journal of Respiratory and Critical Care Medicine. 1997;156(2):341–57.
48. Kirkham PA, Barnes PJ. Oxidative stress in COPD. Chest. 2013;144(1):266–73.
49. Marginean C, Popescu MS, Vladaia M, Tudorascu D, Pirvu DC, Petrescu F. Involvement of oxidative stress in COPD. Current Health Sciences Journal. 2018;44(1):48.
50. Barnes PJ. Oxidative stress-based therapeutics in COPD. Redox Biology. 2020;33:101544.
51. Manca ML, Ferraro M, Pace E, Di Vincenzo S, Valenti D, Fernàndez-Busquets X, et al. Loading of beclomethasone in liposomes and hyalurosomes improved with mucin as effective approach to counteract the oxidative stress generated by cigarette smoke extract. Nanomaterials. 2021;11(4):850.
52. Barnes PJ, Celli BR. Systemic manifestations and comorbidities of COPD. European Respiratory Journal. 2009;33(5):1165–85.
53. Negewo NA, McDonald VM, Gibson PG. Comorbidity in chronic obstructive pulmonary disease. Respiratory Investigation. 2015;53(6):249–58.
54. Nussbaumer-Ochsner Y, Rabe KF. Systemic manifestations of COPD. Chest. 2011;139(1):165–73.
55. Ryu JH, Scanlon PD. Obstructive lung diseases: COPD, asthma, and many imitators. Mayo Clinic Proceedings. 2001;76(11):1144–53.
56. Chellappan DK, Pabreja K, Faiyazuddin M, eds. Advanced Drug Delivery Strategies for Targeting Chronic Inflammatory Lung Diseases [Internet]. Singapore: Springer Singapore; 2022. [cited 2024 Apr 25]. Available from: https://link.springer.com/10.1007/978-981-16-4392-7
57. Allemailem KS, Almatroudi A, Alrumaihi F, Aljaghwani A, Alnuqaydan AM, Khalilullah H, et al. Antimicrobial, immunomodulatory and anti-inflammatory potential of liposomal thymoquinone: Implications in the treatment of bacterial pneumonia in immunocompromised mice. Biomedicines. 2021;9(11):1673.
58. Rinaldi F, Hanieh PN, Sennato S, Santis FD, Forte J, Fraziano M, et al. Rifampicin–liposomes for *Mycobacterium abscessus* infection treatment: Intracellular uptake and antibacterial activity evaluation. Pharmaceutics. 2021;13(7):1070.
59. Ambati S, Ellis EC, Pham T, Lewis ZA, Lin X, Meagher RB. Antifungal liposomes directed by dectin-2 offer a promising therapeutic option for pulmonary aspergillosis. MBio. 2021;12(1):10–128.
60. Zhang T, Chen Y, Ge Y, Hu Y, Li M, Jin Y. Inhalation treatment of primary lung cancer using liposomal curcumin dry powder inhalers. Acta Pharmaceutica Sinica B. 2018;8(3):440–8.

8 Polymeric Nanoparticles for Targeted Therapy of Chronic Obstructive Pulmonary Disease

Nitin Verma, Komal Thapa, Neha Kanojia, and Prarit Chandel
School of Pharmacy, Chitkara University
Baddi, Himachal Pradesh, India

Monica Arora
Faculty of Health Sciences, Villa College, QI Campus
Rahdhebai Hingun, Malé, Maldives

Rajnish Srivastava
School of Pharmacy, Chitkara University
Baddi, Himachal Pradesh, India

8.1 INTRODUCTION

Increases in neutrophil and macrophage concentrations are part of the inflammation response, a component of chronic obstructive pulmonary disease (COPD), which leads to tissue remodelling [1, 2]. Each characteristic plays a role in the breathing problems (dyspnea) experienced by COPD patients. According to data from the World Health Organization, tobacco use accounts for more than 70% of COPD cases in higher-income nations, compared to lower- to middle-income nations. Thirty to forty percent of COPD cases are caused by tobacco use [3]. Chronic airway inflammation, remodelling, and irreversible lung parenchymal damage are clinical aspects associated with COPD that led to mucus retention and airflow restriction. This results in an increase in the diameter of the chest wall, coughing, wheezing, dyspnea, and permanent sensitivity of the airways. Hospitalisations due to COPD exacerbations often result in increased financial burdens for both the patient and the healthcare system. Smoking is the major risk factor for COPD progression because it includes hundreds of pro-oxidant and pro-inflammatory chemicals that cause airway inflammation [4]. Reactive oxygen species (ROS) production and removal are out of balance, which leads to oxidative stress. One biological mediator that scavenges ROS is glutathione. Furthermore, as a cellular detoxification mechanism, the carcinogenic chemicals in tobacco are conjugated with glutathione [5, 6]. The majority of people affected with COPD are middle-aged and older, and the condition is more common as people age. The population of Aboriginal and Torres Strait Islanders (ATSI) aged 45 and over had a higher prevalence of COPD in 2018–2019, at 10%; in comparison, this meant that ATSI people were 2.3 times more likely to develop COPD. In the reported demographic, people under 70 years old and those living in low- to middle-income nations accounted for about 90% of COPD-related deaths. Globally speaking, COPD has been found to be the seventh most common cause of disease [7]. Inhaled corticosteroids (ICS), long-acting muscarinic antagonists (LAMA), short-acting beta-2-agonists (SABA) and long-acting beta-2-agonists (LABA) are some of the current

DOI: 10.1201/9781003498926-8

pharmacological therapies for COPD. Usually, the symptoms experienced by the individual and the severity of their COPD are taken into consideration when choosing which of these drugs to take alone or in combination [8]. These treatments have significant negative effects even when they can simply relieve symptoms. To manage COPD without the use of pharmaceuticals, one should quit smoking, receive pulmonary rehabilitation, immunisations, and eat a healthy diet [9–12]. The traditional pharmaceutical therapies and their corresponding dose forms have drawbacks, such as their limited capacity to treat symptoms alone and their significant side effect profiles, which are partially brought on by the patient's systemic exposure to the therapeutic agents. Systems of delivery based on nanoparticles (NPs) may improve patient outcomes. Physicochemical features of NPs generally make them extremely desired and have shown therapeutic potential in the treatment of a variety of chronic respiratory morbidities, including COPD. Additionally, it has been shown that organic NPs can enhance the bioavailability of medications and exhibit a high degree of penetration selectivity, rendering them an appropriate drug delivery method for a range of medical problems [13, 14]. To demonstrate the effectiveness and potential of these cutting-edge, NP-based drug delivery systems as therapeutic approaches, we shall examine them all in this review.

8.2 PATHOPHYSIOLOGY OF COPD

Chronic inflammation and production of free radicals, such as oxidative stress, are key factors in the pathophysiology of COPD. The pathophysiology of COPD is shown in Figure 8.1.

8.2.1 Chronic Inflammation

Pollutants, smoke, and biomass all cause the activation of toll-like receptor-mediated signalling, which in turn increases the expression of pro-inflammatory mediators to start an inflammatory response. The transcription factor NF-κB and p38 map kinase (MAPK) tightly regulate the expression of these mediators [15–17]. Innate immune cells, such as neutrophils, macrophages, and lymphocytes, quickly infiltrated as a result. Subsequently, there is an increase in the quantity and migration of adaptive immune cells, such as CD8+ cytotoxic T cells and CD4+ T helper cells, towards the lungs, which increases the cellular inflammation [18, 19]. Asthma-related features such

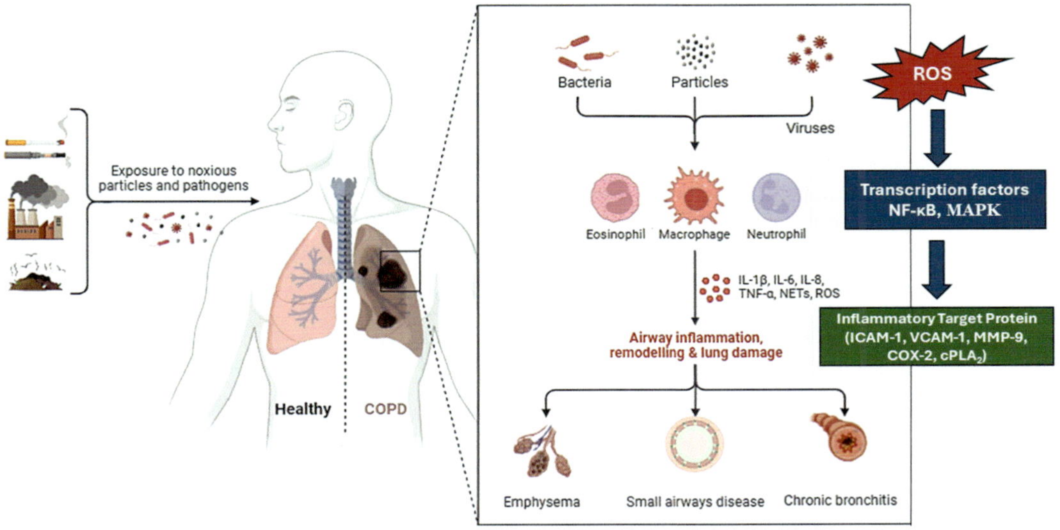

FIGURE 8.1 Role of reactive oxygen species (ROS) in the pathogenesis of COPD.

as increased eosinophil counts in the respiratory tract can occasionally be seen in patients [20]. In addition, lung airways have increased expression and production of many proteases that break down elastin fibres, including those produced by neutrophils, macrophages, and epithelial cells. Additionally, in COPD patients, neutrophil elastases promote mucus hypersecretion. Emphysema is characterised by elastolysis in lung parenchymatous tissues, which is further exacerbated by metalloproteinases such as MMp9 and MMp12. When exposed to inhaled smoke irritants, the lung mucosa normally responds by becoming more inflamed in the airways of healthy persons with normal lung functions. Patients with COPD experience increased inflammation, which is associated with the lower respiratory tract colonisation of *Haemophilus influenzae* and *Streptococcus pneumoniae* [21–24]. Due to this, macrophages were unable to properly phagocytose bacteria and apoptotic cells, which hindered the clearance of lung inflammation [25–27]. This is one of the causes of the inflammation's continued presence even after quitting smoking.

8.2.2 Free Radicals

Elevated oxidative stress is the primary cause of the pathophysiology of COPD, leading to a variety of disease progression aspects [28]. Individuals with COPD who are exposed to cigarette smoke have severe oxidative stress, which is exacerbated by neutrophil and macrophage activation. ROS produce a high oxidative burden and hyperinflammation, which complicates the pathophysiology of COPD. For example, ROS trigger the activation of NF-κB and p38MAPK, which in turn increases the expression of genes and proteins linked to the inflammatory process. Moreover, endogenously produced antiprotease α1-antitrypsin is inhibited by ROS, leading to severe elastolysis. Oxidative stress causes DNA damage that is not well repaired by DNA repair machinery; under normal circumstances, DNA repair machinery corrects DNA damage. Failure to repair the double-stranded DNA breaks increases the risk of lung cancer [29, 30]. Because carbonylation of proteins during ROS generation causes autoantibodies in COPD patients, these antibodies move in the bloodstream [31]. Lung tissues undergo fibrosis due to ROS-mediated activation of transforming growth factor-β (TGFβ). The reactivity to corticosteroids is further decreased by oxidative stress, which also lowers histone deacetylase 2 (HD2) levels. Defective ROS burden and internal antioxidant defence lower SIRT1 activity in the lungs of COPD patients, which impairs autophagy and genomic stability and damages the cell senescence and aging defence mechanism [32]. Increased acetylation and decreased HD2 activity are the causes of low NRF2 activation levels under oxidative stress [33].

There is growing evidence that COPD is associated with mitochondrial dysfunction, and that mitochondrial ROS diminish oxidative phosphorylation and reduce endogenous ATP. Epithelial cells from COPD patients have been shown to exhibit cellular senescence and ROS overproduction, which results in mitochondrial fragmentation, when *in vitro* cigarette smoke is mimicked [34]. In lung tissues, cigarette smoke induces mitochondrial autophagy and/or mitophagy in epithelial cells. Necroptosis and mitochondrial insufficiency were eventually caused by PINK1 protein and mitophagy regulator phosphatases [35]. Researchers further showed that long-term exposure to cigarette smoke causes emphysema and mucus secretion due to elevated PINk1 levels in COPD using PINk1 deletion mice models.

8.3 LIMITATIONS OF CURRENT COPD THERAPIES

8.3.1 Advantages and Disadvantages of Conventional Therapies

LABA and LAMA are currently authorised first-line treatments for COPD patients utilised in clinical settings. Smooth bronchial muscle cells relax as a result of beta-2-adrenoreceptors being activated by LABAs such as formoterol and indacaterol [36]. LABAs act for around 12 hours, during which time they have been demonstrated to dramatically lower the frequency of exacerbations and increase lung volume [37]. They do not, however, have any bearing on the patient death rate or the rate at which lung function declines. In addition, these medications cause side effects that include hypokalaemia, somatic tremor, and heart rhythm disruption [38]. In contrast, the Rodrigo 2017 study review casts doubt on the existence of these adverse effects, noting that their extrapolated safety data does not indicate an increase in severe cardiovascular events [. According to the Rodrigo 2017 study, an additional adverse consequence linked to the traditional usage of combined LABA/ICS therapy is a higher risk of pneumonia, which might worsen a patient's condition. In contrast, muscarinic M3 receptors are inhibited by LAMAs, which are inhaled anticholinergics like tiotropium and umeclidinium, in an attempt to reduce bronchoconstriction [39]. It has also been demonstrated that LAMA improves the efficacy of initiatives like pulmonary rehabilitation. Anticholinergics administered by inhalation have a lower absorption rate than when administered systemically, but this has been shown to be advantageous since it limits the severity of any potential systemic side effects. Dry mouth and mouth bitterness are the most common side effects of LAMA that have been documented, especially when using ipratropium [40, 41]. It has been demonstrated that using LABA and LAMA together may reduce severe exacerbations of COPD [42]. Furthermore, there are numerous ways to administer the medication. Alternatively, a number of inhalers with distinct dosing regimens can be used. When taken as anti-inflammatory maintenance therapy, ICS may decrease airway hyper-reactivity in patients with COPD [43]. In patients experiencing severe exacerbations of COPD, ICSs are frequently used in addition to LABA medicines to improve lung function and overall health status [44]. Based on randomised controlled trials (RCTs) spanning 6–12 months, a bronchodilator and an ICS used in combination therapy increased forced expiratory volume (FEV1). RCTs, however, were unable to demonstrate a clinically meaningful impact in trials where the ICS was used as a monotherapy [45]. Individuals with a history of asthma, blood eosinophil counts greater than 300 cells per microliter, and more than two hospitalisations due to COPD exacerbations each year are typically better candidates for ICS treatment. Nonetheless, the patient's therapeutic objectives should guide the drug choice for an effective COPD treatment plan. If survival is the main goal, selecting whether to use an ICS for a patient requires striking a compromise between clinical judgement and a risk-benefit analysis [46]. This raises the possibility that patients will get pneumonia [47]. Patients using ICS are more likely to develop pneumonia if they are older (over age 55), have a FEV1 between 30% and 50%, or have a history of low eosinophil counts [48]. Given that COPD is a condition that primarily affects the elderly, there is an inherent increase in the risk of pneumonia. This is further supported by the Janson observational trial, which found no statistically significant advantage in survival associated with simultaneous LABA/ICS treatment.

8.4 POLYMERIC NANOPARTICLES FOR COPD

Drug molecules are transported via nanoparticulate delivery structures, which developed from platforms based on nanotechnology, using nanocarriers, which are typically smaller than 500 nm. The precursor material for nanocarriers must satisfy the requirements for biodegradability, biocompatibility, and non-immunogenicity [49, 50]. At the moment, there are two types of smart nanostructures: inorganic and organic nanocarriers. Their physicochemical properties are primarily determined by their chemical compositions, which can be inorganic, organic, or hybrid; their sizes, which can range from small to large; their shapes and arrangements, which can include lamellar or multilamellar structures, branched or hyperbranched, star, rod, and sphere; surface charges; and their functional groups and coating materials [51, 52].

A polymer is a macromolecule created when monomers make covalent bonds with one another. Polymers can be classified as either branching or linear chains depending on how their monomeric constituents are arranged. Each monomer has functional groups that govern the connections and, in turn, particular characteristics [53]. Polymers are characterised by their adaptability and ease of customisation to meet the needs of the investigator. The polymeric NPs, such as nanospheres or nanocapsules, vary in size from 1 nm to 1000 nm. Thus, the morphological structures of these polymeric NPs vary. Poly(lactic-co-glycolic acid) (PLGA) and poly (lactic acid) (PLA) copolymers are the most widely utilised biodegradable polymeric materials for drug delivery systems. PLGA NPs are used in many different ways as medication carriers to treat various diseases. We will discuss PLGA NPs and their potential therapeutic use in COPD in the sections that follow [54, 55].

Applications for pulmonary diseases make use of a range of natural (albumin, chitosan, gelatin, alginate, and collagen) and synthetic (PLA, PLGA, polyacrylates, and polyanhydrides) polymers. Because natural polymers release drugs more quickly, they are not frequently employed. Artificial polymers are more effective at releasing the medication over time [56].

The core of PLGA NPs contains encapsulated or adsorbed therapeutic ingredients that are shielded from degradation. The effective release was mediated by the encapsulation, which enabled the active components to pass through cellular barriers. PLGA NPs can be used to codeliver molecular agents and chemotherapy drugs. These NPs have a negatively charged surface and are internalised by cells by fluid-phase pinocytosis or clathrin-coated endocytosis. Catalytic polymers such as chitosan have the ability to modify the negatively charged surface, hence enhancing the adsorption of target molecules [57]. The medications can target sick cells after getting past these defence mechanisms [58]. Use of particle drug delivery devices raises serious concerns regarding chronic lung blockage in patients with COPD. Preclinical studies are presently being conducted on a small number of formulations and synthesis techniques for drug-loaded NPs targeted at particular cell types in the airways [59].

As particulate carriers for pulmonary inflammation and related diseases, PLGA NPs have been employed (Figure 8.2). In mice models of cystic fibrosis (CF), a drug-loaded polylactic acid nanoparticles (PLGA-PEGPS-341) were first employed for delayed and sustainable administration. However, because the polymer is hydrophobic, researchers found that diffusion in bacterial biofilms is reduced [60, 61]. The mobility of polymeric NPs in human lung mucus and sputum has been enhanced by the use of hydrophilic PEG polymer to modify PLGA NPs. Better NP mobility may be controlled by the density and molecular weight of PEG on the particle surface (for example, 200 nm PLGA NPs coated with PEG [Mol. Wt. 2–5 kDa] can pass through cystic fibrosis sputum with ease, whereas PLGA NPs coated with 10 kDa PEG showed mucoadhesion).

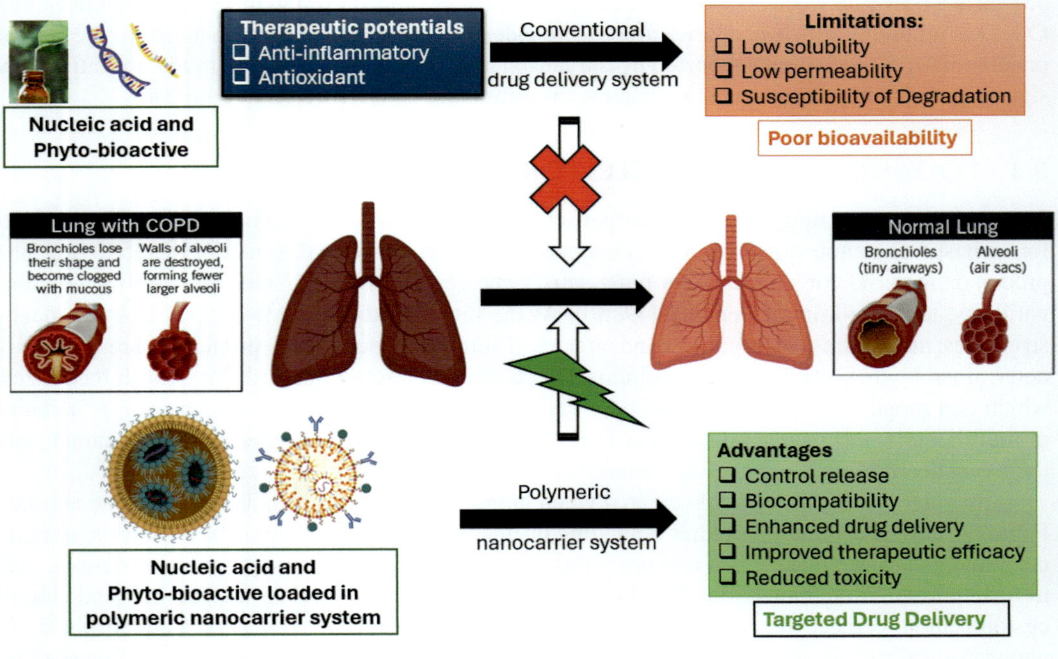

FIGURE 8.2 Advantages of polymer-based drug delivery system tie the effective management of COPD.

8.5 ENGINEERING STRATEGIES FOR POLYMERIC NANOPARTICLES

The lactic and glycolic acids that make up the biodegradable copolymer PLGA are joined via ester bonds. It decomposes readily into harmless components like carbon dioxide and water. Two distinct processes are used to fabricate PLGA NPs: the solvent-evaporation approach, also known as the single or double emulsion method, and the nanoprecipitation method.

8.5.1 Size

Currently, therapeutic NPs should be between 10 nm and 100 nm in size. The immune defence barrier and the mucosal lining can both be penetrated by PLGA NPs that are less than 100 nm. These particles prematurely collapse as a result of being exposed to several barriers during their transit across membranes prior to being taken up by lung cells. The NPs are tiny enough to pass through the layer with little steric inhibition, preventing them from depositing in the airways. Compared to disc- or barrel-shaped NPs, spherical and rod-shaped NPs exhibit superior penetration [62].

8.5.2 Surface Properties

When compared to their bulk components, NPs have a higher surface-to-volume ratio, which regulates how they interact with the biological system. The positive or negative surface of polyethylene glycol (PEG)-stabilised PLGA NPs limits their ability to interact with other particles as well as with themselves [63]. The greater surface charge, whether positive or negative, would facilitate macrophage clearance via the reticuloendothelial system. In addition to regulating interactions, surface charge would also affect distribution and places in the system that are limited in size. Therefore, it is essential to research surface features with size specificities in order to identify target locations for polymeric NPs [64].

8.5.3 Surface Ligands

When administered, targeting ligands improves the contacts between the NP and the cell surface and determines the final cellular position. Therefore, adding the right ligands to change the surface characteristics of NPs will improve their uptake by cells [65].

8.5.4 Solvent Evaporation Method

The three most often utilised organic solvents are ethyl acetate, dichloromethane, and chloroform. It has recently been demonstrated that ethyl acetate is less poisonous and more suitable for use in biological applications. Polyvinyl acetate (PVA) and other surfactants are present in the aqueous phase. At this point, the organic phase is gradually emulsified in a surfactant-containing aqueous phase. Finally, homogenisation or ultrasonication are used to form the emulsion, resulting in the dispersion of nanodroplets. After that, the polymer solvent is removed by magnetic stirring or evaporation at room temperature. Centrifugation can be used to wash and collect solid NPs once the solvent has evaporated. These nanospheres can be stored for an extended time by lyophilisation or freeze drying [66, 67].

8.5.5 Nanoprecipitation

Two miscible solvents are needed for the nanoprecipitation process, which is often referred to as the solvent displacement method. Using a syringe or dropper, the polymer is dissolved in an organic solvent that is miscible, such as acetonitrile or acetone, in the internal phase. The resulting solution is then gradually added to the aqueous phase while being constantly stirred. For the suspension to remain stable during this process, surfactants can also be employed. Because the NPs produced by this process have a well-defined size range of 170 nm to 200 nm and a restricted size distribution, they have an advantage over the emulsion solvent evaporation approach. With the exception of the emulsion approach, these traditional preparation methods have minimal drawbacks because they produce polydisperse, spherical particles with sizes ranging from 150 nm to 300 nm. The polydispersity is influenced by centrifugation and agitation speeds as well. Additionally, these techniques reduce the effectiveness of particles in drug loading and encapsulation. PEG surface coatings of PLGA NPs can increase loading and release rate, expanding their range of dynamic applications. The hydrophilic polymer of choice for surface tailoring in nanomaterials based on PLA and PLGA is PEG. This constitutes an amphiphilic copolymer, in which the hydrophilic PEG shell both prevents undesired protein aggregation in the bloodstream and shields the hydrophobic core carrying medication molecules from phagocytes.

The main justification for employing PLGA NPs as drug delivery system is their versatility in terms of molecular loading efficiency. The lactide-to-glycolide ratio, surface shape, coating, storage conditions, and polymer molecular weight all affect the drug release kinetics from encapsulated PLGA NPs. The monomeric components of PLGA NPs are primarily responsible for their solubility. Glycolic acid concentrations greater than 75% will cause the NPs to dissolve in fluorinated solvents such as hexafluoroisopropanol. PLGA NPs can be characterised physicochemically using a variety of sophisticated methods. Dynamic laser scattering (DLS) can be used to assess size, while atomic force microscopy (AFM), transmission electron microscopy (TEM), and scanning electron microscopy (SEM) can be used to assess morphology.

8.6 FUTURE PERSPECTIVES

Although using NPs for medicinal purposes has many positive aspects, dose-dependent toxicity and related adverse effects provide a challenge. To protect the safety of potential patients, future studies should be specifically designed to evaluate how various NP respond to various environmental circumstances. Furthermore, since kidney function may be harmed by the removal of the agent, it is crucial to assess the level of toxicity in the case of non-biodegradable metal NPs. Therefore, more investigation is needed to evaluate the long-term danger associated with these advanced therapeutic applications and determine which NP drug delivery systems are suitable for which demographic groups with differing degrees of COPD severity.

8.7 CONCLUSION

A significant part of the pulmonary pathophysiology in COPD is attributed to oxidative stress and airway inflammation. Furthermore, the harm done to the airways by COPD is not easily repaired, and the present treatments can only slow the progression of the condition and lower the chance of flare-ups. Infections of the lungs caused by bacteria, fungi, or viruses even worsen symptoms. Enhancing knowledge of the risk factors that contribute to the pathophysiology of COPD would not only lead to better screening practices but also aid in the assessment of more effective preventative measures. Being a less common ailment, COPD requires greater public awareness to enable early patient diagnosis. The care of COPD that is now available, which includes reducing the amount of time that a person is exposed to cigarette smoke, using bronchodilators, and pulmonary restoration, can improve exercise capacity and breathlessness momentarily, but it cannot stop the illness from progressing. Furthermore, the COPD medications that are now on the market have a shorter half-life and restricted absorption. Thus, the urgent need for efficient medication delivery systems to treat COPD arises.

Drug conjugates made of biodegradable polymers may be used to treat respiratory conditions like COPD by inhaling the medication. When compared to drug administration alone, polymer-conjugated medicines have a better pharmacokinetic profile, which leads to a sustained release of the drug. It has been discovered that using nanotechnology to build medication delivery systems is more effective in combating oxidative stress and inflammation; however, further research is needed to fully understand the process of choosing materials and creating NPs to achieve the appropriate level of bioefficacy.

Despite their many benefits, nanomedicines conjugated with polylactic acid (PLGA) face significant obstacles such as narrow therapeutic effectiveness and drug toxicity. Therefore, it is imperative to find new medications and improved PLGA-drug conjugates to provide enhanced efficacy with reduced toxicity for the successful treatment of COPD. This will facilitate the successful treatment of COPD in clinics using PLGA-drug conjugates. Research aimed at clarifying the molecular impact of drug-loaded PLGA NPs on immune cell elicitation is another topic of great importance.

REFERENCES

1. J. Saxena, M. Bisen, A. Misra, V.K. Srivastava, S. Kaushik, A.J. Siddiqui, N. Mishra, A. Singh, A. Jyoti, Targeting COPD with PLGA-based nanoparticles: Current status and prospects, BioMed. Res. Int. 2022 (2022) 5058121. https://doi.org/10.1155/2022/5058121.
2. N. Vij, T. Min, M. Bodas, A. Gorde, I. Roy, Neutrophil targeted nano-drug delivery system for chronic obstructive lung diseases, Nanomed. Nanotechnol. Biol. Med. 12 (8) (2016) 2415–2427. https://doi.org/10.1016/j.nano.2016.06.008.
3. World Health Organization, Chronic Obstructive Pulmonary Disease (COPD), 2023. https://www.who.int/news-room/fact-sheets/detail/chronic-obstructive-pulmonary-disease-(copd)#:~:text=Chronic%20obstructive%20pulmonary%20disease%20(COPD)%20is%20the%20third%20leading%20cause,%2Dincome%20countries%20(LMIC).&text=Tobacco%20smoking%20accounts%20for%20over,cases%20in%20high%2Dincome%20countries.
4. G. De Rubis, K.R. Paudel, B. Manandhar, S.K. Singh, G. Gupta, R. Malik, J. Shen, A. Chami, R. MacLoughlin, D.K. Chellappan, B.G. Oliver, P.M. Hansbro, K. Dua, Agarwood oil nanoemulsion attenuates cigarette smoke-induced inflammation and oxidative stress markers in BCi-NS1.1 airway epithelial cells, Nutrients 15 (4) (2023).
5. Australian Institute of Health and Welfare, Chronic Respiratory Conditions, 2023. https://www.aihw.gov.au/reports/chronic-respiratory-conditions/chronic-respiratory-conditions/contents/chronic-obstructive-pulmonary-disease.
6. Global Initiative for Chronic Obstructive Lung Disease (GOLD) guidelines for chronic obstructive pulmonary disease, 2025. https://goldcopd.org/2025-gold-report/. https://tgldcdp-tg-org.au.ezproxy.lib.uts.edu.au/viewTopic?etgAccess=true&guidelinePage=Respiratory&topicfile=chronic-obstructive-pulmonary-disease.
7. S. Kumari, A. Goyal, E. Sönmez Gürer, E. Algın Yapar, M. Garg, M. Sood, R.K. Sindhu, Bioactive loaded novel nano-formulations for targeted drug delivery and their therapeutic potential, Pharmaceutics. 14 (5) (2022) 1091.
8. J.E. Hartman, J.B.A. Welling, K. Klooster, O.A. Carpaij, S.W.S. Augustijn, D.J. Slebos, Survival in COPD patients treated with bronchoscopic lung volume reduction, Respir. Med. 196 (2022) 106825. https://doi.org/10.1016/j.rmed.2022.106825.
9. P. Sharma, M. Mehta, D.S. Dhanjal, S. Kaur, G. Gupta, H. Singh, L. Thangavelu, S. Rajeshkumar, M. Tambuwala, H.A. Bakshi, D.K. Chellappan, K. Dua, S. Satija, Emerging trends in the novel drug delivery approaches for the treatment of lung cancer, Chem. Biol. Interact. 309 (2019) 108720. https://doi.org/10.1016/j.cbi.2019.06.033.
10. M. Imran, S.K. Jha, N. Hasan, A. Insaf, J. Shrestha, J. Shrestha, H.P. Devkota, S. Khan, N. Panth, M.E. Warkiani, K. Dua, P.M. Hansbro, K.R. Paudel, Y. Mohammed, Overcoming multidrug resistance of antibiotics via nanodelivery systems, Pharmaceutics 14 (3) (2022), https://doi.org/10.3390/pharmaceutics14030586.
11. R. Awasthi, B. Bhushan, G.T. Kulkarni, Chapter 9 - Concepts of nanotechnology in nanomedicine: From discovery to applications, in: K. Dua, P.M. Hansbro, R. Wadhwa, M. Haghi, L.G. Pont, K.A. Williams (Eds.), Targeting Chronic Inflammatory Lung Diseases Using Advanced Drug Delivery Systems, Academic Press, 2020, pp. 171–209, https://doi.org/10.1016/B978-0-12-820658-4.00009-1
12. N.S. Skolnik, T.S. Nguyen, A. Shrestha, R. Ray, T.C. Corbridge, S.A. Brunton, Current evidence for COPD management with dual long-acting muscarinic antagonist/long-acting β2-agonist bronchodilators, PGM (Postgrad. Med.). 132 (2) (2020) 198–205. https://doi.org/10.1080/00325481.2019.1702834.

13. S. David, C.W. Edwards, Forced Expiratory Volume, StatPearls Publishing, 2022. https://www.ncbi.nlm.nih.gov/books/NBK540970/.
14. Global Initative for Chronic Obstructive Lung Disease, GOLD Report, 2023, p. 2023. https://goldcopd.org.
15. W. Szafranski, A. Cukier, A. Ramirez, et al., Efficacy and safety of budesonide/formoterol in the management of chronic obstructive pulmonary disease, Eur. Respir. J. 21 (2003) 74–81.
16. J. Christer, Inhaled corticosteroids in COPD: Risk and benefits, Thorax 77 (6) (2022) 530–531. https://doi.org/10.1136/thoraxjnl-2021-217930.
17. A. Sharafkhaneh, J.G. Southard, M. Goldman, et al., Effect of budesonide/formoterol pMDI on COPD exacerbations: A double-blind, randomized study, Respir. Med. 106 (2012) 257–268.
18. A. Agusti, L.M. Fabbri, D. Singh, J. Vestbo, B. Celli, F.M.E. Franssen, K.F. Rabe, A. Papi, Inhaled corticosteroids in COPD: Friend or foe? Eur. Respir. J. 52 (6) (2018) https://doi.org/10.1183/13993003.01219-2018.
19. P. Calverley, R. Pauwels, J. Vestbo, et al., Combined salmeterol and fluticasone in the treatment of chronic obstructive pulmonary disease: A randomised controlled trial, Lancet 361 (2003) 449–456.
20. P.M. Calverley, W. Boonsawat, Z. Cseke, et al., Maintenance therapy with budesonide and formoterol in chronic obstructive pulmonary disease, Eur. Respir. J. 22 (2003) 912–919.
21. J. Christer, W. Fredrik, T. Gunilla, S. Georgios, S. Hanna, High use of short-acting beta-2-agonists in COPD is associated with an increased risk of exacerbations and mortality, ERJ Open Research (2023) 722–2022, https://doi.org/10.1183/23120541.00722-2022.
22. M. Doroudian, M.E. Armstrong, S.C. Donnelly, Nano-Based Therapies for Acute and Chronic Lung Diseases, Springer Nature Singapore Pte Ltd, 2023, https://doi.org/10.1007/978-981-19-8342-9_12.
23. V. Puri, A. Sharma, P. Kumar, I. Singh, Thiolation of biopolymers for developing drug delivery systems with enhanced mechanical and mucoadhesive properties: A review, Polymers. 12 (8) (2020) 1803.
24. O. Afzal, A.S.A. Altamimi, M.S. Nadeem, S.I. Alzarea, W.H. Almalki, A. Tariq, B. Mubeen, B.N. Murtaza, S. Iftikhar, N. Riaz, I. Kazmi, Nanoparticles in drug delivery: From history to therapeutic applications, Nanomaterials 12 (24) (2022), https://doi.org/10.3390/nano12244494.
25. K.R. Paudel, N. Panth, B. Manandhar, S.K. Singh, G. Gupta, P.R. Wich, S. Nammi, R. MacLoughlin, J. Adams, M.E. Warkiani, D.K. Chellappan, B.G. Oliver, P.M. Hansbro, K. Dua, Attenuation of cigarette-smoke-induced oxidative stress, senescence, and inflammation by berberine-loaded liquid crystalline nanoparticles: In vitro study in 16HBE and RAW264.7 cells, Antioxidants 11 (5) (2022), https://doi.org/10.3390/antiox11050873.
26. K.S. Konduri, S. Nandedkar, N. Düzgünes, V. Suzara, J. Artwohl, R. Bunte, P.R. Gangadharam, Efficacy of liposomal budesonide in experimental asthma, J. Allergy Clin. Immunol. 111 (2) (2003) 321–327. https://doi.org/10.1067/mai.2003.104.
27. Y. Matsuo, T. Ishihara, J. Ishizaki, K. Miyamoto, M. Higaki, N. Yamashita, Effect of betamethasone phosphate loaded polymeric nanoparticles on a murine asthma model, Cell. Immunol. 260 (1) (2009) 33–38. https://doi.org/10.1016/j.cellimm.2009.07.004.
28. Y.J. Oh, J. Lee, J.Y. Seo, T. Rhim, S.-H. Kim, H.J. Yoon, K.Y. Lee, Preparation of budesonide-loaded porous PLGA microparticles and their therapeutic efficacy in a murine asthma model, J. Contr. Release. 150 (1) (2011) 56–62. https://doi.org/10.1016/j.jconrel.2010.11.001.
29. M. Govender, S. Indermun, P. Kumar, Y.E. Choonara, V. Pillay, Chapter 3 - Advanced drug delivery systems for respiratory diseases, in: K. Dua, P.M. Hansbro, R. Wadhwa, M. Haghi, L.G. Pont, K.A. Williams (Eds.), Targeting Chronic Inflammatory Lung Diseases Using Advanced Drug Delivery Systems, Academic Press, 2020, pp. 41–55. https://doi.org/10.1016/B978-0-12-820658-4.00003-0.
30. A. Mohamed, N.K. Kunda, K. Ross, G.A. Hutcheon, I.Y. Saleem, Polymeric nanoparticles for the delivery of miRNA to treat chronic obstructive pulmonary disease (COPD), Eur. J. Pharm. Biopharm. 136 (2019) 1–8. https://doi.org/10.1016/j.ejpb.2019.01.002.
31. X. Guo, X. Zuo, Z. Zhou, Y. Gu, H. Zheng, X. Wang, G. Wang, C. Xu, F. Wang, PLGA-based micro/nanoparticles: An overview of their applications in respiratory diseases, Int. J. Mol. Sci. 24 (5) (2023). https://doi.org/10.3390/ijms24054333.
32. R.M. Séeguin, N. Ferrari, Emerging oligonucleotide therapies for asthma and chronic obstructive ulmonary disease, Expet Opin. Invest. Drugs. 18 (10) (2009) 1505–1517. https://doi.org/10.1517/13543780903179294.
33. W. Liao, J. Dong, H.Y. Peh, L.H. Tan, K.S. Lim, L. Li, W.F. Wong, Oligonucleotide therapy for obstructive and restrictive respiratory diseases, Molecules 22 (1) (2017), https://doi.org/10.3390/molecules22010139.

34. M. Mehta, D. Deeksha, G. Tewari, R. Gupta, H. Awasthi, P. Singh, D.K. Pandey, R. Chellappan, T. Wadhwa, P.M. Collet, S.R. Hansbro, L. Kumar, P. Thangavelu, K. Negi, S. Dua, Satija, Oligonucleotide therapy: An emerging focus area for drug delivery in chronic inflammatory respiratory diseases, Chem. Biol. Interact. 308 (2019) 206–215. https://doi.org/10.1016/j.cbi.2019.05.028.
35. National Library of Medicine (U.S.), (2012, December 3). Safety, Tolerability, Pharmacodynamics and Pharmacokinetics of Inhaled TPI 1100 in Healthy Volunteers, 2009, June 5. Identifier NCT00914433, https://clinicaltrials.gov/ct2/show/NCT00914433?cond=TPI+1100&draw=2&rank=1.
36. C.E. Guerrero-Beltrán, J. Bernal-Ramírez, O. Lozano, Y. Oropeza-Almazán, C. Castillo, J.R. Garza, N. García, J. Vela, A. García-García, E. Ortega, G. TorreAmione, N. Ornelas-Soto, G. García-Rivas, Silica nanoparticles induce cardiotoxicity interfering with energetic status and Ca2+ handling in adult rat cardiomyocytes, Am. J. Physiol. Heart Circ. Physiol. 312 (4) (2017) H645–H661. https://doi.org/10.1152/ajpheart.00564.2016.
37. J.R. Baker, C. Vuppusetty, T. Colley, S. Hassibi, P.S. Fenwick, L.E. Donnelly, K. Ito, P.J. Barnes, MicroRNA-570 is a novel regulator of cellular senescence and inflammaging, Faseb J. 33 (2) (2019) 1605–1616. https://doi.org/10.1096/fj.201800965R.
38. G. Kim, C. Piao, J. Oh, M. Lee, Combined delivery of curcumin and the heme oxygenase-1 gene using cholesterol-conjugated polyamidoamine for antiinflammatory therapy in acute lung injury, Phytomedicine. 56 (2019) 165–174. https://doi.org/10.1016/j.phymed.2018.09.240.
39. G.J. Rodrigo, D. Price, A. Anzueto, D. Singh, P. Altman, G. Bader, F. Patalano, R. Fogel, K. Kostikas, LABA/LAMA combinations versus LAMA monotherapy or LABA/ICS in COPD: A systematic review and meta-analysis, Int J Chron Obstruct Pulmon Dis. 12 (2017) 907–922. https://doi.org/10.2147/copd.S130482.
40. M. Di Francesco, R. Primavera, M. Summa, M. Pannuzzo, V. Di Francesco, D. Di Mascolo, R. Bertorelli, P. Decuzzi, Engineering shape-defined PLGA microPlates for the sustained release of anti-inflammatory molecules, J. Contr. Release. 319 (2020) 201–212. https://doi.org/10.1016/j.jconrel.2019.12.039.
41. H.O. Alsaab, F.D. Alharbi, A.S. Alhibs, N.B. Alanazi, B.Y. Alshehri, M.A. Saleh, F.S. Alshehri, M.A. Algarni, T. Almugaiteeb, M.N. Uddin, R.M. Alzhrani, PLGAbased nanomedicine: History of advancement and development in clinical applications of multiple diseases, Pharmaceutics 14 (12) (2022), https://doi.org/10.3390/pharmaceutics14122728.
42. K. Dua, P.M. Hansbro, R. Wadhwa, M. Haghi, L.G. Pont, Targeting Chronic Inflammatory Lung Diseases Using Advanced Drug Delivery Systems, Elsevier Science, 2020.
43. D. Ribeiro de Araujo, M. Carneiro-Ramos, Biotechnology Applied to Inflammatory Diseases: Cellular Mechanisms and Nanomedicine, Springer, 2023.
44. B. Mishra, J. Singh, Chapter 4 - Novel drug delivery systems and significance in respiratory diseases, in: K. Dua, P.M. Hansbro, R. Wadhwa, M. Haghi, L.G. Pont, K.A. Williams (Eds.), Targeting Chronic Inflammatory Lung Diseases Using Advanced Drug Delivery Systems, Academic Press, 2020, pp. 57–95. https://doi.org/10.1016/B978-0-12-820658-4.00004-2.
45. A.S. Mishra, A. Sugumaran, N. Damodharan, Nanoparticulate delivery systems for the effective treatment of COPD, Annals of the Romanian Society for Cell Biology. 25 (4) (2021) 13892–13904.
46. O.F. Khan, E.W. Zaia, S. Jhunjhunwala, W. Xue, W. Cai, D.S. Yun, C.M. Barnes, J.E. Dahlman, Y. Dong, J.M. Pelet, M.J. Webber, J.K. Tsosie, T.E. Jacks, R. Langer, D.G. Anderson, Dendrimer-inspired nanomaterials for the in vivo delivery of siRNA to lung vasculature, Nano Lett. 15 (5) (2015) 3008–3016. https://doi.org/10.1021/nl5048972.
47. E. Vasile, A. Serafim, D. Petre, D. Giol, P. Dubruel, H. Iovu, I.C. Stancu, Direct synthesis and Morphological Characterization of gold-dendrimer nanocomposites Prepared using PAMAM Succinamic acid dendrimers: Preliminary study of the calcification potential, Sci. World J. (2014) 103462, https://doi.org/10.1155/2014/103462, 2014.
48. J. Liu, Y. Xu, M. Yan, Y. Yu, Y. Guo, 18β-glycyrrhetinic acid suppresses allergic airway inflammation through NF-κB and Nrf2/HO-1 signaling pathways in asthma mice, Sci. Rep. 12 (1) (2022) 3121. https://doi.org/10.1038/s41598-022-06455-6.
49. H.P. Devkota, A. Adhikari-Devkota, K.R. Paudel, N. Panth, G. Gupta, D.K. Chellappan, P.M. Hansbro, K. Dua, in: D.K. Chellappan, K. Pabreja, M. Faiyazuddin (Eds.), Phytochemicals and Their Nanoformulations Targeted for Pulmonary Diseases, 2022.
50. A. Bohr, N. Tsapis, C. Foged, I. Andreana, M. Yang, E. Fattal, Treatment of acute lung inflammation by pulmonary delivery of anti-TNF-α siRNA with PAMAM dendrimers in a murine model, Eur. J. Pharm. Biopharm. 156 (2020) 114–120. https://doi.org/10.1016/j.ejpb.2020.08.009.

51. S. Castellani, A. Trapani, A. Spagnoletta, L. di Toma, T. Magrone, S. Di Gioia, D. Mandracchia, G. Trapani, E. Jirillo, M. Conese, Nanoparticle delivery of grape seed-derived proanthocyanidins to airway epithelial cells dampens oxidative stress and inflammation, J. Transl. Med. 16 (1) (2018) 140. https://doi.org/10.1186/s12967-018-1509-4.
52. M. Habib, et al., Heart Biotech Pharma Ltd. Nanoparticle Drug Delivery, 2018 US10034837B2.
53. W. Chin, et al., Nanoparticle Based Therapy for Dispersing Mucin, University of California, 2014. US20140186290.
54. J. Geiger, et al., Ethris GmbH. Pulmonary Delivery of Messenger RNA, US20150126589, 2015.
55. M. Hartman, et al. A.G. Novartis, Respirable Glomerates of Porous Carrier Particles and Micronised Drug, 2017 US20140302147.
56. L.T. Ferguson, X. Ma, J.W. Myerson, J. Wu, P.M. Glassman, M.E. Zamora, E.D. Hood, M. Zaleski, M. Shen, E.-O. Essien, V.V. Shuvaev, J.S. Brenner, Mechanisms by which liposomes improve inhaled drug delivery for alveolar diseases, Advanced NanoBiomed Research. 3 (3) (2023) 2200106. https://doi.org/10.1002/anbr.202200106.
57. V. Subbiah, J.E. Grilley-Olson, A.J. Combest, N. Sharma, R.H. Tran, I. Bobe, A. Osada, K. Takahashi, J. Balkissoon, A. Camp, A. Masada, D.J. Reitsma, L.A. Bazhenova, Phase Ib/II trial of NC-6004 (nanoparticle Cisplatin) plus gemcitabine in patients with advanced solid tumors, Clin. Cancer Res. 24 (1) (2018) 43–51. https://doi.org/10.1158/1078-0432.Ccr-17-1114.
58. H. Almanza-Reyes, S. Moreno, I. Plascencia-Lopeź, M. Alvarado-Vera, L. Patron-Romeró, B. Borrego, A. Reyes-Escamilla, D. Valencia-Manzo, A. Brun, A. Pestryakov, N. Bogdanchikova, Evaluation of silver nanoparticles for the prevention of SARS-CoV-2 infection in health workers: In vitro and in vivo, PLoS One. 16 (8) (2021) e0256401. https://doi.org/10.1371/journal.pone.0256401.
59. A.H.Y. Chan, R.A. Pleasants, R. Dhand, S.L. Tilley, S.A. Schworer, R.W. Costello, R. Merchant, Digital inhalers for asthma or chrochronic obstructive pulmonary disease: A scientific perspective, Pulmonary Therapy. 7 (2) (2021) 345–376. https://doi.org/10.1007/s41030-021-00167-4.
60. A. Lewis, S. Torvinen, P.N.R. Dekhuijzen, H. Chrystyn, A.T. Watson, M. Blackney, A. Plich, The economic burden of asthma and chronic obstructive pulmonary disease and the impact of poor inhalation technique with commonly prescribed dry powder inhalers in three European countries, BMC Health Serv. Res. 16 (1) (2016) 251. https://doi.org/10.1186/s12913-016-1482-7.
61. P. Rogliani, L. Calzetta, A. Coppola, F. Cavalli, J. Ora, E. Puxeddu, M.G. Matera, M. Cazzola, Optimizing drug delivery in COPD: The role of inhaler devices, Respir. Med. 124 (2017) 6–14. https://doi.org/10.1016/j.rmed.2017.01.006.
62. M.B. Dolovich, R. Dhand, Aerosol drug delivery: Developments in device design and clinical use, Lancet. 377 (9770) (2011) 1032–1045. https://doi.org/10.1016/s0140-6736(10)60926-9.
63. D. Guzman-Villanueva, I.M. El-Sherbiny, D. Herrera-Ruiz, A.V. Vlassov, H.D. Smyth, Formulation approaches to short interfering RNA and MicroRNA: Challenges and implications, J. Pharm. Sci. 101 (11) (2012) 4046–4066.
64. K.S. Babu, J.B. Morjaria, Emerging therapeutic strategies in COPD, Drug Discov. Today. 20 (3) (2015) 371–379.
65. N. Vij, T. Min, R. Marasigan, C.N. Belcher, S. Mazur, H. Ding, K.T. Yong, I. Roy, Development of PEGylated PLGA nanoparticle for controlled and sustained drug delivery in cystic fibrosis, J. Nanobiotechnology. 8 (2010) 1–18.
66. S. Fekri Aval, A. Akbarzadeh, M.R. Yamchi, F. Zarghami, K. Nejati-Koshki, N. Zarghami, Gene Silencing effect of SiRNA-magnetic modified with biodegradable copolymer nanoparticles on hTERT gene expression in lung cancer cell line, Artif. Cells Nanomed. Biotechnol. 44 (1) (2016) 188–193.
67. I. d'Angelo, B. Casciaro, A. Miro, F. Quaglia, M.L. Mangoni, F. Ungaro, Overcoming barriers in *Pseudomonas aeruginosa* lung infections: Engineered nanoparticles for local delivery of a cationic antimicrobial peptide, Colloids Surf. B Biointerfaces. 135 (2015) 717–725.

9 Hybrid Nanocarriers for the Delivery of Therapeutics in Chronic Obstructive Pulmonary Disease

Flavia C. Zacconi
Facultad de Química y de Farmacia,
Pontificia Universidad Católica de Chile
Santiago, Chile
Institute for Biological and Medical Engineering,
Schools of Engineering, Medicine and Biological
Sciences, Pontificia Universidad Católica de Chile
Santiago, Chile

9.1 INTRODUCTION

Hybrid nanocarriers represent an innovative approach in drug delivery systems, particularly in the context of treating chronic obstructive pulmonary disease (COPD)[1–4]. These nanocarriers integrate multiple materials, often from different classes, to harness synergistic properties that enhance therapeutic efficacy and address specific challenges associated with COPD treatment[5, 6].

This chapter provides a comprehensive overview of the fundamental definition and concept of hybrid nanocarriers specifically designed for COPD therapy, emphasizing their transformative potential in revolutionizing drug delivery strategies for respiratory disorders, including COPD. By exploring the key concepts and principles that form the foundation of hybrid nanocarriers in the context of COPD therapy, this chapter emphasizes their significant role in addressing the complex challenges associated with delivering therapeutics to the lungs effectively. Through a comprehensive exploration of the development and application of hybrid nanocarriers, this chapter highlights the innovative approaches and promising solutions that these advanced drug delivery systems offer in the realm of COPD treatment, paving the way for enhanced therapeutic outcomes and improved management of this chronic respiratory condition.

9.2 HYBRID NANOCARRIERS FOR COPD THERAPY

9.2.1 Definition and Concept

The concept of hybrid nanocarriers in COPD therapy involves combining various materials like lipids, polymers, inorganic nanoparticles, and organic compounds to create versatile drug delivery systems. By integrating these diverse components, hybrid nanocarriers can address limitations of individual materials, such as poor stability, low drug loading capacity, and inadequate targeting capabilities. This strategic combination allows for the development of advanced drug delivery platforms that can effectively overcome challenges in delivering therapeutics to the lungs. Hybrid nanocarriers offer a multifaceted approach to enhancing therapeutic efficacy and targeted drug delivery in COPD, leveraging the strengths of different materials to optimize treatment outcomes[7, 8].

DOI: 10.1201/9781003498926-9

The integration of these diverse materials within a hybrid nanocarrier platform allows for the development of innovative and multifunctional drug delivery systems that can effectively overcome the unique challenges encountered in delivering therapeutics to the lungs of COPD patients. The synergistic combination of the strengths and properties of different materials, such as the biocompatibility of lipids, the controlled release capabilities of polymers, and the unique functionalities of inorganic nanoparticles, enables hybrid nanocarriers to offer a comprehensive and multifaceted approach to enhancing the therapeutic efficacy and targeted delivery of drugs in the management of COPD.

This strategic approach to combining various materials within a hybrid nanocarrier system provides a versatile platform that can be tailored and optimized to address the specific needs and challenges associated with COPD treatment, ultimately leading to improved patient outcomes and a more effective management of this chronic respiratory condition.

9.2.2 Rationale for Using Hybrid Nanocarriers in COPD

The utilization of hybrid nanocarriers in the treatment of COPD offers several compelling advantages over conventional drug delivery systems[9–13].

- ***Enhanced drug delivery***: hybrid nanocarriers can improve drug delivery efficiency by overcoming barriers in the lungs, such as mucus layers and cellular uptake mechanisms, leading to increased drug bioavailability at the target site.
- ***Targeted delivery***: the design of hybrid nanocarriers allows for targeted delivery of therapeutics to specific regions of the lungs affected by COPD, minimizing systemic side effects and maximizing drug efficacy.
- ***Sustained release***: hybrid nanocarriers can provide sustained release of drugs, ensuring a prolonged therapeutic effect and reducing the frequency of dosing, which is beneficial for managing chronic conditions like COPD.
- ***Customization and flexibility***: the versatility of hybrid nanocarriers enables customization of drug release profiles, incorporation of multiple drugs for combination therapy, and modification of surface properties for enhanced cellular uptake.
- ***Improved pulmonary retention and targeting***: the unique physicochemical properties of hybrid nanocarriers enable prolonged residence time and enhanced targeting specificity within the lungs. Surface modifications, such as ligand conjugation and polymer coatings, facilitate efficient deposition and retention of nanocarriers in the diseased airways, minimizing systemic exposure and maximizing local drug concentrations at the site of action.
- ***Reduced systemic toxicity***: the localized delivery of therapeutics using hybrid nanocarriers reduces systemic exposure to drugs, thereby minimizing off-target effects and mitigating systemic toxicity. This targeted approach is particularly advantageous in COPD, where chronic administration of medications can lead to systemic side effects and complications.
- ***Enhanced drug loading and encapsulation efficiency***: Hybrid nanocarriers allow for the combination of different materials with complementary properties, enabling improved drug loading capacity and encapsulation efficiency. By integrating lipids, polymers, and inorganic nanoparticles, hybrid nanocarriers can accommodate a wider range of therapeutic agents, including small molecules and biologics, thereby enhancing the efficacy of COPD treatments.
- ***Synergistic therapeutic effects***: hybrid nanocarriers offer the opportunity to exploit synergistic interactions between different materials to enhance therapeutic outcomes. By combining lipids for improved biocompatibility, polymers for sustained release kinetics, and inorganic nanoparticles for targeted delivery or imaging capabilities, hybrid nanocarriers can exert multifaceted effects on COPD pathophysiology, including anti-inflammatory, bronchodilator, and antioxidant activities.

- ***Adaptability to personalized medicine***: hybrid nanocarriers offer a versatile platform for personalized medicine approaches in COPD therapy. Their tunable physicochemical properties allow for tailored drug delivery strategies based on individual patient characteristics, disease severity, and treatment response, thereby optimizing therapeutic outcomes and minimizing adverse effects.

The rationale for employing hybrid nanocarriers in the treatment of COPD stems from their unique properties that address the challenges associated with conventional drug delivery systems.

9.3 DESIGN CONSIDERATIONS FOR HYBRID NANOCARRIERS IN COPD THERAPY

The design of hybrid nanocarriers for COPD therapy involves several key considerations to optimize their performance and therapeutic efficacy. These design aspect plays a crucial role in enhancing drug delivery, targeting, and controlled release in the management of chronic respiratory diseases. Furthermore, designing hybrid nanocarriers for COPD therapy involves careful consideration of various parameters to optimize their efficacy and safety including surface functionalization, size and shape optimization, and controlled release mechanisms, with examples highlighting their importance in enhancing therapeutic outcomes.

9.3.1 Surface Functionalization

The surface of hybrid nanocarriers can be functionalized with various moieties to improve their targeting capabilities, cellular uptake, and drug release kinetics in COPD. Surface functionalization plays a crucial role in determining the biocompatibility, stability, and targeting capabilities of hybrid nanocarriers. By modifying the surface of nanocarriers with ligands, polymers, or targeting moieties, it is possible to improve their interaction with biological components and enhance their specificity towards diseased tissues[14–16]. Examples of surface functionalization strategies include:

1. ***Targeting ligands***: incorporation of ligands, such as antibodies, peptides, or small molecules, that can bind to specific receptors or markers expressed on the surface of lung cells or the extracellular matrix in COPD[17].
2. ***Penetration enhancers***: coating the nanocarrier surface with mucolytic agents or cell-penetrating peptides to facilitate transport through the mucus layer and enhance cellular internalization[18].
3. ***Stimuli-responsive moieties***: attachment of pH-sensitive, redox-sensitive, or enzyme-cleavable groups to trigger drug release in response to specific environmental cues within the lungs[19].

For instance, conjugation of targeting ligands, such as antibodies or peptides, to the surface of hybrid nanocarriers can facilitate active targeting of inflammatory cells or specific receptors in the diseased lungs of COPD patients. For instance, PEGylation of liposomal nanoparticles decorated with arginine–glycine–aspartic acid (RGD) peptides has been shown to enhance their accumulation in inflamed lung tissues, thereby improving the therapeutic efficacy of anti-inflammatory drugs.

9.3.2 Size and Shape Optimization

The size and shape of hybrid nanocarriers play a crucial role in their deposition, distribution, and cellular interactions within the lungs. The size and shape of hybrid nanocarriers are critical parameters that influence their pharmacokinetics, biodistribution, and cellular uptake. Size optimization is essential to ensure efficient pulmonary deposition and penetration into the affected regions of the lungs, while shape optimization can further enhance cellular internalization and therapeutic payload delivery. Optimizing these parameters can enhance the therapeutic efficacy in COPD:

1. ***Aerodynamic diameter***: designing hybrid nanocarriers with an optimal aerodynamic diameter, typically between 1–5 µm, to ensure efficient lung deposition and deep alveolar penetration[20].
2. ***Spherical vs. non-spherical shapes***: exploring non-spherical geometries, such as rod-like or disc-shaped nanocarriers, to improve lung retention and cellular uptake compared to traditional spherical designs[21].
3. ***Multistage delivery systems***: developing hierarchical or multi-layered hybrid nanocarriers that can undergo size reduction or shape transformation upon reaching the target site in the lungs[22].

Engineering hybrid nanocarriers with a size range optimized for deep lung penetration (e.g., 100–200 nm) can facilitate their deposition in the alveolar and bronchial regions, where COPD pathology primarily manifests. Moreover, incorporating a rod-like or filamentous shape into nanocarriers can promote mucoadhesion and facilitate mucus penetration, thereby improving their retention and efficacy in mucus-rich airways[13].

9.3.3 Controlled Release Mechanisms

Hybrid nanocarriers can be engineered to provide controlled and sustained drug release in the lungs, addressing the challenges associated with conventional COPD therapies. Controlled release mechanisms enable sustained and targeted delivery of therapeutics, allowing for optimal drug concentrations to be maintained at the site of action over an extended period. Various strategies, including stimuli-responsive systems and matrix-based formulations, can be employed to achieve controlled release kinetics tailored to the dynamics of COPD progression and patient-specific needs. Examples of controlled release mechanisms include:

1. ***Polymer-based release***: incorporating biodegradable polymers, such as poly(lactic-co-glycolic acid) (PLGA) or chitosan, to enable gradual drug diffusion or polymer erosion-mediated release[23].
2. ***Lipid-based release***: utilizing lipid-based hybrid nanocarriers, such as liposomes or lipid-polymer hybrids, to modulate drug release through membrane permeability and lipid bilayer disruption[24, 25].
3. ***Stimuli-responsive release***: designing hybrid nanocarriers that can respond to specific environmental triggers (e.g., pH, redox potential, enzymes) within the lungs to trigger on-demand drug release[26].

Considering the information, incorporating stimuli-responsive polymers, such as pH-sensitive or enzyme-sensitive hydrogels, into hybrid nanocarriers can enable triggered release of drugs in response to disease-specific cues, such as acidic pH or elevated protease levels in inflamed lung tissues. This spatiotemporal control over drug release can minimize systemic exposure and maximize therapeutic efficacy while reducing the frequency of dosing and associated side effects.

9.4 EXAMPLES OF HYBRID NANOCARRIERS

Some examples of different types of hybrids nanocarriers for the delivery of therapeutics in COPD are as follows considering the reported and current composition and materials.

9.4.1 Lipid-Polymer Hybrid Nanoparticles

- ***Lipid-polymer hybrid nanoparticles for roflumilast delivery in COPD***: Novel lipid-polymer hybrid nanoparticles (LPHNPs) targeted to lung macrophages were developed as carriers for the delivery of roflumilast, a phosphodiesterase-4 inhibitor, in the management of COPD[27].
- ***LPHNPs for siRNA delivery in COPD***: Lipid-polymer hybrid nanoparticles were used for the pulmonary delivery of siRNA to treat obstructive airway diseases, including COPD[28].
- ***LPHNPs for miRNA delivery in COPD***: Nebulized lipid-polymer nanoparticles (LPNs) were engineered to deliver miR-17, a potent inhibitor of interleukin (IL)-8, to bronchial epithelial cells for the treatment of inflammatory lung diseases like COPD[29].

9.4.2 Polymer-Inorganic Hybrid Nanoparticles

4. ***Polymer-inorganic hybrid nanoparticles for antioxidant delivery in COPD***: Antioxidant-loaded polymer-inorganic hybrid nanoparticles were developed as nanomedicines for the treatment of lung disorders, including COPD[30].
5. ***Polymer-inorganic hybrid nanoparticles for mitochondrial dysfunction in COPD***: Polymer-inorganic hybrid nanoparticles were investigated as potential carriers for the delivery of therapeutics targeting mitochondrial dysfunction in COPD[6].
6. ***Polymer-inorganic hybrid nanoparticles for combination therapy in COPD***: Multifunctional polymer-inorganic hybrid nanoparticles were engineered to co-deliver combination therapies for the management of COPD[31].

9.4.3 Lipid-Inorganic Hybrid Nanoparticles

7. ***Lipid-inorganic hybrid nanoparticles for imaging in COPD***: Lipid-inorganic hybrid nanoparticles were explored for imaging applications in the diagnosis and monitoring of COPD[32].
8. ***Lipid-inorganic hybrid nanoparticles for targeted drug delivery in COPD***: Lipid-inorganic hybrid nanoparticles were designed to target specific cell types and pathways involved in the pathogenesis of COPD[33].
9. ***Lipid-inorganic hybrid nanoparticles for stimuli-responsive drug release in COPD***: Lipid-inorganic hybrid nanoparticles were engineered to respond to environmental cues within the lungs, triggering on-demand drug release for the treatment of COPD[7].

9.5 COMBINATION THERAPY USING HYBRID NANOCARRIERS IN COPD AD THEIR CHALLENGES

Combination therapy, which involves the simultaneous administration of multiple drugs with complementary mechanisms of action, has emerged as a promising approach for enhancing the efficacy of treatment in COPD. Hybrid nanocarriers offer a versatile platform for delivering combination therapy, allowing for the co-encapsulation of multiple therapeutic agents within a single carrier system. The use of hybrid nanocarriers offers the potential to deliver combination therapies for the management of COPD. Targeting a single pathway or mechanism may not be sufficient to effectively manage the disease. COPD is a complex and heterogeneous disease characterized by multiple pathological processes, including airway inflammation, oxidative stress, mucus hypersecretion, and airflow limitation. Given the multifactorial nature of COPD, single-agent therapies often provide suboptimal efficacy in controlling symptoms and preventing disease progression[34,35]. Combination therapy addresses this challenge by targeting multiple disease pathways simultaneously, resulting in synergistic effects and improved therapeutic outcomes and it can provide the following advantages:

1. ***Targeting multiple pathways***: Hybrid nanocarriers can be designed to co-deliver a combination of therapeutics, such as bronchodilators, anti-inflammatory agents, and antioxidants, to address the various underlying mechanisms involved in COPD pathogenesis.
2. ***Synergistic effects***: The simultaneous delivery of complementary drugs can lead to synergistic therapeutic effects, potentially enhancing the overall efficacy in COPD management.
3. ***Reduced side effects***: Combination therapy using hybrid nanocarriers can allow for the use of lower individual drug doses, potentially reducing the risk of adverse effects compared to high-dose monotherapy.
4. ***Simultaneous delivery of multiple drugs***: Hybrid nanocarriers enable the co-encapsulation of multiple drugs with different physicochemical properties, allowing for precise control over drug ratios and release kinetics. This facilitates synergistic interactions between therapeutic agents and enhances their efficacy in treating COPD.
5. ***Targeted delivery to diseased airways***: Surface functionalization of hybrid nanocarriers with targeting ligands allows for site-specific delivery of combination therapy to inflamed airways or specific cell types implicated in COPD pathogenesis. This minimizes off-target effects and maximizes therapeutic efficacy while reducing systemic toxicity.
6. ***Improved drug stability and bioavailability***: Encapsulation of drugs within hybrid nanocarriers protects them from degradation and enhances their stability during storage and administration. This ensures consistent drug release and bioavailability, resulting in more predictable pharmacological effects and improved patient compliance.
7. ***Potential for personalized medicine***: Hybrid nanocarriers offer a versatile platform for personalized medicine approaches in COPD therapy. By tailoring the composition and properties of nanocarriers to individual patient characteristics and disease phenotypes, it is possible to optimize treatment regimens and maximize therapeutic benefits.

However, the development of combination therapy using hybrid nanocarriers in COPD presents several challenges, including:

- ***Formulation complexity***: Incorporating multiple therapeutic agents within a single hybrid nanocarrier system requires careful optimization of the composition, drug loading, and release kinetics.
- ***Pharmacokinetic interactions***: The co-delivery of multiple drugs may lead to potential pharmacokinetic interactions, which need to be thoroughly investigated and addressed.
- ***Regulatory considerations***: The clinical translation of combination therapy using hybrid nanocarriers may face additional regulatory hurdles compared to single-agent formulations, requiring comprehensive evaluation of safety and efficacy.
- ***Optimization of drug ratios and release profiles***: Achieving optimal drug ratios and release profiles within hybrid nanocarriers is essential for maximizing therapeutic efficacy while minimizing adverse effects. This requires careful consideration of drug physicochemical properties, pharmacokinetics, and synergistic interactions.
- ***Biocompatibility and safety***: Ensuring the biocompatibility and safety of hybrid nanocarriers is paramount for clinical translation. Comprehensive preclinical studies are needed to assess the long-term effects of combination therapy on lung function, systemic toxicity, and immune response.
- ***Regulatory considerations***: Regulatory approval for combination therapy using hybrid nanocarriers requires rigorous evaluation of efficacy, safety, and quality control. Meeting regulatory requirements for drug approval and manufacturing processes is essential for successful clinical translation and commercialization.

Combination therapy using hybrid nanocarriers has several potential applications in COPD treatment, including:

- Simultaneous delivery of bronchodilators and anti-inflammatory agents to control symptoms and prevent exacerbations.
- Co-encapsulation of antioxidants and mucoactive agents to reduce oxidative stress and mucus hypersecretion.
- Combined delivery of small molecule drugs and biologics for targeted therapy of specific COPD phenotypes, such as eosinophilic inflammation or emphysema.

The use of hybrid nanocarriers for combination therapy in COPD can provide benefits such as targeting multiple disease mechanisms, achieving synergistic effects, and reducing the risk of side effects, ultimately leading to improved treatment outcomes for COPD patients[36–38].

In conclusion, combination therapy using hybrid nanocarriers holds great promise for improving the efficacy and safety of COPD treatment. Tackling the complex nature of the disease and leveraging the benefits of nanotechnology, personalized medicine, and targeted drug delivery, hybrid nanocarriers present innovative solutions to improve patient outcomes and quality of life for those with COPD.

9.6 ONGOING CLINICAL TRIALS AND FUTURE PERSPECTIVES

Ongoing clinical trials represent a crucial step toward the translation of hybrid nanocarriers into clinical practice for the treatment of COPD. This section provides an overview of current clinical trials involving hybrid nanocarriers in COPD therapy and discusses future perspectives for the development and optimization of these innovative drug delivery systems.

1. ***Lipid-polymer hybrid nanoparticles***: Several clinical trials are evaluating the safety, efficacy, and pharmacokinetics of lipid-polymer hybrid nanoparticles in COPD patients. These trials aim to assess the therapeutic benefits of targeted drug delivery and controlled release formulations compared to conventional treatments. For example, a phase II trial (NCT03635450) is investigating the use of inhalable lipid-polymer hybrid nanoparticles loaded with bronchodilators and corticosteroids for the management of COPD exacerbations[39].
2. ***Lipid-polymer hybrid nanoparticles for roflumilast delivery in COPD***: A phase II clinical trial (NCT04321420) is currently evaluating the safety, tolerability, and efficacy of inhaled lipid-polymer hybrid nanoparticles loaded with the phosphodiesterase-4 inhibitor roflumilast for the management of COPD[27].
3. ***Polymer-inorganic hybrid nanoparticles for antioxidant delivery in COPD***: A phase I/II clinical trial (NCT03511391) is investigating the use of antioxidant-loaded polymer-inorganic hybrid nanoparticles as a nanomedicine approach for the treatment of lung disorders, including COPD[6].
4. ***Lipid-inorganic hybrid nanoparticles for imaging in COPD***: A pilot study (NCT03481556) is exploring the potential of lipid-inorganic hybrid nanoparticles as imaging agents for the diagnosis and monitoring of COPD progression[6].

As the research in this field continues to evolve, several future perspectives and areas of focus for hybrid nanocarriers in COPD therapy can be identified:

- ***Personalized medicine approaches***: Future developments in hybrid nanocarriers for COPD therapy are likely to focus on personalized medicine approaches, tailoring drug delivery strategies to individual patient characteristics, disease phenotypes, and treatment responses. By integrating patient-specific data, such as genetic markers, biomarker profiles, and imaging findings, it will be possible to optimize the design and dosing regimens of hybrid nanocarriers for enhanced therapeutic efficacy and minimized side effects.

- ***Multifunctional nanocarriers***: Advancements in nanotechnology and materials science may lead to the development of multifunctional hybrid nanocarriers capable of simultaneous drug delivery, imaging, and diagnostic monitoring. By incorporating theranostic components, such as imaging contrast agents, biosensors, and stimuli-responsive nanoparticles, hybrid nanocarriers can offer real-time feedback on treatment response and disease progression, enabling precision medicine approaches in COPD management.
- ***Targeted drug delivery strategies***: Future research efforts will likely focus on refining targeted drug delivery strategies using hybrid nanocarriers to enhance their specificity toward diseased tissues and minimize off-target effects. By incorporating advanced targeting ligands, such as antibodies, peptides, or aptamers, hybrid nanocarriers can selectively accumulate in inflamed airways or specific cell types implicated in COPD pathogenesis, improving therapeutic outcomes, and reducing systemic toxicity.
- ***Combination therapy delivery***: Developing multifunctional hybrid nanocarriers capable of co-delivering a combination of therapeutics, such as bronchodilators, anti-inflammatory agents, and antioxidants, to address the complex pathophysiology of COPD.
- ***Targeted delivery to specific cell types***: Enhancing the targeting capabilities of hybrid nanocarriers to selectively deliver drugs to specific cell types, such as alveolar macrophages or airway epithelial cells, involved in the pathogenesis of COPD.
- ***Stimuli-responsive drug release***: Designing hybrid nanocarriers that can respond to environmental cues within the lungs, such as pH changes or oxidative stress, to trigger on-demand drug release and improve therapeutic efficacy.
- ***Improved aerodynamic properties***: Optimizing the size, shape, and surface properties of hybrid nanocarriers to enhance their aerodynamic performance and improve lung deposition, thereby increasing the therapeutic efficacy in COPD.
- ***Optimization of formulation and manufacturing processes***: Future research efforts should focus on optimizing the formulation and manufacturing processes of hybrid nanocarriers to improve scalability, reproducibility, and regulatory compliance. This includes exploring novel fabrication techniques, such as microfluidics and spray drying, to enhance the uniformity and stability of hybrid nanocarriers.
- ***Integration of advanced targeting strategies***: Advances in targeting ligands, such as antibodies, peptides, and aptamers, offer opportunities to enhance the specificity and efficacy of hybrid nanocarriers in COPD therapy. Future studies should explore the integration of advanced targeting strategies to selectively deliver therapeutics to inflamed airways or specific cell types implicated in COPD pathogenesis.
- ***Development of multifunctional theranostic platforms***: The development of multifunctional hybrid nanocarriers capable of simultaneous drug delivery, imaging, and diagnostic monitoring holds promise for personalized medicine approaches in COPD management. Future research should focus on integrating theranostic components, such as imaging contrast agents and biosensors, into hybrid nanocarriers to enable real-time monitoring of treatment response and disease progression.
- ***Translation to clinical practice***: Translating hybrid nanocarriers from preclinical studies to clinical practice requires rigorous evaluation of safety, efficacy, and pharmacokinetics in human subjects. Future clinical trials should focus on assessing the therapeutic benefits of hybrid nanocarriers in large patient cohorts, with an emphasis on patient-reported outcomes and long-term safety monitoring.

As these clinical trials and future research directions unfold, hybrid nanocarriers hold great promise in revolutionizing the delivery of therapeutics for the management of chronic obstructive pulmonary disease, potentially leading to improved patient outcomes and quality of life. Hybrid nanocarriers represent a promising avenue for advancing the delivery of therapeutics in COPD therapy.

9.7 CONCLUSIONS

The development of hybrid nanocarriers has emerged as a promising approach for the delivery of therapeutics in the management of COPD. These advanced drug delivery systems combine the advantages of different materials, such as lipids, polymers, and inorganic compounds, to address the unique challenges associated with COPD treatment.

Throughout this chapter, we have explored the role of hybrid nanocarriers as innovative drug delivery systems for COPD treatment. We discussed the rationale for using hybrid nanocarriers in COPD therapy, highlighting their ability to overcome challenges such as poor drug solubility, limited bioavailability, and off-target effects. By combining different materials, including lipids, polymers, and inorganic nanoparticles, hybrid nanocarriers offer multifunctional platforms for delivering a wide range of therapeutic agents to the lungs. We examined various types of hybrids nanocarriers, including lipid-polymer hybrids, polymer-inorganic hybrids, and lipid-inorganic hybrids, each with unique advantages in terms of drug loading capacity, controlled release kinetics, and targeting specificity. Examples of ongoing clinical trials and future perspectives were also discussed, illustrating the translational potential of hybrid nanocarriers in COPD therapy.

Considering the current level of development of hybrid nanocarriers and their challenges, some important points to conclude this chapter are:

- Hybrid nanocarriers offer enhanced drug encapsulation, controlled release kinetics, improved stability, and targeted delivery to the affected lung tissues in COPD.
- Lipid-polymer hybrid nanoparticles, polymer-inorganic hybrid nanoparticles, and lipid-inorganic hybrid nanoparticles have demonstrated their potential in preclinical studies for COPD therapy.
- These hybrid nanocarriers can be engineered to incorporate various functionalities, including targeted delivery, stimuli-responsive behavior, and the ability to co-deliver combination therapies.
- Ongoing clinical trials are evaluating the safety, tolerability, and efficacy of hybrid nanocarrier-based formulations for the delivery of drugs like roflumilast, antioxidants, and imaging agents in COPD.

REFERENCES

1. Agustí, A., Celli, B.R., Criner, G.J., Halpin, D., Anzueto, A., Barnes, P., Bourbeau, J., Han, M.K., Martinez, F.J., Montes de Oca, M., Mortimer, K., Papi, A., Pavord, I., Roche, N., Salvi, S., Sin, D.D., Singh, D., Stockley, R., López Varela, M.V., Wedzicha, J.A., Vogelmeier, C.F. Global initiative for chronic obstructive lung disease 2023 report: GOLD executive summary. *European Respiratory Journal*, doi: 10.1183/13993003.00239-2023
2. Sengel-Turk, C.T., Paksoy, A.O., Alpturk, O. The state of the art in core–shell-type lipid–polymer hybrid nanocarriers and beyond. *Polymer Bulletin*, 2024, 81, 4771–4800. https://doi.org/10.1007/s00289-023-04951-x
3. Sharma, M., Shah, R., Saraf, A., Kumar, R., Maheshwari, R., Balakrishnan, K., Nair, A., Kumar, R., Gupta, P. Current advances in the therapeutic potential of nanomedicines for pulmonary disorders. *Emergent Materials*, (2024). https://doi.org/10.1007/s42247-024-00708-y
4. Huayamares, S.G., Zenhausern, R., Loughrey, D. Nanocarriers for inhaled delivery of RNA therapeutics. *Current Research in Biotechnology*, 2024, 100200. https://doi.org/10.1016/j.crbiot.2024.100200

5. Liu, C., Xi, L., Liu, Y., Mak, J.C.W., Mao, S., Wang, Z., Zheng, Y. An inhalable hybrid biomimetic nanoplatform for sequential drug release and remodeling lung immune homeostasis in acute lung injury treatment. *ACS Nano*, 2023, 17, 11626–11644. https://doi.org/10.1021/acsnano.3c02075
6. Sato, K., Kawakami, H. Mitochondrial dysfunction and nanocarrier-based treatments in chronic obstructive pulmonary disease (COPD). *Oxygen*, 2023, 3, 394–406. https://doi.org/10.3390/oxygen3040026
7. Maurya, P., Singh, S., Saraf, S.A. Inhalable hybrid nanocarriers for respiratory disorders. Targeting Chronic Inflammatory Lung Diseases Using Advanced Drug Delivery Systems, 2020, 281–302. doi: 10.1016/B978-0-12-820658-4.00013-3
8. Wadhwa, A., Bobak, T.R., Bohrmann, L., Geczy, R., Sekar, S., Sathyanarayanan, G., Kutter, J.P., Franzyk, H., Foged, C., Saatchi, K., Häfeli, U.O. Pulmonary delivery of siRNA-loaded lipid-polymer hybrid nanoparticles: Effect of nanoparticle size. *OpenNano*, 2023, 13, 100180. https://doi.org/10.1016/j.onano.2023.100180
9. Kaushik, N., Borkar, S.B., Nandanwar, S.K., Panda, P.K., Choi, E.H., Kaushik, N.K. Nanocarrier cancer therapeutics with functional stimuli-responsive mechanisms. *Journal of Nanobiotechnology*, 2022, 20, 152. doi: 10.1186/s12951-022-01364-2
10. Gajbhiye, K.R., Salve, R., Narwade, M. et al. Lipid polymer hybrid nanoparticles: A custom-tailored next-generation approach for cancer therapeutics. *Molecular Cancer*, 2023, 22, 160. https://doi.org/10.1186/s12943-023-01849-0
11. Abdelaziz, H.M., Gaber, M., Abd-Elwakil, M.M., Mabrouk, M.T., Elgohary, M.M., Kamel, N.M., Kabary, D.M., Freag, M.S., Samaha, M.W., Mortada, S.M., Elkhodairy, K.A., Fang, J.Y., Elzoghby, A.O. Inhalable particulate drug delivery systems for lung cancer therapy: Nanoparticles, microparticles, nanocomposites and nanoaggregates. *Journal of Control Release*, 2018, 10, 374–392. doi: 10.1016/j.jconrel.2017.11.036
12. Saxena, J., Bisen, M., Misra, A., Srivastava, V.K., Kaushik, S., Siddiqui, A.J., Mishra, N., Singh, A., Jyoti, A. Targeting COPD with PLGA-based nanoparticles: Current status and prospects. *Biomed Research International*, 2022, 5058121. doi: 10.1155/2022/5058121
13. Pontes, J.F., Grenha, A. Multifunctional nanocarriers for lung drug delivery. *Nanomaterials*, 2020, 10, 183. https://doi.org/10.3390/nano10020183
14. Li, X., Peng, X., Zoulikha, M. et al. Multifunctional nanoparticle-mediated combining therapy for human diseases. *Signal Transduction and Targeted Therapy*, 2024, 9, 1. https://doi.org/10.1038/s41392-023-01668-1
15. Dhoble, S., Patravale, V., Weaver, E., Lamprou, D.A., Patravale, T. Comprehensive review on novel targets and emerging therapeutic modalities for pulmonary arterial hypertension. *International Journal of Pharmaceutics*, 2022, 621, 121792. https://doi.org/10.1016/j.ijpharm.2022.121792
16. Yanar, F., Carugo, D., Zhang, X. Hybrid nanoplatforms comprising organic nanocompartments encapsulating inorganic nanoparticles for enhanced drug delivery and bioimaging applications. *Molecules*, 2023, 28, 5694. https://doi.org/10.3390/molecules28155694
17. Cheng, X., Xie, Q., Sun, Y. Advances in nanomaterial-based targeted drug delivery systems. *Frontiers in Bioengineering and Biotechnology*, 2023, 11, 1177151. doi: 10.3389/fbioe.2023.1177151
18. Marcello, E., Chiono, V. Biomaterials-enhanced intranasal delivery of drugs as a direct route for brain targeting. *International Journal of Molecular Sciences*, 2023, 24, 3390. doi: 10.3390/ijms24043390
19. Fatima, M., Almalki, W.H., Khan, T., Sahebkar, A., Kesharwani, P. Harnessing the power of stimuli-responsive nanoparticles as an effective therapeutic drug delivery system. *Advanced Materials*, 2024, 2312939. https://doi.org/10.1002/adma.202312939
20. Chaurasiya, B., Zhao, Y.Y. Dry powder for pulmonary delivery: A comprehensive review. *Pharmaceutics*, 2020, 13, 31. doi: 10.3390/pharmaceutics13010031
21. Sayed, F.A., Eissa, N.G., Shen, Y., Hunstad, D.A., Wooley, K.L., Elsabahy, M. Morphologic design of nanostructures for enhanced antimicrobial activity. *Journal of Nanobiotechnology*, 2022, 20, 536. doi: 10.1186/s12951-022-01733-x
22. Kapate, N., Clegg, J.R., Mitragotri, S. Non-spherical micro- and nanoparticles for drug delivery: Progress over 15 years. *Advanced Drug Delivery Reviews*, 2021, 177, 113807. https://doi.org/10.1016/j.addr.2021.05.017
23. Tsung, T.-H., Tsai, Y.-C., Lee, H.-P., Chen, Y.-H., Lu, D.-W. Biodegradable polymer-based drug-delivery systems for ocular diseases. *International Journal of Molecular Sciences*, 2023, 24, 12976. https://doi.org/10.3390/ijms241612976
24. Lu, H., Zhang, S., Wang, J., Chen, Q. A review on polymer and lipid-based nanocarriers and its application to nano-pharmaceutical and food-based systems. *Frontiers in Nutrition*, 2021, 8, 783831. doi: 10.3389/fnut.2021.783831

25. Jain, S., Kumar, M., Kumar, P., Verma, J., Rosenholm, J.M., Bansal, K.K., Vaidya, A. Lipid–Polymer hybrid nanosystems: A rational fusion for advanced therapeutic delivery. *Journal of Functional Biomaterials*, 2023, 14, 437. https://doi.org/10.3390/jfb14090437
26. Xiao, R., Zhou, G., Wen, Y., Ye, J., Li, X., Wang, X. Recent advances on stimuli-responsive biopolymer-based nanocomposites for drug delivery. *Composites Part B: Engineering*, 2023, 266, 111018. https://doi.org/10.1016/j.compositesb.2023.111018
27. Craparo, E.F., Cabibbo, M., Scialabba, C., Giammona, G., Cavallaro, G. Inhalable formulation based on lipid-polymer hybrid nanoparticles for the macrophage targeted delivery of Roflumilast. *Biomacromolecules*, 2022, 23, 3439–3451. https://doi.org/10.1021/acs.biomac.2c00576
28. Wadhwa, A., Bobak, T.R., Bohrmann, L., Geczy, R., Sekar, S., Sathyanarayanan, G., Kutter, J.P., Franzyk, H., Foged, C., Saatchi, K., Häfeli, U.O. Pulmonary delivery of siRNA-loaded lipid-polymer hybrid nanoparticles: Effect of nanoparticle size. *OpenNano*, 2023, 13, 100180. https://doi.org/10.1016/j.onano.2023.100180
29. Vencken, S., Foged, C., Ramsey, J.M., Sweeney, L., Cryan, S.A., MacLoughlin, R.J., Greene, C.M. Nebulised lipid-polymer hybrid nanoparticles for the delivery of a therapeutic anti-inflammatory microRNA to bronchial epithelial cells. *ERJ Open Research*, 2019, 5, 00161–2018. doi: 10.1183/23120541.00161-2018
30. Yazdi, M.E.T., Qayoomian, M., Beigoli, S., Boskabady, M.H. Recent advances in nanoparticle applications in respiratory disorders: A review. *Frontiers in Pharmacology*, 2023, 14, 1059343. doi: 10.3389/fphar.2023.1059343
31. Passi, M., Shahid, S., Chockalingam, S., Sundar, I.K., Packirisamy, G. Conventional and nanotechnology based approaches to combat chronic obstructive pulmonary disease: Implications for chronic airway diseases. *International Journal of Nanomedicine*, 2020, 15, 3803–3826. doi: 10.2147/IJN.S242516
32. Uskoković, V. Lessons from the history of inorganic nanoparticles for inhalable diagnostics and therapeutics. *Advances in Colloid and Interface Science*, 2023, 315, 102903. https://doi.org/10.1016/j.cis.2023.102903
33. Alfahad, A.J., Alzaydi, M.M., Aldossary, A.M., Alshehri, A.A., Almughem, F.A., Zaidan, N.M., Tawfik, E.A. Current views in chronic obstructive pulmonary disease pathogenesis and management. *Saudi Pharmaceutical Journal*, 2021, 29, 1361–1373. doi: 10.1016/j.jsps.2021.10.008
34. Donohue, J.F. Combination therapy for chronic obstructive pulmonary disease: Clinical aspects. *Proceedings of the American Thoracic Society*, 2005, 2, 272–81, discussion 290–1. doi: 10.1513/pats.200505-047SR
35. British Columbia. BC Guidelines, Chronic Obstructive Pulmonary Disease (COPD): Diagnosis and Management, 2024. https://www2.gov.bc.ca/gov/content/health/practitioner-professional-resources/bc-guidelines/copd
36. Prada, B.S., Jadhav, U., Ghewade, B., Wagh, P., Karnan, A., Ledwani, A. Comparing glycopyrronium/formoterol combination therapy with monotherapy in moderate-to-severe chronic obstructive pulmonary disease (COPD): A narrative review. *Cureus*, 2024, 16, e58633. doi: 10.7759/cureus.58633
37. Virmani, T., Kumar, G., Virmani, R., Sharma, A., Pathak, K. Nanocarrier-based approaches to combat chronic obstructive pulmonary disease. *Nanomedicine*, 2022, 17, 1833–1854.
38. Xie, L., Xie, D., Du, Z., Xue, S., Wang, K., Yu, X., Liu, X., Peng, Q., Fang, C. A novel therapeutic outlook: Classification, applications and challenges of inhalable micron/nanoparticle drug delivery systems in lung cancer (Review). *International Journal of Oncology*, 2024, 64, 38. doi: 10.3892/ijo.2024.5626
39. Serrenho, I., Rosado, M., Dinis, A., Cardoso, C., Grãos, M., Manadas, B., Baltazar, G. Stem cell therapy for neonatal hypoxic-ischemic encephalopathy: A systematic review of preclinical studies. *International Journal of Molecular Sciences*, 2021, 22, 3142. https://doi.org/10.3390/ijms22063142

10 Dendrimers as Emerging Drug Carriers in Chronic Obstructive Pulmonary Disease

Shraddha Chugh
Department of Chemistry, University of Petroleum & Energy Studies
Dehradun, Uttarakhand, India

Mousmee Sharma
Department of Chemistry, Uttaranchal University
Dehradun, Uttarakhand, India

Parteek Prasher
Department of Chemistry, University of Petroleum & Energy Studies
Dehradun, Uttarakhand, India

10.1 INTRODUCTION

Chronic obstructive pulmonary disease (COPD), marked by progressive airflow limitation and persistent respiratory symptoms, poses a substantial global health concern. Despite advancements in therapeutic interventions, the management of COPD remains complex, often hindered by the limitations of conventional drug delivery systems. There is an increased interest in devising novel drug delivery approaches to improve the effectiveness and safety of treatments for COPD [1, 2]. COPD patients face chronic respiratory issues marked by airflow limitation, continuous sputum, dyspnea, and cough. The complex pathophysiology of COPD, involves various factors such as genetic predisposition, environmental exposures like cigarette smoke and air pollution, and host characteristics. A key aspect of COPD's development is the onset of chronic inflammation in the airways, lung tissues, and pulmonary blood vessels. When exposed to harmful particles and gases, notably from cigarette smoke, immune cells like macrophages, neutrophils, and T lymphocytes become activated, which leads to the release of various pro-inflammatory substances like cytokines, reactive oxygen species (ROS), and chemokines. This sustained inflammatory response contributes to airway remodeling, mucus hypersecretion, and the destruction of lung parenchyma, leading to progressive airflow limitation and impaired gas exchange [3].

Destruction of balance between ROS production and the body's antioxidant mechanism cause oxidative stress, which is a central player in the development of COPD. Increased ROS production, largely driven by activated inflammatory cells and dysfunctional mitochondria, surpasses the capacity of natural antioxidant mechanisms, leading to oxidative injury to lipids, proteins, and DNA in the pulmonary tissues. This oxidative damage amplifies inflammation, promotes tissue injury, and contributes to the development of COPD phenotypes, including emphysema and chronic bronchitis [4, 5].

Dendrimers are hyperbranched, three-dimensional macromolecules characterized by a highly symmetrical, well-defined structure. They are composed of a central core, branching units extending

 DOI: 10.1201/9781003498926-10

from the core called dendrons, and an outer shell containing functional groups. These structures can be meticulously crafted with precise control over their molecular weight, dimensions, and surface properties, rendering them appealing options for drug delivery purposes [6, 7].

A dendrimer's characteristic feature is to encapsulate drug molecules within its interior cavities or on its surface. This encapsulation can enhance the solubility, stability, and bioavailability of poorly water-soluble drugs, thereby improving its pharmacokinetic and pharmacodynamic profiles. Additionally, dendrimers can be tailored to modulate drug release kinetics, allowing for sustained or targeted drug delivery to specific tissues or cells. The surface functionalization of dendrimers with targeting ligands, such as peptides, antibodies, or small molecules, further enhances their utility as drug carriers. By decorating dendrimer surfaces with ligands that selectively bind to receptors or biomarkers overexpressed in diseased tissues, targeted drug delivery can be achieved, minimizing off-target effects and maximizing therapeutic efficacy [8, 9]. In the context of COPD treatment, dendrimers offer unique advantages as drug carriers for delivering anti-inflammatory and antioxidant agents to the lungs. Their ability to encapsulate drugs, modulate drug release kinetics, and target specific cell types or tissues holds promise for enhancing the efficacy and safety of COPD therapies. By harnessing the properties of dendrimers, researchers have explored novel strategies to address the underlying pathways driving COPD pathogenesis, ultimately improving treatment outcomes for patients [10–12].

Dendrimers serve as versatile nanocarriers in respiratory medicine, offering innovative solutions for drug delivery and gene therapy. Below is a summary of key mechanisms and their advantages in dendrimer-based drug delivery systems for various classes of drugs. These processes, comprising encapsulation, regulated release, precise targeting, and gene transfer, empower accurate and efficient therapeutic approaches for respiratory ailments like asthma, COPD, and lung cancer. Additionally, the integration of therapeutic and diagnostic functionalities within dendrimers fosters the development of theragnostic approaches, paving the way for personalized medicine in respiratory care. Table 10.1 shows mechanisms and advantages of different dendrimer-based drug delivery system for an overview.

This chapter focuses on different drug delivery systems supported by dendrimers for efficient drug delivery in treatment of COPD targeting various anti-inflammatory, bronchodilator, antioxidant and gene targeting drug pathways (Figure 10.1).

TABLE 10.1
Mechanisms and Advantages of Dendrimer-based Drug Delivery Systems for Chronic Obstructive Pulmonary Disease Therapy

Mechanism	Description	Advantages	Relevant Drug Classes	References
Encapsulation	Drug molecules are encapsulated within the dendritic architecture or interior cavities of dendrimers.	Enhanced solubility, stability, and bioavailability of hydrophobic drugs. Protection from degradation.	Corticosteroids, nonsteroidal anti-inflammatory drug (NSAIDs), bronchodilators, antioxidants	[6–9]
Controlled release	Dendrimers can modulate the release kinetics of encapsulated drugs, enabling sustained or controlled drug release.	Prolonged therapeutic effect, reduced dosing frequency, improved patient compliance.	Corticosteroids, bronchodilators, NSAIDs	[13–16]
Targeted delivery	Surface modification of dendrimers with targeting ligands or moieties facilitates selective delivery to specific cells or tissues.	Enhanced drug accumulation at the site of action, reduced off-target effects, improved therapeutic efficacy.	Corticosteroids, NSAIDs, antioxidants, gene therapy agents	[17–19]

(Continued)

TABLE 10.1 (Continued)
Mechanisms and Advantages of Dendrimer-based Drug Delivery Systems for Chronic Obstructive Pulmonary Disease Therapy

Mechanism	Description	Advantages	Relevant Drug Classes	References
Improved solubility	Dendrimers can enhance the solubility of poorly water-soluble drugs through encapsulation or conjugation.	Improved bioavailability and pharmacokinetic profiles of hydrophobic drugs.	Corticosteroids, NSAIDs, antioxidants	[20–22]
Gene delivery	Dendrimers can complex with genetic materials (e.g., small interfering RNA [siRNA], pDNA) and facilitate their cellular uptake and delivery.	Targeted gene therapy, modulation of disease pathways, improved treatment outcomes.	Gene therapy agents	[23–29]
Combination therapy	Dendrimers can co-deliver multiple therapeutic agents simultaneously, enabling synergistic effects.	Targeting multiple disease pathways, enhanced therapeutic efficacy, reduced drug resistance.	Corticosteroids, bronchodilators, antioxidants, gene therapy agents	[30–34]
Theragnostic	Dendrimers can integrate therapeutic agents with diagnostic imaging modalities or reporter molecules.	Simultaneous therapy and disease monitoring, personalized medicine, improved treatment outcomes.	Corticosteroids, antioxidants, gene therapy agents	[32]

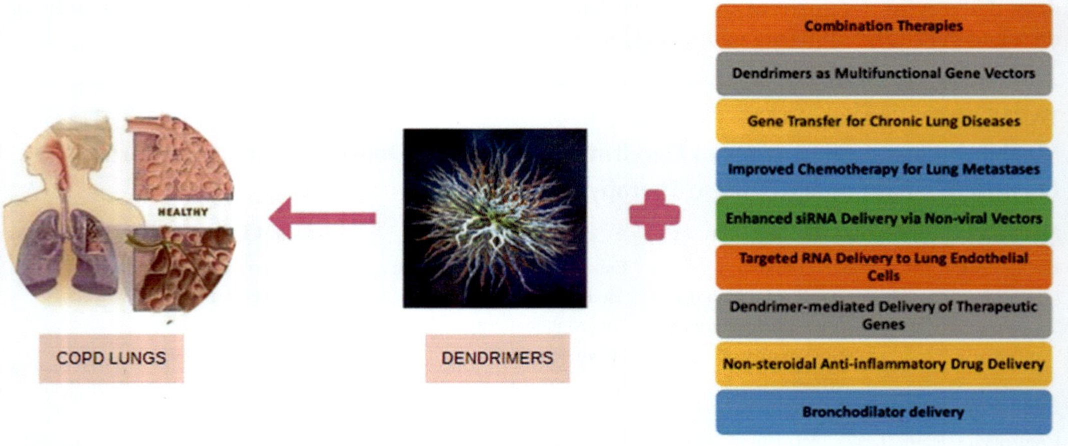

FIGURE 10.1 Strategies for chronic obstructive pulmonary disease treatment by dendrimers.

10.2 ANTI-INFLAMMATORY DRUG DELIVERY BY DENDRIMERS IN COPD

COPD, marked by intricate inflammatory mechanisms, leads to structural changes in airways and alveoli, causing respiratory impairment. Managing COPD effectively involves addressing this chronic inflammation. The inflammatory response in COPD initiates with the infiltration and activation of various immune cells like neutrophils, macrophages, and lymphocytes in lung tissue. These cells release a range of substances including cytokines, chemokines, and enzymes, perpetuating inflammation and causing ongoing tissue damage [35, 36].

Neutrophils, upon activation, release proteolytic enzymes like elastase and matrix metalloproteinases, as well as ROS, which can degrade the lung parenchyma and airway structures. The excessive recruitment and activation of neutrophils in the lungs is a hallmark of COPD [37]. Macrophages play a central role in driving the inflammatory process as they release chemokines and cytokines, leading to activation of other inflammatory cells, thus sustaining the inflammatory cascade. Additionally, lymphocytes, specifically CD8+ T cells, contribute to the damage of lung tissue by releasing cytotoxic substances and triggering apoptosis in structural cells [38].

This persistent and unresolved inflammatory response leads to structural changes in the airways and alveoli, including airway remodeling, mucus hypersecretion, and emphysematous destruction of the lung parenchyma. These structural alterations result in airflow limitation, air trapping, and reduced gas exchange, ultimately contributing to the characteristic symptoms of COPD, such as dyspnea, cough, and exercise intolerance [39].

Conventional anti-inflammatory therapies, such as corticosteroids and bronchodilators, have demonstrated limited efficacy in controlling chronic inflammation and preventing disease progression in COPD. Furthermore, these therapies are often associated with systemic side effects, particularly with long-term use, highlighting the need for more effective and targeted treatment approaches [40].

Dendrimers, highly branched, monodisperse, and nanoscale polymeric macromolecules, offer promising opportunities for the lungs-targeting efficient delivery of anti-inflammatory drugs. Their unique structural features, including a well-defined architecture, multivalency, and tailorable surface chemistry, make them attractive carriers for enhancing the solubility, stability, and targeted delivery of therapeutic agents [13–16].

Dendrimer-based drug delivery systems offer a versatile platform for encapsulating or conjugating anti-inflammatory drugs. This protects the drugs from degradation and allows for controlled release specifically at the intended target site. Additionally, the surface modification of dendrimers with specific targeting ligands or moieties can facilitate selective delivery to inflamed tissues or specific cell types involved in the inflammatory cascade, such as macrophages, neutrophils, or epithelial cells [17–19].

By leveraging the unique properties of dendrimers, researchers have explored various strategies to improve the targeted delivery of anti-inflammatory drugs to the lungs. These strategies aim to enhance therapeutic efficacy while minimizing systemic exposure and associated side effects. This targeted and controlled delivery approach has the potential to address the limitations of conventional therapies and provide more effective treatment options for COPD patients.

10.2.1 Corticosteroid Delivery

10.2.1.1 Enhancing Corticosteroid Delivery in COPD with Dendrimer-Based Formulations

Corticosteroids are crucial anti-inflammatory agents in COPD treatment, yet their efficacy is hindered by challenges in delivery and pharmacokinetics. Poor solubility, rapid clearance, and systemic side effects limit their therapeutic potential. To overcome these hurdles, researchers have turned to dendrimer-based formulations for targeted pulmonary delivery of corticosteroids [20].

Dendrimers, emerging as nanoscale drug carriers, provide several advantages, including enhanced solubility, controlled release, and targeted delivery to the lungs. In a research study assessing methylprednisolone (MP) linked to polyamidoamine (PAMAM) dendrimers, significant improvements in airway delivery were observed in a mouse model of pulmonary inflammation due to allergen exposure [21, 22]. The conjugation of MP PAMAM Generation 4 Hydroxyl-terminated dendrimers with 12 MP per dendrimer enhanced solubility. Daily treatments with MP-dendrimer conjugate resulted in a remarkable decrease in allergen-induced inflammation compared to free MP alone. Notably, the conjugate further diminished airway and tissue eosinophil recoveries, indicating improved therapeutic efficacy.

Moreover, dendrimer-mediated delivery exhibited minimal additional lung inflammation and decreased baseline phagocytic cell recoveries compared to free MP. The study highlights the potential of dendrimer-based formulations to enhance corticosteroid retention in the lung and mitigate inflammatory responses associated with conditions like asthma.

These findings underscore the promising role of dendrimers in revolutionizing corticosteroid therapy for COPD by offering a targeted and efficient delivery platform, thus addressing the limitations of conventional corticosteroid treatment strategies [29].

Chawla et al. (2022) developed complexes of roflumilast (RofA) with dendrimers for sustained delivery. Roflumilast, a specific inhibitor of phosphodiesterase-4 (PDE4), is employed for the treatment of COPD. Unlike corticosteroids, commonly employed in COPD therapy for their anti-inflammatory properties, roflumilast acts via a distinct mechanism by selectively inhibiting PDE4 enzyme activity. This inhibition elevates cyclic adenosine monophosphate (cAMP) levels within cells, resulting in anti-inflammatory and bronchodilatory effects. The study thoroughly investigated the encapsulation potency, loading capacity, in vitro drug release behavior, and stability of these complexes. Results indicated increased solubility of RofA when complexed with dendrimers of various generations and under different pH conditions. Moreover, both RofA and RofA-PAMAM complexes exhibited significant anti-inflammatory activity. The drug release kinetics from the complexes followed a non-Fickian diffusion pattern ($n < 0.89$), indicating a sustained release profile. Additionally, the study explored the impact of adjunct antibiotic therapy using a cytotoxicity assay, providing insights into potential combination therapies for enhanced therapeutic outcomes in COPD treatment [41].

The dendrimer nanocarriers loaded with fluticasone showed improved anti-inflammatory properties in laboratory tests. This was evident in a notable decrease in the generation of inflammatory signaling molecules such as interleukin (IL)-8, tumor necrosis factor-alpha (TNF-α), and monocyte chemoattractant protein-1 (MCP-1) within human lung epithelial cells. Additionally, in vivo studies using a mice model of acute lung inflammation revealed that the dendrimer nanocarriers were more effective in reducing neutrophil infiltration and pro-inflammatory cytokine levels compared to conventional fluticasone formulations [42]. Importantly, the dendrimer-based fluticasone formulation showed reduced systemic exposure and associated side effects compared to the free drug. This outcome can be attributed to the controlled release and targeted delivery of fluticasone to the lungs, minimizing systemic absorption and potential adverse effects. Liu et al. (1996) have also investigated dendrimer-based formulations for the delivery of corticosteroids in COPD. Researchers created a nanocarrier system based on dendrimers to deliver beclomethasone dipropionate to the lungs, which is another corticosteroid used in COPD therapy. In this study, beclomethasone was attached to PAMAM dendrimers through a biodegradable ester connection, leading to prolonged release and increased anti-inflammatory effectiveness both in laboratory experiments and in living organisms [43]. In general, employing dendrimers as carriers for delivering corticosteroids in COPD shows potential for enhancing treatment effectiveness and reducing the systemic side effects linked with traditional corticosteroid treatments.

10.2.2 Non-steroidal Anti-inflammatory Drug Delivery

While corticosteroids are the mainstay of anti-inflammatory therapy in COPD, nonsteroidal anti-inflammatory drug (NSAIDs) have also been explored as an alternative or complementary treatment approach. NSAIDs work to reduce inflammation by blocking the activity of cyclooxygenase (COX) enzymes, which are essential for the production of prostaglandins and other inflammatory substances.

However, the systemic administration of NSAIDs is often associated with gastrointestinal and cardiovascular side effects, limiting their long-term use in COPD patients. These adverse effects are primarily attributed to the non-specific distribution and systemic exposure of NSAIDs when administered orally or parenterally.

To overcome these challenges, scientists have explored dendrimer-based formulations as carriers for delivering NSAIDs to the lungs. Utilizing the pulmonary route presents several benefits, such as direct delivery to the lung tissue, thereby minimizing systemic exposure and potentially lowering the risk of gastrointestinal and cardiovascular side effects linked with systemic administration of NSAIDs.

In one of the pioneering studies in this field, a dendrimer-based formulation for the pulmonary delivery of indomethacin was developed, a commonly used NSAID, where conjugation of indomethacin polyamidoamine (PAMAM) dendrimers through a cleavable ester linkage, forming dendrimer-drug conjugates was carried. In comparison with free drugs, the dendrimer-indomethacin conjugates exhibited improved solubility and sustained release kinetics. Laboratory experiments conducted with human lung epithelial cells showed that the dendrimer-indomethacin combinations maintained their ability to reduce inflammation, as indicated by a notable decrease in the production of pro-inflammatory cytokines like IL-8 and TNF-α [44]. Importantly, the dendrimer-indomethacin conjugates showed enhanced pharmacokinetic properties, with a prolonged half-life and increased residence time in the lungs compared to free indomethacin. In vivo studies in a rat model of acute lung injury (ALI) further demonstrated the superior anti-inflammatory activity of the dendrimer-indomethacin conjugates, as evidenced by reduced neutrophil infiltration, pro-inflammatory cytokine levels, and lung injury scores.

Building upon these findings, development of ibuprofen-loaded PAMAM dendrimer nanocarriers for pulmonary delivery was carried. Ibuprofen is another widely used NSAID with potent anti-inflammatory properties. Ibuprofen offers pain relief and anti-inflammatory effects in lung diseases like pleurisy, bronchitis, and asthma. Ibuprofen, encapsulated within the dendritic structure of PAMAM dendrimers, resulted in nanocarriers with heightened solubility and sustained release characteristics. These ibuprofen-loaded dendrimer nanocarriers displayed prolonged release profiles and increased anti-inflammatory efficacy in vitro. This was demonstrated by notable reductions in the secretion of pro-inflammatory cytokines and chemokines, including IL-8, TNF-α, and MCP-1, in human lung epithelial cells [45, 46].

An inventive nanoplatform employing generation 5 poly(amidoamine) dendrimers entrapping gold nanoparticles (Au DENPs) has been developed for the concurrent delivery of dexamethasone (Dex) and a microRNA-155 inhibitor (miR-155i) to target ALI. These Au DENPs, featuring 10 Dex molecules per dendrimer and a 2.1 nm gold core diameter, effectively transport miR-155i to lipopolysaccharide-activated alveolar macrophages. Within these macrophages, miR-155i facilitates the upregulation of anti-inflammatory IL-10 and suppressor of cytokine signaling 1, while concurrently suppressing the expression of pro-inflammatory cytokines TNF-α, IL-1β, and IL-6. In parallel, Dex acts to inhibit the secretion of pro-inflammatory cytokines by downregulating COX-2 expression. This dual approach of combining chemotherapy with gene therapy effectively reduces the expression of pro-inflammatory cytokines, leading to the restoration of damaged lung tissues in an ALI model. Histological examinations and decreased cytokine levels confirm the efficacy of this strategy. By precisely targeting the inflammatory processes within the lungs, this dendrimer-based nanotechnology offers considerable promise in the treatment of inflammatory lung diseases such as ALI. It provides a potential solution for addressing the complex underlying mechanisms of these conditions and ultimately improving patient outcomes [30].

Thus, pulmonary delivery of NSAIDs using dendrimer-based carriers has the potential to minimize systemic exposure and associated side effects.

10.3 BRONCHODILATOR DRUG DELIVERY

By relaxing the smooth muscle cells in the airways, bronchodilators provide symptomatic relief and improve lung function in COPD patients [47]. However, conventional inhalation formulations of bronchodilators often face challenges such as poor solubility, rapid mucociliary clearance, and suboptimal lung deposition, limiting their therapeutic efficacy [48]. To overcome these challenges, pharmaceutical strategies such as modifying the drug formulation, particle size, and delivery device can be used to extend pulmonary exposure and improve therapeutic efficacy. For example, lipid-based formulations can be used to enhance the solubility and lung deposition of lipid-soluble drugs, and dry powder inhalers (DPIs) can be used to deliver drugs directly to the lungs with minimal mucociliary clearance [49]. Dendrimers have emerged as promising nanocarriers for the pulmonary delivery of bronchodilators, offering several advantages over traditional formulations. Dendrimers can encapsulate or conjugate bronchodilator drugs within their unique architecture, improving drug solubility, stability, and bioavailability [50]. Studies have demonstrated that positively charged dendrimers can form complexes with bronchodilators, leading to increased drug bioavailability and efficacy in preventing conditions like deep vein thrombosis after pulmonary administration. Ray et al.'s (2011)study reveals PAMAM dendrimers as promising carriers for delivering the poorly soluble anti-asthma drug beclomethasone dipropionate (BDP) to the lungs. BDP's known benefits in COPD management include reducing airway inflammation and improving lung function. The study demonstrates that BDP complexed with dendrimers exhibits sustained release over eight hours, with higher dendrimer generations and elevated pH media enhancing drug solubility. While nebulizer type affects aerosol performance more than dendrimer generation, air-jet and vibrating-mesh nebulizers show higher aerosol output, highlighting the potential of BDP-dendrimers for pulmonary inhalation using specific nebulizer types. This innovative approach not only ensures targeted drug delivery to affected lung tissues but also facilitates convenient administration through nebulizers, aligning with BDP's therapeutic benefits in COPD management. By enhancing BDP's solubility and enabling sustained release, dendrimer-based delivery systems offer a promising strategy for optimizing treatment outcomes in COPD patients, addressing the specific needs of COPD management effectively [51]. Similarly Malkoch et al. (2005) developed/synthesized triphenylphosphonium (TPP)-modified generation 4 poly(amidoamine) (PAMAM) dendrimers (G4NH2-TPP) for bronchodilator delivery in COPD. These dendrimers enhance small interfering RNA (siRNA) delivery to pulmonary epithelium via aerosol formulations in pressurized metered dose inhalers (pMDIs) and DPIs. Complexes of siRNA and G4NH2-TPP showed increased transfection efficiency with higher TPP density and N/P ratios. G4NH2-12TPP dendriplexes at N/P ratio 30 exhibited the highest gene knockdown efficiency in vitro. Micron particle technologies incorporating 12TPP-dendriplexes in mannitol microparticles achieved efficient aerosolization for deep lung deposition with no impact on gene knockdown efficiency. This approach demonstrated the potential of TPP-conjugated dendrimers for siRNA delivery to COPD-afflicted lungs via portable inhalers [52]. Kim et al. (2010) developed/synthesized positively charged dendrimers to enhance pulmonary delivery of enoxaparin, a low-molecular weight heparin (LMWH), for potential use in COPD treatment. Their study revealed that these dendrimers formed complexes with enoxaparin, increasing its relative bioavailability by 40% through ionic interactions. Administering enoxaparin-dendrimer formulations into the lungs of rats showed efficacy in preventing deep vein thrombosis comparable to subcutaneous injection of enoxaparin. Importantly, these formulations did not adversely affect mucociliary transport rate or cause extensive lung damage. The positively charged dendrimers were identified as suitable carriers for pulmonary delivery of enoxaparin, potentially due to their ability to reduce the negative surface charge density of the drug molecule [53]. Numerous studies have explored the potential of dendrimers in delivering various classes of bronchodilators, including beta-2 (β2). Bronchodilators, particularly β2-agonists, represent a mainstay of COPD therapy by relaxing airway smooth muscle and promoting bronchodilation. However, conventional inhaler-based delivery suffers from limitations, including poor drug deposition in the lower airways, systemic side effects due to oropharyngeal deposition, and dose-related complications. Dendrimers may offer a promising strategy for targeted and enhanced pulmonary delivery of β2-agonists in COPD patients.

β2-agonists, such as salbutamol, formoterol, and salmeterol, work by binding to β2-adrenergic receptors on the smooth muscle cells, inducing bronchodilation. Dendrimers can encapsulate β2-agonists like salbutamol and formoterol. Dendrimers have the ability to encapsulate a variety of bioactive substances, including drugs, due to their large number of hydrophobic pockets. The encapsulation of β2-agonists within dendrimers can be achieved through non-covalent interactions, where dendrimers wrap drug molecules within their interior and protect them, the release rate of β2-agonists from dendrimers can be controlled by adjusting the dendrimer-to-drug molar ratio. Dendrimer-mediated transdermal delivery of β2-agonists has shown a 2–4.5 times enhancement factor in steady-state flux compared to the pure drug [54–56].

Anticholinergic drugs, used in bronchodilator delivery include muscarinic receptor antagonists, which are used in the treatment of COPD and asthma. These drugs work by blocking the muscarinic effects of acetylcholine, which causes bronchoconstriction. The use of muscarinic receptor antagonists has been shown to improve lung function, reduce exacerbations, and improve quality of life in patients with COPD and asthma [57]. Dendrimers can be used as a drug delivery system for anticholinergic drugs, such as ipratropium bromide, a short-acting muscarinic receptor antagonist (SAMA) [58]. Ipratropium bromide has been encapsulated in dendrimers, which has shown generation-dependent drug loading and controlled release [59]. The use of dendrimers as a drug delivery system for ipratropium bromide has been shown to improve the efficacy and safety of the drug, as it can target specific cells or tissues in the body and reduce systemic side effects.

Antioxidant therapy represents a promising avenue for managing COPD by targeting the oxidative stress at its core. Dendrimers, complex polymer structures, have emerged as effective carriers for delivering antioxidant drugs, offering improved bioavailability, minimal toxicity, and precise delivery profiles [60]. Recent advancements in dendrimer-based formulations show considerable potential in COPD treatment, harnessing the antioxidant properties of polyphenols and other bioactive compounds [61]. Dendrimers can encapsulate antioxidant molecules within their interior spaces or on their surfaces, safeguarding them from degradation and ensuring sustained release for prolonged therapeutic effects.

For instance, caffeic acid, a plant-derived antioxidant, demonstrates promise in mitigating oxidative stress in lung diseases like asthma and COPD, thereby reducing inflammation and shielding against tissue damage [62]. When conjugated with carbosilane dendrimers, polyphenols like caffeic acid exhibit enhanced biocompatibility and antioxidant efficacy, protecting against oxidative hemolysis and reducing ROS production in human fibroblasts [62].

Moreover, dendrimers functionalized with TEMPO (2,6,6-Tetramethylpiperidine-1-oxy) radicals show potent scavenging activity against nitric oxide, 2,2-diphenyl-1-picrylhydrazyl and hydrogen peroxide radicals, offering a versatile platform for mitigating oxidative stress in COPD and potentially alleviating associated inflammation and airway remodeling [63].

Similarly, Coenzyme Q10 (CoQ10) shows promise for COPD management due to its antioxidative and anti-inflammatory properties. CoQ10-loaded dendrimers, formulated with sugary maize dendrimer-like glucan (SMDG) as a carrier, demonstrate improved solubility, cellular uptake, and protective effects against oxidative damage, highlighting their therapeutic potential [64].

Polyphenol-based antioxidant dendrimers synthesized with electron-donating groups exhibit potent antioxidant activities without pro-oxidant effects, providing targeted antioxidant therapy for COPD and minimizing systemic toxicity [65]. Additionally, dendrimers functionalized with metal-organic frameworks and hyaluronic acid offer a promising strategy for targeted drug delivery in COPD, enhancing therapeutic efficiency while minimizing adverse effects [66].

Furthermore, curcumin-conjugated PAMAM dendrimers demonstrate enhanced bioavailability and therapeutic potential for COPD management, delivering this potent antioxidant to target sites within the lungs to mitigate oxidative stress and inflammation [67].

In summary, dendrimer-based drug delivery systems offer a versatile platform for delivering antioxidants to the lungs, optimizing therapeutic outcomes, and minimizing systemic side effects in COPD management. While challenges remain in targeted antioxidant delivery, ongoing research continues to explore innovative approaches to address the complex pathophysiology of COPD [68].

10.4 GENE THERAPY BY DENDRIMERS

While conventional therapeutics have made modest strides in alleviating the symptomatic burden of COPD, this multifactorial disease continues to exert a formidable toll on patient morbidity and mortality. The advent of gene therapy has unveiled an enticing strategy to subvert the relentless pathophysiological processes underpinning COPD progression at their molecular roots. However, the clinical translation of gene-based therapeutics has been paradoxically constrained by the inadequacies of conventional viral and non-viral vectors. Enter dendrimers – hyperbranched, structurally precise macromolecules that have emerged as an avant-garde class of gene delivery vectors with immense potential for pulmonary applications [69, 70].

10.4.1 Dendrimer-Mediated Delivery of Therapeutic Genes

In the pursuit of enhancing therapeutic outcomes for various pulmonary conditions, recent advancements in dendrimer-based delivery systems offer promising avenues for targeted gene therapy.

10.4.2 Targeted RNA Delivery to Lung Endothelial Cells

Chemically modified dendrimer nanomaterials show significant promise in addressing inflammation-related lung disorders such as chronic asthma and obstructive pulmonary disease. Utilizing innovative combinatorial strategies, dendrimers are tailored for the precise delivery of siRNA to the lung vasculature, particularly targeting Tie2-expressing lung endothelial cells. Importantly, these modified materials demonstrate minimal elevation of proinflammatory cytokines and toxicity, indicating their potential as effective agents for pulmonary RNA therapeutics [23, 71].

10.4.3 Enhanced siRNA Delivery via Non-viral Vectors

Overcoming the challenges of poor siRNA bioavailability, researchers investigate the efficacy of a TPP modified poly(amidoamine) dendrimer in enhancing siRNA transfection efficiency in pulmonary epithelial models. By modulating complexation efficiency through varying TPP densities and N/P ratios, dendriplexes exhibit significant gene knockdown in lung alveolar epithelial cells. Moreover, aerosol formulations utilizing micron particle technologies demonstrate potential for deep lung deposition, paving the way for local siRNA delivery via portable inhalers [24].

10.4.4 Improved Chemotherapy for Lung Metastases

In response to the challenges posed by low drug concentration and systemic toxicity in the treatment of lung metastases, researchers have turned to nanocarrier-based systems for localized delivery of doxorubicin (DOX). By conjugating DOX to poly(amidoamine) dendrimers, therapeutic efficacy is enhanced, and systemic toxicity is reduced. This is evidenced by a decrease in tumor burden and improved survival rates observed in a lung metastasis mouse model. Pulmonary administration of the dendrimer-DOX conjugate extends lung accumulation while limiting distribution to other tissues, underscoring its potential as a targeted chemotherapy approach [25].

10.4.5 Gene Transfer for Chronic Lung Diseases

Highlighting the importance of gene transfer in chronic lung diseases like cystic fibrosis, researchers investigate the gene delivery efficiency of cationic polymers such as polyethylenimine (PEI) and fractured polyamidoamine dendrimers in murine lungs. Additionally, the influence of surfactants on PEI-mediated gene transfer is examined, shedding light on factors affecting in vivo gene delivery efficiency. Collectively, these studies underscore the versatility and efficacy of dendrimer-based delivery systems in addressing diverse challenges associated with pulmonary gene therapy, offering novel strategies for targeted treatment of lung diseases [26].

10.4.6 Dendrimers as Multifunctional Gene Vectors

Moving beyond their conventional role as mere carriers for gene delivery, dendrimers have evolved into multifunctional gene vectors that integrate various therapeutic and diagnostic features into a unified nanoarchitecture. This convergence of functionalities has given rise to dendrimer-based theranostic gene vectors, which synergistically combine gene therapy with complementary imaging and therapeutic capabilities. Gene transfer to the lung is facilitated through the airway and pulmonary vasculature.

Kukowska-Latallo and colleagues explored gene transfer through intravascular and endobronchial routes using DNA combined with G9 PAMAM dendrimers or naked plasmid DNA. When dendrimer-complexed pCF1CAT plasmid was administered intravenously, there was substantial lung transgene expression at 12 hours and 24 hours, with another peak observed at 3 days to 5 days afterward. Lung-specific COPD assessment test (CAT) expression was evident, contrasting with minimal levels seen after intravenous delivery of naked plasmid DNA. Repeated intravascular doses of dendrimer-complexed plasmid maintained expression at 15–25% of peak levels. Endobronchial delivery of naked pCF1CAT plasmid also resulted in significant lung CAT protein expression. Intratracheal and intranasal routes yielded similar CAT expression levels in the lungs and trachea, though intranasal delivery of dendrimer-complexed DNA exhibited lower expression compared to naked DNA. Localization of CAT enzymatic activity suggested that vascular delivery achieved lung parenchymal expression, while endobronchial delivery targeted bronchial epithelium. These findings highlight the efficacy of G9 dendrimers as carriers for pulmonary gene transfer, with repeated administration extending transgene expression [27].

Labhasetwar (2003) synthesized and characterized dendritic poly(l-lysine) vectors of varying generations for gene transfection. Gel retardation assays confirmed the binding of plasmid DNA by dendrimers, with higher generation dendrimers showing increased stability of the complexes. DNase I assays revealed better protection of plasmid DNA by G5 dendrimers compared to PEI. Atomic force microscopy showed consistent particle sizes (100–200 nm) across all generations, while zeta potential measurements indicated higher positive charges with increasing N/P ratio, facilitating interaction with cell membranes. Both in vitro and in vivo cytotoxicity evaluations demonstrated the biocompatibility of dendrimers at different N/P ratios. In vitro gene transfection tests demonstrated higher efficiency and resistance to serum degradation of G5 dendrimers compared to other generations, suggesting their potential as promising candidates for gene delivery. G5 dendrimers showed comparable transfection efficiency to PEI but with lower toxicity, indicating their potential superiority for gene delivery applications. Overall, these studies highlight the versatility and effectiveness of dendrimers in gene delivery, offering insights into their potential for targeted and sustained gene expression in therapeutic approaches [28].

10.5 COMBINATION THERAPIES

Dendrimers play a pivotal role in combinational therapies for COPD treatment by serving as versatile drug delivery platforms. Their unique properties, such as precise molecular architecture, tunable size, and high surface functionality, enable the encapsulation and controlled release of multiple therapeutic agents simultaneously. In COPD, dendrimer-based combinational therapies can involve the co-delivery of anti-inflammatory drugs, bronchodilators, antioxidants, and gene therapy agents to target various aspects of the disease pathophysiology, including airway inflammation, oxidative stress, and tissue remodeling. By encapsulating and delivering these agents directly to the affected lung tissue, dendrimers enhance drug efficacy, reduce systemic side effects, and enable synergistic therapeutic effects. Additionally, dendrimers can facilitate targeted delivery to specific cell types or regions within the lungs, further optimizing treatment outcomes. Overall, dendrimers serve as versatile nanocarriers in combinational therapies for COPD, offering the potential for enhanced therapeutic efficacy and improved disease management.

DOX, a potent chemotherapeutic agent, is utilized in lung cancer treatment to impede tumor growth by disrupting DNA synthesis and prompting cancer cell demise. Moghimi et al. (2001) conducted studies aimed at bolstering the therapeutic effectiveness of lung-resident cancers through the application of dendrimer-based drug delivery systems. Their research centered on a 56 kDa PEGylated polylysine dendrimer linked to DOX, showcasing regulated and extended exposure of lung tumors to cytotoxic drugs [31]. Expanding on this notion, Danhier et al. (2012) delved into the possibilities of dendrimer-based nanoplatforms for amalgamated chemotherapy and gene therapy in ALI. Severe exacerbations of COPD may culminate in acute respiratory failure, mirroring the manifestations of acute respiratory distress syndrome/injury. Dexamethasone, a potent anti-inflammatory corticosteroid, is used to alleviate airway inflammation and symptoms in COPD. MicroRNAs, small RNA molecules, are implicated in COPD pathogenesis, regulating inflammatory processes and potentially influencing treatment response, by entrapping dexamethasone and microRNA inhibitors within generation 5 poly(amidoamine) dendrimer-entrapped Au DENPs, the study achieved efficient delivery to lipopolysaccharide-activated alveolar macrophages, leading to downregulation of pro-inflammatory cytokines and enhanced repair of damaged lung tissues [32]. Expanding the scope of dendrimer-based combination therapies, Petros and DeSimone (2010) investigate their utility in combating lung metastases. Through the conjugation of DOX to carboxyl-terminated poly(amidoamine) dendrimers, the study demonstrated a significant decrease in tumor burden and increased survival rates in a lung metastasis mouse model, highlighting the potential of dendrimer-based local lung delivery in improving therapeutic outcomes [33]. Furthermore, Cheng et al. delve into the application of dendrimers for improving airway delivery and reducing lung inflammation in pulmonary disorders. By conjugating MP to PAMAM dendrimers, the study observed enhanced efficacy in decreasing allergen-induced inflammation in a murine model, potentially mitigating the systemic side effects associated with traditional MP administration. Collectively, these studies underscore the versatility and potential of dendrimer-based drug delivery systems for combination therapies in lung diseases, offering targeted delivery, enhanced therapeutic efficacy, and reduced systemic toxicity. The findings pave the way for future research aimed at optimizing dendrimer formulations, elucidating their mechanisms of action, and translating these innovative approaches into clinical practice for improved management of respiratory conditions such as COPD [34].

10.6 CONCLUSION

Utilizing dendrimer-based drug delivery systems in managing COPD offers a multifaceted approach to improving treatment effectiveness while minimizing adverse effects. By encapsulating or linking drugs like corticosteroids, bronchodilators, and antioxidants, dendrimers enhance their solubility, stability, and availability in COPD therapy. An essential advantage is dendrimers' ability to precisely target delivery to specific lung cells or tissues, reducing systemic exposure and related side effects. This targeted delivery enhances treatment efficacy while optimizing drug concentration at the treatment site. Moreover, dendrimers serve as versatile carriers for combination therapies, enabling the simultaneous delivery of multiple agents like anti-inflammatory drugs, bronchodilators, and gene therapy tools. Combining these agents within dendrimer-based formulations can produce synergistic therapeutic effects, leading to improved disease management and patient outcomes. Additionally, dendrimers show promise as theranostic platforms, merging gene therapy with diagnostic imaging to monitor disease progression and sustain gene expression. While preclinical studies highlight dendrimer-based therapy's potential benefits in COPD, further research is necessary to refine formulations, understand the mechanisms, and validate these approaches through robust clinical trials. Overall, dendrimer-based drug delivery systems offer a hopeful avenue for advancing COPD treatment, delivering improved efficacy, safety, and targeted therapy to enhance patient well-being and long-term prognosis.

REFERENCES

1. Vargason, A. M., Anselmo, A. C., & Mitragotri, S. (2021). The evolution of commercial drug delivery technologies. Nature Biomedical Engineering, 5(9), 951–967.
2. Browning, R. J., Reardon, P. J. T., Parhizkar, M., Pedley, R. B., Edirisinghe, M., Knowles, J. C., & Stride, E. (2017). Drug delivery strategies for platinum-based chemotherapy. ACS Nano, 11(9), 8560–8578.
3. Manzari, M. T., Shamay, Y., Kiguchi, H., Rosen, N., Scaltriti, M., & Heller, D. A. (2021). Targeted drug delivery strategies for precision medicines. Nature Reviews. Materials, 6(4), 351–370. doi: 10.1038/s41578-020-00269-6.
4. Gundry, S. (2019). COPD 1: Pathophysiology, diagnosis and prognosis. Nursing Times, 116, 27–30.
5. Agarwal, A. K., Raja, A., & Brown, B. D. Chronic Obstructive Pulmonary Disease. [Updated 2023 Aug 7]. In: StatPearls [Internet]. Treasure Island (FL): StatPearls Publishing; 2024 Jan.
6. Wang, J., Li, B., & Qiu, L., et al. (2022). Dendrimer-based drug delivery systems: History, challenges, and latest developments. Journal of Biological Engineering, 16, 18.
7. Nikzamir, M., Hanifehpour, Y., & Akbarzadeh, A., et al. (2021). Applications of dendrimers in nanomedicine and drug delivery: A review. Journal of Inorganic and Organometallic Polymers and Materials, 31, 2246–2261.
8. Duncan, R., & Izzo, L. (2005). Dendrimer biocompatibility and toxicity. Advanced Drug Delivery Reviews, 57(15), 2215–2237.
9. Patel, H. N., & Patel, P. M. (2013). Dendrimer applications–a review. International Journal of Pharma and Bio Science, 4(2), 454–463.
10. Rao, C., & Tam, J. P. (1994). Synthesis of peptide dendrimer. Journal of the American Chemical Society, 116(15), 6975–6976.
11. Majoros, I. J., Williams, C. R., & Baker, J. R., Jr. (2008). Current dendrimer applications in cancer diagnosis and therapy. Current Topics in Medicinal Chemistry, 8(14), 1165–1179.
12. Tomalia, D. A. (1995). Dendrimer molecules. Scientific American, 272(5), 62–66.
13. Jessamine, V., Mehndiratta, S., De Rubis, G., Paudel, K. R., Shetty, S., Suares, D., ... & Dua, K. (2024). The application of nanoparticles as advanced drug delivery systems in attenuating COPD. Heliyon, 10(3), e25393.
14. Xu, Y., Lv, L., Wang, Q., Yao, Q., Kou, L., & Zhang, H. (2024). Emerging application of nanomedicine-based therapy in acute respiratory distress syndrome. Colloids and Surfaces B: Biointerfaces, 237, 113869.
15. Choi, Myoungjee, Thuy L. T., Lee, Y., Piao, C., Choi, J. S., & Lee, M. (2021). Dual-functional dendrimer micelles with glycyrrhizic acid for anti-inflammatory therapy of acute lung injury. ACS Applied Materials & Interfaces, 13(40), 47313–47326.
16. Cojocaru, F. D., Botezat, D., Gardikiotis, I., Uritu, C. M., Dodi, G., Trandafir, L., ... & Mihai, C. T. (2020). Nanomaterials designed for antiviral drug delivery transport across biological barriers. Pharmaceutics, 12(2), 171.
17. Pedziwiatr-Werbicka, E., Milowska, K., Dzmitruk, V., Ionov, M., Shcharbin, D., & Bryszewska, M. (2019). Dendrimers and hyperbranched structures for biomedical applications. European Polymer Journal, 119, 61–73.
18. Abd-El-Aziz, A. S., Benaaisha, M. R., Abdelghani, A. A., Bissessur, R., Abdel-Rahman, L. H., & Fayez, A. M., & El-ezz, D. (2021). Aspirin-based organoiron dendrimers as promising anti-inflammatory, anticancer, and antimicrobial drugs. Biomolecules 11(11), 1568.
19. Kang, S., & Kim, J. (2016). Design of hyaluronic acid-based nanocarriers for cancer therapy. Progress in Polymer Science, 60, 86–128.
20. Karimi, M., Ghasemi, A., Sahandi Zangabad, P., Rahighi, R., Moosavi Basri, S. M., Mirshekari, H., ... & Hamblin, M. R. (2016). Smart micro/nanoparticles in stimulus-responsive drug/gene delivery systems. Chemical Society Reviews, 45(5), 1457–1501.

21. Cai, L., Qiu, N., Li, X., Jiang, X., Liu, Q., Hou, W., … & Huang, W. (2018). Design of high-efficient drug delivery systems based on ROS-sensitive prodrug/polymer nanomedicine conjugates. Biomaterials, 171, 219–231.
22. Farokhzad, O. C., & Langer, R. (2009). Impact of nanotechnology on drug delivery. ACS Nano, 3(1), 16–20.
23. Malam, Y., Loizidou, M., & Seifalian, A. M. (2009). Liposomes and nanoparticles: Nanosized vehicles for drug delivery in cancer. Trends in Pharmacological Sciences, 30(11), 592–599.
24. Dinarvand, R., Sepehri, N., Manoochehri, S., Rouhani, H., & Atyabi, F. (2011). Polylactide-co-glycolide nanoparticles for controlled delivery of anticancer agents. International Journal of Nanomedicine, 6, 877–895.
25. Torchilin, V. P. (2005). Recent advances with liposomes as pharmaceutical carriers. Nature Reviews Drug Discovery, 4(2), 145–160.
26. Alexis, F., Pridgen, E., & Molnar, L. K. (2008). Factors affecting the clearance and biodistribution of polymeric nanoparticles. Molecular Pharmaceutics, 5(4), 505–515.
27. Parveen, S., & Sahoo, S. K. (2008). Polymeric nanoparticles for cancer therapy. Journal of Drug Targeting, 16(2), 108–123.
28. Sahoo, S. K., & Labhasetwar, V. (2003). Nanotech approaches to drug delivery and imaging. Drug Discovery Today, 8(24), 1112–1120.
29. Peer, D., Karp, J. M., Hong, S., & Farokhzad, O. C. (2007). Nanocarriers as an emerging platform for cancer therapy. Nature Nanotechnology, 2(12), 751–760.
30. Svenson, S., & Tomalia, D. A. (2005). Dendrimers in biomedical applications—Reflections on the field. Advanced Drug Delivery Reviews, 57(15), 2106–2129.
31. Moghimi, S. M., Hunter, A. C., & Murray, J. C. (2001). Long-circulating and target-specific nanoparticles: Theory to practice. Pharmacological Reviews, 53(2), 283–318.
32. Danhier, F., Ansorena, E., & Silva, J. M. (2012). PLGA-based nanoparticles: An overview of biomedical applications. Journal of Controlled Release, 161(2), 505–522.
33. Petros, R. A., & DeSimone, J. M. (2010). Strategies in the design of nanoparticles for therapeutic applications. Nature Reviews Drug Discovery, 9(8), 615–627.
34. Cheng, Z., Al Zaki, A., & Hui, J. Z. (2012). Multifunctional nanoparticles: Cost versus benefit of adding targeting and imaging capabilities. Science, 338(6109), 903–910.
35. Forrester, J. S., & Bick-Forrester, J. (2005). Persistence of inflammatory cytokines cause a spectrum of chronic progressive diseases: Implications for therapy. Medical Hypotheses, 65(2), 227–231. doi: 10.1016/j.mehy.2005.03.010.
36. Yu, C. H., Suh, B., Shin, I., Kim, E. H., Kim, D., Shin, Y. J., Chang, S. Y., Baek, S. H., Kim, H., & Bae, O. N. (2019). Inhibitory effects of a novel chrysin-derivative, CPD 6, on acute and chronic skin inflammation. International Journal of Molecular Science, 20(11), 2607. doi: 10.3390/ijms20112607.
37. Jasper, A. E., McIver, W. J., Sapey, E., & Walton, G. M. (2019). Understanding the role of neutrophils in chronic inflammatory airway disease. F1000Res, 8, F1000 Faculty Rev-557. doi: 10.12688/f1000research.18411.1.
38. Burger, D., & Dayer, J. M. (2002). The role of human T-lymphocyte-monocyte contact in inflammation and tissue destruction. Arthritis Research & Therapy, 4, 1–8.
39. Gea, J., Pascual, S., Casadevall, C., Orozco-Levi, M., & Barreiro, E. (2015). Muscle dysfunction in chronic obstructive pulmonary disease: Update on causes and biological findings. Journal of Thoracic Disease, 7(10), E418.
40. Papadopoulou, E., Safar, S. B., Khalil, A., Hansel, J., Wang, R., Corlateanu, A., … & Mathioudakis, A. G. (2024). Inhaled versus systemic corticosteroids for acute exacerbations of COPD: A systematic review and meta-analysis. European Respiratory Review, 33(171), 230151.
41. Chawla, M., Kaushik, R. D., Malik, M. K., Pundir, V., Singh, J., & Rehmaan, H. (2022). Development and optimization of RofA-PAMAM dendrimer complex materials for sustained drug delivery. Materials Today Communications, *33*, 104881.
42. Klajnert, B., Bryszewska, M., & Palecz, B. (2006). Interactions between dendrimers and biological structures. Biochemical Journal, 399(2), 321–329.
43. Liu, M., & Fréchet, J. M. (1999). Designing dendrimers for drug delivery. Pharmaceutical Science & Technology Today, 2(10), 393–401.
44. Hong, S., Leroueil, P. R., Majoros, I. J., Orr, B. G., Baker, J. R., Jr, & Banaszak Holl, M. M. (2007). The binding avidity of a nanoparticle-based multivalent targeted drug delivery platform. Chemistry & Biology, 14(1), 107–115.

45. Menjoge, A. R., Kannan, R. M., & Tomalia, D. A. (2010). Dendrimer-based drug and imaging conjugates: Design considerations for nanomedical applications. Drug Discovery Today, 15(5–6), 171–185.
46. Fréchet, J. M., & Tomalia, D. A. (2001). Dendrimers and other dendritic polymers. John Wiley & Sons.
47. Ziemba, B., Janaszewska, A., Ciepluch, K., Appelhans, D., Voit, B., & Klajnert-Maculewicz, B. (2015). Influence of dendrimer's structure on its activity against amyloid fibril formation. Biochimica et Biophysica Acta (BBA)-General Subjects, 1850(5), 995–1005.
48. Majoros, I. J., Thomas, T. P., Mehta, C. B., & Baker, J. R., Jr. (2005). Poly (amidoamine) dendrimer-based multifunctional engineered nanodevice for cancer therapy. Journal of Medicinal Chemistry, 48(19), 5892–5899.
49. Tomalia, D. A., & Frechet, J. M. (2002). Discovery of dendrimers and dendritic polymers: A brief historical perspective. Journal of Polymer Science Part A: Polymer Chemistry, 40(16), 2719–2728.
50. Janaszewska, Anna, Lazniewska, J., Trzepiński, P., Marcinkowska, M., & Klajnert-Maculewicz, B. (2019). Cytotoxicity of dendrimers. Biomolecules, **9**(8), 330.
51. Sadekar, S., Ray, A., Janat-Amsbury, M., Peterson, C. M., & Ghandehari, H. (2011). Comparative biodistribution of PAMAM dendrimers and HPMA copolymers in ovarian-tumor-bearing mice. Biomacromolecules, 12(1), 88–96.
52. Wu, P., Malkoch, M., & Hunt, J. N. (2005). Acid-mediated amidation of poly (propylene imine) dendrimers. Tetrahedron Letters, 46(1), 45–48.
53. Kim, Y., Klutz, A. M., & Jacobson, K. A. (2010). Systematic investigation of polyamidoamine dendrimers surface-modified with poly (ethylene glycol) for drug delivery applications: Synthesis, characterization, and evaluation of cytotoxicity. Bioconjugate Chemistry, 21(11), 1890–1900.
54. Shah, S., Solanki, K., & Vyas, S. P. (2010). Preparation and evaluation of polymer–dendrimer nanoparticle for oral delivery of anticancer drug. European Journal of Pharmaceutical Sciences, 39(5), 336–345.
55. Mintzer, M. A., & Grinstaff, M. W. (2011). Biomedical applications of dendrimers: A tutorial. Chemical Society Reviews, 40(1), 173–190.
56. Kannan, R. M., Nance, E., Kannan, S., & Tomalia, D. A. (2014). Emerging concepts in dendrimer-based nanomedicine: From design principles to clinical applications. Journal of Internal Medicine, 276(6), 579–617.
57. Tripathi, P. K., & Khopade, A. J. (2011). N-, O-, S-containing dendrimers and triazine dendrimers: Synthesis and applications as amphiphilic polymers. Polymer Reviews, 51(1), 23–70.
58. Nanjwade, B. K., Bechra, H. M., Derkar, G. K., Manvi, F. V., & Nanjwade, V. K. (2009). Dendrimers: Emerging polymers for drug-delivery systems. European Journal of Pharmaceutical Sciences, 38(3), 185–196.
59. Etemadzade, M., Majidi, S., & Tayefi-Nasrabadi, H. (2009). Synthesis, characterization, and biological evaluation of a new generation of dendrimers based on melamine as a core for the delivery of 5-fluorouracil anticancer drug. Biopolymers, 91(4), 291–301.
60. Kannan, S., & Kolhe, P. (2004). Delivery of neurotrophic factors across the blood-brain barrier using dendrimer nanocarriers. Journal of Drug Targeting, 12(9–10), 651–657.
61. Han, S., Liu, Y., Nie, X., Xu, Q., & Jiao, F. (2007). Oligonucleotides conjugated with short polyethylene glycol-containing CpG motifs: A promising strategy to attenuate systemic inflammatory responses. Bioconjugate Chemistry, 18(3), 663–670.
62. Lee, H., Larson, R. G., & Lawrence, D. A. (2009). Biomolecular coating of platinum nanoparticles for targeted delivery of gemcitabine. Oncology Research, 18(1), 33–38.
63. Lee, C. C., Gillies, E. R., & Fox, M. E. (2006). A single dose of doxorubicin-functionalized bow-tie dendrimer cures mice bearing C-26 colon carcinomas. Proceedings of the National Academy of Sciences, 103(45), 16649–16654.
64. Vaidya, B., Nayak, M. K., Dash, D., Agrawal, G. P., & Vyas, S. P. (2009). Development and characterization of protein-loaded PLGA nanoparticles for sustained delivery of the anti-inflammatory molecule curcumin. Nanomedicine, 5(4), 439–449.
65. Aryal, S., Hu, C. M., & Zhang, L. (2011). Polymeric nanoparticles with precise ratiometric control over drug loading for combination therapy. Molecular Pharmaceutics, 7(4–
66. Alex, S. A., Mukerjee, A., & Agrawal, M. (2009). Preparation and evaluation of nanoparticulate carrier system for hydrochlorothiazide delivery: Effect of formulation variables. AAPS PharmSciTech, 10(1), 169–176.
67. Trivedi, R., & Kompella, U. B. (2010). Nanomicellar formulations for sustained drug delivery: Strategies and underlying principles. Nanomedicine, 5(3), 485–505.

68. Park, K. (2012). Controlled drug delivery systems: Past forward and future back. Journal of Controlled Release, 1(64), 3–20.
69. Gref, R., Lück, M., & Quellec, P. (2000). 'Stealth'corona-core nanoparticles surface modified by polyethylene glycol (PEG): Influences of the corona (PEG chain length and surface density) and of the core composition on phagocytic uptake and plasma protein adsorption. Colloids and Surfaces B: Biointerfaces, 18(3–4), 301–313.
70. Elzoghby, A. O., Samy, W. M., & Elgindy, N. A. (2012). Protein-based nanocarriers as promising drug and gene delivery systems. Journal of Controlled Release, 161(1), 38–49.
71. Ganta, S., Devalapally, H., & Amiji, M. (2008). Curcumin enhances oral bioavailability and anti-tumor therapeutic efficacy of paclitaxel upon administration in nanoemulsion formulation. Journal of Pharmaceutical Sciences, 97(8), 3179–3189.

11 Biodegradable Polymers for Controlled Release of Drugs in Chronic Obstructive Pulmonary Disease

Rabab Fatima and Parteek Prasher
Department of Chemistry, UPES, Energy Acres
Dehradun, Uttarakhand, India

Mousmee Sharma
Department of Chemistry, Uttaranchal University
Dehradun, Uttarakhand, India

11.1 INTRODUCTION

Chronic obstructive pulmonary disease (COPD) emerges as a significant contributor to mortality and disability, yet its exploration from a cellular and molecular standpoint has only recently gained momentum. Within the pathophysiology of COPD, persistent inflammation triggers the irreversible constriction of small airways and the destruction of alveolar walls, a process known as emphysema. This pathological cascade is further characterized by increased neutrophils, alveolar macrophages, and cytotoxic T lymphocytes, accompanied by numerous inflammatory mediators including lipids, cytokines, chemokines, and growth factors. Moreover, oxidative stress is prominently implicated in COPD, potentially exacerbating the inflammatory milieu. Enhanced elastolysis and the likely involvement of matrix metalloproteinases further contribute to tissue remodeling. Importantly, the inflammatory response and proteolytic activity observed in COPD represent an augmentation of the typical response owing to cigarette smoke [1]. Unlike asthma, this inflammatory process appears refractory to corticosteroid therapy, prompting the exploration of novel anti-inflammatory modalities aimed at halting the inexorable progression of the disease.

For the treatment of infectious lung disorders, pulmonary drug administration is a very appealing issue. Since the pulmonary route of drug delivery has great bioavailability and no first-pass effect, it is a valuable way to deliver drugs directly to the deliberated site. Millions of individuals worldwide suffer from pulmonary diseases, prompting continuous advancements in drug delivery systems based on natural polymers. Researchers and physicians are working worldwide in the fields of pharmaceutical technology, new drug development, computational therapies, biomimetics, nanotechnology, and bionanotechnology to try and find an early diagnosis and treatment for these chronic illnesses. It is now understood that a major pathophysiological characteristic of many people with cystic fibrosis, asthma, and COPD is airway mucus hypersecretion. Therefore, it's critical to design drugs that prevent these vulnerable patients' mucus hypersecretion. Conventional treatments, such as corticosteroids, mucolytics, β2-adrenoceptor agonists, anticholinergics, and macrolide antibiotics, are less successful in treating COPD than asthma and have inconsistent efficacy in reducing airway mucus hypersecretion [2]. Pharmacotherapeutic targets undergoing strategic

DOI: 10.1201/9781003498926-11

investigation encompass novel nerve activity inhibitors, such as retinoic acid receptor-α antagonists; tachykinin receptor antagonists; inducers of epoxygenase activity like benzafibrate; inhibitors targeting mucin exocytosis, including anti-myristoylated alanine-rich C kinase substrate (MARCKS) peptides and Munc-18B blockers [3]. However, it's crucial to note that the mucus hypersecretory phenotype observed in asthma differs from that seen in COPD, owing to both factual and theoretical distinctions. Further data on these variations is needed to pinpoint particular therapeutic targets, which should then result in the logical development of anti-hypersecretory drugs to treat COPD and COPD mucus hypersecretion [4]. Natural polymers sourced from plants, animals, and microbes are preferred for their biocompatibility, biodegradability, and lower toxicity compared to synthetic counterparts [5].

Pharmaceutical formulations are designed to deliver therapeutic agents, and their effectiveness is directly influenced by their pharmacokinetic and pharmacodynamic behaviors in the body, depending crucially on plasma-protein interaction, solubility, and Log P index [6]. Polymers execute a crucial role in maintaining or modifying the inherent physicochemical aspects of intended therapeutic agents to achieve the desired response while minimizing toxicity [7]. Polymers, whether of natural or synthetic origin, consist of high molecular-weight macromolecules formed by repetitive monomeric units. Natural polymers, also known as biopolymers, are derived from various biological sources such as plants, bacteria, algae, and animals. Within the realm of active pharmaceutical ingredient delivery, natural polymers have played a pivotal role as they possess inherent biocompatibility and biodegradability properties, are readily available, and are cost-effective. As a result, natural polymers have garnered significant attention from formulation scientists in the development of effective pulmonary drug delivery systems [8]. Their versatility and favorable properties make them attractive candidates for enhancing drug delivery efficacy while minimizing adverse effects. Key natural polymers utilized in developing advanced pulmonary drug delivery systems for treating pulmonary diseases such as COPD such as dextran, chitosan, alginate, hyaluronic acid, poly(lactic-co-glycolic acid) (PLGA), natural gums, starch, cellulose and poly lactic acid (PLA) are widely employed in the form of nanoparticles, micelles, liposomes, and solid lipid nanoparticles (SLNs) along with administration of drugs into the pulmonary system because of their favorable characteristics such as high hydrophobicity, low toxicity, additional mucociliary clearance, mucoadhesivity, and biological activity [9]. These attributes enable precise drug targeting to the designated site. This section explores contemporary advancements in naturally derived polymer-based nanomedicines (NMCs), encompassing their transit and mode of operation within the respiratory and bronchiolar domains, alongside their diverse physicochemical attributes, comprising surface charge, particle dimensions, surface alterations, and *in vitro* safety profiles [10]. This chapter also offers a summary of the various natural biodegradable polymers and covered more specifics on how they are utilized in controlled delivery of drugs used in management of COPD (Figure 11.1).

11.1.1 Mechanisms of Drug Release from Biodegradable Polymers

The primary objective of regulating drug release kinetics is to maintain a consistent concentration of the drug in the bloodstream within the therapeutic range, spanning from the minimum toxic concentration (MTC) to the minimum effective concentration (MEC) [11]. Zero-order release kinetics refers to the continuous release of cargo drug per unit time from the intended carriers that occurs within this range and is both well-tolerated and effective. The properties and ratios of the drug, excipients, and polymers as well as additives, the interactions between the various components physically and/or chemically, and even the formulation techniques all have an impact on the drug release profile [12]. They can be categorized into solvent-controlled drug release, degradation-controlled drug release, and diffusion-controlled drug release [13].

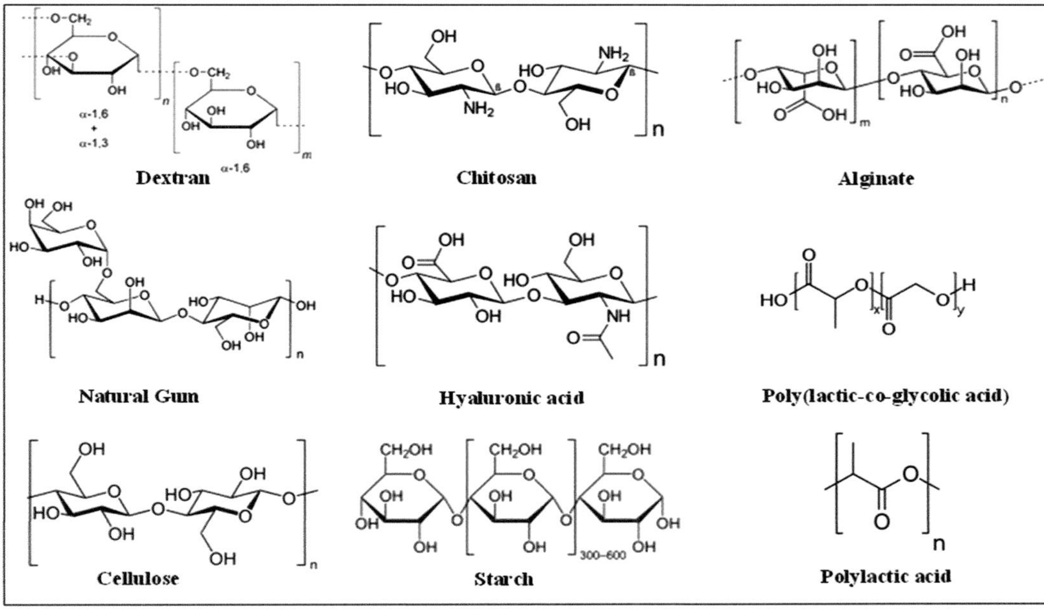

FIGURE 11.1 Chemical structure of biodegradable polymers in contemporary pulmonary drug delivery against chronic obstructive pulmonary disease.

In a standard drug release mechanism, the drug molecule migrates from the interior of the polymer matrix to the exterior of the particle, and ultimately disperses into the surrounding environment [14]. During the initial phase, the active pharmaceutical ingredient diffuses out of the particle through water-filled pores, mostly propelled by the osmotic pressure or chemical potential gradient [15]. But once polymer breakdown starts, erosion of the polymer might happen on the surface of the matrix (surface erosion) or inside the matrix (bulk erosion) when water seeps into the polymer bulk [16]. While certain cases of drug release from biodegradable polymers are linked to diffusion-controlled release only [17] the majority of examples involve the simultaneous combination of diffusion and degradation/erosion [18].

Drug release from biodegradable polymers often exhibits bi-phasic and tri-phasic characteristics [19], yet achieving zero-order from polymeric particles is preferable to attain the desired release profile of drug [20]. A first burst release can be observed in the tri-phasic release phase, wherein the medication that is not encapsulated or that is encapsulated close to the polymer surface is rapidly released through hydration [21]. Driven by erosion and polymer breakdown, the second and third phases may proceed slowly or quickly [22]. Remarkably, the first burst release phase is not always required, and the release velocities of the subsequent phases might alternate under certain conditions [23]. Biodegradable polymers on the other hand offer a versatile platform for drug delivery due to their appreciable surface area-to-volume ratio and tunable properties. Several biodegradable polymers have been extensively investigated herein for nanoparticle-based drug delivery systems targeting pulmonary disorders such as COPD (Figure 11.2).

11.2 DEXTRAN

The hydrophilic polysaccharide dextran is mostly made up of α-1,6 linked glucopyranoside residues for the parent chain and α-1,2/3/4 linked residues for the side chain. Dextran is an invaluable option for the development of excipients and drug delivery vehicles because of its high biocompatibility, stability in slightly acidic and basic environments, water solubility, lack of immunogenicity,

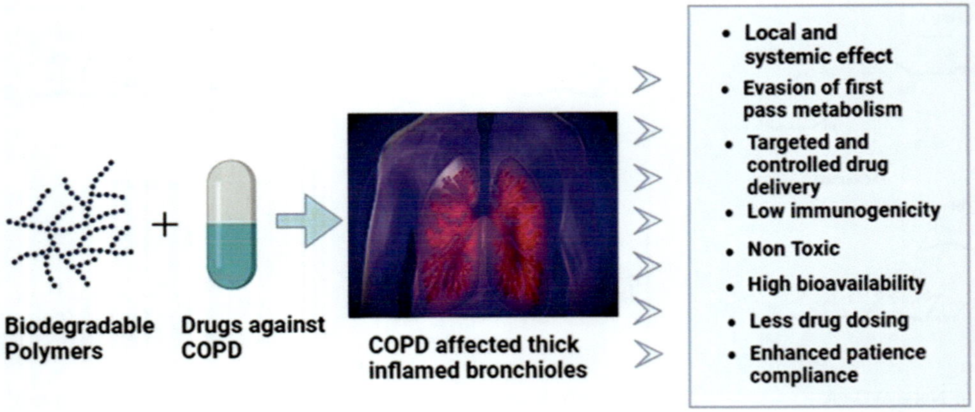

FIGURE 11.2 Potential benefits of biodegradable polymers in management of chronic obstructive pulmonary disease-affected lungs.

and chemically adjustable –OH groups. Improved chain mobility due to α-1,6 links in the parent chain determines dextran's aqueous solubility, and its biocompatibility is confirmed by the digestive enzymes' conversion into physiologically innocuous breakdown products [24]. In COPD, excessive mucus production and its viscosity exacerbate airway constriction, impeding breathing. Dextran, with its mucolytic properties, has the potential to ameliorate airway clearance and mitigate symptoms such as coughing and excessive sputum production. Furthermore, dextran's anti-inflammatory attributes offer promise in attenuating symptoms and retarding disease progression by curbing the inflammatory response, which is pivotal in the pathogenesis of COPD. The immunomodulatory potential of dextran is being explored in COPD management, as dysregulated immune responses contribute to lung inflammation and tissue damage in this condition. Leveraging nebulization as a delivery method facilitates targeted administration directly to the lungs, thereby enhancing dextran's therapeutic efficacy in COPD.

As drug delivery vehicles, excipients, and inhalable aerosolized nanoformulations, dextran-based nanomaterials have significant potential for the effective modulation of inflammatory lung disorders such as COPD. It is possible to construct controlled release and stimuli-responsive release systems since natural dextran is enzymatically degradable [25]. Yet, native dextran's characteristics can be adjusted through chemical alteration [26]. The use of nanocomposite microparticle (nCmP) systems in therapeutics for pulmonary delivery in COPD shows great promise. In this context, a study aimed to determine the 22 best spray drying condition(s) to manufacture nanocmP with desired drug delivery characteristics was conducted. Curcumin served as the model drug with mannitol as excipient in formulation of 26 nCmP, and acetalated dextran (Ac-Dex) was utilized to create nanoparticles. The Design-Expert software was utilized to apply Box-Behnken design for nCmP 27 parameter optimization. Two important factors impacting the aerodynamics of nCmP systems are the feed concentration (Fc) and the nanoparticle ratio (NP%). nCmP systems with greater than 431 NP% had higher drug loading, higher Fine particle fraction (FPF), higher Respirable fraction (RF), and lower Mass median aerodyanamic diameter(MMAD). In the 434 design range, tin has no effect on the nCmP properties however, in higher ranges, tin might prevent nanoparticles from being redispersed into the nCmP [27]. Similarly, budesonide-loaded nanoparticles having Microparticulate system (MPS), Polydispersity index (PDI), and Zeta Potential (ZP) of 160.8 ± 0.27 nm, 0.36 ± 0.04, and 13 ± 0.894 mV were studied in a COPD *in vitro* model. All of the batches' drug loading percentages were found to be between 10% and 16%. Using High Performance Liquid Chromatography (HPLC), the amount of medication released in the target location (alveoli) was determined to be 18.26%.

It was discovered that the best *in vitro* aerodynamic performance was exhibited by nanodispersion. The MPS, PDI, ZP, and % Cumulative Drug Release (CDR) did not significantly alter after three months of storage, according to the stability study results. Altogether, the cross-linking of dextran sulfate nanoparticles with chitosan exhibited characteristics that were appropriate for nebulized dispersion of drug, drug release, and mitigating the first-pass effect [28].

11.3 CHITOSAN

Chitosan, a natural polysaccharide derived from chitin, exhibits excellent biocompatibility, biodegradability, and mucoadhesive properties. Chitosan nanoparticles can efficiently encapsulate drugs and facilitate sustained release, making them suitable for targeted delivery to the lungs. Chitosan possesses positively charged amino groups that interact with negatively charged mucin molecules present in the mucus layer lining the respiratory tract. This mucoadhesive property allows chitosan-based formulations to adhere to the mucosal surfaces of the lungs, prolonging residence time and facilitating sustained drug release. Enhanced mucoadhesion promotes better drug absorption and penetration into lung tissues, improving therapeutic efficacy. In rats with COPD, the isoquinoline alkaloid berberine (Ber) has demonstrated therapeutic potential. However, Ber's poor absorption and aqueous solubility limit how effective it can be as a medication. To create an effective nanocarrier platform for the oral delivery of Ber, chitosan and SLNs were combined. The developed Ber-encapsulated nanoparticles were then assessed for their potential to reduce airway inflammation in rats with COPD that had been exposed to cigarette smoke. *In vitro* tests using simulated digestive fluids of variable pH revealed that ber-encapsulated SLE-chitosan had a regulated slow-release profile and good stability at nanoscale particle sizes. When compared to pure Ber, intake of Ber-loaded SLE-chitosan nanoparticles could dramatically reduce levels of inflammatory cytokines and inflammatory cells (macrophages and neutrophils) in Bronchoalveolar lavage fluid (BALF). Chitosan decorated SLN nanoparticles provide an outstanding oral delivery system for controlled release of berberine, enhancing its aqueous solubility and oral bioavailability to further support its pharmacological actions [29].

In terms of pharmaceutical therapy, a class of drugs known as inhaled corticosteroids (ICS) is thought to lessen the long-term inflammatory effects in the bronchial region, improving restricted airflow in COPD [30]. Synthetic trifluorinated glucocorticoids, such as fluticasone propionate, are used as topical eosinophil inhibitors to prevent the release of pro-inflammatory substances in lung epithelium and to treat and manage a variety of inflammatory conditions over the long term [31]. Controlled drug delivery and chemical modification of chitosan is a useful method for creating novel matrices with more functional groups and enhanced characteristics, like increased hydrophilicity and swelling rate. The alkaloid theophylline (1,3-dimethylxanthine) (THP), which is present in *Camellia sinensis* leaves, is frequently employed in clinical practice to treat chronic obstructive lung disorders by acting as a bronchodilator. Because THP has a short duration of action (six hours), it is necessary to provide the typical dose forms three to four times a day to prevent significant changes in plasma concentration. THP has therapeutic benefits at plasma concentrations between 5 μg/mL and 10 μg/mL, whereas hazardous effects often occur at dosages greater than 20 μg/mL [32]. Sustained release dosage formulations for THP reduce patient therapeutic problems by maintaining desired serum concentrations for extended periods of time without frequent dosing as THP dissolves in both alkaline and acidic media [33].

Chitosan and chitosan/montmorillonite crosslinked with glutaraldehyde are the basis of polymeric materials that were developed as novel THP release carriers. White Swiss mice were given intraperitoneal injections of polymeric matrices as suspensions. The life rate of the mice was used to test the samples' biocompatibility and toxicity; human blood hemolysis was used to assess the samples' biocompatibility, and the chemiluminescence assay was used to test the samples' cell viability. THP's sustained release was demonstrated by the *in vivo* release profiles and the drug's presence was detected for up to 50 hours after injection. For the matrix containing montmorillonite, an increase in the half-release time (t1/2) from 12 hours to 14 hours was observed suggesting that the patient may benefit from a lower dosage frequency but with an extended action period [34]. Chitosan possesses immunomodulatory properties that can influence the local immune response in the lungs. It has been reported to stimulate mucosal immunity, enhance antigen presentation, and modulate inflammatory processes, making it potentially useful for the treatment of respiratory infections, COPD, and other inflammatory lung conditions. Additionally, chitosan-based formulations may also promote wound healing and tissue regeneration in damaged pulmonary tissues, further supporting their therapeutic utility in pulmonary drug delivery.

11.4 ALGINATE

In terms of efficacy, safety, pharmacokinetic/pharmacodynamic properties, and physicochemical properties, alginates are superior to conventional polysaccharides. They find extensive use in NMCs, including dendrimers, nanocrystals, emulsions, liposomes, SLNs, micelles, and polymeric nanoparticles. In the past ten years, there has been a lot of interest in drug delivery strategies based on biodegradable polymers for a variety of life-threatening conditions, including COPD, owing to newer methods and developments [35].

A strong inhibitor of phosphodiesterases-4 (PDE4), which is the primary isoenzyme present in COPD-affected cells, is roflumilast. Therefore, in addition to its anti-inflammatory properties, roflumilast is given to reduce the frequency of COPD symptoms [30]. The 500 μg roflumilast pills have been widely utilized for the treatment of COPD as the largest amount taken orally in one dose 1000 μg. Divalent cations like calcium form a suitable material for immobilized matrix formation since it can easily transition from liquid to gel form in the presence of cations [36]. The calcium ion source, calcium beta-glycerophosphate, was dissolved in the oil phase and subsequently distributed throughout the alginate aqueous solution to create an oil/water (o/w) emulsion. This was then dried using a nano-spray dryer to create cross-linked emulsified spray dried alginate particles. In comparison to microparticles made without a cross-linking agent, the cross-linked spray-dried alginate particles had greater values of both particle size and span. The formulation with the biggest particle size and the most regulated drug released pattern was from the alginate drug formulation, which was made with β-cyclodextrin as a carrier [37]. A drug delivery system using microspheres of rifampicin-PLGA and sodium alginate with a 1.748 μm diameter yielded a 91.83% cumulative drug release, which was slower than the from the alginate-free microspheres (97.36% release). The drug containing polymer gel stuck to the alveoli of lungs in less than a day, and the microspheres remained there for ~21 days, causing a delayed and sustained release. Following five hours of dosing, the drug's highest concentration in the lungs was observed at 48.60 ± 15.67 g ml^{-1} [38].

Moreover, it has been witnessed that the alginate microparticles loaded with Paclitaxel helped achieve the drug's non-invasive pulmonary administration method in COPD. Paclitaxel-loaded alginate microparticles were delivered endotracheally, which successfully concentrated the drug's cargo in the lung tissues and extended the drug's residence time relative to free-paclitaxel. The local drug activity in the lungs was further prolonged when mucoadhesive polymers like chitosan and alginate are used to form solid lipid microparticles (SLMs). To acquire mucoadhesive properties, anionic alginate creates hydrogen bonds with the mucosal layer, while the amino $-NH_2$ group of chitosan engaged in electrostatic interactions with the anionic groups of mucins. The cellulose membrane, saturated with mucin, served as a surrogate pulmonary epithelial surface to evaluate the bioadhesion of microparticles [39].

Solid lipid Paclitaxel-loaded alginate microparticles observed a protracted delivery of the corticosteroid fluticasone propionate directly into bronchial system, as demonstrated by a mere 60% drug release after two days at normal pH. Less than 10% of the drug was released throughout the course of the next few hours after the SLMs showed an initial 40% burst drug release [40]. Additionally, the loading of beta (β)-agonists on alginate mucoadhesive systems, like salmeterol xinafoate, demonstrated how well these systems work to increase the cargo drug's efficacy and retention duration in patients with secondary COPD inflamed bronchi. The microparticles size of 3.5–4 µm seemed ideal for pulmonary delivery, preventing any negative effects related to the drug's systemic absorption in alveolar capillariesMucoadhesive solid lipid microparticles (SLMs) incorporating the long-acting β2-agonist Salmeterol Xinafoate (SX) were developed, characterized for size, zeta potential, aerodynamic diameter, turbidity, drug loading, and encapsulation efficiency, and finally evaluated using a bronchial epithelial cell model. The results signify that encapsulating SX within the SLMs produced inhalable particles that were more effective than the free drug in enhancing cAMP expression in bronchial epithelial cells in vitro. The aerodynamic properties and mucoadhesive characteristics of the SLMs further suggest their potential to extend SX retention time in lung epithelium and enhance its therapeutic efficacy, making them an ideal and promising approach for treating COPD [41]. These studies signify the possibility of a variety of mucolytic cargo drugs and gene or nucleic acid transfer techniques that can be used with different drug delivery methods to treat respiratory conditions like cystic fibrosis and COPD. To confirm the benefit of these cutting-edge drug delivery systems like nanoparticles over conventional drug delivery strategies, more clinical trials as well as animal studies are required.

11.5 NATURAL GUMS

Xanthan gum is a naturally occurring microbial polysaccharide that possesses all the previously stated properties. It can be broken down into smaller molecules or eliminated by the kidneys directly. With varied degrees of efficacy and unique rheological characteristics, this natural excipient has been studied as a matrix for solid dosage forms (tablets, capsules), hydrogels, nanoparticles, microparticles, and buccal/transdermal patches [42].

Utilizing natural gums for the manufacture of sustained-release (SR) oral dosage forms in contrast to cellulosic polymers was assessed as a means of addressing persistent pulmonary issues associated with COPD and promoting patient compliance while reducing the frequency of doses. Through the use of cellulosic polymers, such as hydroxypropyl methylcellulose (H) or carboxymethyl cellulose (C), and natural gums, such as guar (G) or xanthan (X), SR tablets containing 7.5 mg of TBS were created. Various ratios, such as C:H, G:X, X:H, H:G, or 20:80, 40:60, and 60:40, were employed. Release rates were compared with conventional tablets such as Bricanyl® SR in phosphate basic buffer solution of pH 7.4 following the assessment of the tablets' physical attributes. Every tablet fulfilled the required physical requirements. Lower dissolution efficiency (DE8%) and longer mean dissolution time (MDT) were observed in tablets containing an 80:20 polymer ratio of H8X2 and G8X2. The formulations with the highest MDT and the smallest DE8% were H4X6 and H6X4, in the ratios of 60:40 and 40:60. H6X4 followed a Higuchi pattern, similar to a conventional tablet, whereas formulation H4X6 released the drug with zero-order kinetics. We conclude that the formulation most similar to the standard was G8X2, which contained guar and xanthan following a zero-order drug release kinetic [43].

Natural gums were used to make THP-controlled release matrix tablets, which were then assessed for inflammation associated with COPD *in vitro* models where a drug to polymer ratio of 1:0.5 regulated release was achieved by combining guar gum and xanthan gum. The manufactured pills' reliability was well within tolerable bounds. Between average and individual weight, there was no discernible difference in weight. After a 24-hour period, 97.28% of the medication was released. The Fourier transform infrared spectroscopy (FTIR) was also used for the compatibility study. The drug alone and drug with excipient FTIR graphs were obtained. Since no other peaks were discovered, it was determined that the excipients were drug-compatible [44]. Using an SR salbutamol sulphate tablet formulation made by direct compression, the prospect of cross-linked *Acacia seyal* gum as a matrix forming was assessed. A fixed amount of salbutamol sulphate (8 mg) was used in the formulation of the tablets, and varying percentages of 30%, 40%, and 50% were used for the matrix material. This resulted in formulations T1, T2, and T3 that had EG, T4, T5, and T6 that contained (chitosan gum) CG, and T7, T8, and T9 that contained hydroxypropyl methylcellulose (HPMC). Weight, diameter, tensile strength, thickness, disintegration time, friability, crushing strength, and drug content of the tablets were assessed; these parameters fell within acceptable limits for SR tablets weighing between 131 mg and 325 mg. The time for 50% drug release (T50%) was shown to be longer for the CG matrices based on the cumulative percent drug released calculation. The release of 76.78% was seen at the eighth hour with Formulation T4 (30% CG percentage), which was the closest to the theoretical release of 66.66%. T4 was determined to be the best formulation in this investigation to maintain the release of salbutamol sulphate for about 12 hours based on this and the outcomes of the statistical analysis of variance ($p < 0.05$). Therefore, it is reasonable to say that cross-linked *A. seyal* gum can be employed as a substitute matrix material to maintain the release of drugs that are soluble in water [45].

The hydrogel of guargum with appreciable mucoadhesiveness connected multiple micelles of chitosan-g-polycaprolactone, which function as a vehicle for the inhalation of drugs that are insoluble in water, like rifampicin. The carrier system demonstrated an impressive 98% encapsulation effectiveness and produced a drug release that was pH-responsive at the slightly acidic pH 5.5, which is where micelle destruction occurs. Just 10% of the drug that is encapsulated leaves the micelles in 12 days at the physiological pH of 7.4, demonstrating the stability and longevity of micelles in physiological settings. The *in vitro* investigation showed that micelles were retained in THP-1 cells even after 24 hours, indicating that the micelles had a controlled and targeted release characteristic in COPD condition [46].

11.6 HYALURONIC ACID

A linear mucopolysaccharide, hyaluronic acid is made up of numerous units of N-acetyl-d-glucosamine and d-glucuronic acid joined by β-1–4 and β-1–3 glycosidic bonds. Hyaluronic acid is a highly desired drug delivery vehicle because of its inherent biocompatibility, viscoelasticity, biodegradability, and the capacity to specifically recognize overexpressed tumor receptors [47]. Although hyaluronic acid after thiolation increases its mucoadhesive function and delays the biodegradation to produce a controlled drug release, hyaluronic acid is primarily modified chemically at the -COOH and -OH groups [48]. The synthesis of hyaluronic acid to produce thiolated byproducts that have a ~6.5-fold longer mucosal layer adhesion period than unmodified hyaluronic acid regulated the rate of thiolated hyaluronic acid's biodegradation by hyaluronidase, which was slowed down by the *in situ* disulfide bonds resulting in a prolonged drug release lasting more than 12 hours [49].

Dry powder inhalers and pressurized metered-dose are two methods of delivering salbutamol that guarantee a focused lung administration, elevated local drug concentration, and slower systemic exposure in COPD [50]. On the other hand, salbutamol delivery by metered-dose inhaler results in a rapid absorption into the systemic circulation within 10–15 minutes [51], Salbutamol's high water solubility and membrane permeability reduce both the drug's duration of action and its local concentration at the intended site. Moreover, the drug delivered encounters airway clearance mechanisms that shorten the duration of drug retention at the target site, including coughing, mucociliary clearance, and alveolar macrophage clearance [52]. These variables make it necessary to reduce the dosing frequency and boost patient compliance. Combining sodium ascorbyl phosphate and urea-crosslinked hyaluronic acid is a promising treatment option for inflammatory lung conditions like asthma, emphysema, and COPD [53]. The sample's ideal viscosity, pH, and osmolality for the airways were demonstrated, which supports the sample's use in pulmonary delivery. Significantly, the formulation of urea-crosslinked sodium ascorbyl phosphate and hyaluronic acid inhalable dry powder showed excellent efficacy as adjuvant therapy for respiratory disorders [54]. Hyaluronic acid's significance in creating bioadhesive drug delivery systems is further validated in rapid surface adsorption of fluticasone propionate ensuring a mucoadhesive as well as sustained pulmonary drug delivery [55].

Commercially available medicines using hyaluronic acid as their active ingredient, such as Hyaneb® and Yabro®, effectively cure hyper-reactivity of bronchi in reaction to several irritants and decrease mucus viscosity in this disease, respectively. In comparison to simple spray-dried salbutamol powder, loading microparticles of hyaluronic on surface of salbutamol improved the drug's mucoadhesive activity, reduced the systemic exposure, and extended the lung retention duration from two hours to eight hours. Significantly, the maximum salbutamol concentration dropped significantly. Hyaluronic acid microparticles release salbutamol in a 25% burst, then the drug is continuously released over the course of the next 20 hours until it is completely depleted. Interestingly, in just ten minutes, the residual salbutamol content dropped sharply from the mucosa to 40%. However, a significantly smaller 12% loss over the course of 60 minutes of flushing indicated that salbutamol loading on hyaluronic acid effectively sustained its mucosal retention [56].

11.7 POLY(LACTIC-CO-GLYCOLIC ACID) OR PLGA

PLGA holds significant promise for pulmonary drug delivery due to its unique properties and versatility. PLGA is a biodegradable copolymer derived from lactic acid and glycolic acid, both of which are metabolizable byproducts. The role of PLGA in pulmonary drug delivery can be explained scientifically through its key characteristics and mechanisms. PLGA undergoes hydrolytic degradation in the body, breaking down into lactic acid and glycolic acid. These metabolites are easily eliminated through the citric acid cycle, minimizing systemic toxicity and enhancing biocompatibility. In pulmonary drug delivery, this biodegradability ensures that the polymer matrix gradually dissolves, releasing the encapsulated drug payload over time. PLGA nanoparticles can be precisely engineered to encapsulate drugs and modulate their release kinetics. By adjusting factors such as polymer composition, molecular weight, and particle size, controlled and sustained drug release profiles tailored to the specific requirements of pulmonary therapeutics can be achieved. This controlled release mechanism enhances drug bioavailability and reduces dosing frequency, improving patient compliance and therapeutic outcomes. PLGA nanoparticles can be surface-modified or functionalized with targeting ligands to enhance their specificity for pulmonary tissues or cells. Functionalization enables active targeting of diseased sites or specific cell receptors, minimizing off-target effects and maximizing therapeutic efficacy. In pulmonary drug delivery, targeted PLGA nanoparticles can efficiently navigate the complex lung microenvironment and penetrate deep into lung tissues, enhancing drug deposition at the desired site of action. PLGA nanoparticles provide a protective encapsulation environment for drugs, shielding them from degradation or premature clearance in the body. This protection ensures the stability and integrity of the encapsulated drug during transit to the lungs, preserving its pharmacological activity and enhancing therapeutic efficacy. PLGA is inherently biocompatible and widely used in biomedical applications, making it suitable for pulmonary drug delivery without eliciting significant immune responses or adverse reactions. The biodegradable nature of PLGA further contributes to its safety profile, as it minimizes the accumulation of non-biodegradable polymer residues in the body.

The two main mechanisms of PLGA disintegration are autocatalytic and hydrolytic ester bond breakdown. PLGA hydrolyzes and breaks down in the presence of water, which penetrates the polymer matrix's amorphous areas by rupturing hydrogen bonds and van der Waals forces. The polymers then break down more quickly into water-soluble pieces due to reduced polymeric molecular weight, which increases their hydrophilicity. After being hydrolyzed into Glycolic acid (GA) and lactic acid (LA), the resulting fragments are subsequently converted into energy, CO_2, and water by regular metabolic pathways. An autocatalytic degradation process is possible for PLGA. The organism may degrade more quickly if the acidic breakdown products it produces during that process, create a somewhat acidic local microenvironment for the organism [57].

Currently, antioxidant, anti-inflammatory, and corticosteroid drugs are the mainstays of COPD treatment; however, these medicines come with substantial side effects and increased dosage requirements [58]. In most inflammatory conditions, the extracellular protein small molecule heat shock protein α,β-crystallin (HSPB5) possesses an anti-inflammatory effect [59]. Even yet, the body naturally produces anti-HSPB5 antibodies to prevent it from attaching to the receptor. Noort et al., additionally discovered that encases HSPB5 in porous PLGA microparticles preventing neutralization of the protein while also encouraging macrophage phagocytosis of HSPB. PLGA microparticles with HSPB5 administered via the pulmonary route where alveolar macrophages in COPD mice preferentially absorbed them, thereby greatly limiting lung invasion by inflammatory cells as opposed to ineffective free HSPB5 dose [60].

Marcianes et al., treated COPD by treating macrophages with gatifloxacin and surface-modifying PLGA microparticles using labrafil. They discovered that surface-modified microparticles can be phagocytosed quickly in the investigation of macrophage phagocytic behavior, and that following a 48-hour co-incubation period, macrophages survived and the microparticles were neatly visible inside the alveolar architecture [61]. These microparticles boost the drug's therapeutic window and mitigate the drawback of overly rapid drug release by improving their targeting from macrophages. While additional research is necessary to fully understand the pharmacokinetic characteristics of PLGA microparticles in the respiratory system, these particles show great potential as an inhaled delivery strategy for the treatment of respiratory disorders. For the controlled-release drug administration of pharmaceuticals against COPD, peptides, proteins, and other compounds such as poly lactic acid (PLA) has been recognized as a suitable dispersion disintegrant due to its exceptional mechanical characteristics, biodegradability, and biocompatibility. Moreover, PLGA's properties are greatly influenced by its molecular weight, characteristic viscosity, and monomer makeup. When PLGA is utilized as a drug-delivery matrix, these variables also affect the drug-loading capacity and drug-release properties. It is feasible to enhance biodistribution in the intended tissues and organs and restrict access into healthy cells by selectively modifying the polymeric microparticle/nanoparticle surface. A more effective treatment for many difficult-to-treat diseases such as COPD is also possible with the use of the PLGA matrix as a carrier of the contrast agent, which can greatly improve the accuracy of major disease diagnosis.

11.8 CELLULOSE

Cellulose plays a significant role in the controlled delivery of drugs for COPD and can be utilized as a carrier or matrix material in various drug delivery systems such as nanoparticles, microparticles, or films. These cellulose-based delivery systems can be engineered to release drugs in a controlled manner, ensuring sustained therapeutic levels in the lungs over an extended period. Additionally, cellulose derivatives can be modified to enhance mucoadhesion, promoting prolonged contact with the lung epithelium and improving drug retention and efficacy. THP and roxithromycin (ROX) in this regard, demonstrated synergistic relationship to treat COPD. The preparation of concurrent drug-loaded microcapsules (CDM) of ROX and TH involved spray drying, using sodium alginate (Alg) and amphiphilic methylcellulose (MC) as the wall components. With the most optimal preparation circumstances and wall material components, MC/Alg/MC exhibited the highest encapsulation efficiency. The drug loading of concurrent drug microcapsules was higher than that of single drug microcapsules when the wall material was MC/Alg/MC. CDM demonstrated clear pH-sensitive controlled-release characteristics. Following a 12-hour period of release, the total release of TH and ROX at pH 1.5 was 22.31% and 21.07%, and at pH 7.4, 73.14% and 97.52%, respectively. The fact that CDM accomplished controlled-release was further demonstrated by the antibacterial activity of both CDM and pure drugs. Neutrophil cytotoxicity tests verified that CDM exhibited lower cellular toxicity compared to unmodified drugs [62].

Acebrophylline is a bronchodilator that is widely used to treat asthma and COPD, as it has mucosecretolytic and anti-inflammatory properties. To reduce the dosing frequency and increase patient turnover, a recent study optimized and characterized the once-daily sustained release formulation of acebrophylline. Matrix tablets of acebrophylline were manufactured through the wet granulation method, employing hydroxypropyl methylcellulose (HPMC K4M & HPMC K15M) and a hydrophobic polymer like ethyl cellulose (EC). Utilizing a USP (United states pharmacopeia) paddle device, an *in vitro* release investigation was carried out. The formulation ASRF10, which had a sustained release effect of up to 12 hours and an initial release of 16%, was referred to as an optimized formulation. It comprised 20% HPMC K15M and 5% EC. Further *in vitro* release data of the ASRF10 was assessed using various kinetic models to determine the drug release mechanism. Since the sustained release effect of the acebrophylline enhances its bioavailability and may be therapeutically beneficial for long-term asthma, the acebrophylline slow and sustained release solid dispersion has been shown to have promising potential as an alternative to the traditional dosage form. Nevertheless, additional clinical research is required to evaluate the effectiveness of this technique for asthmatic patients [63].

For the treatment of COPD, matrix tablets of a salbutamol sulphate sustained release dosage form were created by wet granulation process employing varied concentration of EC and hydroxypropylmethyl cellulose (HPMS K100M), which were also subsequently tested for thickness, hardness, percentage drug content, friability, weight fluctuation, and swelling index. The tablets underwent *in vitro* drug release for the first two hours in acidic pH and the final ten hours in phosphate buffer (pH 6.8), as evaluated by kinetic models where the Higuchi equation suggested that erosion and diffusion might be the mechanisms underlying drug release. Specifically, formulation F2, which contains HPMC K100M, was shown to have the best release profile and to be able to maintain drug release for an extended amount of time [64].

In terms of developing extended bronchodilation for the management of excruciating bronchospasm associated with COPD, THP and salbutamol sulphate demonstrated a healthy synergistic effect. THP and salbutamol sulphate displayed relative short half-lives, which enhanced the drug frequency as an immediate-release dose form but renders them as excellent choices for prolonged release tablet formulation [65]. By using the direct compression technique, prolonged release matrix tablets containing THP and salbutamol sulphate were developed. The formulations were coded as F1–F7, based on the ratios of modified polymers. Controlled drug release from the core tablets occurs upon contact with an aqueous medium, facilitated by varying ratios of hydrophilic polymers. Calibration samples of pure drugs were prepared using different solvents, including 0.1N HCl, phosphate buffer (pH 6.8), and distilled water, for conducting *in vitro* dissolution studies on simulated tablets. Formulations named F1–F7, comprising different concentrations of EC and HPMC K100M, demonstrate extended drug release for up to eight hours. Among these formulations, F6 exhibits ~95% release of salbutamol and 94% for THP at the end of 8 hours. This outcome underscores the potential of extended-release tablets containing a specific ratio of EC and HPMC K100M as rate controlling polymers for effectively treating COPD, highlighting that a specific range of concentrations of the cellulosic formulation can deliver prolonged drug release [66].

11.9 STARCH

Starch is a naturally occurring polysaccharide that is present in plant seeds and tubers. Natural starch has undergone several chemical changes to improve its physicochemical properties, making it suitable for use as a novel drug delivery excipient. A novel inhaled anticholinergic drug called tiotropium, which is utilized in Spiriva, has been approved for the once-daily management of bronchospasm related to COPD. The drug provided appreciable 24-hour bronchodilator therapy after a once-daily dose by binding to all muscarinic receptors within human bronchi with demonstrated toxicity [67]. In this regard, a novel tiotropium inhalant formulation was created using a combination of biodegradable materials including starch manifesting much longer-acting therapeutic activity for longer time period following an initial single dose. An o/w emulsion preparation or basic stirring were used to prepare the formulations. Tiotropium deposition and inhalation efficacy studies were primarily done *in vivo* to evaluate therapeutic potential of these formulations. The lung epithelium of BALB/c mice exhibited 90% inhalation efficiency for all formulations following initial dosing.

Remarkably, the inhalation formulation comprising tiotropium with PLGA demonstrated extended drug residency in contrast to other formulations. By preserving the drug's therapeutic impact for a longer amount of time after the initial inhalation, this microparticle technology is therefore anticipated to increase the effectiveness of treatment for COPD patients [68].

To tackle the severity of COPD, different formulations of THP microbeads were created by combining pregelatinized breadfruit starch with chitosan and alginate at different polymer:drug ratios. For optimal pharmacological therapy, microbead formulations offering regulated THP distribution offer superior options compared to traditional dose forms. The ordered granule distribution of this form of starch was disturbed during gelatinization, according to FTIR and XRD (X Ray Diffraction) spectra. The pregelatinized starch's viscosity and fluidity were improved. The size range of THP microbeads was 1.09 ± 0.672 to 1.58 ± 0.54 mm, with a nearly spherical form. With a decrease in the polymer:drug ratio, microsphere size and swelling rose and entrapment and dissolution time (t50) decreased. The efficiency of trapping varied between 30.99 ± 1.32 and $78.50 \pm 2.37\%$. The starch:alginate ratio of 3:1 at a polymer:drug ratio of 2:1 resulted in an extended dissolving period ($t50 = 8.40 \pm 1.20$ h) with an optimized formulation. For the controlled release of THP in microbeads, breadfruit starch was thus an appropriate copolymer that might replace synthetic polymers in the delivery of drugs in COPD [68].

THP and other xanthine derivatives are known to cause cardiovascular side effects because of their lower affinities for adenosine A1 and A2 receptors. Doxofylline, a new generation xanthine derivative, functions by inhibiting phosphodiesterase activities without causing these side effects. Doxofylline is an antitussive and bronchodilator that is metabolized by the liver to a degree of 80–90%. Doxofylline sublingual tablets containing super disintegrants such sodium starch glycolate and croscarmellose were evaluated to achieve rapid disintegration under the tongue, allowing the oral mucosa to absorb the active ingredient directly and demonstrated an effective slow and sustained release of drug improving the therapeutic value and toxicity [69].

11.10 POLY LACTIC ACID (PLA)

PLA is a biodegradable polymer derived from renewable resources such as corn starch or sugarcane. Its role in pulmonary drug delivery can be understood scientifically through several key mechanisms. PLA undergoes hydrolytic degradation in the body, breaking down into lactic acid, a naturally occurring metabolite that can be metabolized and eliminated. This biodegradation process minimizes the accumulation of non-degradable polymer residues in the lungs, reducing the risk of tissue damage or inflammation. Additionally, PLA is biocompatible and well-tolerated by the body, further enhancing its suitability for pulmonary drug delivery applications [70]. The small size and tunable properties of PLA nanoparticles enable efficient deposition in the lungs following inhalation. These nanoparticles can penetrate deep into the pulmonary airways, reaching target cells or tissues with high precision. PLA nanoparticles provide a protective encapsulation environment for drugs, shielding them from degradation or premature clearance in the lungs. This protection ensures the stability and integrity of the encapsulated drug during transit to the target site, preserving its pharmacological activity and enhancing therapeutic efficacy [71]. When baicalein is inhaled through muco-penetrating polymeric nanoformulations made of PLA and PLGA, it enters the mucosa more deeply and is better absorbed by the target cells, increasing its local distribution. This is in contrast to when baicalein is delivered through mucoadhesive systems made of chitosan. According to the drug release data of the nanoparticles, unbound baicalein released more than 80% of its total amount in less than four hours. On the other hand, within the same time period, the mucoadhesive delivery with PLA showed a 23.4% cumulative drug release [72].

Mucus-penetrating particles with a size of 200 nm and an initial loading efficiency of between 30% and 35% based on poly (lactide) prolonged the lung residence length of fluticasone propionate by preventing the loss of drug via the mucociliary elimination mechanism. According to clinical research, using a dry powder inhaler with fluticasone propionate can help COPD patients experience fewer exacerbations [73]. *In vivo* delivery of fluticasone propionate to the lungs via mucus-penetrating particles significantly enhanced local drug exposure by 60% compared to mucoadhesive formulations. Additionally, this innovative delivery method extended the efficacy of a single dosage to 16 hours in comparison to the administration of free drug alone. The polymeric coatings used in this approach were developed from hydrophilic, neutrally charged peptides, which were identified through advanced techniques such as next-generation sequencing (NGS) and combinatorial peptide-presenting phage libraries and successfully permeated and evenly spread within the dense sputum of COPD patients, demonstrating enhanced persistence within the airways [74]. When compared to the PEGylated coating, the peptide coatings showed mucus penetration behavior and cellular uptake that were two times better. With better clinical results, these peptides may be used as conjugates or surface modifications to increase the quantity of cargo medication delivered to COPD-mutated epithelia [75]. Because of their biocompatibility and biodegradability PLA nanoparticles are suitable for mucus-permeating and controlled drug delivery when loaded with the phospholipid complex of baicalein, the respiratory tract's tissue distribution, transepithelial transport, and cargo drug's interactions with the mucus layer may significantly improve [76].

11.11 CONCLUSION AND FUTURE PROSPECTS

Natural polymers have garnered a lot of interest since they are frequently investigated as vehicles for the delivery of various phytoconstituents and pharmaceuticals. Natural polymers are very versatile in drug administration due to their inherent advantages in controlled enzyme degradation, exceptional biocompatibility, specific interactions with several pulmonary biomolecules, and ease of modification. Because natural polymers have advantages in terms of patient compliance, prolonged drug release, and targeted efficacy, they can be used as drug delivery systems to improve the efficacy of many medicines in lung illnesses.

Recent years have seen a considerable development in the study of pulmonary diseases attributable to biodegradable polymers, thanks to the increased demand and ongoing progress in the field of biodegradable materials. Basic research on the material–organism interaction might be carried out in the future to evaluate potential advantages and disadvantages, which would boost their application in clinical therapy. Sustained bronchodilation remains the main goal of COPD care because airway inflammation in COPD does not respond well to the anti-inflammatory medicines that are now on the market. Moreover, compared to pharmaceuticals integrated into biodegradable polymers, those encapsulated in biodegradable polymers may have a slower and sustained release rate and a longer duration of action. Pro-drug conjugation on the other hand also offers a potential mechanism for controlled release of drugs for either localized or systemic pulmonary sites in COPD, whereas the pro-drug strategy has been successful in developing long-lasting bronchodilators. Reducing the drug's aqueous solubility or co-precipitating relatively insoluble components with aqueously soluble drugs might also result in sustained release in the lungs.

There is mounting evidence that starting long-acting bronchodilator therapy early in the course of a COPD patient's illness can help to improve their quality of life and decrease the disease's progression. To further improve therapy effectiveness, bronchodilators with various pharmacological characteristics, along with natural polymers, may be combined to prevent future rises in the socioeconomic burden of COPD, all of these strategies ought to be implemented. Furthermore, the availability of defined pharmacopoeia procedures to assess the pharmacokinetics, clearance, and deposition of controlled release systems, along with biodegradable polymers, needs clinical validation. The absence of appropriate animal models and the undetermined long-term adverse effects of drug carriers are the main reasons for the numerous constraints in current research on pulmonary delivery systems for COPD. Thus, the development of efficient controlled drug delivery systems intended for lung administration in this disease require more interdisciplinary research.

REFERENCES

1. M. Passi, S. Shahid, S. Chockalingam, I.K. Sundar, G. Packirisamy, Conventional and nanotechnology based approaches to combat chronic obstructive pulmonary disease: Implications for chronic airway diseases, Int. J. Nanomedicine. 15 (2020) 3803–3826. https://doi.org/10.2147/IJN.S242516.
2. A.J. Alfahad, M.M. Alzaydi, A.M. Aldossary, A.A. Alshehri, F.A. Almughem, N.M. Zaidan, E.A. Tawfik, Current views in chronic obstructive pulmonary disease pathogenesis and management, Saudi Pharm. J. 29 (2021) 1361–1373. https://doi.org/10.1016/j.jsps.2021.10.008.
3. C. Wang, J. Zhou, J. Wang, S. Li, A. Fukunaga, J. Yodoi, H. Tian, Progress in the mechanism and targeted drug therapy for COPD, Signal Transduct. Target. Ther. 5 (2020). https://doi.org/10.1038/s41392-020-00345-x.
4. D.F. Rogers, P.J. Barnes, Treatment of airway mucus hypersecretion, Ann. Med. 38 (2006) 116–125. https://doi.org/10.1080/07853890600585795.
5. E. Rytting, J. Nguyen, X. Wang, T. Kissel, Biodegradable polymeric nanocarriers for pulmonary drug delivery, Expert Opin. Drug Deliv. 5 (2008) 629–639. https://doi.org/10.1517/17425247.5.6.629.
6. A. Balde, S.-K. Kim, S. Benjakul, R.A. Nazeer, Pulmonary drug delivery applications of natural polysaccharide polymer derived nano/micro-carrier systems: A review, Int. J. Biol. Macromol. 220 (2022) 1464–1479. https://doi.org/10.1016/j.ijbiomac.2022.09.116.
7. S. Senapati, A.K. Mahanta, S. Kumar, P. Maiti, Controlled drug delivery vehicles for cancer treatment and their performance, Signal, Transduct. Target. Ther. 3 (2018) 7. https://doi.org/10.1038/s41392-017-0004-3.
8. J.M. Dang, K.W. Leong, Natural polymers for gene delivery and tissue engineering, Adv. Drug Deliv. Rev. 58 (2006) 487–499. https://doi.org/10.1016/j.addr.2006.03.001.
9. I.A. Prestisya, A. Miatmoko, M. Rahmadi, D.M. Hariyadi, Dry powder inhalation microparticles (Alginate, Carrageenan, Chitosan, and combination polymers): A review on characteristics and in vivo activity, Egypt. J. Chem. 65 (2022) 181–206. https://doi.org/10.21608/EJCHEM.2022.119143.5364.
10. X.M. Zeng, G.P. Martin, C. Marriott, The controlled delivery of drugs to the lung, Int. J. Pharm. 124 (1995) 149–164. https://doi.org/10.1016/0378-5173(95)00104-Q.
11. P. Prasher, R. Fatima, M. Sharma, Therapeutic delivery with V-amylose, Drug Dev. Res. 82 (2021) 727–729. https://doi.org/10.1002/ddr.21804.
12. X. Chen, W. Han, G. Wang, X. Zhao, Application prospect of polysaccharides in the development of anti-novel coronavirus drugs and vaccines, Int. J. Biol. Macromol. 164 (2020) 331–343. https://doi.org/10.1016/j.ijbiomac.2020.07.106.
13. R. Fatima, M. Sharma, A. Dhiman, A. Arora, H. Mudila, P. Prasher, Targeted delivery of fenamates with aminated starch, Ther. Deliv. 14 (2023) 183–192. https://doi.org/10.4155/tde-2023-0013.
14. B. Layek, S. Mandal, Natural polysaccharides for controlled delivery of oral therapeutics: A recent update, Carbohydr. Polym. 230 (2020) 115617. https://doi.org/10.1016/j.carbpol.2019.115617.
15. R. Fatima, P. Prasher, M. Sharma, D.K. Chellappan, G. Gupta, S.K. Singh, V.B. Patravale, K. Dua, Aminated polysaccharides: Unveiling a new frontier for enhanced therapeutic efficacy, J. Drug Deliv. Sci. Technol. 89 (2023) 105090. https://doi.org/10.1016/j.jddst.2023.105090.
16. H.A. Yassin, M.A. Sharaf, H.A. El-Ghamry, A.E. Abdelaziz, Effect of different types of polymers as well as different preparation techniques on the in-vitro release of dyphylline controlled release matrix tablets, Res. Sq. (2021). https://doi.org/10.21203/rs.3.rs-211212/v1.
17. F. Farshi Azhar, A. Olad, A study on sustained release formulations for oral delivery of 5-fluorouracil based on alginate-chitosan/montmorillonite nanocomposite systems, Appl. Clay Sci. 101 (2014) 288–296. https://doi.org/10.1016/j.clay.2014.09.004.
18. B. Çelik, Risperidone mucoadhesive buccal tablets: Formulation design, optimization and evaluation, Drug Des. Devel. Ther. 11 (2017) 3355–3365. https://doi.org/10.2147/DDDT.S150774.
19. S. Fredenberg, M. Wahlgren, M. Reslow, A. Axelsson, The mechanisms of drug release in poly(lactic-co-glycolic acid)-based drug delivery systems—A review, Int. J. Pharm. 415 (2011) 34–52. https://doi.org/10.1016/j.ijpharm.2011.05.049.
20. S.S. Shah, Y. Cha, C.G. Pitt, Poly (glycolic acid-co-dl-lactic acid): Diffusion or degradation controlled drug delivery? J. Control. Release. 18 (1992) 261–270. https://doi.org/10.1016/0168-3659(92)90171-M.
21. Y. Fu, W.J. Kao, Drug release kinetics and transport mechanisms of non-degradable and degradable polymeric delivery systems, Expert Opin. Drug Deliv. 7 (2010) 429–444. https://doi.org/10.1517/17425241003602259.
22. I.S. Bayer, Controlled drug release from nanoengineered polysaccharides, Pharmaceutics 15 (2023). https://doi.org/10.3390/pharmaceutics15051364.

23. M. Zhu, A.K. Whittaker, F.Y. Han, M.T. Smith, Journey to the market: The evolution of biodegradable drug delivery systems, Appl. Sci. 12 (2022). https://doi.org/10.3390/app12020935.
24. Z. Wang, S.K. Gupta, S.A. Meenach, Development and physicochemical characterization of acetalated dextran aerosol particle systems for deep lung delivery, Int. J. Pharm. (2017). https://doi.org/10.1016/j.ijpharm.2017.04.052.
25. S. Suarez, A. Muñoz, A. Mitchell, R.L. Braden, C. Luo, J.R. Cochran, A. Almutairi, K.L. Christman, Degradable acetalated dextran microparticles for tunable release of an engineered hepatocyte growth factor fragment, ACS Biomater. Sci. Eng. 2(2) 2016) 197–204. https://doi.org/10.1021/acsbiomaterials.5b00335.
26. S. Wang, F. Fontana, M.A. Shahbazi, H.A. Santos, Acetalated dextran based nano- and microparticles: Synthesis, fabrication, and therapeutic applications, Chem. Commun. 57 (2021) 4212–4229. https://doi.org/10.1039/d1cc00811k.
27. Z. Wang, S.A. Meenach, Optimization of acetalated dextran–based nanocomposite microparticles for deep lung delivery of therapeutics via spray-drying, J. Pharm. Sci. 106 (2017) 3539–3547. https://doi.org/10.1016/j.xphs.2017.07.022.
28. D.S. Bodas, P.P. Ige, Central Composite rotatable design for optimization of budesonide-loaded cross-linked chitosan–dextran sulfate nanodispersion: Characterization, in vitro diffusion and aerodynamic study, Drug Dev. Ind. Pharm. 45 (2019) 1193–1204. https://doi.org/10.1080/03639045.2019.1606823.
29. H. Liu, Y. Li, X. Zhang, M. Shi, D. Li, Y. Wang, Chitosan-coated solid lipid nano-encapsulation improves the therapeutic antiairway inflammation effect of berberine against COPD in cigarette smoke-exposed rats, Can. Respir. J. 2022 (2022) 1–13. https://doi.org/10.1155/2022/8509396.
30. J.G. Montana, H.J. Dyke, Update on the therapeutic potential of PDE4 inhibitors, Expert Opin. Investig. Drugs. 11 (2002) 1–13. https://doi.org/10.1517/13543784.11.1.1.
31. A. Markham, B. Jarvis, Inhaled Salmeterol/Fluticasone propionate combination, Drugs. 60 (2000) 1207–1233. https://doi.org/10.2165/00003495-200060050-00012.
32. W.F. Zhang, X.G. Chen, P.W. Li, Q.Z. He, H.Y. Zhou, Preparation and characterization of theophylline loaded chitosan/β-cyclodextrin microspheres, J. Mater. Sci. Mater. Med. 19 (2008) 305–310. https://doi.org/10.1007/s10856-006-0021-1.
33. V.S. Mastiholimath, P.M. Dandagi, S.S. Jain, A.P. Gadad, A.R. Kulkarni, Time and pH dependent colon specific, pulsatile delivery of theophylline for nocturnal asthma, Int. J. Pharm. 328 (2007) 49–56. https://doi.org/10.1016/j.ijpharm.2006.07.045.
34. C. Cheaburu-Yilmaz, R. Dumitriu, M.-T. Nistor, C. Lupusoru, M. Popa, L. Profire, C. Silvestre, C. Vasile, Biocompatible and biodegradable chitosan/clay nanocomposites as new carriers for theophylline controlled release, Br. J. Pharm. Res. 6 (2015) 228–254. https://doi.org/10.9734/bjpr/2015/16525.
35. F. Abedini, M. Ebrahimi, A.H. Roozbehani, A.J. Domb, H. Hosseinkhani, Overview on natural hydrophilic polysaccharide polymers in drug delivery, Polym. Adv. Technol. 29 (2018) 2564–2573. https://doi.org/10.1002/pat.4375.
36. A. Bianchera, R. Bettini, Polysaccharide nanoparticles for oral controlled drug delivery: The role of drug–polymer and interpolymer interactions, Expert Opin. Drug Deliv. 17 (2020) 1345–1359. https://doi.org/10.1080/17425247.2020.1789585.
37. A.A. Mahmoud, N.A. Elkasabgy, A.A. Abdelkhalek, Design and characterization of emulsified spray dried alginate microparticles as a carrier for the dually acting drug roflumilast, Eur. J. Pharm. Sci. 122 (2018) 64–76. https://doi.org/10.1016/j.ejps.2018.06.015.
38. I.R. Scolari, X. Volpini, M.L. Fanani, B. De La Cruz-Thea, L. Natali, M.M. Musri, G.E. Granero, Exploring the toxicity, lung distribution, and cellular uptake of rifampicin and ascorbic acid-loaded alginate nanoparticles as therapeutic treatment of lung intracellular infections, Mol. Pharm. 18 (2021) 807–821. https://doi.org/10.1021/acs.molpharmaceut.0c00692.
39. J.R. Kanwar, G. Mahidhara, R.K. Kanwar, Novel alginate-enclosed chitosan-calcium phosphate-loaded iron-saturated bovine lactoferrin nanocarriers for oral delivery in colon cancer therapy, Nanomedicine. 7 (2012) 1521–1550. https://doi.org/10.2217/nnm.12.29.
40. M.H. Desoqi, H.S. El-Sawy, E. Kafagy, M. Ghorab, S. Gad, Fluticasone propionate–loaded solid lipid nanoparticles with augmented anti-inflammatory activity: Optimisation, characterisation and pharmacodynamic evaluation on rats, J. Microencapsul. 38 (2021) 177–191. https://doi.org/10.1080/02652048.2021.1887383.
41. E. Amore, M.L. Manca, M. Ferraro, D. Valenti, V. La Parola, S. Di Vincenzo, M. Gjomarkaj, G. Giammona, M.L. Bondì, E. Pace, Salmeterol Xinafoate (SX) loaded into mucoadhesive solid lipid microparticles for COPD treatment, Int. J. Pharm. 562 (2019) 351–358. https://doi.org/10.1016/j.ijpharm.2019.03.059.

42. T. Virmani, G. Kumar, R. Virmani, A. Sharma, K. Pathak, Xanthan gum-based drug delivery systems for respiratory diseases, in: Natural Polymeric Materials Based Drug Delivery Systems in Lung Diseases, Springer Nature Singapore, Singapore, 2023: pp. 279–295. https://doi.org/10.1007/978-981-19-7656-8_16.
43. J. Varshosaz, J. Emami, E. Jaffari, Comparison of hydrophilic natural gums and cellulosic polymers in formulation of sustained-release matrix tablets of terbutalin sulfate, Reasearch Pharm. Sci. 1 (2006) 30–39.
44. P.D. Nithya, S. Brahmani, A. Sriram, N.P. Jaya Krishna, P. Koteswari, P.S. Babu, Formulation and evaluation of theophylline controlled release matrix tablets by using natural gums, Int. J. Pharm. Sci. Rev. Res. 27 (2014) 183–187.
45. A.B. James, A.B. Isah, A.K. Olowosulu, Acacia seyal gum cross-links as matrix for sustained release salbutamol sulphate tablet formulation, J. Pharm. Dev. Ind. Pharm. 4 (2022) 1–15.
46. X. Yuan, R. Amarnath Praphakar, M.A. Munusamy, A.A. Alarfaj, S. Suresh Kumar, M. Rajan, Mucoadhesive guargum hydrogel inter-connected chitosan-g-polycaprolactone micelles for rifampicin delivery, Carbohydr. Polym. 206 (2019) 1–10. https://doi.org/10.1016/j.carbpol.2018.10.098.
47. F. Acarturk, Mucoadhesive vaginal drug delivery systems, Recent Pat. Drug Deliv. Formul. 3 (2009) 193–205. https://doi.org/10.2174/187221109789105658.
48. J. Griesser, G. Hetényi, A. Bernkop-Schnürch, Thiolated hyaluronic acid as versatile mucoadhesive polymer: From the chemistry behind to product developments—What are the capabilities? Polymers (Basel). 10 (2018) 243. https://doi.org/10.3390/polym10030243.
49. K. Kafedjiiski, R. Jetti, F. Foger, H. Hoyer, M. Werle, M. Hoffer, A. Bernkop-Schnurch, Synthesis and in vitro evaluation of thiolated hyaluronic acid for mucoadhesive drug delivery, Int. J. Pharm. 343 (2007) 48–58. https://doi.org/10.1016/j.ijpharm.2007.04.019.
50. P. Mehta, S. Kadam, A. Pawar, C. Bothiraja, Dendrimers for pulmonary delivery: Current perspectives and future challenges, New J. Chem. 43 (2019) 8396–8409. https://doi.org/10.1039/c9nj01591d.
51. C. Lafortuna, F. Fazio, Acute effect of inhaled salbutamol on mucociliary clearance in health and chronic bronchitis, Respiration. 45 (1984) 111–123. https://doi.org/10.1159/000194607.
52. N. Ullmann, S. Caggiano, R. Cutrera, Salbutamol and around, Ital. J. Pediatr. 41 (2015) A74. https://doi.org/10.1186/1824-7288-41-S2-A74.
53. A. Fallacara, E. Baldini, S. Manfredini, S. Vertuani, Hyaluronic acid in the third millennium, Polymers (Basel). 10 (2018) 701. https://doi.org/10.3390/polym10070701.
54. A. Fallacara, L. Busato, M. Pozzoli, M. Ghadiri, H.X. Ong, P.M. Young, S. Manfredini, D. Traini, Co-spray-dried urea cross-linked hyaluronic acid and sodium ascorbyl phosphate as novel inhalable dry powder formulation, J. Pharm. Sci. 108 (2019) 2964–2971. https://doi.org/10.1016/j.xphs.2019.04.015.
55. K. Madhusudana Rao, S.V. Krishna Rao, P. Sudhakar, K. Chowdoji Rao, S. Subha, Synthesis and characterization of biodegradable poly (vinyl caprolactam) grafted on to sodium alginate and its microgels for controlled release studies of an anticancer drug, J. Appl. Pharm. Sci. 3 (2013) 61–69.
56. Y. Li, M. Han, T. Liu, D. Cun, L. Fang, M. Yang, Inhaled hyaluronic acid microparticles extended pulmonary retention and suppressed systemic exposure of a short-acting bronchodilator, Carbohydr. Polym. 172 (2017) 197–204. https://doi.org/10.1016/j.carbpol.2017.05.020.
57. J. Siepmann, K. Elkharraz, F. Siepmann, D. Klose, How autocatalysis accelerates drug release from PLGA-based microparticles: A quantitative treatment, Biomacromolecules. 6 (2005) 2312–2319. https://doi.org/10.1021/bm050228k.
58. J. Saxena, M. Bisen, A. Misra, V.K. Srivastava, S. Kaushik, A.J. Siddiqui, N. Mishra, A. Singh, A. Jyoti, Targeting COPD with PLGA-based nanoparticles: Current status and prospects, Biomed Res. Int. 2022 (2022) 1–13. https://doi.org/10.1155/2022/5058121.
59. S.T. Erni, G. Fernandes, M. Buri, M. Perny, R.J. Rutten, J.M. van Noort, P. Senn, D. Grandgirard, M. Roccio, S.L. Leib, Anti-inflammatory and oto-protective effect of the small heat shock protein alpha B-crystallin (HspB5) in experimental pneumococcal meningitis, Front. Neurol. 10 (2019). https://doi.org/10.3389/fneur.2019.00570.
60. J.M. van Noort, M. Bsibsi, P.J. Nacken, W.H. Gerritsen, S. Amor, I.R. Holtman, E. Boddeke, I. van Ark, T. Leusink-Muis, G. Folkerts, W.E. Hennink, M. Amidi, Activation of an immune-regulatory macrophage response and inhibition of lung inflammation in a mouse model of COPD using heat-shock protein alpha B-crystallin-loaded PLGA microparticles, Biomaterials. 34 (2013) 831–840. https://doi.org/10.1016/j.biomaterials.2012.10.028.
61. P. Marcianes, S. Negro, E. Barcia, C. Montejo, A. Fernández-Carballido, Potential active targeting of gatifloxacin to macrophages by means of surface-modified PLGA microparticles destined to treat tuberculosis, AAPS PharmSciTech. 21 (2020) 15. https://doi.org/10.1208/s12249-019-1552-3.

62. Y. Zhou, B. Qiu, X. Yin, H. Liu, L. Zhu, Concomitant drugs-loaded microcapsules of roxithromycin and theophylline with pH-sensitive controlled-releasing properties, Int. J. Polym. Mater. Polym. Biomater. 69 (2020) 679–690. https://doi.org/10.1080/00914037.2019.1596917.
63. S. Ahmed, N. Panda, M. Ansari, A. Kauser, Development and in vitro evaluation of acebrophylline sustained release matrix tablets employing different grade of HPMC and ethyl cellulose, J. Glob. Trends Pharm. Sci. 6 (2015) 2716–2727.
64. P.B. Mote, P.K.R.K. Rawat, S.K. Singh, N.S. Zadbuke, A.A. Salunke, V.B. Rajendra, Formulation and evaluation of sustained release matrix tablets of anti-asthmatic agent using various polymers, J. Drug Deliv. Ther. 3 (2013). https://doi.org/10.22270/jddt.v3i2.440.
65. D. Sharma, Formulation development and evaluation of fast disintegrating tablets of salbutamol sulphate for respiratory disorders, ISRN Pharm. 2013 (2013) 1–8. https://doi.org/10.1155/2013/674507.
66. N. Goyal, A. Kumar, Formulation and in vitro evaluation of salbutamol sulphate and theophylline extended-release tablets using modified polymers, Int. J. Pharm. Pharm. Sci. 10 (2018) 67. https://doi.org/10.22159/ijpps.2018v10i8.27865.
67. S. Shetty, D. Suares, S. Shetty, Starch-based drug delivery systems in lung disorders, in: Natural Polymeric Materials Based Drug Delivery Systems in Lung Diseases, Springer Nature Singapore, Singapore, 2023: pp. 115–127. https://doi.org/10.1007/978-981-19-7656-8_6.
68. N.M. Oh, K.T. Oh, Y.S. Youn, D.-K. Lee, K.-H. Cha, E.S. Lee, Development of tiotropium inhalation formulations for the treatment of chronic obstructive pulmonary disease, J. Pharm. Investig. 43 (2013) 71–74. https://doi.org/10.1007/s40005-013-0054-7.
69. Y.K. Reddy, S.S. Ahmed, Formulation and in vitro evaluation of doxophylline Oral disintegrating tablets, Res. J. Pharm. Technol. 13 (2020) 2811. https://doi.org/10.5958/0974-360X.2020.00500.4.
70. T.G. Barclay, C.M. Day, N. Petrovsky, S. Garg, Review of polysaccharide particle-based functional drug delivery, Carbohydr. Polym. 221 (2019) 94–112. https://doi.org/10.1016/j.carbpol.2019.05.067.
71. P. Jana, M. Shyam, S. Singh, V. Jayaprakash, A. Dev, Biodegradable polymers in drug delivery and oral vaccination, Eur. Polym. J. 142 (2021) 110155. https://doi.org/10.1016/j.eurpolymj.2020.110155.
72. W. Dong, J. Ye, J. Zhou, W. Wang, H. Wang, X. Zheng, Y. Yang, X. Xia, Y. Liu, Comparative study of mucoadhesive and mucus-penetrative nanoparticles based on phospholipid complex to overcome the mucus barrier for inhaled delivery of baicalein, Acta Pharm. Sin. B. 10 (2020) 1576–1585. https://doi.org/10.1016/j.apsb.2019.10.002.
73. J. Kurczewska, Recent reports on polysaccharide-based materials for drug delivery, Polymers (Basel). 14 (2022). https://doi.org/10.3390/polym14194189.
74. A. Papi, D. Dokic, W. Tzimas, I. Mészáros, A. Olech-Cudzik, Z. Koroknai, K. McAulay, S. Mersmann, P. Dalvi, T. Overend, Fluticasone propionate/formoterol for COPD management: A randomized controlled trial, Int. J. Chron. Obstruct. Pulmon. Dis. 12 (2017) 1961–1971. https://doi.org/10.2147/COPD.S136527.
75. Q. Zhong, O.M. Merkel, J.J. Reineke, P. Da Rocha, Effect of the route of administration and PEGylation of poly(amidoamine) dendrimers on their systemic and lung cellular biodistribution, Mol. Pharm. 13 (2016) 1866–1878. https://doi.org/10.1021/acs.molpharmaceut.6b00036.
76. F. Emami, S.J. Mostafavi Yazdi, D.H. Na, Poly(lactic acid)/poly(lactic-co-glycolic acid) particulate carriers for pulmonary drug delivery, J. Pharm. Investig. 49 (2019) 427–442. https://doi.org/10.1007/s40005-019-00443-1.

12 Nano Approaches for the Delivery of Biologics in Chronic Obstructive Pulmonary Disease

Gaurav Bisht, Shivam Pandey, Ritu Painuli, and Ajay Singh
Department of Chemistry, School of Applied
and Life Sciences, Uttaranchal University,
Dehradun, Uttarakhand, India

Chetan Kumar
School of Pharmaceutical and Populations Health
Informatics, Faculty of Pharmacy, DIT University
Dehradun, Uttarakhand, India

Waseem Ahmad
Department of Chemistry, Graphic Era (Deemed to Be University)
Dehradun, Uttarakhand, India

12.1 INTRODUCTION

Chronic bronchitis and emphysema are among the respiratory disorders that make up chronic obstructive pulmonary disease (COPD), a progressive lung illness marked by persistent inflammation of the airways and blockage of airflow (Hussain et al., 2021). Currently, the fourth leading cause of death worldwide is COPD. Up to 2.8 million people died from COPD globally in 2010, while 175 million people were estimated to be have the disease in 2015. Despite advancements in therapeutic options to help control symptoms and prevent complications, COPD is incurable. Therefore, innovative approaches are badly needed to reduce mortality and delay or even reverse the progression of this disease (Shukla et al., 2020). The primary cause of COPD is prolonged exposure to irritants such as dust, gaseous pollutants, air pollution, and cigarette smoke. These allergens cause inflammation and affect the lungs' tiny airways, air sacs, or alveoli. Breathing becomes difficult and airflow restricted because of the air sacs losing their mobility and the airways thickening and narrowing with time.

Patients with severe forms of COPD and asthma experience frequent exacerbations and subsequent emergency room visits throughout the year. Patients with any form of COPD display a wide range of symptoms, with breathlessness or breathe shortness during daily activities being the significant indication that breaks down over time. This is brought on by either chronic bronchitis, which is more frequent in smokers and ex-smokers with COPD, or partial alveolar damage (emphysema), which is the result of excess mucus and inflammatory cells forming in the bronchioles. These disorders prevent airflow and reduce the lungs' ability to interact with gases, which reduces blood oxygen levels (hypoxemia) and ultimately results in respiratory failure (Joshi, 2018). A chronic cough (with or without mucus), fatigue, wheezing, and tightness in the chest are common symptoms

 DOI: 10.1201/9781003498926-12

that people with any kind of COPD may experience; these symptoms are sometimes misdiagnosed as age-related. Some people may not exhibit any symptoms at all until their illness worsens.

Despite decades of intensive research, there is currently no treatment to repair the damaged tissue and restore lung function. In addition, COPD is a degenerative condition that deteriorates with age. The global burden of disease research, according to the World Health Organization (WHO), found that there were 251 million cases of COPD worldwide, with 90% of those instances occurring in low- and middle-income nations. In 2015 alone, there were an estimated 3.17 million fatalities, or 5% of all deaths, which is an increase of 11.6% from 1990 (Beck-Broichsitter et al., 2012).

12.2 TYPES OF CHRONIC OBSTRUCTIVE PULMONARY DISEASE

There are mainly two types of COPD—chronic bronchitis and emphysema, which often coexist in patients. Discussed next is a detailed explanation of each (see Figure 12.1):

12.2.1 Chronic Bronchitis

Inflammation and constriction of the bronchial tubes, which transport air to and from the lungs, are hallmarks of chronic bronchitis. Coughing and excessive mucus production are caused by inflammation. A productive cough that persists for two consecutive years for at least three months is the primary sign of chronic bronchitis. Chest tightness, exhaustion, wheezing, and shortness of breath are possible additional symptoms (Kim & Criner, 2013). Chronic bronchitis mainly occurs due to cigarette smoking, though other lung irritants like dust, air pollution, and chemical fumes can also cause it. Prolonged exposure to irritants causes the bronchial tubes to become chronically inflamed, which triggers the production of excessive mucus, which clogs the airways and makes breathing difficult (Wilson et al., 2000). A patient's medical history, symptoms, physical examination, and results of pulmonary function tests—such as spirometry—are used to make the diagnosis. Lung damage can also be evaluated using imaging tests such as computerized tomography (CT) scans or chest X-rays.

12.2.2 Emphysema

The lung disease known as emphysema is characterized by the destruction of the alveoli, or air sacs, in the lungs. Breathing becomes more difficult because of this injury, which decreases the surface

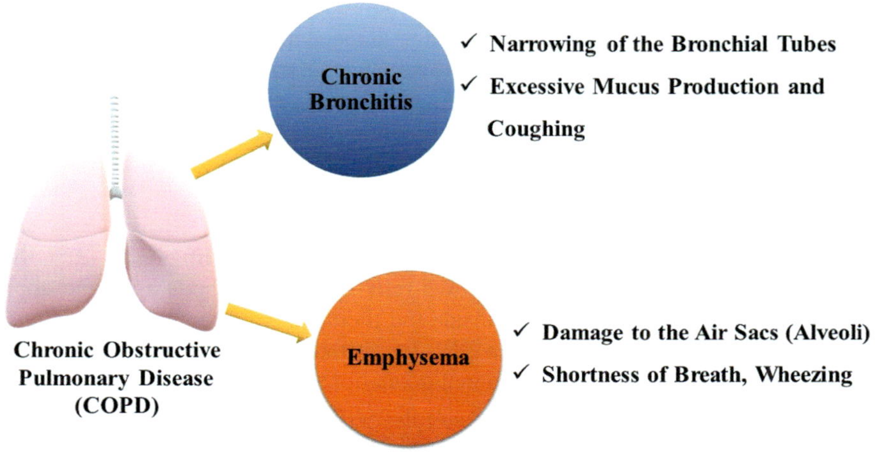

FIGURE 12.1 Types of chronic obstructive pulmonary disease.

area accessible for gas exchange. Emphysema symptoms usually include exhaustion, wheezing, shortness of breath, and a persistent cough (Jelic, 2019). Patients may have weight loss and respiratory infections as the condition worsens. Smoking cigarettes is the main cause of emphysema. It can also develop as a result of prolonged exposure to other lung irritants such as industrial dust, air pollution, and chemical fumes. Emphysema results in damage to the alveolar walls, which causes the tiny airways to collapse during expiration and lose their suppleness. Breathlessness and other symptoms are brought on by the lungs' decreased ability to exchange gases efficiently (Sharafkhaneh et al., 2008). Emphysema is diagnosed using a multimodal approach that includes imaging scans, lung function testing, physical examination, and medical history, like chronic bronchitis.

12.3 BIOLOGICS FOR COPD TREATMENT

The chronic airflow limitation associated with COPD occurs due to a combination of tiny airway blockage and fibrosis, as well as the loss of parenchymal assistance given by alveolar affixations to the tiny airways. Thus, structural changes reduce the tendency to keep the airways open during inhalation and restriction of airflow. Since it is produced not fully sensitive to bronchodilators, resulting in a bronchodilator associated Forced expiratory volume in one second (FEV1)/ Forced vital capacity (FVC) that is regularly less than 0.7, spirometry is often employed to identify airflow limitation (Pavord, 2018).

Currently, available pharmaceutical types for maintaining stable COPD are bronchodilators (such as beta (β)2 agonists, antimuscarinics, and methylxanthines) and anti-inflammatories (like corticosteroids) (Yousuf & Brightling, 2018). To enhance overall health and exercise tolerance, pharmacologic treatment for COPD aims to lessen symptoms as well as the frequency and intensity of exacerbations. A variety of clinical conditions, the cost and accessibility of pharmaceuticals, and the anticipated adverse effects of treatment all influence the selection of a particular treatment (Wechsler, 2018). Bronchodilator therapy, which includes long-acting β2 agonists and long-acting antimuscarinic drugs, generally exerts proportional effects on symptoms and airflow restriction. Conversely, inhaled corticosteroids (ICS) exhibit a marginally differentiating effect on the likelihood of exacerbations and the likelihood of subsequent exacerbations. Present practices for therapy states that the pharmaceutical regimen should be customized according to the patient's symptoms and the risk of exacerbations, which is assessed by examining previous episodes of severe or recurrent exacerbations (Seshadri & Ramamurthi, 2018).

The investigation of biological therapeutics that specifically target inflammatory pathways is warranted, considering the prevalence of airway inflammation. The current mediocre outcomes of the research align with the overarching challenge associated with the development of innovative COPD medications. Numerous instances of this may have transpired because of the treatment trait's lack of precise specification. Interventions, which aim to reduce type-2 or elevated levels of eosinophilic airway inflammation, are a notable exception (Yadav et al., 2023).

12.4 NANO APPROACHES FOR COPD

To improve COPD management, novel drug delivery strategies, testing tools, and therapeutic approaches are created using nanotechnology. We refer to these methods as "nano approaches". By extending the duration of the treatment inside the therapeutic window and reducing systemic exposure right after inhalation, nanoparticles (NP) for pulmonary drug delivery have the potential to improve upon the shortcomings of conventional inhalational drug administration. As a result, this approach can lower the necessary dosage, frequency of use, and systemic adverse effects (Yhee et al., 2016). Polylactic-co-glycolic acid (PLGA), approved by the U.S. Food and Drug Administration (FDA), is the most widely used biodegradable polymer for drug delivery in NPs. Biodegradable polymers are frequently utilized as carriers for embedding medicines in NPs.

One benefit of locally administered nanomaterial-based frameworks is the ability to reduce side effects by reducing the dose and consistency. Other benefits include protecting cargo drugs in opposed to deterioration or deactivation and having the potential to overcome types of intracellular and extracellular limitations due to their nano-dimension size (Vij, 2017). Since PLGA is a malleable polymer, it has been used to encapsulate proteins, nucleic acids, and other small-molecule drugs in particle carriers. PLGA progressively bioerodes *in vivo* and splits into its monomeric constituents (lactic acid and glycolic acid) by the hydrolysis of the ester backbone. Since these byproducts are typical metabolites of the body, PLGA is frequently tolerable when used *in vivo* for biomedical applications. Here are some various nanotherapies for COPD:

- **Lipid nanoparticles:**
  Medication for COPD, such as anti-inflammatory drugs and bronchodilators, can be encapsulated in lipid-based NPs like liposomes and solid lipid nanoparticles (SLNs). By preventing drug breakdown, increasing the solubility, and enabling targeted distribution to the lungs, these NPs maximize medication effectiveness while reducing systemic adverse effects (Liu et al., 2022). Liposomes are the most often used lipid NPs for pulmonary administration since they primarily consist of cholesterol and phospholipids. The main structural components of the skeleton are phospholipids, which may be obtained from synthetic sources (dipalmitoyl phosphatidylethanolamine, or DPPE) and natural foods (lecithin and soy lecithin). One material that is crucial for regulating the membrane mobility of liposomes is cholesterol (Doroudian et al., 2021). The self-assembled liposomes exhibit an amphiphilic domain because the phospholipid molecules have a lipophilic tail and a hydrophilic head. This makes it possible to load the drugs with different solubilities and use the co-solubility effect to increase solubility. Solid colloid drug delivery systems (SLNs) are another kind of lipid-based NPs made of natural or synthesized lipids.

  Active substances can be fully deposited in the lungs thanks to SLNs, which are more stable during atomization and may break down to produce supersonic droplets more readily than liposomes. Longer administration periods and improved patient compliance can result from the adhesion, buildup, long-term discharge, and retention of SLNs in the lung, all of which have the potential to improve and maintain the therapeutic effect on respiratory conditions. Finally, SLNs with strong biodegradability will tolerate well in the airway (Bian et al., 2021).
- **Polymeric nanoparticles:**
  Biodegradable polymers can be utilized to create polymeric NPs that can hold COPD drugs and release them gradually. These NPs can reduce the frequency of dosages and enhance medicine absorption by specifically targeting certain cells or tissues in the lungs. Both synthetic (Percutaneous Endoscopic Gastrotomy (PEG), PLGA, polyethylene, and polyacrylate) and natural (chitosan, alginate, and cyclodextrin) polymer materials are often used (Chikuma et al., 2020). PLGA is a biodegradable polymer that may hydrolyze to produce lactic and glycolic acids when consumed by cells. Because of its great biocompatibility, the FDA has approved PLGA-based formulations for use in clinical settings.

  To make NPs for the administration of pulmonary medicines, PLGA is used frequently. Still, the mild dosage of PLGA NPs frequently restricts their use. PEG surface modification not only reduces the amount of PLGA NP cleared by the reticuloendothelial system, but also makes it possible for PEG to swiftly traverse the mucus barrier and airway defense. Researchers added antibodies that target neutrophils, which are capable of delivering ibuprofen to the lung, to PLGA NPs by modifying them with a copolymer of distearoylphosphatidylethnaolamine (DSPE)-PEG2000/DSPE-mPEG2000. Furthermore, several polymer materials (Table 12.1) have specific therapeutic benefits on respiratory disorders in addition to being utilized as drug carriers to deliver therapeutic medications (Gaul et al., 2018).

TABLE 12.1
Different Nano Approaches for the Treatment of COPD and Their Route of Administration

Nanocarrier	Preparation Method	Route of Administration	Reference
Solid-lipid nanoparticles	Precipitation	Dry inhaled powder	(Sharma et al., 2016)
PLGA-Large Porous Microparticles (LPMs)	Freeze-drying	Aerosol	(Yadollahi et al., 2015)
PEG/Phosphatidylethanolamine (PEG-DSPE)	Coprecipitation and reconstitution	Dry inhaled powder	(Jacobs & Muller, 2002)
Polyaspartamide	Spray and freeze-drying	Dry inhaled powder	(Craparo et al., 2017)
PEG@CS/BPQDs-AM NPs	liquid Exfoliation	Dry inhaled powder	(Li et al., 2020)
Chitosan/Alginate-LPMs	Ethanolic Precipitation	Aerosol	(Amore et al., 2017)
PEGylated immuno-conjugated NPs	Coprecipitation and reconstitution	Dry inhaled powder	(Vij et al., 2016)
TPGS-PLGA hybrid NPs	Nanoprecipitation	Aerosol	(Wan et al., 2019)
PEG-modified silica NPs	Precipitation Technique	Dry inhaled powder	(Guo et al., 2023)
TPCA-1-PMs	Dialysis method	Dry inhaled powder	(Zhang et al., 2019)

12.5 NANOPARTICLES FOR GENE THERAPY

Gene therapy vectors can be transported by NPs, allowing for the precise delivery of therapeutic genes to lung cells. By modifying gene expression, gene therapy has the potential to cure the underlying molecular causes of COPD, including oxidative stress and inflammation. Therapeutic genes can be selectively delivered by designed NPs to the lung's afflicted cells, such as alveolar macrophages or bronchial epithelial cells. The therapy's effectiveness is increased, and off-target effects are reduced because of this tailored administration. Genetic material (such as DNA or RNA) must be delivered to the target cells for gene therapy to work. These delicate molecules can be shielded by NPs from the body's enzymes and other biological processes, maintaining their integrity and functioning until they reach the target cells (Gill et al., 2007).

12.6 NANOPARTICLE-BASED IMAGING AND DIAGNOSTICS

- **Nanoparticle contrast agent:**
  NPs can be functionalized with contrast agents for imaging modalities such as positron emission tomography (PET), magnetic resonance imaging (MRI), and CT (Wadhwa et al., 2019). Enhancing the sensitivity, specificity, and adaptability of imaging methods utilized in COPD diagnosis and treatment has enormous potential thanks to NP contrast agents. To maximize the design, targeting, and clinical translation of NP-based contrast agents for COPD imaging applications, further research and development in this field are required. These contrast-enhanced NPs make it possible to visualize lung anatomy and function non-invasively, which helps with COPD progression monitoring and diagnosis (Bahmanpour et al., 2020).
- **Biosensor:**
  Biosensors that can identify certain biomarkers linked to COPD, such as reactive oxygen species (ROS), inflammatory cytokines (like interleukin-8), and specific proteins (like C-reactive protein), have been made possible by nanotechnology. These indicators can help with the diagnosis and tracking of COPD development by offering important information about the degree of oxidative stress and inflammation in the lungs. Graphene, metallic NPs, and carbon nanotubes are examples of nanomaterials with special electrical, optical, and catalytic capabilities that can be used to improve the sensitivity and specificity of biosensors for COPD biomarker detection. This makes it possible to find biomarkers

in complicated biological materials like sputum or exhaled breath condensate, even at extremely low quantities (Dong et al., 2020).

12.7 NANOPARTICLE-BASED THERAPEUTIC AGENT

Antioxidant-loaded NPs, including those containing catalase or superoxide dismutase (SOD), can scavenge ROS and lower oxidative stress in COPD patients' lungs. These antioxidant NPs may lessen lung damage brought on by COPD, lessen inflammation, and lessen exacerbations. Anti-inflammatory drugs can be specifically delivered to COPD patients' irritated lung tissues using NP engineering. These NPs may decrease inflammation and maintain lung function by targeting certain inflammatory pathways linked to the pathophysiology of COPD (Xu et al., 2020).

12.8 DRUG NANOCRYSTALS

The use of drug nanosuspensions (NS) and nanocrystals (NC) has increased as a means of addressing the shortcomings of poorly soluble pharmaceuticals. Medicated NC particles, which are less than a micron in size, include solely the medication and a stabilizing surfactant. By improving the volume-to-surface area ratio and, hence, the solubility of the medication and dissolving rate, the process of nanonization enhances the drug's pharmacokinetics and bioavailability. There are two main approaches for producing nanocomposites, bottom-up, where the medication becomes saturated, and top-down, where the large insoluble particles are physically broken up by mechanical techniques like homogenization or media milling. But mainly, only top-down methods are applied at the time to create commercial items. High drug-loading efficiency, scalability, and flexibility to various routes of administration are only a few advantages of drug NC. Lower regulatory barriers to the production of items utilizing nanotechnology are also a consequence of the industry's well-established practices and the composition's simplicity. While many NC formulations are being studied in preclinical settings, a few are now approved for use in clinical settings (Guan et al., 2022).

12.9 APPLICATION OF NANOTECHNOLOGY IN DIFFERENT TYPES OF LUNG DISEASE

12.9.1 Targeted Drug Delivery

Over the past ten years, the relatively young field of multidisciplinary research of nanomedicine's use for lung illnesses has expanded quickly. According to findings of a study, nanomedicine has the potential to completely transform the medical field through the creation of novel therapeutic agent delivery strategies, vaccine development, and nanotechnology-based medical diagnostics (Doroudian et al., 2019). Targeted drug delivery, made possible by nanotechnology, ensures that drugs more efficiently reach certain cells or tissues in the lungs while reducing systemic adverse effects. Therapeutic medicines like antibiotics, anti-inflammatory medications, or gene treatments can be delivered straight to the damaged lung tissues via NP engineering. This method can lower the necessary dose while increasing therapy effectiveness (Chan et al., 2021) (see Figure 12.2).

12.9.2 Treatment of Pulmonary Fibrosis

Antifibrotic medications can be delivered using NP design to treat pulmonary fibrosis, a disorder marked by an excessive buildup of fibrous tissue in the lungs. The high related morbidity and death rate of pulmonary fibrosis is a result of the absence of appropriate medications. Patients usually have an average lifespan of 40 years. Making muco-inert nanomaterials that can enter cystic fibrosis (CF) sputum and get to the target epithelial cells seems to be beneficial. Within the area of 60–300 nm, the overall mesh size in CF sputum remained 140 ± 50 nm (Suk et al., 2009). Additionally, they

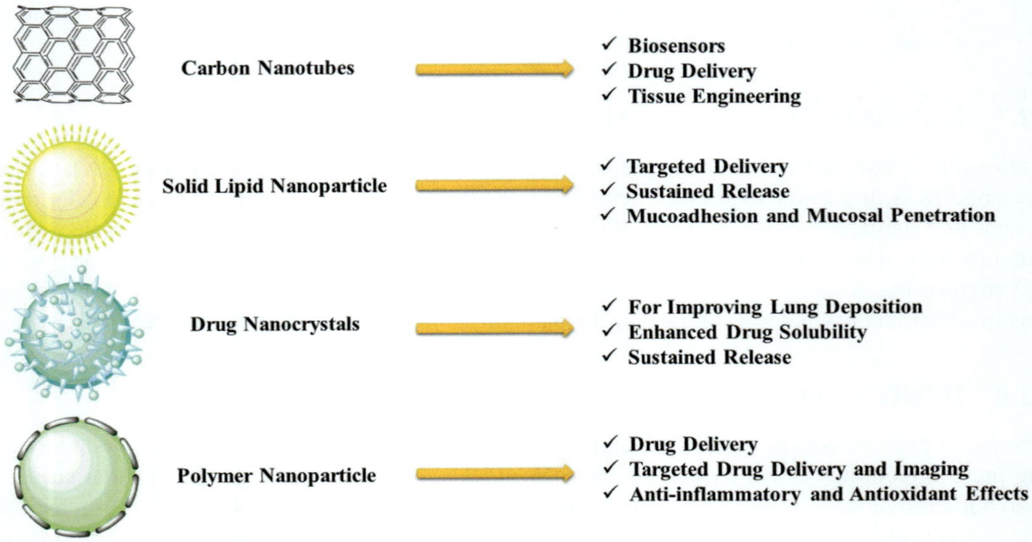

FIGURE 12.2 Different nano approaches and their application in chronic obstructive pulmonary disease.

displayed that PEG-dense particles with low molecular mass up to 200 nm in diameter may pass through CF sputum up to 90 times easier than an uncovered particle of the same size. The mobility of these NPs after processing CF sputum with N-acetylcysteine (NAC) was examined in a different investigation (Suk et al., 2011). These NPs may be able to lessen symptoms and maybe even stop the progression of the disorder by specifically targeting fibrotic tissue.

12.9.3 Lung Cancer Disorder

Targeting lung disorders, particularly cancer, has been a new priority for nanotechnology throughout the past ten years. The goal is to integrate the pulmonary/inhalational form of administration with therapeutic delivery based on nanotechnology (Zhang et al., 2008). In lung cancer treatment, nanotechnology offers targeted drug delivery and imaging capabilities. NPs can carry chemotherapeutic agents directly to tumor cells, reducing systemic toxicity and improving therapeutic outcomes (Sharma et al., 2019). Additionally, nanomaterials can be used for imaging purposes, enabling early detection and precise monitoring of tumor growth and response to treatment.

12.9.4 Respiratory Infection

Due mostly to intrahospital infections, which more readily impact patients with impaired immune systems, respiratory infections pose a substantial cost to health systems around the world. Furthermore, as a result of the HIV/AIDS epidemic, tuberculosis (TB), an infectious illness that is widespread in many impoverished countries, has returned to prominence in the industrialized world (Ibarra-Sánchez et al., 2022). Drug delivery systems based on nanotechnology (nano-DDS) have shown promise in addressing drug resistance and overcoming the drawbacks of traditional formulations. This has created the possibility of further advancements in this field. Antimicrobial medicines can be delivered via NPs to fight respiratory illnesses including TB and pneumonia. By more efficiently targeting pathogens and penetrating bacterial biofilms, these NPs can overcome antibiotic resistance and improve treatment efficacy (Andrade et al., 2013).

TABLE 12.2
Nanoparticles and Their Applications in Different Types of Lung Disease

Type of Nanoparticle Used	Pharmaceutical Ingredient Used	Application in Treatment	Reference
Polymeric NPs	Aminophylline	COPD	(Varshosaz et al., 2013)
ZnO and AgO NPs	Rifampicin	Tuberculosis	(Ellis et al., 2018)
Polymeric NPs	Azithromycin	Pneumonia	(Ahmad et al., 2009)
Nano (SBS)	Salbutamol	Asthma	(Gao et al., 2020)
Polymeric NPs	Bavachinin	Asthma	(Wang et al., 2018)
GNPs/Polymeric	Small interfering RNA	Lung cancer	(Liu et al., 2019)
Polymeric/silica NPs	Small interfering RNA	Lung cancer	(Bregoli et al., 2016)
Liposome	miR-660 (CCL660)	Lung cancer	(Moro et al., 2019)
Polymeric NPs	IFN-γ-pDNA	Asthma	(Kuzmov & Minko, 2015)
Liposome/Polymeric	MnPD and pHDAC2	COPD	(Chikuma et al., 2020)
Se NP	Isoniazid	Tuberculosis	(Pi et al., 2020)
Polymeric NPs	Gene encoding CFTR	Cystic Fibrosis	(Konstan et al., 2004)
Polymeric NPs	Pentoxifylline	Lung cancer	(Gupta et al., 2014)
Liposome	miR-146a	COPD	(Mohamed et al., 2019)
Polymeric NPs	Tobramycin	Cystic Fibrosis	(Ernst et al., 2018)

Abbreviations: NPs: Nanoparticles, ZnO: Zinc Oxide, AgO: Silver Oxide, Se: Selenium, SBS: Solution Blow Spinning Fibers, GNPs: Gold Nanoprisms, CFTR: Cystic Fibrosis Transmembrane Regulator, MnPD: Mn-porphyrin dimer, pHDAC2: HDAC2-encoding plasmid DNA

12.9.5 Regenerative Medicinal Application

Numerous studies have shown that the use of nanotechnology improves the regeneration of many types of tissue, including bone, vascular, heart, cartilage, bladder, and brain. Numerous metals (alloys) and nanostructured materials have been studied for their bio- (and cyto-) compatibility characteristics. Nanomaterials can be incorporated into scaffolds that replicate the native extracellular matrix and promote cell growth, differentiation, and tissue regeneration in diseases such as acute respiratory distress syndrome (ARDS) or COPD (Khang et al., 2010). For other applications see Table 12.2.

12.10 PROMINENT LITERATURE ON USE OF NANO APPROACHES FOR COPD

Nano delivery strategies for long-term respiratory disorders are hindered by airway defense, severe inflammation, and mucus hypersecretion. Asthma, COPD, and CF are characterized by continuous respiratory inflammation and mucus discharge. Different pathophysiologies produce inflammation and mucous hypersecretion, which is aggravated by infection or tobacco smoking. Corticosteroids, antibiotics, and other drugs on the market reduce chronic inflammation, but getting them to target cells or tissues over time may be difficult. COPD requires ongoing NP-targeted medicine to control fibrosis, protease-mediated chronic emphysema, inflammatory cell chemotaxis, and/or chronic lung blockage. In addition, targeted epithelial administration and inflammatory cell dispersion are necessary for treating CF and other chronic inflammatory lung diseases. Nanotechnology is being used to investigate a potential lung disease treatment for COPD and CF (Vij, 2012).

Yoo and Colleagues presented a novel class of biodegradable polyoxalate (HPOX) combined with hydroxybenzyl alcohol (HBA) as a potential airway inflammation treatment. Peroxalate ester linkages in HPOX can react with H2O2 and HBA, making it antioxidant and anti-inflammatory. HPOX NPs effectively reduce intracellular oxidative stress, scavenge H2O2, and suppress pro-inflammatory mediators in activated macrophages, including cyclooxygenase-2 (COX-2 production-2), inducible nitric oxide synthase (iNOS), and interleukin (IL), resulting in strong antioxidant and anti-inflammatory effects. A mouse treated with asthma medication was tested for HPOX NPs' anti-asthmatic properties. Intratracheal HPOX NPs reduced IL-4, iNOS, and inflammatory cell recruitment. HPOX NPs show antioxidant, anti-inflammatory, and anti-asthmatic capabilities, suggesting they could be used as therapeutic agents and drug carriers for airway inflammatory diseases including asthma (Yoo et al., 2013).

Pulmonary diseases' therapeutic effectiveness is frequently restricted, and drug delivery technologies provide novel approaches to clinical issues. Since solid lipid microparticles (SLMs) can deposit into secondary airways due to their size, they are recommended as methods for delivering medicines to the lungs. In recent research, two novel SLMs inhibit inflammatory mechanisms and senescence in isolates from cigarette smoke by combining alginate and chitosan, like polymers with mucoadhesive properties. The biocompatibility of these SLMs and their efficacy in comparison to the free drug were also examined. According to recent research, SLMs loaded with fluticasone propionate (FP) are superior to FP alone in reducing oxidative stress. Treating chronic inflammatory lung illnesses using FP-loaded microparticles appears to be a potential therapeutic method (Amore et al., 2017).

Nanomaterials may be used to alter the tumor microenvironment, thereby improving lung cancer patient outcomes. Nanomaterials provide more precise tumor targeting while reducing or even eliminating negative effects through more persistent drug delivery. Small-molecule inhibitors may be administered with the help of nanomaterials to patients suffering from COPD and non-small cell lung cancer (NSCLC). This might prolong the course of treatment and improve patient outcomes. The goal of the study was to emphasize the role that nanomedicines play in improving and reversing negative outcomes in the tumor microenvironment (Seshadri & Ramamurti, 2018).

Because COPD is an incurable illness defined by a sticky mucus layer, therapeutic medication delivery using nanoagents that can penetrate mucus is made possible. Optimizing drug release from nanovehicles is necessary to improve the therapeutic benefits of COPD treatment, though. In this instance, PEGylated chitosan nanospheres containing amikacin (referred to as PEG@CS/BPQDs-AM NPs) are mixed with black phosphorus quantum dots (BPQDs). PEG's hydrophilicity and chitosan's positive charge allow nanovehicles to deeply enter the layer of mucus that sticks to the mucous membrane, therefore serving as a drug delivery method. Furthermore, the inside BPQDs break down quickly into harmless PO_4^{3-} and acidic H^+, which encourages PEGylated CS nanosphere dissociation, speeds up the release of medicinal AM (amikacin), and ultimately weakens biofilm viability for simple removal. The findings showed that using BPQDs to mediate drug delivery is a viable and attractive approach for precision medicine, with great potential for COPD clinical management. (Li et al., 2020)

Oxidative stress affects COPD development and etiology. Combining antioxidants or nutrients helps COPD animal models. The benefits of exogenous antioxidants may not have been duplicated in clinical settings due to their limited half-life, concentration, and absorption. For COPD oxidative stress, novel medicine delivery techniques are needed. The study covers the latest innovations in medicine administration and research on oxidative stress in COPD and technology to cure it. Researchers have solved various problems and improved drug pharmacokinetics and bioavailability since nanotechnology. Large hollow small particles and enclosed hollow microparticles are promising carriers for regulated medication release and lung deposition of inhaled therapy. In pulmonary clinical settings, drug delivery system translation is still in its infancy (Xu et al., 2020).

Larsson et al., examined compliance with COPD medication recommendations in Swedish primary and secondary healthcare facilities using data from the Swedish National Airway Register (SNAR). SNAR collected data on the evidence (COPD Assessment Test (CAT), Medical Research Council dyspnea scale (mMRC)), lung function, pharmacological medication, and history of exacerbations of 15,595 COPD patients from 853 primary care and 125 secondary care medical centers over 1.5 years. Individuals with asthma co-diagnosis were rejected. Patients were placed in four therapy groups. Long-acting bronchodilators were prescribed to 78% of patients and ICS to 46%. Thirty three percent, 46%, 55%, and 71% of patients received any combination of ICS and therapy, while 19%, 34%, 39%, and 61% received triple therapy. Twenty-one percent, 11%, 11%, and 5% of groups A, B, C, and D received no inhaler treatment. In persons without ICS symptoms, drug combinations with ICS were common (Larsson et al., 2021).

Many millions of individuals worldwide suffer from chronic respiratory disorders, which are the leading cause of death, morbidity, social and economic pressures, and disability-adjusted life-years (DALYs). Chronic respiratory problems affect the parenchyma, pulmonary vasculature, and airways. Currently, the medicinal material is not delivered to the target location for optimal care and/or prevention; it largely relieves symptoms. Recent research focuses on parenchymal rehabilitation, oxidative stress, airway inflammation, and systemic inflammation. Neutrophils, macrophages, T cells, smooth muscle cells, endothelium, epithelium, and fibroblasts are active actors in chronic respiratory diseases. Oxidants, growth factors, chemokines, and cytokines are involved in the pathophysiology of each chronic respiratory illness. Prescription drugs now reduce airway tightness and irritation. Every investigation shows that cytokines promote long-term airway remodeling and inflammation. Due to the limitations of traditional medicines, this study summarizes chronic airway disease knowledge and discusses traditional treatments in greater detail. It also highlights cutting-edge IL-targeting pharmaceutical delivery methods for Chronic Respiratory Diseases (CRDs) (Hussain et al., 2021).

Targeting lung macrophages, new lipid–polymer hybrid nanoparticles (LPHNPs) were identified as viable delivery systems for roflumilast, a medication used to treat COPD. These have a core-shell shape, 50% phospholipid content, colloidal size, negative potential, and slowly release the medication that is entrapped in a physiological fluid simulation. Additionally, their large PEG density on the surface, which bestows mucus-piercing qualities, was shown by the surface examination. Man@LPHFNPs-Roflumilast can match with the bronchial of human epithelial, which actively targets and absorbs them via mannose. Man@LPHFNPs-Roflumilast was encapsulated using the nano-into-micro technique in poly(vinyl alcohol)/leucine-based microparticles by spray-drying to produce an inhalable formulation (Craparo et al., 2022).

12.11 CONCLUSION AND FUTURE PROSPECTIVE

Biologics therapy might be revolutionized by using nano techniques in the ongoing fight against COPD. Future COPD therapy paradigms might be redefined by the confluence of biologics and nanotechnology, which would provide individualized and tailored therapies that address the complex molecular causes behind this crippling illness. NPs and nanofibers, for example, are powerful vehicles for targeted medication delivery that minimize systemic adverse effects while guaranteeing ideal therapeutic concentrations at the sick location. Furthermore, platforms provided by nanotechnology can surmount biological barriers present in the lungs, hence augmenting the bioavailability and effectiveness of biological agents. Bioengineered nanostructures that enable the continuous release of anti-inflammatory cytokines and gene therapy using NPs to improve lung tissue regeneration are just two examples of how nanobiologic advancements could shape the landscape of COPD treatments in the future. With the advancement of research, there is hope that personalized nanobiology interventions catered to specific patient profiles will one day enable COPD management to surpass its current constraints and usher in a new era of increased effectiveness, decreased side effects, and better quality of life for millions of people globally.

REFERENCES

Ahmad, F. J., Mittal, G., Jain, G. K., Malhotra, G., Khar, R. K., & Bhatnagar, A. (2009). Nano salbutamol dry powder inhalation: A new approach for treating broncho-constrictive conditions. *European Journal of Pharmaceutics and Biopharmaceutics*, *71*(2), 282–291.

Amore, E., Ferraro, M., Manca, M. L., Gjomarkaj, M., Giammona, G., Pace, E., & Bondì, M. L. (2017). Mucoadhesive solid lipid microparticles for controlled release of a corticosteroid in the chronic obstructive pulmonary disease treatment. *Nanomedicine*, *12*(19), 2287–2302.

Andrade, F., Rafael, D., Videira, M., Ferreira, D., Sosnik, A., & Sarmento, B. (2013). Nanotechnology and pulmonary delivery to overcome resistance in infectious diseases. *Advanced Drug Delivery Reviews*, *65*(13–14), 1816–1827.

Bahmanpour, A. H., Ghaffari, M., Ashraf, S., & Mozafari, M. (2020). Nanotechnology for Pulmonary and Nasal Drug Delivery. In M. Mozafari (ed.), *Nanoengineered Biomaterials for Advanced Drug Delivery* (pp. 561–579). Elsevier.

Beck-Broichsitter, M., Merkel, O. M., & Kissel, T. (2012). Controlled pulmonary drug and gene delivery using polymeric nano-carriers. *Journal of Controlled Release*, *161*(2), 214–224.

Bian, S., Cai, H., Cui, Y., Liu, W., & Xiao, C. (2021). Nanomedicine-based therapeutics to combat acute lung injury. *International Journal of Nanomedicine*, 2247–2269.

Bregoli, L., Movia, D., Gavigan-Imedio, J. D., Lysaght, J., Reynolds, J., & Prina-Mello, A. (2016). Nanomedicine applied to translational oncology: A future perspective on cancer treatment. *Nanomedicine: Nanotechnology, Biology and Medicine*, *12*(1), 81–103.

Chan, Y., MacLoughlin, R., Zacconi, F. C., Tambuwala, M. M., Pabari, R. M., Singh, S. K., … Dua, K. (2021). Advances in nanotechnology-based drug delivery in targeting PI3K signaling in respiratory diseases. *Nanomedicine*, *16*(16), 1351–1355.

Chikuma, K., Arima, K., Asaba, Y., Kubota, R., Asayama, S., Sato, K., & Kawakami, H. (2020). The potential of lipid-polymer nanoparticles as epigenetic and ROS control approaches for COPD. *Free Radical Research*, *54*(11–12), 829–840.

Craparo, E. F., Cabibbo, M., Scialabba, C., Giammona, G., & Cavallaro, G. (2022). Inhalable formulation based on lipid–polymer hybrid nanoparticles for the macrophage targeted delivery of roflumilast. *Biomacromolecules*, *23*(8), 3439–3451.

Craparo, E. F., Ferraro, M., Pace, E., Bondì, M. L., Giammona, G., & Cavallaro, G. (2017). Polyaspartamide-based nanoparticles loaded with fluticasone propionate and the in vitro evaluation towards cigarette smoke effects. *Nanomaterials*, *7*(8), 222.

Dong, T., Santos, S., Yang, Z., Yang, S., & Kirkhus, N. E. (2020). Sputum and salivary protein biomarkers and point-of-care biosensors for the management of COPD. *Analyst*, *145*(5), 1583–1604.

Doroudian, M., MacLoughlin, R., Poynton, F., Prina-Mello, A., & Donnelly, S. C. (2019). Nanotechnology based therapeutics for lung disease. *Thorax*, *74*(10), 965–976.

Doroudian, M., O'Neill, A., Mac Loughlin, R., Prina-Mello, A., Volkov, Y., & Donnelly, S. C. (2021). Nanotechnology in pulmonary medicine. *Current Opinion in Pharmacology*, *56*, 85–92.

Ellis, T., Chiappi, M., Garcia-Trenco, A., Al-Ejji, M., Sarkar, S., Georgiou, T. K., … Ryan, M. P. (2018). Multimetallic microparticles increase the potency of rifampicin against intracellular Mycobacterium tuberculosis. *ACS Nano*, *12*(6), 5228–5240.

Ernst, J., Klinger-Strobel, M., Arnold, K., Thamm, J., Hartung, A., Pletz, M. W., … Fischer, D. (2018). Polyester-based particles to overcome the obstacles of mucus and biofilms in the lung for tobramycin application under static and dynamic fluidic conditions. *European Journal of Pharmaceutics and Biopharmaceutics*, *131*, 120–129.

Gao, Y., Wang, J., Chai, M., Li, X., Deng, Y., Jin, Q., & Ji, J. (2020). Size and charge adaptive clustered nanoparticles targeting the biofilm microenvironment for chronic lung infection management. *ACS Nano*, *14*(5), 5686–5699.

Gaul, R., Ramsey, J. M., Heise, A., Cryan, S.-A., & Greene, C. M. (2018). Nanotechnology Approaches to Pulmonary Drug Delivery: Targeted Delivery of Small Molecule and Gene-Based Therapeutics to the Lung. In A. M. Grumezescu (ed.), *Design of Nanostructures for versatile Therapeutic Applications* (pp. 221–253). Elsevier.

Gill, S., Löbenberg, R., Ku, T., Azarmi, S., Roa, W., & Prenner, E. J. (2007). Nanoparticles: Characteristics, mechanisms of action, and toxicity in pulmonary drug delivery—a review. *Journal of Biomedical Nanotechnology*, *3*(2), 107–119.

Guan, W., Liu, Y., Ding, S., & Wang, Y. (2022). Ciprofloxacin nanocrystals and N-acetylcysteine co-solidified powders for pulmonary drug delivery: Development and in vitro and in vivo characterization. *Journal of Nanoparticle Research*, *24*(2), 41.

Guo, Y., Ma, Y., Chen, X., Li, M., Ma, X., Cheng, G., … Sun, B. (2023). Mucus penetration of surface-engineered nanoparticles in various pH microenvironments. *ACS Nano*, *17*(3), 2813–2828.

Gupta, N., Hatoum, H., & Dy, G. K. (2014). First line treatment of advanced non-small-cell lung cancer–specific focus on albumin bound paclitaxel. *International Journal of Nanomedicine*, 209–221.

Hussain, M. S., Sharma, P., Dhanjal, D. S., Khurana, N., Vyas, M., Sharma, N., … Sohal, S. S. (2021). Nanotechnology based advanced therapeutic strategies for targeting interleukins in chronic respiratory diseases. *Chemico-Biological Interactions*, *348*, 109637.

Ibarra-Sánchez, L. Á., Gámez-Méndez, A., Martínez-Ruiz, M., Nájera-Martínez, E. F., Morales-Flores, B. A., Melchor-Martínez, E. M., … Iqbal, H. M. (2022). Nanostructures for drug delivery in respiratory diseases therapeutics: Revision of current trends and its comparative analysis. *Journal of Drug Delivery Science and Technology*, *70*, 103219.

Jacobs, C., & Müller, R. H. (2002). Production and characterization of a budesonide nanosuspension for pulmonary administration. *Pharmaceutical Research*, *19*, 189–194.

Jelic, T. M. (2019). Emphysema *Update in Respiratory Diseases*: IntechOpen.

Joshi, N. (2018). Nanotechnology enabled inhalation of bio-therapeutics for pulmonary diseases: Design considerations and challenges. *Current Pathobiology Reports*, *6*, 225–231.

Khang, D., Carpenter, J., Chun, Y. W., Pareta, R., & Webster, T. J. (2010). Nanotechnology for regenerative medicine. *Biomedical Microdevices*, *12*, 575–587.

Kim, V., & Criner, G. J. (2013). Chronic bronchitis and chronic obstructive pulmonary disease. *American Journal of Respiratory and Critical Care Medicine*, *187*(3), 228–237.

Konstan, M. W., Davis, P. B., Wagener, J. S., Hilliard, K. A., Stern, R. C., Milgram, L. J., … Gedeon, C. R. (2004). Compacted DNA nanoparticles administered to the nasal mucosa of cystic fibrosis subjects are safe and demonstrate partial to complete cystic fibrosis transmembrane regulator reconstitution. *Human gene Therapy*, *15*(12), 1255–1269.

Kuzmov, A., & Minko, T. (2015). Nanotechnology approaches for inhalation treatment of lung diseases. *Journal of Controlled Release*, *219*, 500–518.

Larsson, K., Ekberg-Jansson, A., Stridsman, C., Hanno, M., & Vanfleteren, L. E. (2021). Adherence to treatment recommendations for chronic obstructive pulmonary disease-results from the swedish national airway register. *International Journal of Chronic Obstructive Pulmonary Disease*, 909–918.

Li, Z., Luo, G., Hu, W. P., Hua, J. L., Geng, S., Chu, P. K., … Yu, X. F. (2020). Mediated drug release from nanovehicles by black phosphorus quantum dots for efficient therapy of chronic obstructive pulmonary disease. *Angewandte Chemie*, *132*(46), 20749–20757.

Liu, B., Cao, W., Qiao, G., Yao, S., Pan, S., Wang, L., … Cui, D. (2019). Effects of gold nanoprism-assisted human PD-L1 siRNA on both gene down-regulation and photothermal therapy on lung cancer. *Acta Biomaterialia*, *99*, 307–319.

Liu, D., Long, M., Gao, L., Chen, Y., Li, F., Shi, Y., & Gu, N. (2022). Nanomedicines targeting respiratory injuries for pulmonary disease management. *Advanced Functional Materials*, *32*(22), 2112258.

Mohamed, A., Kunda, N. K., Ross, K., Hutcheon, G. A., & Saleem, I. Y. (2019). Polymeric nanoparticles for the delivery of miRNA to treat chronic obstructive pulmonary disease (COPD). *European Journal of Pharmaceutics and Biopharmaceutics*, *136*, 1–8.

Moro, M., Di Paolo, D., Milione, M., Centonze, G., Bornaghi, V., Borzi, C., … Ponzoni, M. (2019). Coated cationic lipid-nanoparticles entrapping miR-660 inhibit tumor growth in patient-derived xenografts lung cancer models. *Journal of Controlled Release*, *308*, 44–56.

Pavord, I. D. (2018). Biologics and chronic obstructive pulmonary disease. *Journal of Allergy and Clinical Immunology*, *141*(6), 1983–1991.

Pi, J., Shen, L., Yang, E., Shen, H., Huang, D., Wang, R., … Cai, J. (2020). Macrophage-targeted isoniazid–selenium nanoparticles promote antimicrobial immunity and synergize bactericidal destruction of tuberculosis bacilli. *Angewandte Chemie*, *132*(8), 3252–3260.

Seshadri, D. R., & Ramamurthi, A. (2018). Nanotherapeutics to modulate the compromised micro-environment for lung cancers and chronic obstructive pulmonary disease. *Frontiers in Pharmacology*, *9*, 340406.

Sharafkhaneh, A., Hanania, N. A., & Kim, V. (2008). Pathogenesis of emphysema: From the bench to the bedside. *Proceedings of the American Thoracic Society*, *5*(4), 475–477.

Sharma, A., Mehta, V., Parashar, A., Patrwal, R., & Malairaman, U. (2016). Solid lipid nanoparticle: Fabricated through nanoprecipitation and their physicochemical characterization. *International Journal of Pharmacy and Pharmaceutical Sciences*, *8*(10), 144–148.

Sharma, P., Mehta, M., Dhanjal, D. S., Kaur, S., Gupta, G., Singh, H., … Bakshi, H. A. (2019). Emerging trends in the novel drug delivery approaches for the treatment of lung cancer. *Chemico-Biological Interactions*, *309*, 108720.

Shukla, S. D., Vanka, K. S., Chavelier, A., Shastri, M. D., Tambuwala, M. M., Bakshi, H. A., … O'toole, R. F. (2020). Chronic Respiratory Diseases: An Introduction and Need for Novel Drug Delivery Approaches In K. Dua, P. M. Hansbro, R. Wadhwa, M. Haghi, L. G. Pont & K. A. Williams (eds.), *Targeting Chronic Inflammatory Lung Diseases Using Advanced Drug Delivery Systems* (pp. 1–31). Elsevier.

Suk, J. S., Lai, S. K., Boylan, N. J., Dawson, M. R., Boyle, M. P., & Hanes, J. (2011). Rapid transport of muco-inert nanoparticles in cystic fibrosis sputum treated with N-acetyl cysteine. *Nanomedicine*, *6*(2), 365–375.

Suk, J. S., Lai, S. K., Wang, Y.-Y., Ensign, L. M., Zeitlin, P. L., Boyle, M. P., & Hanes, J. (2009). The penetration of fresh undiluted sputum expectorated by cystic fibrosis patients by non-adhesive polymer nanoparticles. *Biomaterials*, *30*(13), 2591–2597.

Varshosaz, J., Ghaffari, S., Mirshojaei, S., Jafarian, A., Atyabi, F., Kobarfard, F., & Azarmi, S. (2013). Biodistribution of amikacin solid lipid nanoparticles after pulmonary delivery. *BioMed Research International*, *2013*.

Vij, N. (2012). Synthesis and evaluation of airway targeted PLGA nanoparticles for drug delivery in obstructive lung diseases. *Nanoparticles in Biology and Medicine: Methods and Protocols*, 303–310.

Vij, N. (2017). Nano-based rescue of dysfunctional autophagy in chronic obstructive lung diseases. *Expert Opinion on Drug Delivery*, *14*(4), 483–489.

Vij, N., Min, T., Bodas, M., Gorde, A., & Roy, I. (2016). Neutrophil targeted nano-drug delivery system for chronic obstructive lung diseases. *Nanomedicine: Nanotechnology, Biology and Medicine*, *12*(8), 2415–2427.

Wadhwa, R., Aggarwal, T., Thapliyal, N., Chellappan, D. K., Gupta, G., Gulati, M., … Hansbro, P. M. (2019). Nanoparticle-based drug delivery for chronic obstructive pulmonary disorder and asthma: Progress and challenges. *Nanotechnology in Modern Animal Biotechnology*, 59–73.

Wan, F., Bohr, S. S.-R., Kłodzińska, S. N., Jumaa, H., Huang, Z., Nylander, T., … Sternberg, C. (2019). Ultrasmall TPGS–PLGA hybrid nanoparticles for site-specific delivery of antibiotics into *Pseudomonas aeruginosa* biofilms in lungs. *ACS Applied Materials & Interfaces*, *12*(1), 380–389.

Wang, K., Feng, Y., Li, S., Li, W., Chen, X., Yi, R., … Hong, Z. (2018). Oral Delivery of bavachinin-loaded PEG-PLGA nanoparticles for asthma treatment in a murine model. *Journal of Biomedical Nanotechnology*, *14*(10), 1806–1815.

Wechsler, M. E. (2018). Current and emerging biologic therapies for asthma and COPD. *Respiratory Care*, *63*(6), 699–707.

Wilson, L., Devine, E., & So, K. (2000). Direct medical costs of chronic obstructive pulmonary disease: Chronic bronchitis and emphysema. *Respiratory Medicine*, *94*(3), 204–213.

Xu, Y., Liu, H., & Song, L. (2020). Novel drug delivery systems targeting oxidative stress in chronic obstructive pulmonary disease: A review. *Journal of Nanobiotechnology*, *18*, 1–25.

Yadav, A. K., Gu, W., Zhang, T., Xu, X., & Yu, L. (2023). Current perspectives on biological therapy for COPD. *COPD: Journal of Chronic Obstructive Pulmonary Disease*, *20*(1), 197–209.

Yadollahi, R., Vasilev, K., & Simovic, S. (2015). Nanosuspension technologies for delivery of poorly soluble drugs. *Journal of Nanomaterials*, *2015*, 1–13.

Yhee, J. Y., Im, J., & Nho, R. S. (2016). Advanced therapeutic strategies for chronic lung disease using nanoparticle-based drug delivery. *Journal of Clinical Medicine*, *5*(9), 82.

Yoo, D., Guk, K., Kim, H., Khang, G., Wu, D., & Lee, D. (2013). Antioxidant polymeric nanoparticles as novel therapeutics for airway inflammatory diseases. *International journal of pharmaceutics*, *450*(1–2), 87–94.

Yousuf, A., & Brightling, C. E. (2018). Biologic drugs: A new target therapy in COPD? *COPD: Journal of Chronic Obstructive Pulmonary Disease*, *15*(2), 99–107.

Zhang, C. Y., Lin, W., Gao, J., Shi, X., Davaritouchaee, M., Nielsen, A. E., … Wang, Z. (2019). pH-responsive nanoparticles targeted to lungs for improved therapy of acute lung inflammation/injury. *ACS Applied Materials & Interfaces*, *11*(18), 16380–16390.

Zhang, L.-J., Xing, B., Wu, J., Xu, B., & Fang, X.-L. (2008). Biodistribution in mice and severity of damage in rat lungs following pulmonary delivery of 9-nitrocamptothecin liposomes. *Pulmonary Pharmacology & Therapeutics*, *21*(1), 239–246.

13 Nanotheranostics for Chronic Obstructive Pulmonary Disease (COPD)

Status and Implications

Richa Joshi and Hem C. Joshi
Department of Botany, D.S.B. Campus, Kumaun University
Nainital, Uttarakhand, India

13.1 INTRODUCTION

Chronic obstructive pulmonary disease (COPD) is a common lung disease causing restricted airflow and breathing problems (WHO, 2023). COPD comprises two main conditions, emphysema and chronic bronchitis, and is an inflammatory condition that involves airways, lung parenchyma, and pulmonary vasculature. It is characterized by airflow limitation in a progressive manner and is not fully reversible. COPD has resulted in early mortality, high death rates, and significant cost to health systems (Raherison, 2009). People with COPD have their lungs damaged or clogged with phlegm. Cough accompanied with phlegm, difficulty in breathing, wheezing, and tiredness are common symptoms seen in people with COPD. Wheezing and fatigue are also common symptoms in people with COPD. The main cause of COPD is long-term or active smoking and air pollution, along with occupational exposure coming from exposure to dust, chemicals, and fumes. Because of the weakening of the overall immunity, people with COPD are at higher risk of other health problems. Although COPD is not curable, symptoms can be improved by avoiding smoking and long-term exposure to air pollution and getting vaccinated to prevent infections. COPD can be treated with medications, oxygen, and pulmonary rehabilitation. With the increasing global population, the risk of COPD is predicted to increase in the coming years. Studies has also found that the disease shows considerable variability among populations and can affect each population group differently (Mannino, 2007). Lung infections, heart disease, fatigue, and anxiety along with depression are complications involved in COPD.

The term theranostics was used by John Funkhouser in 1988. It is made up of combination of two terms, *therapeutics* and *diagnostics* technology using suitable combination of drugs (Gilham, 2002). One technique involved the use of a radioactive drug to diagnose and a second radioactive drug to deliver the therapy. It is an image-guided therapy and combines a few forms of diagnostic testing to identify the presence of a molecular target having a specific drug (Lee, 2011). Molecular imaging is used for diagnosis, providing an essential means for detecting diseases non-invasively. One of the main reason for an emerging trend of theranostics is the revolutionary approach that will provide precise therapy selection on the basis of specific molecular features of disease, advanced ways to monitor therapy response and predictive adverse effects being the basis of personalized medicine (Lee, 2011). Theranostics includes imaging and targeting. Radionuclide imaging techniques such as positron emission tomography (PET) and radioactive tracer as a drug are used for treatment. These allows personalized treatment based on patient characteristics.

DOI: 10.1201/9781003498926-13

Nanotechnology refers to the technology that operates at the nanoscale. It is the understanding and control of matter at dimensions roughly 1 nm to 100 nm, where unique phenomena enable novel applications. Nanotechnology includes the forming and use of materials, structures, devices, and systems that have unique properties because of their small size as well as control of materials at nanoscale (Allhoff, 2009). Theranostics nanoparticles with both therapeutic and imaging abilities have the promise to revolutionize diagnosis, therapy, and prognosis. Nanomedicine is extensively applied to diagnose, treat, and prevent pulmonary diseases. Nanomedicine offers a unique opportunity to develop nanopackaged therapeutics in an aerosolized format for delivery to the lung. It provides the opportunity to target COPD through the fabrication of various nanocarriers. The application of theranostics nanoparticles that combine both therapeutic and diagnostic capability in a single dose has shown a promising future in the biomedical field toward personalized medicine (Kelkar, 2011).

COPD is a leading cause death globally every year. It is a respiratory disease that causes obstruction in airflow, leading to significant damage to the lungs. COPD includes chronic bronchitis and emphysema. Chronic bronchitis is the inflammation and narrowing of airway tubes called bronchi, associated with mucus build-up, and emphysema is the damage to air sacs present in the lungs, interrupting the outflow of air from the lungs (Haas, 2008).

Causes and risk factors:

1. ***Active smoking***: Cigarette smoking is the most common cause of COPD. Long-term smoking damages the lungs and airways.
2. ***Air pollution***: Long-term exposure to air pollution is also one of the rising factors that leads to COPD.
3. ***Occupational exposure***: Exposure to chemical fumes, dust, smoke from cooking, and heating fuels without ventilation and secondhand smoke in the air from other people smoking.
4. ***Asthma***: A lung disease that causes swelling and narrowing of the airways. People that have asthma are also at high risk of developing COPD.
5. ***Early childhood infections***: Infections related to respiratory tracts and lungs at an early age are also common causes that lead to COPD later.
6. ***Genetics***: Conditions like alpha-1 antitrypsin (AAT) deficiency that runs in families. AAT deficiency leads to a high risk of developing COPD at an early age. Vitamin D deficiency and mutations in CHRNA are other genetic conditions that cause COPD.

COPD is a life-threatening condition that causes significant damage to airways and lungs. Pathophysiology is the understanding of adverse functional changes associated with diseases (Braun, 2007). In people with COPD, this starts with damage to the airways and tiny air sacs in the lungs. The disease progresses from a cough with mucus to an extreme difficulty in breathing. In COPD, less air enters and comes out of lungs because the airways and air sacs lose their elasticity, walls between several air sacs are destroyed, airways walls become thick and inflamed, and airways make more mucus than natural and often become clogged. Two main conditions of COPD are chronic bronchitis and emphysema, affecting different parts of lungs. When breathed, air travels down the bronchial tubes that branch into thousands of small thin tubes called bronchioles, which in turn end in bunches of tiny round air sacs called alveoli, which are elastic. Alveoli contains capillaries at their end. Through capillaries, oxygen moves from the lungs to the bloodstream and carbon dioxide moves from the blood into the capillaries and finally into the lungs before it is exhaled. Emphysema directly affects the alveoli, damaging the fibers that make wall of the alveoli, resulting in loss of their elasticity and ability to coil during exhale, making it difficult to exhale carbon dioxide out of the lungs. Emphysema is a structural change seen in COPD, and an irritant often leads to an inflammatory response due to recruitment of neutrophils and macrophages. They release multiple inflammatory mediators.

Bronchitis results in inflammation of the lung airways, resulting in subsequent mucus production. Prolonged bronchitis results in chronic bronchitis. Until the disease is more advanced, serious symptoms of COPD are not usually seen. Being a respiratory disease, a difficulty in breathing that gradually gets worse is the most common symptom. Wheezing, tightness of breath during physical activity, tiredness, frequent infection of lungs, and weight loss are other common symptoms of COPD.

13.2 CURRENT TREATMENT OPTIONS AND LIMITATIONS

The damage done by COPD is not reversable. However, there are some treatments and medications to control the symptoms of COPD and improve quality of life. Diagnosis of COPD involves tests that evaluate the conditions of the lungs and severity of the disease. Diagnosis is usually done by:

a. **Complete blood count,** to check infection causing the symptoms. Due to infection, the white blood cell (WBC) count is high.
b. **Arterial blood gas test,** to measure oxygen levels in the blood.
c. **X-ray** of chest is done to examine any abnormality causing the symptoms.
d. **Pulmonary function test (PFT),** to check functioning of lungs and severity of COPD.

Treatment:

The main aim of treatment is to control symptoms, reduce risk factors, delay progression through medications, physical therapy, and future complications.

Smoking cessation and pulmonary rehabilitation are non-medicinal ways to control COPD.

Medications

a. **Bronchodilators** are used to relax the muscles of the airways and improve breathing. They are given through inhalers. Albuterol, levalbuterol, glycopyrronium, tiotropium, aclidinium, vilanterol, ipratropium, and umeclidinium are few drugs used for treatment.
b. **Glucocorticosteriods** are used to reduce inflammation in the airways. They are used as inhalers. Beclomethasone, betamethasone, cortisone, budesonide, and dexamethasone are common drugs used for this purpose.
c. **Mucolytics** used in cases of severe COPD attack. N-acetylcysteine is used for this purpose.

Therapy

a. **Oxygen therapy** is performed during exacerbations.
b. **Pulmonary rehabilitation** involves learning, practicing, and improving breathing and physical activities with the help of a group of specialists.
c. **Surgery** is performed to replace severely damaged lungs with healthy lungs from a donor.

Limitations

Damage caused by COPD cannot be undone but can be managed through treatments and medication. However, even after management of symptoms, several complications are observed such as:

a. Acute or chronic respiratory failure.
b. Pulmonary hypertension.
c. Acute exacerbation of COPD.
d. Cor pulmonale.
e. Weight loss.
f. Bacterial infections.
g. Adverse reaction to glucocorticoids.

13.3 THERANOSTICS: A NOVEL APPROACH

The word theranostics is a combination of the two words *therapeutics* and *diagnostics* for use as an image-guided therapy (Thanou, 2018). Diagnostic testing is combined with concerning COPD. Typically, diagnostic testing for COPD is combined with clinical assessments, imaging studies (like chest X-rays or CT scans), and pulmonary function tests (such as spirometry).. This is achieved through molecular imaging. It is a personalized approach that uses both diagnosis and therapy tools as a part of treatment. PET scan imaging is used to identify specific targets. After confirmation of the presence of target through the scan, a radioactive drug is used for further treatment. The drug is designed in such a way that it attacks only specific targets while avoiding healthy areas. The radioactive drug that does not reach the target is quickly passed out of the body. Theranostics involve radionuclide diagnostics procedures and radionuclide therapies using appropriate radiopharmaceutics and treatments that target specific receptors (Królicki, 2022). Theranostics has provided a transition from conventional medicine to personalized medicine. It is based upon the development of personalized medical treatment plan based on specific requirements of individuals, which results in providing the right drug for the right patient at the right time (Chan, 2011). Genetics is a significant part of theranostics, while pharmacogenetics, proteomics, and biomarker profiling form the fundamental part of theranostics. It provides a cost-effective specific treatment plan. Being a proposed process of diagnostic therapy, it provides an opportunity for the development of a personalized treatment plan on the basis of test results based on taking new medicine.

13.3.1 Advantages of Theranostics Over Conventional Treatments

Theranostics is an emerging targeted, safe, and efficient pharmacotherapy that focuses on patient-centered care (Shrivastava, 2019). Innovation in science and technology has led to many breakthroughs and successes, resulting in advancement and improvement in quality of life. Traditional medicine been based on rationale history taking requires several clinical examinations to arrive at the cause of problem. All of this was initially done with the help of herbs and, with advancement in technology, it progressed to chemical products of plants and other different sources. However, this practice has not been able to treat certain conditions, as the treatments are based on a holistic approach, not focusing on individual requirements and suitability. The arrival of diagnosis and treatment that is completely focused on an individual patient's medical history, symptoms, personal and family history, and laboratory analysis along with imaging and scanning can help diagnose disease and develop patient-centered treatment plans and medication (Jeelani, 2014). Theranostics is one such approach that is a combination of both therapeutics and diagnosis. It has transformed conventional treatment method into a more personalized treatment. Patient-centered care has become a priority and widely used treatment for treating patients with drugs. It has become an important transition from universal medicine, which emphasizes a "one medicine fits all" practice, to a more personalized and patient-centered practice (Jeelani, 2014).

Unlike conventional treatment, it has become a unique approach for custom made drug therapy that is based on patient's response to drugs. Thus, theranostics has led to a transition from trial and error medicine and one medicine suits all practice to more personalized medicine (Vogenberg, 2010). The theranostics approach is considered as "P4 medicine", predictive, preventive, personalized, and participatory (Assadi, 2014).

13.4 ROLE OF NANOTECHNOLOGY IN ADVANCING THERANOSTICS

Nanotechnology is an approach that operates at nanoscale. Theranostic nanomedicine is an emerging field in therapeutics and nanocarriers have made a revolutionary change in healthcare systems. Micelles, liposomes, polymers, dendrimers, metallic and non-metallic inorganic nanoparticles, nanobubbles, carbon nanotubes (CNTs) and other nanocarriers provide a specific and effective target-specific diagnosis and drug-delivery system. Apart from improved drug diagnosis and delivery, they can also monitor therapeutic responses to medication.

Quantum dots, iron oxide nanoparticles (IONPs), CNTs, and gold/silica nanoparticles are some nanostructures that have precise and remarkable surface characteristics that can perform target-specific actions, thus improving the approach of personalized medicine (Xie, 2010). Nanoparticles can be engineered into different sizes and shapes. Nanomedicine thus forms an important part of theranostics. Some nanostructures used in the field of theranostics are described in detail next:

Iron oxide nanoparticles: IONPs have a diameter in the range of 1–100 nm. They can bind to proteins, drugs, antibodies and nucleotides. They are targeted toward a cell, tissue, or organ with the help of external magnetic field.

Gold nanoparticles: Synthesized by utilizing several reducing reagents through their controlled reduction of aqueous solutions. Citrate is the most common reducing agent used for this purpose because of its ability to generate gold nanospheres of monodisperse nature (Tyagi, 2016).

Carbon nanotubes: CNTs have optical, chemical, and electrical properties. They are cylindrical in shape and this makes them widely used for drug delivery. CNTs consist of graphene sheets rolled in a cylindrical manner, with an open or capped end (Passi, 2020). When made up of single graphene sheets, they turn into a monolayered nanotubes, while many layered graphene sheets result in the formation of multiwalled CNTs. Having a large surface area makes them widely preferred for therapeutic uses.

Liposomal drug delivery: Liposomes are spherical in shape and contain one lipid layer. Phospholipid phosphatidylcholine is also found in liposomes. They also include other lipids such as phosphatidylethanolamine. Liposomes are nano drug carriers with higher accuracy and reduced toxic effects.

13.5 NANOTECHNOLOGY IN COPD THERANOSTICS

13.5.1 Nanoparticle-Based Drug Delivery Systems

The use of nanoparticles has led theranostics into new advancements. Nanoparticles have different chemical properties, surface properties, nature, composition, and characteristics that can be engineered into required shapes and sizes. Nanotheranostic medicine is widely preferred over other theranostic means because its properties are offered in a single platform (Muthu, 2014). Nanomedicine is applied to solve emerging respiratory diseases. Enhanched drug accumulation at the target site, extended circulation time for drugs, reduced drug degradation, and precise administration of a drug at the target site are some of the advantages that nanocarrier drug delivery offers.

13.5.1.1 Types of Nanoparticles used in COPD Theranostics

COPD is one of the most life threating disease affecting people annually all over the world. Nanotechnology has offered opportunities for manipulations of many nanoparticle vehicles to tackle COPD (Hussain, 2021). Nanoparticles used in COPD posses several unique properties and have advanced the field of therapeutics. Several nanoparticles are engineered as nanocarriers for diagnosis and treatment of chronic lung diseases (Yhee, 2016). These nanoparticles are divided into various groups based on their component element and characteristic dimensionalities. They are:

Liposomes: Widely used lipid based carriers. They consist of a single lipid bilayer or multiple lipid bilayers. Liposomes are produced using amphiphilic phospholipids and cholesterol. They are small, spherical in shape, and in the size range of 400 nm or less. Nanosized diagnostic substances such as IONPs, quantum dots, and gold nanoparticles are enclosed within theranostic liposomes for effective drug delivery. The external surface of liposomes is modified with the help of polymers such as polyethylene glycol (PEG), which helps in prolonging the blood circulation time of the liposomes. Lipid-based nanoparticles are highly suitable for inhalation delivery of various pulmonary drugs.

Solid lipid nanoparticles: Lipid-based nanoparticles are also called nanostructural lipid carriers. They can be engineered easily from the lipid-like substances such as bicompatible cholesterol and phosphatidylcholine. They can carry large amounts of drugs and facilitate easy cellular uptake. Examples include, dipalmitoyl phosphatidylcholine (DPPC) and dipalmitoyl phosphatidylethanolamine-methoxy (polyethene glycol) largely used for drug delivery to the lungs. Solid lipid nanoparticles are less than 100 nm in size.

Polymeric/magnetic nanoparticles: Made from biodegradable polymers. They are used in theranostics for their specific targeted delivery of diagnostic and therapeutic agents. Use of biodegradable polymers is made to provide biological compatibility with less cytotoxicity. Poly (D,L,-lactide-co-glycolide) (PLGA) and poly (D,L,-lactide-co-glycolide)-polyethylene glycol (PLGA-PEG) are most commonly used biodegradable polymers for the fabrication of theranostic polymeric nanoparticles.

Magnetic nanoparticles are IONPs comprised of magnetite or hematite.

Dendrimers: Being synthetic nanoparticles, they possess high branching in their structure and are generally spherical in shape. They usually range from 10 nm to 100 nm in size. They are extensively employed in drug delivery. They briskly penetrate the circulatory system and reach the lungs. They are generally monodispersed and their size and formulation can be controlled.

Nanospheres: Nanocarriers of varying size, ranging from 10–200 nm. They are specifically designed to deliver drugs and genes. Amorphous and crystalline in nature, they have ability to provide protection to drugs from the action of enzyme and chemical degradation. Albumin nanospheres, gelatin nanospheres, dextran nanospheres, and polylactic nanospheres are some common biodegradable nanospheres.

Inorganic nanoparticles: Substances such as metal, metal oxide, which include gold, silver, silica cerium oxide, and iron oxide, are largely used in nanotherapeutics. Unique plastic and magnetic properties of gold and IONPs make them suitable for use as imaging contrast agents in several techniques such as computed tomography (CT) scan, magnetic resonance imaging (MRI), and PET.

Gold nanoparticles: They are based on a gold core and extensively used in nanotheranostics in COPD. They are made up of a 1.5–10 nm core size, which provides a large surface area for drug and ligand conjugations. They are synthesized by the chemical treatment of hydrogen tetrachloroaurate. They are target specific and interact either through non-covalent interaction or covalent interactions. Gold particles are known for their monodispersity, diagnostic characteristics, controllable core size, low toxicity, large surface area to volume ratio, ability to bind to molecules through gold–sulfur bonds, and easy fabrication.

Carbon nanomaterials: Classified into sp2 nanocarbon materials which include zero-dimensional (0D) fullerene, one-dimensional (1D) CNTs, two-dimensional (2D) graphene, and carbon dots (i.e., nanoclusters with <10 nm size). They make ideal material for use in theranostic due to their large surface area, ability to encapsulate therapeutic agents and surface modification suitability. They are generally cylindrical in shape and classified into single-walled CNTs or multiwalled CNTs based upon the number of layers of graphene sheets.

13.6 MECHANISMS OF TARGETED DRUG DELIVERY

At present, there is no specific treatment mechanism for COPD. The only way through which the symptoms are controlled is improving airflow limitation. Generally, drug intervention is not followed in COPD under stable periods, but under severe attack conditions, antibiotics, inhaled corticosteroids, bronchodilators, and some other drugs are widely used in clinical practice (Qaseem, 2011). Targeted drug delivery emphasizes the delivery of drugs directly to specific sites. This approach has enhanced therapeutic efficacy and reduced side effects associated with conventional drug delivery. Targeted drug delivery aims to providing successful direction to the therapeutic agent and its accumulation at the targeted site (Bae, 2011). For this purpose, the agent-loaded system has to be accumulated in the physiological system for preferred time. Within this duration, it has to evade the immunological system, target specific cells or tissues, and release loaded therapeutic agents.

13.7 IMAGING MODALITIES IN COPD DIAGNOSIS AND MONITORING

Imaging modalities are a useful tool in the diagnosis of COPD. Advancement in imaging techniques such as densitometry, MRI, CT, PET, single photon emission computed tomography, micro-CT, optical coherence tomography (OCT), and electrical impedance tomography (EIT) have been used for diagnosis and monitoring of COPD.

13.8 OVERVIEW OF IMAGING TECHNIQUES

Magnetic resonance imaging: The use of MRI to study lung disease was limited by poor spatial resolution and the low signal-to-noise ratio of lung parenchyma (Biederer, 2012). This technique involves large magnetic fields that influence and alter the behavior of individual atoms. It utilizes nuclear spin properties of hydrogen atoms present in abundance in water. In a natural state, hydrogen ions spin in random orientation on an axis. With the application of a strong and uniform magnetic field, they become aligned parallel and anti-parallel to the field; when the magnetic field in switched off, the spinning protons return to their natural state with a release/burst of radiofrequency energy. This energy is detected by the MRI scanner and is used to construct the image. This technique offers excellent anatomical resolution in tissue with high water content, but as the lungs are mostly filled with air, they have a low density of hydrogen ions and offer low signal intensity and a poor signal-to-noise ratio.

Computed tomography: CT provides insight into pulmonary structure. It involves an X-ray beam source with detectors in a row placed opposite to the source. Detectors are assembled in a circular arrangement that rotates around the patient through 180 degrees. The X-ray beams are measured as they pass through tissues of varying densities (Milne, 2014). This allows the construction of a 2D axial slice through the body. The digitized image slices comprise pixels with the relative densities that are measured in Hounsfield units (HU), representing the tissue density of that particular location. High-resolution CT is obtained using thinner detectors that allow the reduced thickness of the axial slices, around 1mm. The physical size of the X-ray detectors is one of the determinants of resolution and scanning is another more important determinant of spatial resolution. At present, all the CT scans are performed in helical or spiral mode.

Micro-CT: Used to examine the microstructures of lungs. Through micro-CT, *ex vivo* identification and examination of individual alveoli and terminal bronchioles takes placed, which can not be done with standard Multidetector Computed Tomography (MDCT) (Milne, 2014). Just like conventional CT, it provides full volumetric data and fascinating insights into the structural changes in COPD.

Positron emission tomography: A 3D nuclear imaging technique. It utilizes radioisotopes that emit positrons as they decay. A positron is a subatomic particle presents in the nucleus. It encounters free electrons as it moves out of the nucleus. They combine and annihilate each other, emitting two identical beams of gamma radiation at 180 degrees to each other. As they arise from the same source, the location of the source can be detected geometrically through them, the scattering of beams, random coincidences, and attenuation as the beams travel through tissues of different densities (Milne, 2014). This image noise is corrected during image processing through techniques like the CT scan. Fluorine-18, which has a life of approximately 110 minutes, is a commonly used PET radioisotope.

It is attached to fluorodeoxyglucose (FDG) to form the radiotracer 18F-FDG. This radiotracer is then picked up by metabolically active tissues.

Optical coherence tomography: The field of interventional pulmonology has given rise to OCT. It involves the measurement of the lung structure from an endobronchial approach. OCT uses light waves and involves insertion of a near-infrared optical probe into the airway along with sensors to detect back-scattered and reflected light waves (Milne, 2014). It provides enough resolution to distinguish between different types of tissues with the airways. OCT is an important imaging technique used to examine the anatomical properties of small airways that are limited only by the physical reach of the probe. It provides more precise measurement of airway size and is a useful tool for analyzing structural and functional changes in COPD *in vivo* with high resolution.

13.9 INTEGRATION OF IMAGING WITH THERAPEUTIC NANOPARTICLES

Nanoparticles play a major role in imaging and drug delivery. Nanotheranostics combines both imaging and targeted drug delivery in one nanoparticle for both disease diagnosis and therapy. It is also used for image-guided drug delivery. The advancement in nanotechnology and biotechnology at present has led to development and advancement of nanoparticles with multifunctionality in nanomedicine. Multimodal imaging and theranostics are two highly advanced techniques where the advancement of nanoparticles is maximized (Liu, 2020). Some of the highly precise molecular imaging agents are developed through the use of nanotechnology. Integration of several imaging agents having different properties into multifunctional nanoparticles allows precise and quick diagnosis though multimodal imaging (Lee, 2012). Thus, nanotheranostics helps in the development of highly significant contrasting agents for most imaging techniques, as nanoparticles have less toxicity, enhanced permeability, and a high retention effect (Siafaka, 2021). Distribution, blood circulation, half-life, cellular uptake, penetration, and the targeting of tissues are influenced by the size of nanoparticles. Some of the nanoparticles use as contrast agents are:

a. Perfluorocarbon, used in imaging angiogenesis, cancer metastases, and blood clots.
b. Fullerenes, used in MRI to enhance contrast.
c. Quantum dots, used in specific cellular imaging.
d. Silicon nanoparticles, usually coated with a conductive layer for enhancing MRI.
e. Gold nanoparticles (core shell structure), used in ultrasound and also in X-ray and CT scans.
f. Liposomes, inorganic nanoparticles, and superparamagnetic iron-oxide nanoparticles coated with dextran are used for MRI imaging.

13.10 CURRENT STATUS OF NANOTHERANOSTICS FOR COPD

Nanotheranostics is the combination of therapeutics and diagnosis in a single approach. It offers more personalized medicine and targeted drug delivery. Nanotheranostics has been used widely in several medications including COPD, cancer research, cardiovascular diseases, etc. Recent studies in theranostics are focused on developing more personalized medicines that are patient-centric. Personalized medicine has been proven to be revolutionary in the treatment of COPD as the disease symptoms vary from person to person. Researchers are also focused on improving the efficacy and efficiency of nanoparticles used for targeted drug delivery. Imaging approaches like MRI, CT, PET, SPEC (Single Photon Emission Computed Tomography), and PAT (Photoacoustic Tomography) are also being advanced with recent technologies for early diagnosis. Continued research and trials are important for developing new techniques to control the spread of the disease and to ensure that nanoparticles and other novel approaches being used in treatment of COPD through nanotheranostics have minimum side effects and no toxicity related. It is also essential to develop more advanced treatment methodology over conventional treatment methods.

13.11 EFFICACY AND SAFETY CONSIDERATION

Airway defense, severe inflammation, and mucous hypersecretion are some of the major challenges involved in the delivery and therapeutic efficacy of the nanodelivery technique. Controlling chronic inflammation is the root of treatments required for COPD. Nanotechnology allows for the fabrication of various nanocarriers that provide an effective means for targeting COPD. Toxicity assessment becomes essential when nanoparticles are used for the purpose of treatment, as only limited research has been conducted to assess the toxic effects of the nanotheranostics. Efficacy of a targeted drug delivery system is an important integral part of theranostics. It allows precise drug delivery to the targeted cell or tissue, but it may also result in potential side effects in the system. Education of the patients and general public and awareness regarding the safety and potential effects of the treatments are also important. Patients should be made more aware of the symptoms and treatment procedures. They must be encouraged to actively participate in decision-making. Nanoparticles must not be used above their set limit and there must be flexibility in the treatment process.

13.12 CHALLENGES AND FUTURE DIRECTIONS

The concept of nanotheranostics has become an important approach for the treatment of COPD. Targeted drug delivery systems that involve the use of a nanocarrier are very efficient. But there are possible potential systematic side effects. There is a need for more research to assess the toxicity of nanocarriers. Personalized medications prove to be significant for the management of symptoms but require more detailed analysis of patient's conditions, which can take time in cases of severe conditions. Hence, it must offer flexibility of use for patients in whom no positive results are seen with the use of one type of medicine. Nanotheranostics can become more expensive in prolonged use, so there is a need to maintain cost and efficacy of the treatment.

13.13 IMPLICATIONS AND POTENTIAL BENEFITS

COPD can affect people of different ages and has become a major health threat globally. The main cause of COPD is active smoking followed by exposure to air pollution and genetic conditions. Improved early diagnosis and monitoring provides the possibility to slow the progression and reduce the severity of the disease. There is no universal way to identify COPD due to a wide range of symptoms, but it has been found through various studies that small airway diseases indicate the earlier stages of COPD progression and are believed to be a precursor for the development of emphysema. The concept of early diagnosis is emerging as an important concept for the research and development of advanced treatment modalities. Airway obstruction has been an early indication of COPD in early diagnosis (Johns, 2014). Various techniques have emerged recently and proven to be useful for improved diagnosis and monitoring of COPD. Some of these techniques are the forced oscillation technique for analyzing lung functioning during breathing, X-ray contrast for detailed imaging of structures of lung, the impulse oscillometry system using robust artificial intelligence along with machine learning for real-time monitoring of COPD progression, photoacoustic tomography for imaging, and ultrasound computed tomography that provides real-time imaging of lung function.

As a disease that affects the respiratory system, COPD progression is managed through the control of symptoms. This is achieved by the use of nanotheranostics, imaging, and personalized medication according to the patient's requirement. The efficacy of these treatments is directly related to the patient's adherence. Even though management of symptoms is crucial to check progression of COPD, it is also important to ensure patient compliance (Bourbeau, 2008). Patients with higher adherence show better health outcomes; on the other hand, patients with suboptimal adherence show uncontrolled spread of disease in later stages. Various methods have been described to enhance therapeutic outcomes and promote patient compliance. Some of these methods include monitoring tools, personalized treatment plans, adherence awareness and reminders, simple medication regimes, and regular follow-ups. It is important to educate patients regarding possible side effects. It is also crucial to ensure that patients have a supporting social network, counseling, and motivational sessions.

13.13.1 Reduction of Treatment-Related Side Effects

Along with the management of symptoms, it is also important to manage possible treatment-related side effects. Prolonged treatment often results in exacerbation, which influences quality of life adversely in patients dealing with COPD. The methods used to mitigate treatment-related side effects are:

1. ***Selection of medication***: Personalized medication has been proven to improve patient conditions and management of symptoms. It is important to choose the correct medication that has identified and favorable side effects. The use of bronchodilators, inhaled corticosteroids, and therapy combinations are some common methods involved.
2. ***Proper technique for using an inhaler***: It is important to use an inhaler properly to ensure effective drug delivery. This also helps to avoid local side effects.
3. ***Addressing bone health***: Long-term corticosteroids result in bone-related risks, so it is important to examine such risks.
4. ***Cramps***: Use of certain medications that include beta-agonists results in muscle cramps. It is advised to stray hydrated and maintain electrolyte balance.
5. ***Risk of pneumonia***: Seen when patients inhale corticosteroids for a long time, as they are prone to contacting pneumonia more easily.
6. ***Follow up***: Regular follow-up and proper education of patients regarding the disease help reduce possible side effects related to treatments.

13.13.2 Potential for Personalized Medicine Approaches

COPD being a disease that affects the respiratory tract shows a wide range of symptoms that vary from person to person. Hence, the increasing approach toward personalized medicine provides an opportunity in the field of research and treatment to slow the progression of the symptoms and aid in the management of the condition. Personalized medication provides various approaches such as:

- Strategies for treatment adherence.
- Risk prediction methods.
- Imaging.
- Genetic profiling.
- Predictive side effects.
- Risk management.
- Remote patient monitoring

13.14 ETHICAL AND REGULATORY CONSIDERATIONS

13.14.1 Safety Consideration Related to Nanoparticles Use

Advancement in nanotechnology and its use in theranostics has been greatly used in the treatment of COPD. Desirable modification in nanocarriers has led to improvement in efficacy and lowered toxicity. Still there are various safety considerations that should be followed. Despite being commonly used in treatment of COPD, severe health hazards exist, such as respiratory issues, irritation in skin, and toxicity, when exposed to the human body. Nanoparticles can impact the ecosystem and biodiversity considerably. Even though nanoparticles are used for their localized delivery, it is difficult to ensure precise delivery and retention within the lung tissue. It is also important to carry out assessments to analyze toxicity and any possible adverse effects of nanomaterials. There are some policies that ensure the use of nanoparticles under safer limits. Regulatory guidelines provide certain standards to industries and ensure that industries follow the safety protocols while developing, manufacturing, and using nanoparticles. Transparency by making users aware of the risks associated with the use nanoparticles is also ensured by regulatory policies.

13.15 REGULATORY HURDLES AND APPROVAL PROCESSES

Regulatory hurdles in the use of theranostics in COPD are extremely complex. As COPD causes a wide range of symptoms that vary person to person, it becomes challenging to prescribe a standard universal treatment. Hurdles related to radiopharmaceutical approval are important to overcome. This includes ensuring that radiopharmaceuticals are not toxic and biocompatible, particle sizes are balanced, and immunological responses are minimum. Regulatory approval is necessary to ensure that nanoparticles use is safe and effective.

13.15.1 Patient Education and Informed Consent

It is very important to make sure that the patient is aware of proper treatment and medication. It should also be ensured that patients are well informed regarding various available treatment options. They should also be educated regarding the importance of self-management. Patients should be encouraged to actively participate in decision-making. They must be educated and well informed regarding the medications they are given and all possible outcomes and side effects of the treatment. Informed consent is defined as a legal and ethical process where patients voluntarily agree to medical treatments after properly understanding the purpose, risks, and other alternatives. Patients are provided all the related and necessary information regarding the treatment. Their consent should be voluntarily given without any pressure. It must be ensured that patients are given positive support and confidence in the decisions they make.

13.16 CONCLUSION

Nanotheranostics based on precise drug delivery and more patient-based treatments has been proven an emerging and promising approach toward the treatment of COPD. Nanotheranostics provides target drug delivery where nanoparticles are engineered to deliver drugs to specific target tissues. It integrates diagnosis and therapy in a single approach and provides real-time diagnosis and targeted drug delivery, which has been proven to be the best treatment option available. It also offers drug stability and controlled release of drugs with minimum fluctuations. Nanotheranostics offers a more personalized medication approach, which has revolutionized the treatment method for COPD. As symptoms of COPD show great variations, which differ person to person, a more patient-centric treatment plan proves to be more useful and effective. Use of nanoparticles in theranostics has made it possible to modify and use nanomaterials more efficiently with more predictable side effects. However, nanoparticles must be used according to the set criteria in medicine, with all the necessary assessment done to show more transparency and addressing ethical issues. Nanotheranostics is an innovative solution for the treatment of COPD and its management.

Use of nanotechnology in theranostics holds a promising future in health care systems. It provides a more personalized approach for the treatment where medication is adjusted and given according to the individual patient's need and symptoms. As COPD symptoms vary from person to person, this approach has proved to be more effective in the treatment and management of symptoms. A targeted drug delivery system ensures that the required drug reaches the targeted cell or tissue precisely, which reduces the chances of systematic side effects while increasing the positive therapeutic side effects. Further research in personalized medicine will help to develop more effective and efficient medicine involving nanotheranostics. Even though recent advancements in the field of nanotechnology and theranostics are being widely used in the treatment of COPD, there are still limited options available when it comes to nanomaterials. It also become important to assess nanoparticles for their possible toxicity, which requires more deep research. There is also a need to set up more advanced protocols and policies for regulatory approvals. Nanotheranostics should be subjected to more thorough clinical trials in COPD cases. Efficacy and efficiency testing of nanoparticles can also be improved if more research is done. More patient-centric personalized medicine can be developed by keeping patient needs, severity of symptoms, and well-being in mind. Proper education and awareness about the disease and treatment is also essential.

REFERENCES

Allhoff, F. L. (2009). *What Is Nanotechnology and Why Does It Matter?: From Science to Ethics*. John Wiley & Sons.

Assadi, M. (2014). The future of molecular imaging in paradigm shift from reactive to proactive (P4) medicine: Predictive, preventive, personalized and participatory. *Nuclear Medicine Communications*, *35*(12), 1193–1196.

Bae, Y. H. (2011). Targeted drug delivery to tumors: Myths, reality and possibility. *Journal of Controlled Release*, *153*(3), 198–205.

Biederer, J. M. (2012). MRI of the lung (3/3)—Current applications and future perspectives. *Insights into Imaging*, *3*, 373–386.

Bourbeau, J. (2008). Patient adherence in COPD. *Thorax*, *63*(9), 831–838.

Braun, C. A. (2007). *Pathophysiology: Functional Alterations in Human Health*. Lippincott Williams & Wilkins.

Chan, I. S. (2011). Personalized medicine: Progress and promise. *Annual Review of Genomics and Human Genetics*, *12*, 217–244.

Gilham, I. (2002). Theranostics: An emerging tool in drug discovery and commercialisation. *Drug Discovery World*, *6*, 24–32.

Haas, F. (2008). *The Chronic Bronchitis and Emphysema Handbook*. Turner Publishing Company.

Hussain, M. S. (2021). Nanotechnology based advanced therapeutic strategies for targeting interleukins in chronic respiratory diseases. *Chemico-Biological Interactions*, *348*, 109637.

Jeelani, S. R. (2014). Theranostics: A treasured tailor for tomorrow. *Journal of Pharmacy and Bioallied Sciences*, *6*(Suppl 1), S6–S8.

Johns, D. P. (2014). Diagnosis and early detection of COPD using spirometry. *Journal of Thoracic Disease*, *6*(11), 1557.

Kelkar, S. S. (2011). Theranostics: Combining imaging and therapy. *Bioconjugate Chemistry*, *22*(10), 1879–1903.

Królicki, L. (2022). Theranostics–present and future. *Bio-Algorithms and Med-Systems*, *17*(4), 213–220.

Lee, D. E. (2012). Multifunctional nanoparticles for multimodal imaging and theragnosis. *Chemical Society Reviews*, *41*(7), 2656–2672.

Lee, D. Y. (2011). Molecular theranostics: A primer for the imaging professional. *American Journal of Roentgenology*, *197*(2), 318–324.

Liu, M. A. (2020). Recent advances in the development of nanoparticles for multimodality imaging and therapy of cancer. *Medicinal Research Reviews*, *40*(3), 909–930.

Mannino, D. M. (2007). Global burden of COPD: Risk factors, prevalence, and future trends. *Lancet*, *370*(9589), 765–773.

Milne, S. (2014). Advanced imaging in COPD: Insights into pulmonary pathophysiology. *Journal of Thoracic Disease*, *6*(11), 1570.

Muthu, M. S. (2014). Nanotheranostics- application and further development of nanomedicine strategies for advanced theranostics. *Theranostics*, *4*(6), 660.

Passi, M. S. (2020). Conventional and nanotechnology based approaches to combat chronic obstructive pulmonary disease: Implications for chronic airway diseases. *International Journal of Nanomedicine*, 3803–3826.

Qaseem, A. W. (2011). Diagnosis and management of stable chronic obstructive pulmonary disease: A clinical practice guideline update from the American College of Physicians, American College of Chest Physicians, American Thoracic Society, and European Respiratory Society. *Annals of Internal Medicine*, *155*(3), 179–191.

Raherison, C. (2009). Epidemiology of COPD. *European Respiratory Review, 18*(114), 213–221.

Shrivastava, S. J. (2019). A review on theranostics: An approach to targeted diagnosis and therapy. *Asian Journal of Pharmaceutical Research and Development, 7*(2), 63–69.

Siafaka, P. I. (2021). Current update on nanoplatforms as therapeutic and diagnostic tools: A review for the materials used as nanotheranostics and imaging modalities. *Asian Journal of Pharmaceutical Sciences, 16*(1), 24–46.

Thanou, M. (2018). *Theranostics and Image Guided Drug Delivery. Royal Society of Chemistry.*

Tyagi, H. K. (2016). A facile pH controlled citrate-based reduction method for gold nanoparticle synthesis at room temperature. *Nanoscale Research Letters, 11*, 1–11.

Vogenberg, F. R. (2010). Personalized medicine: Part 1: Evolution and development into theranostics. *Pharmacy and Therapeutics, 35*(10), 560.

Xie, J. L. (2010). Nanoparticle-based theranostic agents. *Advanced Drug Delivery Reviews, 62*(11), 1064–1079.

Yhee, J. Y. (2016). Advanced therapeutic strategies for chronic lung disease using nanoparticle-based drug delivery. *Journal of Clinical Medicine, 5*(9), 82.

14 Senolytics

A Potential Therapeutic Approach for COPD Management

Abhilasha Ahlawat and Neeta Solanki
Faculty of Pharmaceutical Sciences, Maharshi Dayanand University
Rohtak, Haryana, India

Sachin Kumar Singh
School of Pharmaceutical Sciences, Lovely Professional University
Phagwara, Punjab, India
Faculty of Health, Australian Research Centre in Complementary and Integrative Medicine, University of Technology Sydney
Ultimo, NSW, Australia

Kamal Dua
Faculty of Pharmaceutical Sciences, Maharshi Dayanand University
Rohtak, Haryana, India
Faculty of Health, Australian Research Centre in Complementary and Integrative Medicine, University of Technology Sydney
Ultimo, NSW, Australia
Discipline of Pharmacy, Graduate School of Health,
Faculty of Health, University of Technology Sydney
Ultimo, NSW 2007, Australia

Harish Dureja
Faculty of Pharmaceutical Sciences, Maharshi Dayanand University
Rohtak, Haryana, India
Faculty of Health, Australian Research Centre in Complementary and Integrative Medicine, University of Technology Sydney
Ultimo, NSW, Australia

14.1 INTRODUCTION

Chronic obstructive pulmonary disease (COPD) is a persistent lung disease that causes lung inflammation, airflow obstruction, and damage to lung tissues (Agustí et al., 2023). It exerts a major healthcare burden by increasing mortality and morbidity and is becoming a global pandemic (Agustí et al., 2023; Jarhyan et al., 2022; Pando-Sandoval et al., 2022). The cost of COPD treatment is expected to reach up to $4.8 trillion USD by 2030 (Bloom et al., 2011; Safiri et al., 2022). COPD is known to affect more than 210 million people globally and it is the third leading cause of mortality worldwide (Bloom et al., 2011; Jarhyan et al., 2022; Safiri et al., 2022). COPD cases are five to six times higher in elderly people, which affects 1 in 10 individuals over 45 years of age (Agustí et al., 2023; Barnes, 2023; Jarhyan et al., 2022; Pando-Sandoval et al., 2022). A major risk factor associated with COPD

 DOI: 10.1201/9781003498926-14

is long-term exposure to smoke and other contributors include polluting agents, industrial exposure to noxious agents, genetic factors, asthmatic history, and respiratory viral infections (RSV) (Agustí et al., 2023; Barnes, 2023; Jarhyan et al., 2022; Pando-Sandoval et al., 2022). Lung aging also plays a major role in the onset and development of the COPD (Brandsma et al., 2017; Jarhyan et al., 2022; MacNee, 2016; Mercado et al., 2015). As COPD is an age-associated lung disorder, its prevalence rate accelerates with aging and pulmonary functions and begins to decline due to decreased elastic recoiling of the lungs and lesser compliance of the chest wall (Barnes, 2023; Jarhyan et al., 2022; MacNee, 2016; Pando-Sandoval et al., 2022; Xie et al., 2020). Age-related structural and functional changes in the lungs and vulnerability to external factors increase the pathogenic susceptibility of older people to COPD and make them more vulnerable to lung injury, leading to the increased burden of COPD (Fukuchi, 2009). In addition, cigarette smoke elevates oxidative stress, causes changes in DNA, and contributes to premature lung aging, leading to senescence (Hara et al., 2013). These senescent cells alter the cellular function and contribute to inflammation by producing various pro-inflammatory cytokines (Tsuji et al., 2010). This chapter highlights the role of senescence in COPD, various pathways responsible for causing senescence in COPD, and the role of senolytic agents in the management of COPD.

14.2 COPD AND SENESCENCE

COPD caused due to accelerated lung aging lead to the aggregation of senescent cells that are unable to undergo the repairing process, and these cells secrete various inflammatory markers in COPD (Baker et al., 2020). Other characteristics of aging and cellular senescence, which are also found in chronic lung diseases, include shortening of telomeres, mitochondrial dysfunction, and defective autophagy (Birch et al., 2018).

When these senescent cells are unable to be identified and removed, cellular dysfunction begins and exerts resistance to apoptosis through various pathways, collectively termed as senescent cell anti-apoptotic pathways (SCAPs) (Hu et al., 2022). SCAPs further lead to aggregation of these cells, which may be deleterious to the microenvironment of senescent cells and their surroundings. These senescent cells secrete pro-inflammatory factors, i.e., senescence associated secretory phenotypes (SASPs), such as cytokines and chemokines. This alters the microenvironment by variations in cytokine level, leading to the disruption of physiological cell signaling networks and consequently accelerates the inflammatory response in the initiation of senescence in surrounding cells in a paracrine manner (Acosta et al., 2013). This phase of senescence stimulation creates destruction on dynamics and cell-rich tissues, especially cells that make up the respiratory tract. The senescence induction cycle can have a devastating effect on the respiratory system.

The SASP response of senescent cells is stimulated by p21CIP1, which further activates the p38 mitogen-activated protein kinase (MAPK) and Janus-activated kinases (JNK). Stimulation of nuclear factor-kB (NF-kB) causes the discharge of pro-inflammatory cytokines, specifically interleukins (IL)-1β and 6, tumor necrosis factor-alpha (TNF-α), vascular endothelial growth factors and transforming growth factor-beta (TGF-β), chemokines (CXCL1 and 8, i.e., chemokine [C-X-C motif] ligand 1 and 8, and CCL2, i.e., chemokine [C-C motif] ligand 2), and matrix metalloproteinases 2 and 9 (MMP-2 and 9). There is a great increase in all of these in age-linked health conditions involving COPD. Binding of CXCL8 to chemokine C-X-C motif receptor 2 (CXCR2) may consequently prompt DNA destruction and senescence (Acosta et al., 2008). Another SASP protein, plasminogen activator inhibitor-1 (PAI-1), is increased in the sputum and alveoli of COPD patients (To et al., 2013). It is suggested that enhanced PAI-I leads to senescence and may contribute to inflammation in COPD. Further, p16INK4a leads to the activation of a reduced form of nicotinamide adenine dinucleotide phosphate oxidase (NADPH oxidase), resulting in oxidative stress and activation of NF-kB (Takahashi et al., 2006). The SASP phenomenon also involves activation of the NOD-like receptor protein 3 (NLRP3) inflammasome, causes release of IL-1β, extending the inflammatory reaction (Acosta et al., 2013). The content of SASP varies according to cell categories

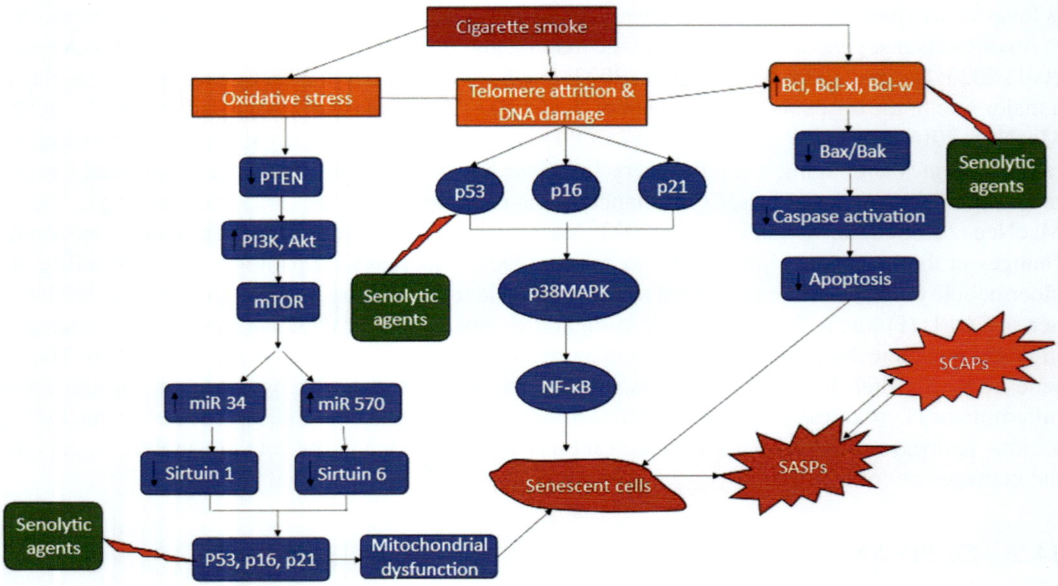

FIGURE 14.1 Cellular senescence in chronic obstructive pulmonary disease.

and inducing elements, but creates significant involvement in the inflammation perceived in numerous chronic lung diseases (Álvarez et al., 2017; Minagawa et al., 2011). Therefore, increased secretion of these inflammation mediators is a typical property of chronic lung diseases including COPD (Figure 14.1). Various pathways responsible for causing cellular senescence and potential targets for senolytics is shown in Figure 14.1.

14.3 TARGETING SENESCENCE THROUGH SENOTHERAPEUTIC AGENTS: A NEW APPROACH TO TARGET COPD

Currently, available treatment options, generally given alone or in combination, only alleviate symptoms and complications accompanying COPD and improve the quality of life. These options comprise smoking cessation, use of bronchodilators, corticosteroid drugs, pulmonary rehabilitation, and lung transplantation in severe conditions (Labaki & Rosenberg, 2020). These treatment options do not involve targeting of the underlying disease process. Recently, various therapies that specifically target senescent cells in various aging-related diseases such as COPD have been explored, including senolytic agents, senostatics, or senomorphic agents (Bateman et al., 2023).

Senostatics/senomorphic agents are drugs that prevent induction of senescence, prevent detrimental cell-extrinsic effects of SCs, and improve cellular repair mechanisms such as DNA damage repair, telomerase action, mitochondrial functioning, and nutrient sensing. Limited research has been conducted to explore the probability of enhancing the DNA damage repair, occurrence of proteostasis, and telomerase activity in the COPD disease (Bodas et al., 2017; Lagoumtzi & Chondrogianni, 2021; Wan et al., 2023). These agents suppress the release of SASPs without cell elimination mainly by neutralizing the activity and functions of SASP elements, including interleukins-1α, 6, and 8 with specific antibodies. These antibodies slow the senescence process without causing cell death (Childs et al., 2015; Childs et al., 2017; Lagoumtzi & Chondrogianni, 2021; Wan et al., 2023). As these agents do not cause straight elimination of senescent cells and are not effective in all types of senescent cells, accordingly, the therapeutic effect may be slow and less marked (Wan et al., 2023). Therefore, these agents may be potentially beneficial for COPD patients (Woldhuis, 2021).

Senomorphics include resveratrol, rapamycin, and metformin (Zhang et al., 2023). Senostatic agents include PI3K-mTOR, p38MAPK, JAK/STAT signaling pathway inhibitors, NF-κB and ataxia telangiectasia mutated (ATM) kinase inhibitors, statins, aspirin, antagomirs of critical microRNAs, and antioxidants (Barnes, 2023; Di Micco et al., 2021).

Another favorable approach is the use of senolytics, which causes elimination of SCs by inducing apoptosis, especially in SCs excluding the functional non-senescent cells (Baker et al., 2011; Woldhuis, 2021). "Senolytics" is formed from the words "senescence" and "lytic" (destroying) (Zhu et al., 2015). Senolytic agents are established as a promising therapeutic approach to eliminate SCs by targeting apoptosis paths and subjecting them to the apoptotic process (Kirkland et al., 2017). During aging, SCs accumulate in tissues and tissue function and repair mechanisms become impaired, resulting in the development of emphysema (Rashid et al., 2018). Removal of SCs and thus reducing their count is recommended to reduce the airway inflammation in COPD and enhance tissue repair by the stem cells and progenitors (Kirkland et al., 2017). Senolytic agents quickly restrict the SCAPs, affecting apoptosis of SC through SASP. It is perceived that, generally, reaccumulation of SCs can take weeks and senolytic agents may be administered periodically – a "hit-and-run" method (Kirkland & Tchkonia, 2020). Preclinical studies have revealed that the use of pharmacological or genetic methods to remove SCs deceased or hindered chronic disorders including several age-linked disorders (Kirkland et al., 2017).

The usage of senolytic drugs may be a favorable way to manage COPD, but the safety and application of the senolytic approach in clinical COPD management needs to be established. Furthermore, the adverse or advantageous effects of senolytics in COPD patients have not been yet assessed (Araya & Kuwano, 2022).

14.4 SENOLYTIC AGENTS

Using a hypothesis-driven methodology, the first senolytic agents, dasatinib (D), quercetin (Q) (both identified in 2015), fisetin, and navitoclax, were identified (Kirkland & Tchkonia, 2020). Various senolytic agents targeting SCs have been identified, which include dasatinib (tyrosine kinase inhibitor), BCL-2 family proteins and histone deacetylase and HSP90 inhibitors (Chang et al., 2016; Fuhrmann-Stroissnigg et al., 2017; Yosef et al., 2016; Zhu et al., 2015; Zhu et al., 2016), UBX101 (MDM2/p53 protein interaction inhibitor) (Jeon et al., 2017), and modified FOXO4-p53 interfering peptide (IP) (Baar et al., 2017). Moreover, targeting SCs having high lysosomal β-galactosidase activity also has been done in practice using cytotoxic medicines entrapped with β(1,4)-galactooligosaccharides (Muñoz-Espín et al., 2018).

A screening of 46 manually selected modulators of anti-apoptotic pathways revealed the senolytic activities of dasatinib and quercetin. Dasatinib was initially discovered for anti-cancerous activity that exhibited senolytic action in preadipocytes and human umbilical vein endothelial cells (HUVEC). Quercetin is a naturally occurring flavonoid and PI3K and serpin inhibitor that showed senolytic effects in the HUVEC assay. Thus, the distinctive activity spectrums of these two agents encouraged their combined usage and is commonly used for targeting in various cells (Power et al., 2023; Zhu et al., 2015).

Other screening agents are piperlongumine (PL) and ABT-263 (navitoclax) (Chang et al., 2016). In addition to balanced screening determinations, initial approaches to detect the new senolytic agents concentrated on targeting the BCL-2 family proteins. These elements are upregulated in SCs and can be deactivated by BH3-only proteins, which favors cell death (Yosef et al., 2016). ABT-263 was identified from a library of 37 independently chosen compounds associated with SC viability and other kind of bioactivities (Chang et al., 2016). The library of compounds involved nutrient metabolism modulators, oxidative stress inducers, anti-apoptotic protein inhibitors, histone deacetylases, and DNA destruction response factors. ABT-263 is established as a significant inhibitor of numerous pro-survival proteins, especially BCL-2, BCL-w and BCL-XL, which is a reliable upregulating agent of anti-apoptotic factors in SCs (Zhu et al., 2015). PL obtained from the long pepper plant has been explored for its anti-cancer properties and capability of modulating the signaling pathways affecting regulation of the cell-cycle, nutrient signaling, immune response (Parama et al., 2021), and inducing apoptosis (Power et al., 2023; Zhang et al., 2018). Further, ABT-737, a chemical mimetic of BH3 protein factors is chemically comparable to ABT-263, has shown senolytic action in IMR-90 fibroblasts induced to senescence involving replicative exhaustion, etoposide treatment, and oncogenic H-Ras expression (Power et al., 2023). Various senolytic agents that may be used for COPD management are listed in Table 14.1.

TABLE 14.1
Senolytic Agents for COPD Management

Class	Agents	Mechanism of Action	Effect in COPD	Reference
		First-Generation Senolytics		
PI3K/Akt inhibitors	Dasatinib (D)	Tyrosine kinase inhibitor	Attenuates the accumulation of various inflammatory markers in lungs	Nishimoto et al., 2019; Wan et al., 2023
	Quercetin (Q)	Targets PI3K/Akt, BCL and HIF-1α pathway, inhibits ferroptosis, decreases polarization of macrophage in airway inflammation due to neutrophils	Quercetin in cigarette smoking (CS)-induced COPD model reduces release of inflammatory mediators, oxidative stress, and inflammation in lungs	Araújo et al., 2022; Shvedova et al., 2022; Wan et al., 2023
	Fisetin	Inhibits PI3K/Akt, BCL, and HIF-1α pathway	Possibly may be effective in COPD	Wan et al., 2023
	ABT-263 (Navitoclax)	Inhibits PI3K/Akt, BCL, and HIF-1α pathway	Lowers the SASPs and ameliorates lung disease. Possibly may be effective in COPD.	Wan et al., 2023
	Azithromycin	Inhibits PI3K/AKT, BCL, and HIF-1α pathway	Suppresses cytokine release in COPD exacerbations	Płusa, 2020
	JN-PK1	Inhibits PI3K/AKT, BCL, and HIF-1α pathway	Possibly may be effective in COPD	Wan et al., 2023
	Dasatinib + Quercetin	Inhibits Src tyrosine kinase and PI3K pathway; inhibits BCL-2- and p53-related SCAPs	Reduces SASPs in bleomycin-induced fibrosis model. Possibly may be effective in COPD.	Kirkland et al., 2017; Russo et al., 2017; Rivera-Torres & San José, 2019
BCL-2, BCL-w, and BCL-XL inhibitors	ABT-737	Inhibits BCL-2, BCL-XL family	Possibly may be effective in COPD	Kirkland et al., 2017; Wan et al., 2023
	ABT-199 (Venetoclax)	Inhibits BCL-2, BCL-Xl family	Possibly may be effective in COPD	Kirkland et al., 2017; Wan et al., 2023
	ABT-263 (Navitoclax)	Inhibits BCL-2- and p53-related SCAPs	Study reported in SaRS CoV-2 pulmonary fibrosis but not in COPD	Kirkland et al., 2017; Wan et al., 2023
	A1331852, A1155463	Inhibits BCL-2, BCL-Xl family	Possibly may be effective in COPD	Delval et al., 2023; Pan et al., 2017

(Continued)

TABLE 14.1 (Continued)
Senolytic Agents for COPD Management

Class	Agents	Mechanism of Action	Effect in COPD	Reference
Cardiac glycosides	Ouabain, Digoxin	Inhibits BCL-2 family proteins, inhibit Na+/K+ ATPase	Affects airway epithelial cells expressing p16, prevent airway inflammation in COPD patients	Broekman et al., 2018; Justice et al., 2019; Rivas et al., 2022
HIF-1 inhibitors	Quercetin (Q)	Targets PI3K/Akt, BCL, and HIF-1 pathways, inhibit ferroptosis, reduces M1 macrophage skewing during airway inflammation due to neutrophils	Reduces inflammation of lungs. Decreases release of inflammatory mediators. Decreases oxidative stress in COPD model.	Araújo et al., 2022; Triantafyllou et al., 2008; Wan et al., 2023
	Fisetin	Inhibits PI3K/AKT, BCL, and HIF-1 pathways	Possibly may be effective in COPD	Triantafyllou et al., 2008
	Vitamin D	Regulates HIF-1α/Notch1 signaling during autophagy	Effective in asthma. Possibly may be effective in COPD.	Huang et al., 2022; Wan et al., 2023
	Epigallocatechin gallate	Suppresses HIF-1α, reduces polarization of macrophages	Effective in asthma. Possibly may be effective in COPD.	Wan et al., 2023; Yang and Li, 2022
	Azithromycin	Inhibits PI3K/Akt/mTOR/ HIF-1α/VEGF pathway	Affect airway remodeling in asthma. Possibly may be effective in COPD.	Wan et al., 2023; Zhao et al., 2018
	Chrysoeriol	Inhibits HIF-1α pathway	Effective in asthma Possibly may be effective in COPD.	Mo et al., 2023; Wan et al., 2023
Immunosuppressors	Rapamycin	Inhibition of mTOR	Reduces inflammation, mean alveolar airspace area, levels of sirtuin-1 and sirtuin-6, and emphysema in CS-induced COPD in mice.	Houssaini et al., 2018; Zhang et al., 2019
FOXO4 modulators	FOXO4 D-retro isoform	Prevents interaction of FOXO4 with p53 and prevents apoptosis	Possibly may be effective in COPD	Baar et al., 2017
HSP 90 inhibitors	17-AAG (Tanespimycin), 17-DMAG (Alvespimycin), Geldanamycin	Targets the BCL-2- and p53-related SCAPs; leads to destabilization of Akt and ERK; promotes apoptosis	Possibly may be effective in COPD	Fuhrmann-Stroissnigg et al., 2017; Kirkland et al., 2017
Calcium channel blockers	Benidipine	Blocks Ca^{2+} channel	Possibly may be effective in COPD	Palazzo et al., 2023
Second-Generation Senolytics				
Vaccines (Nakagami, 2020; Palmer et al., 2019), Immunomodulators (Li et al., 2019), Toxin-loaded nanoparticles preferentially lysed by senescent cells (Chen et al., 2018)				

14.5 CANDIDATES FOR SENOLYTIC THERAPY IN COPD

14.5.1 PI3K/Akt Inhibitors

Dasatinib, mostly employed for the management of leukemia, is a promising protein kinase inhibitor including major action against BCR/ABL and Src family tyrosine kinases (Blake et al., 2008), inflammation-linked cell types, such as T lymphocytes (Schade et al., 2008), pro-fibrotic pathway activated by TGF-β (Distler & Distler, 2010), Syk phosphorylation, and inhibiting the secretion of chemical mediators such as histamine, prostaglandins (PGs) and leukotrienes, and other secretory cytokines by mast cells (Gilfillan & Rivera, 2009). It efficiently diminished the inflammatory and remodeling progressions found in asthmatic lungs, augmenting airway repair and consequently improving lungs function (Da Silva et al., 2016). Additionally, corticosteroid unresponsive airway inflammation is one of foremost obstacles in the active management of chronic lung diseases, such as COPD and asthma. Src is involved in inflammation of airways in mice models of COPD and asthma. Dasatinib has been shown to attenuate the gathering and number of inflammatory cells in airways, and reduces chemokines/cytokines in lipopolysaccharide-induced mice. The targeting of Src offers a therapeutic practice for corticosteroid insensitive pulmonary conditions (Nishimoto et al., 2019).

The senolytic combination of quercetin and dasatinib decreased the SCs and SASP progression in aged mice, and intermittent treatment extended the lifespan and decreased mortality (Xu et al., 2018). The utilization of quercetin and dasatinib also decreased p16, SASPs, and senescent type II pneumocytes in bleomycin-induced fibrosis in mice (Lehmann et al., 2017). The latest open-label pilot investigation of patients with idiopathic pulmonary fibrosis (IPF) indicated that the use of quercetin and dasatinib was well tolerated (Justice et al., 2019). The combination of dasatinib and quercetin was shown to reduce p21WAF1/CIP1 (cyclin dependent kinase inhibitor) (Gomes et al., 2023).

Quercetin, a flavonoid, is an antioxidant, anti-inflammatory substance, and an effective senolytic agent (Kirkland & Tchkonia, 2020; Yamaura et al., 2023). Quercetin has therapeutic uses for various disorders such as cigarette smoking-induced COPD (CS-COPD). Quercetin plays a protecting role in decreasing lung injury produced by long-term cigarette smoking. Moreover, quercetin decreases the inflammation of airway tissues and the liberation of oxidative and inflammation mediators. These outcomes indicate quercetin might be a promising substitute in the prevention of COPD (Araújo et al., 2022).

One *in vitro* study revealed that quercetin in concentration of 25 μM and 50 μM decreased the effects of cigarette smoke (increased the production of nitric oxide and reactive oxygen species) (da Silva Araújo et al., 2020). *In vivo* investigations also confirmed that quercetin reduces the levels of leukocyte, oxidative stress, histological changes of pulmonary parenchyma, and lung functioning variations, signifying that quercetin is an effective agent for managing the effects of cigarette smoke (da Silva Araújo et al., 2020). Further, quercetin is also reported to prevent the progress of lipopolysaccharide (LPS)-induced emphysema in mice and revealed improved elastic recoil and reduced alveolar chord length, reduced lung inflammation and better lung functioning in elastase/LPS-exposed mice (Ganesan et al., 2010).

A randomized Phase I clinical trial evaluated the safety of quercetin supplementation in patients and indicated that COPD patients safely tolerated a quercetin dose up to 2000 mg/day. However, major limitations of the current trial were the smaller sample size and the shorter duration of treatment period with the investigating drug. The safety aspects of quercetin can be explored in a large group of patients including long-term usage of quercetin (Han et al., 2020). Quercetin possesses multiple and wide-ranging protecting properties against CS-COPD. Quercetin stops CS-COPD and stabilizes airway remodeling through antioxidant, immunomodulatory, anti-inflammatory, anti-senescence, autophagy modulation, and gut microbial regulation. Quercetin is a naturally obtained safe constituent, having significant potential in treatment of CS-COPD. Further, studies are needed to test the efficacy of quercetin in combination with beta-agonists and M-receptor antagonists, corticosteroids and roflumilast, antibiotics in CS-COPD and, due to its poor oral bioavailability constraints, entrapment of quercetin in novel drug delivery systems is required (Ding et al., 2023).

Fisetin, a polyphenolic compound found in various fruits and vegetables also possesses senolytic activity. Additionally, it may inhibit PI3K/mTOR signaling (Zhu et al., 2017). The senolytic flavonoid suppresses BCL-2 family members, HIF-1α, and other SCAP network constituents.

14.5.2 Cardiac Glycosides

The electrolytic chemical composition of cells changes when they experience oncogene-induced senescence (Chandeck & Mooi, 2010). In comparison with normal and senescent cells using a fluorescent probe, it was observed that SCs had shown raised intracellular levels of sodium, potassium, and calcium. When these cells were treated with ouabain, the Na+, K+-ATPase inhibitor, BCL-2 proteins, especially protein NOXA, were activated, and senolytic activity was boosted (Guerrero et al., 2019). Furthermore, Triana-Martinez et al., demonstrated that the senolytic action of cardiac glycosides was effective in eliminating lung fibroblasts that were produced by senescence (Triana-Martínez et al., 2019). Digoxin and ouabain have senolytic effects on human airway epithelial cells that express p16, indicating that treating COPD patients with these drugs to target SCs and reduce airway inflammation may be beneficial (Broekman et al., 2018; Justice et al., 2019; Rivas et al., 2022).

14.5.3 B-Cell Lymphoma-2 (BCL-2) Family Protein Inhibitors

BCL-2 plays a role in the pathophysiology of COPD (Zeng et al., 2020). It has been suggested that COPD leads to abnormal inflammation, which is linked to BCL-2 overexpression (Siganaki et al., 2010). BCL-2 exerts a crucial role in homeostasis of airway epithelial cells. Further, BCL-2 may maintain hyperplastic mucus cells, and cigarette smoke may upregulate BCL-2 expression and aid in mucus cell survival (Hussain et al., 2018). The prolonged retention of SCs can be linked to elevated BCL-family protein expression, which can stimulate tissue destruction via SASP. A series of proteins known as BCL-2 inhibitors specifically blocks the pro-survival protein found on the mitochondria's outer membrane (Fan et al., 2020).

In mice, senescent cells can be effectively and selectively eliminated by the BCL-2 family inhibitor ABT-263 (navitoclax) (Chang et al., 2016). In aged hamsters infected with SARS-CoV-2, ABT-263 therapy reduced the senescence-associated signature and partially removed p16-positive cells. Additionally, treatment with ABT-263 improved both early and late lung disease and decreased systemic and pulmonary levels of SASPs (Delval et al., 2023). In addition, quercetin plus navitoclax works well to protect mice from bleomycin-induced lung fibrosis (ILF) (Schafer et al., 2017). ABT-263 decreases lung irradiation-induced senescence by specifically removing type II senescent pneumocytes (Pan et al., 2017). A1331852 and A1155463 are less toxic. A1331852 and A1155463 are selective BCL-XL inhibitors and senolytics *in vitro* that are less toxic and may be more effective for age or senescence-linked disorders (Zhu et al., 2017). Thus, these agents may be further explored for their role in ameliorating exacerbations of COPD.

14.5.4 Immunosuppressants

Normal mouse or human cells undergo senescence when constitutively active serine/threonine kinase Akt is expressed or a phosphatase and tensin homolog (PTEN) is deleted. This process activates mTOR, which is associated with senescence in COPD (Huang & Houghton, 2002; Nogueira et al., 2008). The immunosuppressive drug rapamycin attaches itself to immunophilin FK-binding protein (FKBP12) on mTOR, and the resulting rapamycin-FKBP12 complex subsequently interacts with mTOR to prevent its activity (Huang & Houghton, 2002). Rapamycin is known to inhibit mTOR and decrease pro-inflammatory SASP in lung vascular cells and alveolar epithelial cells, as well as cell senescence in mice. In the COPD model where mice were exposed to smoke, rapamycin was found to be efficient in reducing senescent cells, SASP mediators, and emphysema (Houssaini et al., 2018). Further, rapamycin has also reduced inflammatory cells and mean alveolar airspace area in lungs of mice exposed to cigarette smoke (Zhang et al., 2019). Rapamycin also reduced levels of sirtuin-1 and -6 in COPD (Houssaini et al., 2018; Mitani et al., 2016). Unfortunately, rapamycin has a number of side effects. As a result, less toxic substitutes/rapalogs, such as everolimus and temsirolimus, have been developed. Everolimus at low doses decreases immunosenescence and may lessen respiratory infections in the elderly (Mannick et al., 2014; Mannick et al., 2018). As low doses are both efficacious and often well tolerated in the elderly, this may be a viable long-term COPD treatment option. Since P70 ribosomal S6 kinase is linked to senescence and is the downstream of mTORC1, its inhibition may be more selective than that of mTOR. Various such inhibitors have been generated and need to be evaluated in COPD (Fenton & Gout, 2011).

14.5.5 Hypoxia-Inducible Factor (HIF)– 1α (HIF-1α) Inhibitors

HIF-1α is found to be upregulated under hypoxia and plays a crucial role in regulation of senescence (Alique et al., 2020). Under hypoxic conditions, HIF-1α is persistently activated to sustain or improve mitochondrial biogenesis, which impedes cellular senescence (Liu et al., 2021). Therefore, it is possible to effectively induce cellular senescence and treat aging-related comorbidities by decreasing HIF-1α activity (Chen et al., 2020). HIF-1α is known to play a crucial role in the formation of SASPs and is also involved in senescence of endothelial cells *via* p16, p53, lamin B1, and cyclin D1 (Alique et al., 2019; Giménez-Bastida et al., 2019). HIF-1α is in hypoxic environment due to COPD. Various studies have demonstrated that HIF-1α is overexpressed in lungs of COPD patients and its levels are higher in these patients due to activation of the EGFR/PI3K/AKT pathway (Rong et al., 2018; Shukla et al., 2020; Zhang et al., 2018).

Azithromycin has shown to affect the PI3K/Akt/mTOR/HIF-1α/VEGF pathway to affect airway remodeling in asthma (Zhao et al., 2018). In addition to controlling inflammatory reactions, azithromycin is a unique senolytic medication that also cleanses senescent cells in asthmatic lungs (Wan et al., 2023). Long-term administration of azithromycin reduces the release of inflammatory cytokines, boosts macrophage phagocytosis, and promotes the production of anti-inflammatory cytokines and is well researched in patients with COPD exacerbations (Płusa, 2020).

14.5.6 Calcium Channel Blockers

Reactive oxygen species and carcinogens included in cigarette smoke can cause DNA mutations and alter a cell's environment and homeostasis. It causes cellular senescence in the lungs, which leads to SASPs and a sustained proliferative arrest (Palazzo et al., 2023). Therefore, to find possible agents/drugs that may modulate senescence, it is essential to understand how it is controlled at the molecular and cellular level. Recently, it has been demonstrated that the buildup of senescent cells is linked to several lung conditions (Palazzo et al., 2023). Researchers have demonstrated that calcium (Ca^{2+}) and calcium signaling are important controllers of cellular senescence. A crucial second messenger, calcium, is involved in cell motility, secretion, proliferation, and death. Its effects are achieved by controlling its intracellular concentration and fluxes in which numerous proteins, transporters, channels, pumps, and calcium-binding proteins (CBPs) are involved (Martin et al., 2023). Senescent cells have an increase in intracellular Ca^{2+} content, specifically in the cytosol and mitochondria, because of mitochondrial calcium aggregation and alteration in mitochondrial functions (Luo et al., 1997). Exploring Ca^{2+} may be considered as a potential treatment option for targeting senescence. Apart from CBPs, Ca^{2+} channels and other agents targeting calcium signaling pathways may serve as good senescence target as they can manipulate Ca^{2+} levels (Martin et al., 2023).

Recently, a new class of senolytics known as the dihydropyridine family of voltage gated calcium channel blockers, which includes benidipine, cilnidipine, felodipine, and lacidipine, has been identified and showing senolytic activity on cigarette smoke-induced senescent cells, potentially improving lung disorders. Benidipine is one of these blockers that reduces senescent lung cells and improves pulmonary emphysema in a mouse model (Palazzo et al., 2023).

14.5.7 Forkhead Box Protein O4 (FOXO4)-related Peptides

The transcription factor FOXO4 is essential to the survival of senescent cells (Baar et al., 2017) as it is known to interact with the p53 protein (senescence inducer) (Muñoz-Espín & Serrano, 2014). FOXO4 D-retro in verso isoform, a FOXO4-related peptide, has been shown to inhibit PI3K/AKT/p53/p21/serpine SCAP and further prevent the interaction between FOXO4 and p53. This results in apoptosis and the elimination of SCs and less age-associated organ failure in mice (Baar et al., 2017). FOXO4-related peptides further need to be explored in COPD as they show senolytic activity.

It is demonstrated that senescent cells can be targeted and removed in various animal studies and individuals with aging-related disorders. Senescent cell targeting may be a potential treatment option for COPD and other disorders linked to aging. Until now, no clinical study has evaluated senolytic agents in COPD (Easter et al., 2020). Further preclinical and studies should be conducted to analyze the potential effects of senolytic agents in COPD.

14.6 CONCLUSION AND FUTURE PERSPECTIVES

Progression of COPD due to accelerated aging has brought attention to the possibility of therapies that target the molecular mechanisms responsible for cellular senescence. Senolytic agents have been recently seen as a promising treatment option in various animal models of disorders associated with aging, including diabetes, atherosclerosis, IPF, and chronic kidney disease. Since cellular senescence is the common mechanism underlying these accelerated aging disorders and no clinical study has examined the effect of senolytics in COPD, it may be possible to use these agents in COPD patients based on encouraging clinical trial results. As COPD patients also have comorbid conditions that hasten aging, this approach may have additional therapeutic benefits (Baker et al., 2020).

As senolytics are selective in nature to kill specific senescent cells, it may reduce off-target effects on healthy cells and have a rapid therapeutic effects. The main challenge is in the possibility that eradicating SCs may conflict with the beneficial biological functions of these cells. In fact, senescence serves beneficial biological functions in specific situations rather than always being harmful (Wan et al., 2023). Therefore, before research in clinical studies with COPD patients is possible, senolytic agents need to be chemically optimized to target specific diseases and require thorough preclinical investigations so that these agents can be administered safely (Koll, 2022). The treatment options targeting senescence, such as senolytics, hold much potential to manage COPD, but further extensive preclinical and clinical studies are required to develop specific therapeutic approaches to manage COPD.

REFERENCES

Acosta, J. C., Banito, A., Wuestefeld, T., Georgilis, A., Janich, P., Morton, J. P., & Gil, J. (2013). A complex secretory program orchestrated by the inflammasome controls paracrine senescence. *Nature Cell Biology*, *15*(8), 978–990.

Acosta, J. C., O'Loghlen, A., Banito, A., Guijarro, M. V., Augert, A., Raguz, S., & Gil, J. (2008). Chemokine signaling via the CXCR2 receptor reinforces senescence. *Cell*, *133*(6), 1006–1018.

Agustí, A., Celli, B. R., Criner, G. J., Halpin, D., Anzueto, A., Barnes, P., Bourbeau, J., Han, M. K., Martinez, F. J., Montes de Oca, M., et al. (2023). Global initiative for chronic obstructive lung disease 2023 report: GOLD executive summary. *Archivos De Bronconeumologia*, *59*, 232–248.

Alique, M., Bodega, G., Giannarelli, C., Carracedo, J., & Ramírez, R. (2019). MicroRNA-126 regulates hypoxia-inducible factor-1α which inhibited migration, proliferation, and angiogenesis in replicative endothelial senescence. *Scientific Reports*, *9*(1), 7381.

Alique, M., Sánchez-López, E., Bodega, G., Giannarelli, C., Carracedo, J., & Ramírez, R. (2020). Hypoxia-inducible factor-1α: The master regulator of endothelial cell senescence in vascular aging. *Cells*, *9*(1), 195.

Álvarez, D., Cárdenes, N., Sellarés, J., Bueno, M., Corey, C., Hanumanthu, V. S., & Rojas, M. (2017). IPF lung fibroblasts have a senescent phenotype. *American Journal of Physiology-Lung Cellular and Molecular Physiology*, *313*(6), L1164–L1173.

Araújo, N. P. D. S., de Matos, N. A., Oliveira, M., de Souza, A. B. F., Castro, T. D. F., Machado-Júnior, P. A., & Bezerra, F. S. (2022). Quercetin improves pulmonary function and prevents emphysema caused by exposure to cigarette smoke in male mice. *Antioxidants*, *11*(2), 181.

Araya, J., & Kuwano, K. (2022). Cellular senescence—An aging hallmark in chronic obstructive pulmonary disease pathogenesis. *Respiratory Investigation*, *60*(1), 33–44.

Baar, M. P., Brandt, R. M. C., Putavet, D. A., Klein, J. D. D., Derks, K. W. J., Bourgeois, B. R. M., Stryeck, S., Rijksen, Y., van Willigenburg, H., Feijtel, D. A., van der Pluijm, I., Essers, J., van Cappellen, W. A., van IJcken, W. F., Houtsmuller, A. B., Pothof, J., de Bruin, R. W. F., Madl, T., Hoeijmakers, J. H. J., Campisi, J., & de Keizer, P. L. J. (2017). Targeted apoptosis of senescent cells restores tissue homeostasis in response to chemotoxicity and aging. *Cell*, *169*, 132–147.e16.

Baker, D. J., Wijshake, T., Tchkonia, T., LeBrasseur, N. K., Childs, B. G., Van De Sluis, B., & Van Deursen, J. M. (2011). Clearance of p16Ink4a-positive senescent cells delays ageing-associated disorders. *Nature*, *479*(7372), 232–236.

Baker, J. R., Donnelly, L. E., & Barnes, P. J. (2020). Senotherapy: A new horizon for COPD therapy. *Chest*, *158*(2), 562–570.

Barnes, P. J. (2023). Senotherapy for Lung Diseases. In *Advances in Pharmacology* (Vol. 98, pp. 249–271). Academic Press.

Bateman, G., Guo-Parke, H., Rodgers, A. M., Linden, D., Bailey, M., Weldon, S., & Taggart, C. C. (2023). Airway epithelium senescence as a driving mechanism in COPD pathogenesis. *Biomedicines*, *11*(7), 2072.

Birch, J., Barnes, P. J., & Passos, J. F. (2018). Mitochondria, telomeres and cell senescence: Implications for lung ageing and disease. *Pharmacology & Therapeutics*, *183*, 34–49.

Blake, S., Hughes, T. P., Mayrhofer, G., & Lyons, A. B. (2008). The Src/ABL kinase inhibitor dasatinib (BMS-354825) inhibits function of normal human T-lymphocytes in vitro. *Clinical Immunology*, *127*(3), 330–339.

Bloom, D. E., Cafiero, E. T., Jane-Llopis, E., Abrahams-Gessel, S., Bloom, L. R., Fathima, S., Feigl, A. B., Gaziano, T., Mowafi, M., Pandya, A., Prettner, K., Rosenberg, L., Seligman, B., Stein, A., & Weinstein, C. (2011). The global economic burden of non-communicable diseases (Online). https://www.weforum.org/reports/global-economic-burden-non-communicable-diseases [2 October 2020].

Bodas, M., Patel, N., Silverberg, D., Walworth, K., & Vij, N. (2017). Master autophagy regulator transcription factor EB regulates cigarette smoke-induced autophagy impairment and chronic obstructive pulmonary disease–emphysema pathogenesis. *Antioxidants & Redox Signaling*, *27*(3), 150–167.

Brandsma, C. A., de Vries, M., Costa, R., Woldhuis, R. R., Königshoff, M., & Timens, W. (2017). Lung ageing and COPD: Is there a role for ageing in abnormal tissue repair? *European Respiratory Review*, *26*(146), 170073.

Broekman, W., Khedoe, P. P., Schepers, K., Roelofs, H., Stolk, J., & Hiemstra, P. S. (2018). Mesenchymal stromal cells: A novel therapy for the treatment of chronic obstructive pulmonary disease? *Thorax*, *73*(6), 565–574.

Chandeck, C., & Mooi, W. J. (2010). Oncogene-induced cellular senescence. *Advances in Anatomic Pathology*, *17*(1), 42–48.

Chang, J., Wang, Y., Shao, L., Laberge, R. M., Demaria, M., Campisi, J., Janakiraman, K., Sharpless, N. E., Ding, S., Feng, W., Luo, Y., Wang, X., Aykin-Burns, N., Krager, K., Ponnappan, U., Hauer-Jensen, M., Meng, A., & Zhou, D. (2016). Clearance of senescent cells by ABT263 rejuvenates aged hematopoietic stem cells in mice. *Nature Medicine*, *22*(1), 78–83.

Chen, J., Huang, X., Tao, C., Wang, L., Chen, Z., Li, X., & Wu, Z. (2020). Berberine chloride suppresses non-small cell lung cancer by deregulating Sin3A/TOP2B pathway in vitro and in vivo. *Cancer Chemotherapy and Pharmacology*, *86*, 151–161.

Chen, Z., Hu, K., Feng, L., Su, R., Lai, N., Yang, Z., & Kang, S. (2018). Senescent cells re-engineered to express soluble programmed death receptor-1 for inhibiting programmed death receptor-1/programmed death ligand-1 as a vaccination approach against breast cancer. *Cancer Science*, *109*(6), 1753–1763.

Childs, B. G., Durik, M., Baker, D. J., & Van Deursen, J. M. (2015). Cellular senescence in aging and age-related disease: From mechanisms to therapy. *Nature Medicine*, *21*(12), 1424–1435.

Childs, B. G., Gluscevic, M., Baker, D. J., Laberge, R. M., Marquess, D., Dananberg, J., & Van Deursen, J. M. (2017). Senescent cells: An emerging target for diseases of ageing. *Nature Reviews Drug Discovery*, *16*(10), 718–735.

Da Silva, A. L., Magalhaes, R. F., Branco, V. C., Silva, J. D., Cruz, F. F., Marques, P. S., & Rocco, P. R. M. (2016). The tyrosine kinase inhibitor dasatinib reduces lung inflammation and remodelling in experimental allergic asthma. *British Journal of Pharmacology*, *173*(7), 1236–1247.

da Silva Araújo, N. P., de Matos, N. A., Leticia Antunes Mota, S., Farias de Souza, A. B., Dantas Cangussú, S., Cunha Alvim de Menezes, R., & Silva Bezerra, F. (2020). Quercetin attenuates acute lung injury caused by cigarette smoke both in vitro and in vivo. *COPD: Journal of Chronic Obstructive Pulmonary Disease*, *17*(2), 205–214.

Delval, L., Hantute-Ghesquier, A., Sencio, V., Flaman, J. M., Robil, C., Angulo, F. S., & Trottein, F. (2023). Removal of senescent cells reduces the viral load and attenuates pulmonary and systemic inflammation in SARS-CoV-2-infected, aged hamsters. *Nature Aging*, *3*(7), 829–845.

Di Micco, R., Krizhanovsky, V., Baker, D., & d'Adda di Fagagna, F. (2021). Cellular senescence in ageing: From mechanisms to therapeutic opportunities. *Nature Reviews Molecular Cell Biology*, *22*(2), 75–95.

Ding, K., Jiang, W., Zhan, W., Xiong, C., Chen, J., Wang, Y., & Lei, M. (2023). The therapeutic potential of quercetin for cigarette smoking–induced chronic obstructive pulmonary disease: A narrative review. *Therapeutic Advances in Respiratory Disease*, *17*, 1–23.

Distler, J. H. W., & Distler, O. (2010). Tyrosine kinase inhibitors for the treatment of fibrotic diseases such as systemic sclerosis: Towards molecular targeted therapies. *Annals of the Rheumatic Diseases*, *69*(Suppl 1), i48–i51.

Easter, M., Bollenbecker, S., Barnes, J. W., & Krick, S. (2020). Targeting aging pathways in chronic obstructive pulmonary disease. *International Journal of Molecular Sciences*, *21*(18), 6924.

Fan, Y., Cheng, J., Zeng, H., & Shao, L. (2020). Senescent cell depletion through targeting BCL-family proteins and mitochondria. *Frontiers in Physiology*, *11*, 593630.

Fenton, T. R., & Gout, I. T. (2011). Functions and regulation of the 70 kDa ribosomal S6 kinases. *The International Journal of Biochemistry & Cell Biology*, *43*(1), 47–59.

Fuhrmann-Stroissnigg, H., Ling, Y. Y., Zhao, J., McGowan, S. J., Zhu, Y. I., Brooks, R. W., & Robbins, P. D. (2017). Identification of HSP90 inhibitors as a novel class of senolytics. *Nature Communications*, *8*(1), 422.

Fukuchi, Y. (2009). The aging lung and chronic obstructive pulmonary disease: Similarity and difference. *Proceedings of the American Thoracic Society*, *6*(7), 570–572.

Ganesan, S., Faris, A. N., Comstock, A. T., Chattoraj, S. S., Chattoraj, A., Burgess, J. R., & Sajjan, U. S. (2010). Quercetin prevents progression of disease in elastase/LPS-exposed mice by negatively regulating MMP expression. *Respiratory Research*, *11*, 1–15.

Gilfillan, A. M., & Rivera, J. (2009). The tyrosine kinase network regulating mast cell activation. *Immunological Reviews*, *228*(1), 149–169.

Giménez-Bastida, J. A., Ávila-Gálvez, M., Espín, J. C., & González-Sarrías, A. (2019). Conjugated physiological resveratrol metabolites induce senescence in breast cancer cells: Role of p53/p21 and p16/Rb pathways, and ABC transporters. *Molecular Nutrition & Food Research*, *63*(22), 1900629.

Gomes, P., Rimbert, A., Hassibi, S., Baker, J., & Ito, K. (2023). P206 Identification of novel senolytic candidates for the treatment of chronic respiratory diseases with accelerating aging. Thorax 2023;78:A235.

Guerrero, A., Herranz, N., Sun, B., Wagner, V., Gallage, S., Guiho, R., & Gil, J. (2019). Cardiac glycosides are broad-spectrum senolytics. *Nature Metabolism*, *1*(11), 1074–1088.

Han, M. K., Barreto, T. A., Martinez, F. J., Comstock, A. T., & Sajjan, U. S. (2020). Randomised clinical trial to determine the safety of quercetin supplementation in patients with chronic obstructive pulmonary disease. *BMJ Open Respiratory Research*, *7*(1), e000392.

Hara, H., Araya, J., Ito, S., Kobayashi, K., Takasaka, N., Yoshii, Y., & Kuwano, K. (2013). Mitochondrial fragmentation in cigarette smoke-induced bronchial epithelial cell senescence. *American Journal of Physiology-Lung Cellular and Molecular Physiology*, *305*(10), L737–L746.

Houssaini, A., Breau, M., Kebe, K., Abid, S., Marcos, E., Lipskaia, L., & Adnot, S. (2018). mTOR pathway activation drives lung cell senescence and emphysema. *JCI Insight*, *3*(3), e93203.

Hu, L., Li, H., Zi, M., Li, W., Liu, J., Yang, Y., & He, Y. (2022). Why senescent cells are resistant to apoptosis: An insight for senolytic development. *Frontiers in Cell and Developmental Biology*, *10*, 822816.

Huang, C., Peng, M., Tong, J., Zhong, X., Xian, J., Zhong, L., & Huang, Y. (2022). Vitamin D ameliorates asthma-induced lung injury by regulating HIF-1α/Notch1 signaling during autophagy. *Food Science & Nutrition*, *10*(8), 2773–2785.

Huang, S., & Houghton, P. J. (2002). Inhibitors of mammalian target of rapamycin as novel antitumor agents: From bench to clinic. *Current Opinion in Investigational Drugs (London, England: 2000)*, *3*(2), 295–304.

Hussain, S. S., George, S., Singh, S., Jayant, R., Hu, C. A., Sopori, M., & Chand, H. S. (2018). A small molecule BH3-mimetic suppresses cigarette smoke-induced mucous expression in airway epithelial cells. *Scientific Reports*, *8*(1), 13796.

Jarhyan, P., Hutchinson, A., Khaw, D., Prabhakaran, D., & Mohan, S. (2022). Prevalence of chronic obstructive pulmonary disease and chronic bronchitis in eight countries: A systematic review and meta-analysis. *Bulletin of the World Health Organization*, *100*(3), 216–230.

Jeon, O. H., Kim, C., Laberge, R. M., Demaria, M., Rathod, S., Vasserot, A. P., & Elisseeff, J. H. (2017). Local clearance of senescent cells attenuates the development of post-traumatic osteoarthritis and creates a pro-regenerative environment. *Nature Medicine*, *23*(6), 775–781.

Justice, J. N., Nambiar, A. M., Tchkonia, T., LeBrasseur, N. K., Pascual, R., Hashmi, S. K., & Kirkland, J. L. (2019). Senolytics in idiopathic pulmonary fibrosis: Results from a first-in-human, open-label, pilot study. *EBioMedicine*, *40*, 554–563.

Kirkland, J. L., & Tchkonia, T. (2020). Senolytic drugs: From discovery to translation. *Journal of Internal Medicine*, *288*(5), 518–536.

Kirkland, J. L., Tchkonia, T., Zhu, Y., Niedernhofer, L. J., & Robbins, P. D. (2017). The clinical potential of senolytic drugs. *Journal of the American Geriatrics Society*, *65*(10), 2297–2301.

Koll, L. (2022). *Accelerated lung aging in COPD: finding potential anti-aging therapies* (Doctoral dissertation).

Labaki, W. W., & Rosenberg, S. R. (2020). Chronic obstructive pulmonary disease (Japanese version). *Annals of Internal Medicine*, *173*(3), JITC17–JITC32.

Lagoumtzi, S. M., & Chondrogianni, N. (2021). Senolytics and senomorphics: Natural and synthetic therapeutics in the treatment of aging and chronic diseases. *Free Radical Biology and Medicine*, *171*, 169–190.

Lehmann, M., Korfei, M., Mutze, K., Klee, S., Skronska-Wasek, W., Alsafadi, H. N., & Königshoff, M. (2017). Senolytic drugs target alveolar epithelial cell function and attenuate experimental lung fibrosis ex vivo. *European Respiratory Journal*, *50*(2), 1602367.

Li, W., He, Y., Zhang, R., Zheng, G., & Zhou, D. (2019). The curcumin analog EF24 is a novel senolytic agent. *Aging (Albany NY)*, *11*(2), 771–782.

Liu, J., Huang, X., Liu, D., Ji, K., Tao, C., Zhang, R., & Chen, J. (2021). Demethyleneberberine induces cell cycle arrest and cellular senescence of NSCLC cells via c-Myc/HIF-1α pathway. *Phytomedicine*, *91*, 153678.

Luo, Y., Bond, J. D., & Ingram, V. M. (1997). Compromised mitochondrial function leads to increased cytosolic calcium and to activation of MAP kinases. *Proceedings of the National Academy of Sciences*, *94*(18), 9705–9710.

MacNee, W. (2016). Is chronic obstructive pulmonary disease an accelerated aging disease? *Annals of the American Thoracic Society*, *13*(Supplement 5), S429–S437.

Mannick, J. B., Del Giudice, G., Lattanzi, M., Valiante, N. M., Praestgaard, J., Huang, B., & Klickstein, L. B. (2014). mTOR inhibition improves immune function in the elderly. *Science Translational Medicine*, *6*(268), 268ra179.

Mannick, J. B., Morris, M., Hockey, H. U. P., Roma, G., Beibel, M., Kulmatycki, K., & Klickstein, L. B. (2018). TORC1 inhibition enhances immune function and reduces infections in the elderly. *Science Translational Medicine*, *10*(449), eaaq1564.

Martin, N., Zhu, K., Czarnecka-Herok, J., Vernier, M., & Bernard, D. (2023). Regulation and role of calcium in cellular senescence. *Cell Calcium*, *110*, 102701.

Mercado, N., Ito, K., & Barnes, P. J. (2015). Accelerated ageing of the lung in COPD: New concepts. *Thorax*, *70*(5), 482–489.

Minagawa, S., Araya, J., Numata, T., Nojiri, S., Hara, H., Yumino, Y., & Kuwano, K. (2011). Accelerated epithelial cell senescence in IPF and the inhibitory role of SIRT6 in TGF-β-induced senescence of human bronchial epithelial cells. *American Journal of Physiology-Lung Cellular and Molecular Physiology*, *300*(3), L391–L401.

Mitani, A., Ito, K., Vuppusetty, C., Barnes, P. J., & Mercado, N. (2016). Restoration of corticosteroid sensitivity in chronic obstructive pulmonary disease by inhibition of mammalian target of rapamycin. *American Journal of Respiratory and Critical Care Medicine*, *193*(2), 143–153.

Mo, S., Deng, H., Xie, Y., Yang, L., & Wen, L. (2023). Chryseriol attenuates the progression of OVA-induced asthma in mice through NF-κB/HIF-1α and MAPK/STAT1 pathways. *Allergologia et Immunopathologia*, *51*(1), 146–153.

Muñoz-Espín, D., Rovira, M., Galiana, I., Giménez, C., Lozano-Torres, B., Paez-Ribes, M., & Serrano, M. (2018). A versatile drug delivery system targeting senescent cells. *EMBO Molecular Medicine*, *10*(9), e9355.

Muñoz-Espín, D., & Serrano, M. (2014). Cellular senescence: From physiology to pathology. *Nature Reviews Molecular Cell Biology*, *15*(7), 482–496.

Nakagami, H. (2020). Cellular senescence and senescence-associated t cells as a potential therapeutic target. *Geriatrics & Gerontology International*, *20*(2), 97–100.

Nishimoto, Y., Yasuda, H., Masuko, K., Usui, Y., Ueda, K., Kimura, G., & Kizawa, Y. (2019). The involvement of Src in airway inflammation induced by repeated exposure to lipopolysaccharide in mice. *YakugakuZasshi: Journal of the Pharmaceutical Society of Japan*, *139*(9), 1211–1217.

Nogueira, V., Park, Y., Chen, C. C., Xu, P. Z., Chen, M. L., Tonic, I., & Hay, N. (2008). Akt determines replicative senescence and oxidative or oncogenic premature senescence and sensitizes cells to oxidative apoptosis. *Cancer Cell*, *14*(6), 458–470.

Palazzo, A., Makulyte, G., Goerhig, D., Médard, J. J., Gros, V., Trottein, F., & Bernard, D. (2023). Benidipine calcium channel blocker promotes the death of cigarette smoke-induced senescent cells and improves lung emphysema. *Aging (Albany NY)*, *15*(23), 13581.

Palmer, A. K., Gustafson, B., Kirkland, J. L., & Smith, U. (2019). Cellular senescence: At the nexus between ageing and diabetes. *Diabetologia*, *62*, 1835–1841.

Pan, J., Li, D., Xu, Y., Zhang, J., Wang, Y., Chen, M., & Meng, A. (2017). Inhibition of bcl-2/xl with ABT-263 selectively kills senescent type II pneumocytes and reverses persistent pulmonary fibrosis induced by ionizing radiation in mice. *International Journal of Radiation Oncology* Biology* Physics*, *99*(2), 353–361.

Pando-Sandoval, A., Ruano-Ravina, A., Candal-Pedreira, C., Rodríguez-García, C., Represas-Represas, C., Golpe, R., & Pérez-Ríos, M. (2022). Risk factors for chronic obstructive pulmonary disease in never-smokers: A systematic review. *The Clinical Respiratory Journal*, *16*(4), 261–275.

Parama, D., Rana, V., Girisa, S., Verma, E., Daimary, U. D., Thakur, K. K., Kumar, A., & Kunnumakkara, A. B. (2021). The promising potential of piperlongumine as an emerging therapeutics for cancer. *Exploration of Targeted Anti-Tumor Therapy*, *2*, 323–354.

Płusa, T. (2020). Azithromycin in the treatment of patients with exacerbation of chronic obstructive pulmonary disease. *Polski MerkuriuszLekarski: Organ PolskiegoTowarzystwaLekarskiego*, *48*(283), 65–68.

Power, H., Valtchev, P., Dehghani, F., & Schindeler, A. (2023). Strategies for senolytic drug discovery. *Aging Cell*, *22*, e13948.

Rashid, K., Sundar, I. K., Gerloff, J., Li, D., & Rahman, I. (2018). Lung cellular senescence ism independent of aging in a mouse model of COPD/emphysema. *Scientific Reports*, *8*(1), 9023.

Rivas, M., Gupta, G., Costanzo, L., Ahmed, H., Wyman, A. E., & Geraghty, P. (2022). Senescence: Pathogenic driver in chronic obstructive pulmonary disease. *Medicina*, *58*(6), 817.

Rivera-Torres, J., & San José, E. (2019). Src tyrosine kinase inhibitors: New perspectives on their immune, antiviral, and senotherapeutic potential. *Frontiers in Pharmacology*, *10*, 460745, 1011.

Rong, B., Liu, Y., Li, M., Fu, T., Gao, W., & Liu, H. (2018). Correlation of serum levels of HIF-1α and IL-19 with the disease progression of COPD: A retrospective study. *International Journal of Chronic Obstructive Pulmonary Disease*, 13, 3791–3803.

Russo, M., Milito, A., Spagnuolo, C., Carbone, V., Rosén, A., Minasi, P., & Russo, G. L. (2017). CK2 and PI3K are direct molecular targets of quercetin in chronic lymphocytic leukaemia. *Oncotarget*, *8*(26), 42571–42587.

Safiri, S., Carson-Chahhoud, K., Noori, M., Nejadghaderi, S. A., Sullman, M. J., Heris, J. A., & Kaufman, J. S. (2022). Burden of chronic obstructive pulmonary disease and its attributable risk factors in 204 countries and territories, 1990–2019: Results from the global burden of disease study 2019. *BMJ*, *378*, e069679.

Schade, A. E., Schieven, G. L., Townsend, R., Jankowska, A. M., Susulic, V., Zhang, R., & Maciejewski, J. P. (2008). Dasatinib, a small-molecule protein tyrosine kinase inhibitor, inhibits T-cell activation and proliferation. *Blood, The Journal of the American Society of Hematology*, *111*(3), 1366–1377.

Schafer, M. J., White, T. A., Iijima, K., Haak, A. J., Ligresti, G., Atkinson, E. J., & LeBrasseur, N. K. (2017). Cellular senescence mediates fibrotic pulmonary disease. *Nature Communications*, *8*(1), 14532.

Shukla, S. D., Walters, E. H., Simpson, J. L., Keely, S., Wark, P. A., O'Toole, R. F., & Hansbro, P. M. (2020). Hypoxia-inducible factor and bacterial infections in chronic obstructive pulmonary disease. *Respirology*, *25*(1), 53–63.

Shvedova, M., Thanapaul, R. J. R. S., Thompson, E. L., Niedernhofer, L. J., & Roh, D. S. (2022). Cellular senescence in aging, tissue repair, and regeneration. *Plastic and Reconstructive Surgery*, *150*, 4S–11S.

Siganaki, M., Koutsopoulos, A. V., Neofytou, E., Vlachaki, E., Psarrou, M., Soulitzis, N., & Tzortzaki, E. G. (2010). Deregulation of apoptosis mediators' p53 and bcl2 in lung tissue of COPD patients. *Respiratory Research*, *11*, 1–8.

Takahashi, A., Ohtani, N., Yamakoshi, K., Iida, S. I., Tahara, H., Nakayama, K., & Hara, E. (2006). Mitogenic signalling and the p16INK4a–Rb pathway cooperate to enforce irreversible cellular senescence. *Nature Cell Biology*, *8*(11), 1291–1297.

To, M., Takagi, D., Akashi, K., Kano, I., Haruki, K., Barnes, P. J., & Ito, K. (2013). Sputum plasminogen activator inhibitor-1 elevation by oxidative stress-dependent nuclear factor-κB activation in COPD. *Chest*, *144*(2), 515–521.

Triana-Martínez, F., Picallos-Rabina, P., Da Silva-Álvarez, S., Pietrocola, F., Llanos, S., Rodilla, V., & Collado, M. (2019). Identification and characterization of cardiac glycosides as senolytic compounds. *Nature Communications*, *10*(1), 4731.

Triantafyllou, A., Mylonis, I., Simos, G., Bonanou, S., & Tsakalof, A. (2008). Flavonoids induce HIF-1α but impair its nuclear accumulation and activity. *Free Radical Biology and Medicine*, *44*(4), 657–670.

Tsuji, T., Aoshiba, K., & Nagai, A. (2010). Alveolar cell senescence exacerbates pulmonary inflammation in patients with chronic obstructive pulmonary disease. *Respiration*, *80*(1), 59–70.

Wan, R., Srikaram, P., Guntupalli, V., Hu, C., Chen, Q., & Gao, P. (2023). Cellular senescence in asthma: From pathogenesis to therapeutic challenges. *EBioMedicine*, *94*, 104717.

Woldhuis, R. (2021). *The role of accelerated ageing in aberrant lung tissue repair and remodelling in COPD*. [Thesis fully internal (DIV), University of Groningen]. University of Groningen.

Xie, M., Liu, X., Cao, X., Guo, M., & Li, X. (2020). Trends in prevalence and incidence of chronic respiratory diseases from 1990 to 2017. *Respiratory Research*, *21*, 1–13.

Xu, M., Pirtskhalava, T., Farr, J. N., Weigand, B. M., Palmer, A. K., Weivoda, M. M., & Kirkland, J. L. (2018). Senolytics improve physical function and increase lifespan in old age. *Nature Medicine*, *24*(8), 1246–1256.

Yamaura, K., Nelson, A. L., Nishimura, H., Rutledge, J. C., Ravuri, S. K., Bahney, C., & Huard, J. (2023). Therapeutic potential of senolytic agent quercetin in osteoarthritis: A systematic review and meta-analysis of preclinical studies. *Ageing Research Reviews*, *90*, 101989.

Yang, N., & Li, X. (2022). Epigallocatechin gallate relieves asthmatic symptoms in mice by suppressing HIF-1α/VEGFA-mediated M2 skewing of macrophages. *Biochemical Pharmacology*, *202*, 115112.

Yosef, R., Pilpel, N., Tokarsky-Amiel, R., Biran, A., Ovadya, Y., Cohen, S., & Krizhanovsky, V. (2016). Directed elimination of senescent cells by inhibition of BCL-W and BCL-XL. *Nature Communications*, *7*(1), 11190.

Zeng, H., Li, T., He, X., Cai, S., Luo, H., Chen, P., & Chen, Y. (2020). Oxidative stress mediates the apoptosis and epigenetic modification of the Bcl-2 promoter via DNMT1 in a cigarette smoke-induced emphysema model. *Respiratory Research*, *21*, 229.

Zhang, D., Lee, H., Wang, X., Rai, A., Groot, M., & Jin, Y. (2018). Exosome-mediated small RNA delivery: A novel therapeutic approach for inflammatory lung responses. *Molecular Therapy*, *26*(9), 2119–2130.

Zhang, H., Zhou, X., Chen, X., Lin, Y., Qiu, S., Zhao, Y., & Zhong, X. (2019). Rapamycin attenuates Tc1 and Tc17 cell responses in cigarette smoke-induced emphysema in mice. *Inflammation Research*, *68*, 957–968.

Zhang, H. X., Yang, J. J., Zhang, S. A., Zhang, S. M., Wang, J. X., Xu, Z. Y., & Lin, R. Y. (2018). HIF-1α promotes inflammatory response of chronic obstructive pulmonary disease by activating EGFR/PI3K/AKT pathway. *European Review for Medical & Pharmacological Sciences*, 22(18), 6077–6084.

Zhang, L., Pitcher, L. E., Prahalad, V., Niedernhofer, L. J., & Robbins, P. D. (2023). Targeting cellular senescence with senotherapeutics: Senolytics and senomorphics. *The FEBS Journal*, *290*(5), 1362–1383.

Zhang, X., Zhang, S., Liu, X., Wang, Y., Chang, J., Zhang, X., & Zhou, D. (2018). Oxidation resistance 1 is a novel senolytic target. *Aging Cell*, *17*(4), e12780.

Zhao, X., Yu, F. Q., Huang, X. J., Xu, B. Y., Li, Y. L., Zhao, X. Y., & Luan, B. (2018). Azithromycin influences airway remodeling in asthma via the PI3K/Akt/MTOR/HIF-1α/VEGF pathway. *Journal of Biological Regulators and Homeostatic Agents*, *32*(5), 1079–1088.

Zhu, Y., Doornebal, E. J., Pirtskhalava, T., Giorgadze, N., Wentworth, M., Fuhrmann-Stroissnigg, H., & Kirkland, J. L. (2017). New agents that target senescent cells: The flavone, fisetin, and the BCL-XL inhibitors, A1331852 and A1155463. *Aging (Albany NY)*, *9*(3), 955–963.

Zhu, Y., Tchkonia, T., Pirtskhalava, T., Gower, A. C., Ding, H., Giorgadze, N., & Kirkland, J. L. (2015). The Achilles' heel of senescent cells: From transcriptome to senolytic drugs. *Aging Cell*, *14*(4), 644–658.

Zhu, Y. I., Tchkonia, T., Fuhrmann-Stroissnigg, H., Dai, H. M., Ling, Y. Y., Stout, M. B., & Kirkland, J. L. (2016). Identification of a novel senolytic agent, navitoclax, targeting the Bcl-2 family of anti-apoptotic factors. *Aging Cell*, *15*(3), 428–435.

15 Precision Care

Personalized Medicine in Chronic Obstructive Pulmonary Disease Management

Ayeh Bani Saeid and Stewart Yeung
Discipline of Pharmacy, Graduate School of Health,
University of Technology Sydney
Ultimo, NSW, Australia
Faculty of Health, Australian Research Centre in Complementary and Integrative Medicine, University of Technology Sydney
Ultimo, NSW, Australia

Rachelle Catanzariti
Discipline of Pharmacy, Graduate School of Health,
University of Technology Sydney
Ultimo, NSW, Australia

15.1 INTRODUCTION

Chronic obstructive pulmonary disease (COPD) stands as a formidable global health challenge, characterized by its high prevalence and significant burden on individuals and healthcare systems alike [1]. The World Health Organization (WHO) ranks COPD as the third leading cause of death worldwide and it is projected to become the second leading cause of death by 2030, underscoring the urgent need for effective interventions [2]. Despite its profound impact, COPD remains a disease with limited treatment options, emphasizing the pressing necessity for advancements in precision care [1].

Current medications available for managing COPD can alleviate symptoms and slow the progression of the disease in certain cases; however, they fall short of eradicating the disease [1, 3]. This reality reflects the complex nature of COPD, which encompasses diverse phenotypes and endotypes, presenting unique challenges in treatment optimization [4]. Moreover, the evolving understanding of COPD pathogenesis underscores the need for tailored therapeutic approaches that address individual variations in disease manifestation and progression [4, 5].

In response to these challenges, precision care has emerged as a promising paradigm shift in COPD management [6, 7]. Precision medicine in COPD aims to tailor treatment strategies to the specific characteristics of each patient, leveraging advancements in technology and biomedical research to optimize therapeutic outcomes [7, 8]. This evolution in care delivery represents a departure from the traditional one-size-fits-all approach, offering the potential for more effective, personalized interventions that target the underlying mechanisms driving disease progression [8].

In this chapter, we explore the evolution of precision care in COPD, examining the current biomarkers of the disease, the tools facilitating the use of precision medicine in COPD, the limitations of implementation, and the opportunities presented by precision medicine.

DOI: 10.1201/9781003498926-15

15.2 EVOLUTION OF PRECISION CARE IN COPD

Precision or personalized medicine has been defined as treatment strategies based on the needs of individual patients. This individualization can be made based on genetic reasons, biomarkers, phenotypes, or even psychological characteristics that differentiate between two patients with similar clinical manifestations [8, 9]. Contemporary therapy relies heavily on precise diagnosis, particularly evident in advancements in lung cancer classification and targeted therapy [10, 11]. While this approach has proven effective, there's a fine line where further subclassification may blur distinctions between diseases or subtypes. Some argue for a shift away from broad diagnostic labels like "COPD" toward identifying specific "treatable traits" for more tailored treatment [9, 10]. Nonetheless, diagnostic labels remain crucial for patients, shaping their expectations and facilitating communication with clinicians [11]. Precision medicine emphasizes detailed patient characterization to detect phenotypes and causing endotypes, utilizing molecular and genetic techniques without necessarily discarding current diagnostic labels [12].

In the past two decades, COPD management has transitioned from following national and international guidelines toward a precision medicine approach. In 1995, the American Thoracic Society standards initially advocated a simplistic, one-size-fits-all therapy model devoid of patient-specific considerations [13, 14]. Subsequently, the Global Obstructive Lung Disease (GOLD) Initiative in 2004 introduced recommendations based on airflow obstruction severity, albeit still limited in scope [12, 15]. Notably, the English National Institute for Clinical Excellence in 2004 pioneered a multidimensional approach, considering various patient domains like smoking status, breathlessness, and comorbidities to tailor treatment [12, 16]. Over time, GOLD recommendations have evolved to prioritize individualized patient care. In 2013, initial assessments revealed symptom severity, exacerbation risk, spirometric abnormalities, and comorbidities [17]. This approach was further refined in the 2017 report, where patients are classified into GOLD stages (1–4) based on forced expiratory volume (FEV) and GOLD groups (A–D) based on symptom severity and exacerbation risk [17]. These parameters serve as the foundation for determining personalized therapy strategies.

Furthermore, exacerbation frequency stands as a crucial determinant in COPD, significantly impacting patients' health status, lung function, and overall quality of life [18, 19]. The ECLIPSE study underscored that a patient's history of past exacerbations is the most accurate predictor of future exacerbations [20]. GOLD guidelines use a threshold of two moderate COPD flare-ups per year to identify those at heightened risk [21]. Given the gravity of severe exacerbations, even a single hospitalized exacerbation can also categorize patients as high-risk [22]. In its 2019 update, GOLD incorporated blood eosinophil counts into initial pharmacotherapy considerations. Subsequent therapy adjustments are recommended based on ongoing symptoms, exacerbation occurrences, and eosinophil levels, which can be considered as another example of precision medicine in COPD in the early stages [15].

Additionally, the GOLD report and WHO advocate screening all COPD patients for alpha-1 antitrypsin deficiency (AATD), resulting from mutations in the SERPINA1 gene, which increases the risk of emphysema [23–25]. AATD exemplifies precision medicine in COPD due to its clearly identifiable genetic basis, distinct epidemiology, and clinical features [26]. Crucially, detecting this deficiency through identifiable biomarkers allows for targeted therapy including intravenous augmentation treatment, which has been shown to enhance lung density based on CT scans [27]. However, this subtype of patient, attributed to a specific mutation, constitutes only a small portion of COPD cases [28].

Moreover, for patients with severe disease, a precision-medicine approach aids in assessing the necessity for oxygen therapy or ventilatory support. Arterial blood gas measurements, clinical signs of cor pulmonale, and erythrocytosis determine the indication for long-term oxygen therapy (LTOT) [29]. Although LTOT improves survival in severely hypoxemic patients, its benefits are limited in those with moderate desaturation, referring to a condition where oxygen levels in the blood are moderately low, typically indicating less severe oxygen deficiency compared to severe hypoxemia [30].

Domiciliary non-invasive positive pressure ventilation (NPPV) serves as a targeted intervention for a specific subset of patients, such as those with persistent hypercapnia and hypoxemia post-exacerbation discharge [31]. In these cases, combining NPPV with oxygen therapy prolongs the time to readmission or death. Precision medicine in COPD extends to assessing patients for bronchoscopic interventions, such as lung volume reduction techniques, including endobronchial valves, lung volume reduction coils, or vapor ablation therapy [32]. The suitability and selection of the most appropriate procedure hinge upon a thorough evaluation of various factors. These factors include the severity of airflow obstruction, radiographic findings of emphysema extent, degree of hyperinflation and gas transfer impairment on lung function tests, exercise capacity, and psychological assessment [32].

As evident, the evolution of precision medicine in COPD has come a long way, with current guidelines mainly focusing on precise diagnosis and, eventually, more individualized treatments. This can influence a patient's maintenance drug regimen, exacerbation management, and even, in severe stages, the choice of oxygen therapy and surgical interventions.

15.3 BIOMARKERS AND MICROBIOME

Biomarkers play a crucial role in guiding precision medicine approaches for COPD management, with blood eosinophil counts emerging as a key indicator in the GOLD report for guiding pharmacotherapy decisions [33]. These counts not only predict the efficacy of adding inhaled corticosteroids (ICS) to bronchodilator treatment, but also offer insights into the inflammatory profile of COPD patients, paving the way for tailored and more effective therapeutic interventions [33, 34].

Interleukins, signaling proteins involved in the regulation of the immune response, play a significant role in COPD pathogenesis and treatment response [35]. Among them, interleukin 5 (IL-5), 8 (IL-8), 4 (IL-4), and 13 (IL-13) have garnered particular interest in understanding COPD inflammation dynamics and developing targeted therapies [36]. Elevated eosinophil counts in COPD patients correlate with increased lung eosinophil numbers and heightened levels of type-2 inflammation markers [37–39]. Despite similarities with asthma, therapies targeting IL-5 have yielded disappointing results in eosinophilic COPD [38, 40, 41]. However, the association between higher eosinophil counts and elevated IL-13 expression suggests potential pathways linking eosinophilic inflammation to airway remodeling and mucus secretion [42]. A study by Pavord et al., found no overall impact of mepolizumab, an anti-IL-5 antibody targeting eosinophils, on the annual rate of moderate or severe exacerbations [40]. However, in individuals with higher blood eosinophil counts, a lower dose of mepolizumab was associated with fewer exacerbations [40, 43, 44]. Accordingly, prospective clinical trials have established a threshold of around 100 eosinophils per mL, below which ICSs confer no significant benefit, while a continuous relationship exists between higher eosinophil levels and the efficacy of ICS therapy [44]. Whether eosinophil levels alone can effectively identify a distinct treatable trait of COPD needs further longitudinal study [45].

Epigenetic changes, which modify gene expression without altering DNA, commonly involve DNA methylation and histone H4 acetylation. While exercise is crucial for COPD management, its molecular effects remain unclear [46]. In a study conducted by da Silva and colleagues, inflammatory-response cytokines were also linked to epigenetic alterations in COPD patients undergoing exercise [46]. The study analyzed blood samples from 10 patients during a prolonged exercise program, revealing decreased DNA methylation and altered histone H4 acetylation, correlating with IL-4 and IL-8 levels [46]. These correlations suggest a potential connection between epigenetics and COPD progression modulation via exercise. In a recent review, the utilization of CT and MRI biomarkers in precision medicine was also discussed [47]. It has been highlighted that CT-based findings like emphysema [48], small airways diseases [49], and reduced total airway count [50] have been linked to a rapid decline in lung function, particularly in patients with mild COPD. These imaging biomarkers hold promise as prognostic indicators in early COPD stages, potentially identifying individuals prone to rapid disease progression and enabling early intervention for disease management, as previously suggested by Soriano et al. [50]. Understanding this interplay may offer insights into predicting patient outcomes.

In addition to COPD-related biomarkers, the microbiome—defined as the community of microorganisms inhabiting a particular environment—plays a significant role in COPD symptoms and progression [51]. The lung microbiome, once thought sterile, comprises diverse microbial communities crucial for respiratory health. Its role in immunity, inflammation modulation, and pathogen defense highlights its importance in maintaining lung function and overall well-being [51, 52]. Various studies have explored the connection between lung microbiota and diseases like COPD, sparking debates on whether microbiome changes cause exacerbations or reflect disease severity [51]. Dickson et al., proposed a model suggesting that COPD exacerbations result from dysbiosis triggered by inflammatory factors rather than specific microbial imbalances [53]. They argue that dysbiosis, not newly acquired bacteria, contributes to exacerbations, as seen in the differences between bacterial infections and exacerbations [53]. Antibiotic therapy targets acute bacterial infections but has limited efficacy in exacerbations, with the mild relief attributed potentially to its prophylactic effect against future exacerbations. This dysbiosis perpetuates itself through selective microbial growth and killing [53]. This model helps explain why some COPD patients experience frequent exacerbations, potentially due to specific microbiota alterations or lung structural changes [53]. Chronic azithromycin use reduces exacerbation frequency in COPD, possibly by maintaining proper selective pressure on lung microbiota or by increasing anti-inflammatory bacterial metabolites [54]. Additionally, azithromycin shows more efficacy in reducing exacerbations in older, milder COPD patients who have quit smoking [55]. Hence, precision medicine could target the lung microbial environment in treating frequent exacerbators.

15.4 CURRENT TOOLS IN PRECISION MEDICINE

The absence of dependable COPD biomarkers underscores the necessity for extensive data gathering and integration to unveil the disease's molecular mechanisms. MultiOMICs, an emerging analytical approach, integrates data from various omics studies to comprehensively understand complex biological data, potentially addressing the biomarker deficiency in COPD [56]. By quickly and cost-effectively stratifying patient risk based on specific markers, multiOMICs offer promise for precision medicine in COPD [57]. Large-scale studies integrating multiple omics datasets reveal significant alterations in lung tissue transcriptome, proteome, and metabolome profiles, shedding light on potential therapeutic targets and pathways [57]. Another study focusing on next-generation sequencing of microRNAs (miRNAs) in patient blood suggests miR-106b-5p could be a biomarker of COPD severity, though its mechanism requires further elucidation [58]. Understanding the specific roles of molecules like miR-106b-5p in COPD pathogenesis is crucial for their clinical utility in assessing patient outcomes and severity [58].

Additionally, electronic health records (EHRs) serve as valuable tools in precision medicine across various diseases, including COPD, when accompanied by omics data and analytical programs. The ability to process extensive data is particularly crucial for heterogeneous diseases like COPD, which exhibit diverse symptoms [59]. In Canada, primary care practices saw an increase in EHR usage to 53% by 2014 [60]. Proper EHR implementation involves monitoring usage, record maintenance, and security to ensure effectiveness. Identifying at-risk patients involves assessing factors like second-hand smoke exposure, exercise frequency, environmental pollutants, and tobacco usage, along with psychosocial factors influencing medication adherence and treatment outcomes [60]. A UK study demonstrated that targeted health interventions based on EHR data significantly reduced readmission rates for COPD patients, underscoring the potential of thorough EHRs in disease management. Recognizing symptoms, clinical test results, and psychosocial factors within EHRs is crucial for optimizing therapies and slowing disease progression [61]. Organizing EHR information using non-hypothesis-driven modeling, as demonstrated by the Intermountain Risk Score (IMRS), could provide valuable prognostic insights for COPD and other diseases [62].

While most fields of medicine utilize EHRs, balancing their positive and negative aspects is essential to enhance the quality of care. Successful usage requires close monitoring to prevent negative impacts on patient-physician relationships and overall patient satisfaction. Despite the benefits of EHRs in streamlining data recording and storage, maintaining concise and secure records is essential to prevent important data from being misinterpreted [63, 64]. Establishing partnerships between investigators to integrate EHR analysis with clinical research and biomarker identification holds promise for identifying new treatment options for COPD patients.

15.5 CHALLENGES AND FUTURE DIRECTIONS

One significant challenge in implementing precision medicine for COPD lies in addressing comorbidities. Given that the majority of COPD patients have other chronic conditions, it's crucial to consider these alongside the pulmonary disease when formulating individual management plans. Notably, nearly half of COPD patients also suffer from comorbid depression and anxiety, which are associated with worsened outcomes such as increased mortality, frequent exacerbations, and decreased quality of life [65]. Integrating an assessment of these psychological factors into the precision medicine approach is essential, as treating them could potentially improve COPD outcomes [66–70]. Another challenge is the identification and incorporation of palliative interventions into precision COPD care. Despite maximal therapy, many COPD patients continue to experience distressing symptoms like breathlessness, fatigue, insomnia, anxiety, and depression [71]. Palliative approaches to symptom control, such as oxygen therapy, medication management, pulmonary rehabilitation, advance care planning, and emotional support, are crucial for enhancing the quality of life for these patients [72]. Palliative care services are often underutilized in COPD management, with patients receiving aggressive medical interventions focused solely on survival rather than symptom management [72, 73]. Precision medicine must recognize the individual needs of COPD patients and include palliative interventions when appropriate, ensuring a holistic approach to their care that addresses both physical and psychosocial aspects [72, 74].

15.6 CONCLUSIONS

In conclusion, COPD presents a complex and diverse clinical landscape, with current treatments often falling short due to gaps in current understanding of the disease. To enhance assessment, treatment, and outcomes, it is imperative to elucidate the intricate relationship between phenotype, endotype, and the underlying cellular and molecular pathways driving disease progression. Although there have been calls for improved subphenotyping of COPD patients to guide therapy, we are still in the early stages of achieving this objective. Traditional metrics like FEV1/FVC alone are insufficient for predicting disease outcomes; instead, leveraging multiOMICs approaches can provide insights into molecular alterations and their physiological impacts. Recent advancements in omics data collection and storage offer promising avenues for further research, highlighted by the potential of omics data in elucidating the relationship between lung microbiota dysbiosis and exacerbation frequency. Integration of patient records into EHRs facilitates research endeavors and enhances predictive capabilities in clinical care. Clinicians must consider both biological and psychosocial factors at the individual patient level to optimize assessment, treatment, outcomes, and cost-effectiveness in COPD care.

REFERENCES

1. M.G. Matera, M. Cazzola, and C. Page, *Prospects for COPD treatment.* Current Opinion in Pharmacology, 2021. **56**: p. 74–84.
2. World Health Organization, *Chronic obstructive pulmonary disease (COPD).* 2023; Available from: https://www.who.int/news-room/fact-sheets/detail/chronic-obstructive-pulmonary-disease-(copd).

3. C.F. Vogelmeier, et al., *Goals of COPD treatment: Focus on symptoms and exacerbations.* Respiratory Medicine, 2020. **166**: p. 105938.
4. P.J. Barnes, *Endo-phenotyping of COPD patients.* Expert Review of Respiratory Medicine, 2021. **15**(1): p. 27–37.
5. V.K. Sidhaye, K. Nishida, and F.J. Martinez, *Precision medicine in COPD: Where are we and where do we need to go?* European Respiratory Review, 2018. **27**(149): p. 56–58.
6. D.M. Halpin, *Precision medicine in chronic obstructive pulmonary disease.* Chinese Medical Journal, 2022. **135**(10): p. 1156–1162.
7. C.A. Brandsma, et al., *Recent advances in chronic obstructive pulmonary disease pathogenesis: From disease mechanisms to precision medicine.* The Journal of Pathology, 2020. **250**(5): p. 624–635.
8. J.M. Leung, et al., *Introduction to precision medicine in COPD.* European Respiratory Journal, 2019. **53**(4): p. 1802460.
9. N.R. Council, et al., *Toward precision medicine: building a knowledge network for biomedical research and a new taxonomy of disease.* 2011. National Academies Press (US).
10. A.J. Vargas, and C.C. Harris, *Biomarker development in the precision medicine era: Lung cancer as a case study.* Nature Reviews Cancer, 2016. **16**(8): p. 525–537.
11. J.L. Jameson, and D.L. Longo, *Precision medicine—Personalized, problematic, and promising.* Obstetrical & Gynecological Survey, 2015. **70**(10): p. 612–614.
12. L.M. Fabbri, and S. Hurd, *Global strategy for the diagnosis, management and prevention of COPD: 2003 update.* 2003, European Respiratory Journal. **22**(1): p. 1–2.
13. *Standards for the diagnosis and care of patients with chronic obstructive pulmonary disease. American Thoracic Society.* American Journal of Respiratory Critical Care Medicine, 1995. **152**: p. 78–73.
14. D. Halpin, NICE Guidance for COPD. Thorax, 2004. **59**(3): p. 181–182.
15. R.A. Pauwels, et al., *Global strategy for the diagnosis, management, and prevention of chronic obstructive pulmonary disease: NHLBI/WHO global initiative for chronic obstructive lung disease (GOLD) workshop summary.* American Journal of Respiratory and Critical Care Medicine, 2001. **163**(5): p. 1256–1276.
16. D.J. Pierson, *Clinical practice guidelines for chronic obstructive pulmonary disease: A review and comparison of current resources.* Respiratory Care, 2006. **51**(3): p. 277–288.
17. S. Mayor, *GOLD report on the diagnosis and management of COPD.* Prescriber, 2017. **28**(2): p. 28–32.
18. J. Soler-Cataluña, et al., *Severe acute exacerbations and mortality in patients with chronic obstructive pulmonary disease.* Thorax, 2005. **60**(11): p. 925–931.
19. J.J. Soler-Cataluña, and R. Rodriguez-Roisin, *Frequent chronic obstructive pulmonary disease exacerbators: How much real, How much fictitious?* COPD: Journal of Chronic Obstructive Pulmonary Disease, 2010. **7**(4): p. 276–284.
20. J. Hurst, et al., *Susceptibility to exacerbation in chronic obstructive pulmonary disease.* New England Journal of Medicine, 2010. **363**(12): p. 1128–1138.
21. O. Le Rouzic, et al., *Defining the "frequent exacerbator" phenotype in COPD: A hypothesis-free approach.* Chest, 2018. **153**(5): p. 1106–1115.
22. J. Garcia-Aymerich, et al., *Lung function impairment, COPD hospitalisations and subsequent mortality.* Thorax, 2011. **66**(7): p. 585–590.
23. V. Boulyjenkov, *Alpha1-antitrypsin deficiency: Memorandum from a WHO meeting.* Bulletin of the World Health Organization, 1997. **75**: p. 397–415.
24. M. Miravitlles, et al., *European Respiratory Society statement: Diagnosis and treatment of pulmonary disease in α1-antitrypsin deficiency.* European Respiratory Journal, 2017. **50**(5).
25. P. Strnad, N.G. McElvaney, and D.A. Lomas, *Alpha1-antitrypsin deficiency.* New England Journal of Medicine, 2020. **382**(15): p. 1443–1455.
26. C. Strange, and T. Beiko, Treatment of alpha-1 antitrypsin deficiency. Seminars in Respiratory and Critical Care Medicine. 2015. **36**(4): p. 470–477.
27. K.R. Chapman, et al., *Intravenous augmentation treatment and lung density in severe α1 antitrypsin deficiency (RAPID): A randomised, double-blind, placebo-controlled trial.* The Lancet, 2015. **386**(9991): p. 360–368.
28. M. Miravitlles, *Alpha-1-antitrypsin and other proteinase inhibitors.* Current Opinion in Pharmacology, 2012. **12**(3): p. 309–314.
29. J.M. Cranston, et al., *Domiciliary oxygen for chronic obstructive pulmonary disease.* Cochrane Database of Systematic Reviews, 2005(4).

30. R.D. Yusen, et al., *The long-term oxygen treatment trial for chronic obstructive pulmonary disease: Rationale, design, and lessons learned*. Annals of the American Thoracic Society, 2018. **15**(1): p. 89–101.
31. P.B., Murphy, et al., *Effect of home noninvasive ventilation with oxygen therapy vs oxygen therapy alone on hospital readmission or death after an acute COPD exacerbation: a randomized clinical trial*. JAMA, 2017. **317**(21): p. 2177–2186.
32. F.J., Herth, et al., *Endoscopic lung volume reduction: an expert panel recommendation–update 2019*. Respiration, 2019. **97**(6): p. 548–557.
33. S., Pascoe, et al., *Blood eosinophils and treatment response with triple and dual combination therapy in chronic obstructive pulmonary disease: analysis of the IMPACT trial*. The Lancet Respiratory Medicine, 2019. **7**(9): p. 745–756.
34. K.F. Rabe, et al., *Triple inhaled therapy at two glucocorticoid doses in moderate-to-very-severe COPD*. New England Journal of Medicine, 2020. **383**(1): p. 35–48.
35. K. Chung, *Cytokines as targets in chronic obstructive pulmonary disease*. Current Drug Targets, 2006. **7**(6): p. 675–681.
36. E.T. Osei, et al., *Current perspectives on the role of interleukin-1 signalling in the pathogenesis of asthma and COPD*. European Respiratory Journal, 2020. **55**(2).
37. S.A. Christenson, et al., *Asthma–COPD overlap. Clinical relevance of genomic signatures of type 2 inflammation in chronic obstructive pulmonary disease*. American Journal of Respiratory and Critical Care Medicine, 2015. **191**(7): p. 758–766.
38. A. Higham, et al., *Type 2 inflammation in eosinophilic chronic obstructive pulmonary disease*. Allergy, 2021. **76**(6): p. 1861.
39. L. George, et al., *Blood eosinophil count and airway epithelial transcriptome relationships in COPD versus asthma*. Allergy, 2020. **75**(2): p. 370–380.
40. I.D. Pavord, et al., *Mepolizumab for eosinophilic chronic obstructive pulmonary disease*. New England Journal of Medicine, 2017. **377**(17): p. 1613–1629.
41. G.J. Criner, et al., *Benralizumab for the prevention of COPD exacerbations*. New England Journal of Medicine, 2019. **381**(11): p. 1023–1034.
42. H.C. Atherton, G. Jones, and H. Danahay, *IL-13-induced changes in the goblet cell density of human bronchial epithelial cell cultures: MAP kinase and phosphatidylinositol 3-kinase regulation*. American Journal of Physiology-Lung Cellular and Molecular Physiology, 2003. **285**(3): p. L730–L739.
43. E.H. Bel, and A. Ten Brinke, *New anti-eosinophil drugs for asthma and COPD: Targeting the trait!* Chest, 2017. **152**(6): p. 1276–1282.
44. G.A. Hamad, et al., *Eosinophils in COPD: How many swallows make a summer?* European Respiratory Journal, 2018. **51**(1): p. 1702177.
45. K.F. Rabe, B. Beghé, and L.M. Fabbri, Peripheral Eosinophil Count as a Biomarker for the Management of COPD: Not There Yet. European Respiratory Journal, 2017. **50**(5): 1702165.
46. I.R.V. da Silva, et al., *Exercise-modulated epigenetic markers and inflammatory response in COPD individuals: A pilot study*. Respiratory Physiology & Neurobiology, 2017. **242**: p. 89–95.
47. G.R. Washko, and G. Parraga, *COPD biomarkers and phenotypes: Opportunities for better outcomes with precision imaging*. European Respiratory Journal, 2018. **52**(5).
48. R. Yuan, et al., *Prediction of the rate of decline in FEV1 in smokers using quantitative computed tomography*. Thorax, 2009. **64**(11): p. 944–949.
49. S.P. Bhatt, et al., *Association between functional small airway disease and FEV1 decline in chronic obstructive pulmonary disease*. American Journal of Respiratory and Critical Care Medicine, 2016. **194**(2): p. 178–184.
50. M. Kirby, et al., *Total airway count on computed tomography and the risk of chronic obstructive pulmonary disease progression. Findings from a population-based study*. American Journal of Respiratory and Critical Care Medicine, 2018. **197**(1): p. 56–65.
51. R. Faner, et al., *The microbiome in respiratory medicine: Current challenges and future perspectives*. European Respiratory Journal, 2017. **49**(4).
52. J.P. McAleer, and J.K. Kolls, *Contributions of the intestinal microbiome in lung immunity*. European Journal of Immunology, 2018. **48**(1): p. 39–49.
53. R.P. Dickson, F.J. Martinez, and G.B. Huffnagle, *The role of the microbiome in exacerbations of chronic lung diseases*. The Lancet, 2014. **384**(9944): p. 691–702.
54. L.N. Segal, et al., *Randomised, double-blind, placebo-controlled trial with azithromycin selects for anti-inflammatory microbial metabolites in the emphysematous lung*. Thorax, 2017. **72**(1): p. 13–22.

55. M.K. Han, et al., *Predictors of chronic obstructive pulmonary disease exacerbation reduction in response to daily azithromycin therapy*. American Journal of Respiratory and Critical Care Medicine, 2014. **189**(12): p. 1503–1508.
56. S. Kim, et al., *Integrative clustering of multi-level omics data for disease subtype discovery using sequential double regularization*. Biostatistics, 2017. **18**(1): p. 165–179.
57. J. Li, et al., *Integrating 3-omics data analyze rat lung tissue of COPD states and medical intervention by delineation of molecular and pathway alterations*. Bioscience Reports, 2017. **37**(3): p. BSR20170042.
58. R. Wang, et al., *Peripheral leukocyte microRNAs as novel biomarkers for COPD*. International Journal of Chronic Obstructive Pulmonary Disease, 2017: p. 1101–1112.
59. M. Prosperi, et al., *Big data hurdles in precision medicine and precision public health*. BMC Medical Informatics and Decision Making, 2018. **18**: p. 1–15.
60. M. Jones, et al., *Progress in the enhanced use of electronic medical records: Data from the Ontario experience*. JMIR Medical Informatics, 2017. **5**(1): p. e6928.
61. Y. Liu, et al., *Smoking duration, respiratory symptoms, and COPD in adults aged≥ 45 years with a smoking history*. International Journal of Chronic Obstructive Pulmonary Disease, 2015: p. 1409–1416.
62. B.D. Horne, et al., *Exceptional mortality prediction by risk scores from common laboratory tests*. The American Journal of Medicine, 2009. **122**(6): p. 550–558.
63. D.P. Manca, *Do electronic medical records improve quality of care?: Yes.*. Canadian Family Physician, 2015. **61**(10): p. 846–847.
64. C. Ramanan, et al., *The role of exciton delocalization in the major photosynthetic light-harvesting antenna of plants*. Biophysical Journal, 2015. **108**(5): p. 1047–1056.
65. A. Yohannes, et al., *Depression and anxiety in chronic heart failure and chronic obstructive pulmonary disease: Prevalence, relevance, clinical implications and management principles*. International Journal of Geriatric Psychiatry, 2010. **25**(12): p. 1209–1221.
66. J.N. de Voogd, et al., *Depressive symptoms as predictors of mortality in patients with COPD*. Chest, 2009. **135**(3): p. 619–625.
67. T.-P. Ng, et al., *Depressive symptoms and chronic obstructive pulmonary disease: Effect on mortality, Hospital readmission, symptom burden, functional status, and quality of life*. Archives of Internal Medicine, 2007. **167**(1): p. 60–67.
68. M.D. Eisner, et al., *Influence of anxiety on health outcomes in COPD*. Thorax, 2010. **65**(3): p. 229–234.
69. J.K. Quint, et al., *Relationship between depression and exacerbations in COPD*. European Respiratory Journal, 2008. **32**(1): p. 53–60.
70. D. Halpin, et al., *Understanding fear and anxiety in patients at the time of an exacerbation of chronic obstructive pulmonary disease: A qualitative study*. JRSM Open, 2015. **6**(12): p. 2054270415614543.
71. D. Halpin, D. Seamark, and C. Seamark, Palliative and end-of-life care for patients with respiratory disease. in *Respiratory Diseases in the Elderly*. European Respiratory Monograph, 2009. **43**: p. 327–353. European Respiratory Society.
72. P. Yang, *Maximizing quality of life remains an ultimate goal in the era of precision medicine: Exemplified by lung cancer*. Precision Clinical Medicine, 2019. **2**(1): p. 8–12.
73. D.M. Halpin, *Palliative care for chronic obstructive pulmonary disease. Signs of progress, but still a long way to go*. American Thoracic Society, 2018. p. 1356–1358.
74. M.T. Claessens, et al., *Dying with lung cancer or chronic obstructive pulmonary disease: Insights from SUPPORT*. Journal of the American Geriatrics Society, 2000. **48**(S1): p. S146–S153.

16 Digital Health Solutions

Enhancing Medication Adherence in COPD Treatment

Viswapriya Viswalingam
Poona College of Pharmacy, Department of Pharmaceutical Chemistry, Bharati Vidyapeeth (Deemed to be University)
Pune, Maharashtra, India

Dileep Kumar
Department of Pharm Chemistry, Manipal College of Pharmaceutical Sciences, Manipal Academy of Higher Education
Manipal, Karnataka, India

16.1 INTRODUCTION

Chronic obstructive pulmonary disease (COPD) stands as a global public health concern, afflicting an estimated 294 million individuals annually and resulting in more than 3 million deaths, positioning it as the third leading cause of mortality worldwide (Wang et al., 2015). The escalating prevalence of COPD underscores its pressing significance as a critical health challenge. COPD manifests with multifaceted impairments spanning respiratory, physical, social, and emotional domains, thereby compromising the quality of life and constraining daily activities (Abudiab & Fuller-Thomson, 2022; Verma et al., n.d.). Timely and precise diagnosis, coupled with lifestyle interventions, assumes paramount importance in the long-term management of COPD. However, navigating COPD management proves intricate, contingent upon factors such as symptom severity, frequency and intensity of exacerbations, and hospitalizations (Wedzicha & Seemungal, 2007). A notable proportion of patients encounter substantial barriers in accessing reliable information, competent healthcare providers, and efficacious medications (Bollmeier & Hartmann, 2020). Due to the chronic nature of COPD, educating patients, providing skills, and encouraging self-management have become crucial in COPD care. Patients with COPD need to recognize and respond to symptoms and improve their early detection and treatment of acute exacerbations of COPD (AECOPD) (Cravo et al., 2022; Ogunbayo et al., 2017; Zwerink et al., 2014). They must also identify and avoid triggers, understand and stick to treatment plans, and adapt their work and lifestyle around their symptoms. However, delivering this approach to COPD management is complex and causes suboptimal outcomes such as poor adherence to medications and inadequate inhaler technique (Jardim & Nascimento, 2019; Watson & Wilkinson, 2022). Just 40–60% of patients adhere to their prescribed regime, emphasizing the importance of education and inhaler technique training. Patients also face challenges accessing healthcare services, hindering their ability to self-manage effectively (J. Ma et al., 2023).

In addition, Pulmonary Rehabilitation is crucial for COPD treatment, yet issues with access and engagement are common (Deng et al., 2021; Tonga & Oliver, 2023). Healthcare providers play essential roles in supporting COPD patients, including recognizing symptoms, tailoring treatment plans, providing preventive strategies, and inspiring lifestyle modifications. However, these roles

DOI: 10.1201/9781003498926-16

can be time-consuming and resource-intensive, especially with traditional paper-based records and face-to-face visits (Bourbeau et al., 2004; Hudd, 2020). With an aging population and increased pressure on health service resources, innovative solutions are needed.

Low- and middle-income countries face particular challenges in COPD management due to the high COPD burden and under-resourced healthcare systems. Limited access to diagnostic tools, respiratory care specialists, preventive measures, surveillance data, research, and public health policies further exacerbate the issue (Hurst et al., 2021). Scalable, affordable, and inclusive solutions are necessary to address these challenges in low- and middle-income countries.

Utilizing digital technology for the remote delivery of medical and clinical interventions represents a significant advancement in healthcare provision (Haleem et al., 2021; Senbekov et al., 2020). Initially conceived to address the challenges of providing care to individuals residing in remote and rural locales, this approach, commonly referred to as telemedicine, has evolved to encompass various modalities including synchronous communication (such as real-time video conferencing), asynchronous methods (such as remote consultations via email), and remote monitoring (which involves the transmission of COPD symptoms, for instance) (Agrol, 2023; Chakrabarti, 2015).

With the ubiquitous presence of smartphones, a subset of telemedicine known as Mobile Health (mHealth) has emerged, facilitating aspects of COPD care through dedicated applications (Davids & Ashrafian, 2022; Yang et al., 2018). This proliferation of digital health technologies has ushered in a new era, with digital health serving as an overarching term denoting the broad utilization of digital technologies in healthcare. This encompasses telemedicine applications, electronic health records (EHRs), and a spectrum of digital healthcare systems (Stoumpos et al., 2023; Yeung et al., 2023).

Despite the fluidity and interchangeability of terminologies within this domain, digital health remains a pivotal tool for healthcare providers (Abernethy et al., n.d.; Ronquillo et al., 2024). By enabling seamless sharing and analysis of patient data across hospital and primary care settings, digital health platforms empower practitioners to devise personalized care plans and interventions (National Academies of Sciences et al., 2021; Philpot et al., 2023). Moreover, the integration of sophisticated analysis tools and parallel computing systems allows for the extraction of valuable insights from vast datasets, thereby optimizing care delivery and outcomes.

Recognized as a transformative force in healthcare, the digital health paradigm holds promise in fostering patient empowerment, mitigating health disparities, and driving innovation to achieve superior healthcare standards. Digital health encompasses an array of solutions leveraging digital technologies, including health information technology (HIT), telehealth, mHealth, wearable devices, artificial intelligence (AI), machine learning (ML), the Internet of Things (IoT), and digital therapeutics (Kasoju et al., 2023). These technologies harbor the potential to revolutionize disease screening, diagnosis, and monitoring, thereby enabling more precise profiling of disease progression and bolstering management strategies (Li et al., 2021; Shi et al., 2023). Against the backdrop of the COVID-19 pandemic, nations and healthcare systems globally have swiftly embraced digital health, with telehealth emerging as the *de facto* standard (Bouabida et al., 2022; Monaghesh & Hajizadeh, 2020).

This chapter furnishes a comprehensive exploration of the digital health space, elucidating its various modalities and applications in COPD management. Despite the advantages of digital health, several challenges have impeded their development and widespread integration. Consequently, this chapter delves into key impediments and opportunities to foster the widespread adoption and advancement of digital health services aimed at supporting COPD management.

16.2 DIGITAL HEALTH AND ITS IMPACT

Over recent decades, the trajectory of digital technology has experienced a notable evolution, heralding transformative shifts across diverse spheres of human activity. This technological evolution has sparked profound alterations, both advantageous and detrimental, prompting ongoing scrutiny, particularly concerning its implications for human health (Pandya & Lodha, 2021; Thimbleby, 2013). Fundamental to this paradigm shift is the capacity for mechanical and digital mechanisms to record various facets of human existence, including physiological states, lived experiences, and narratives (Haleem et al., 2021; Haleem, Javaid, Qadri, et al., 2022). This capacity has precipitated revolutionary advancements in the management of individual health, the formulation of population-wide health strategies, and the dynamic generation of novel insights and knowledge in real time (Najjar, 2023; Pang et al., 2023). Collectively, these burgeoning digitally mediated capabilities are encapsulated within the construct of digital health.

Digital health, as a multifaceted domain, encompasses an array of pioneering advancements permeating the realms of health, medicine, and biomedical science, thereby reshaping the requisite tools for cultivating a healthier future (Benis et al., 2021; Cano et al., 2014). Innovations spanning cloud computing, AI, ML, blockchain technology, digitally facilitated diagnostics and therapeutics, telehealth modalities, and mHealth applications have now become commonplace within healthcare systems (Amjad et al., 2023; Tagde et al., 2021). These innovations hold the promise of facilitating earlier disease detection, optimizing treatment outcomes, and fostering heightened levels of patient engagement and empowerment.

Enabled by digital technology, effective, safe, accessible, and engaging healthcare solutions have emerged, capable of integration into global healthcare systems, potentially addressing the demand for COPD care needs (van der Heijden et al., 2013; Watson & Wilkinson, 2022). Diverse digital interventions, uninhibited by individual practices or healthcare systems, are available in various forms, including synchronous applications (apps) facilitating real-time video conferencing, telephone calls, emails, smartphone messages, or notifications (Bird et al., 2019; Ignatowicz et al., 2019). Furthermore, remote monitoring akin to traditional telehealth interventions, alongside information-providing devices, and contemporary multitooled health apps sourcing behavioral changes and self-management, contribute to the array of available options (Alshamrani, 2022; Li, 2013). Within this spectrum, digital health apps harbor the potential to offer multifaceted solutions encompassing patient education programs supporting inhaler technique and lifestyle modification, interactive self-management plans, remote monitoring systems, symptom recording and communication tools, and integration of environmental and physiological data for comprehensive disease understanding and personalized risk factor modification, management, and care adaptation (Farzandipour et al., 2017; Kaplan et al., 2023; Puah et al., 2022). Despite significant progress in digital health, there are still substantial obstacles hindering its full utilization. Challenges related to interoperability, usability, and the effectiveness of digital tools remain unresolved, impeding their smooth integration and usage within healthcare systems (Mumtaz et al., 2023; Renukappa et al., 2022). Furthermore, achieving equitable deployment of digital solutions to address urgent public health issues is still in its early stages, highlighting the need for collaborative efforts to unlock their transformative potential (Kickbusch et al., 2021; Neto & Wyl, 2024). Addressing these challenges necessitates sustained endeavors to leverage the full potential of digital health as a force multiplier in addressing prevailing healthcare challenges (Mumtaz et al., 2023; Williamson & Prybutok, 2024). Crucially, fostering collaboration and synergy among healthcare workers, policymakers, and industry stakeholders is imperative to propel the development and deployment of efficacious digital health solutions (Allioui & Mourdi, 2023; Monaco et al., 2021). These capabilities are elaborated upon in subsequent sections.

16.2.1 Health Information Technology

HIT stands as a pivotal component in modern healthcare delivery, revolutionizing both clinical practice and patient engagement. Amidst its promising capabilities, HIT encounters substantial challenges, constraining its full potential realization (Alotaibi & Federico, 2017; Jen et al., 2024). These challenges encompass issues of interoperability, usability, and the efficacy of digital tools, impeding seamless integration within healthcare ecosystems. Moreover, the equitable deployment of HIT solutions to address pressing public health challenges remains in its nascent stages, necessitating concerted efforts to unlock their transformative impact (Reisman, 2017; Walker et al., 2023). This chapter delves into the multifaceted landscape of HIT, exploring its vast potential to aggregate diverse medical, environmental, biological, and social data, streamline care processes, and bolster population health management. However, amid this promise lies the dichotomy of "digital capitalism" within the health sector, where private gain often supersedes public benefit, hindering innovation and interoperability. Additionally, inadequate regulation of digital health entities further impedes the development of solutions crucial for supporting high-quality primary care, underscoring the urgency for comprehensive regulatory frameworks. This introduction sets the stage for a thorough examination of HIT's role in contemporary healthcare, aiming to elucidate its transformative potential and the avenues for overcoming prevailing challenges to foster a more efficient, equitable, and patient-centered healthcare landscape.

16.2.2 Wearable Devices

Driven by advances in sensor technology, wearable devices have emerged as valuable tools capable of continuously collecting and tracking real-time physiological and biochemical data (Guk et al., 2019; Sharma et al., 2022). These devices, encompassing various forms such as wristbands, rings, smartwatches, armbands, and eyewear, facilitate seamless monitoring in natural environments. Particularly for chronic diseases, wearable devices offer longitudinal data to track patient progress, presenting a viable alternative to more complex, uncomfortable, and expensive monitoring techniques (Kang & Exworthy, 2022; Lu et al., 2020). The trend toward smaller, lighter, and less intrusive wearable devices enhances patient adherence and enables sustained monitoring (Masoumian Hosseini et al., 2023; Smith et al., 2023). Moreover, these devices can wirelessly transmit data to mobile integration platforms, empowering patients and healthcare providers to manage chronic conditions and promptly address exacerbations effectively (Long et al., 2023; Mishra & Singh, 2023). For individuals without chronic illnesses, wearable devices play a pivotal role in monitoring health and fitness indicators, thereby promoting overall wellness and facilitating early detection of potential diseases.

Among wearable technologies, sleep-monitoring devices hold significant prominence. While polysomnography remains the gold standard for sleep monitoring, it is labor-intensive and costly, requiring specialized technicians (de Zambotti et al., 2019; Guillodo et al., 2020). In contrast, sleep-monitoring wearables offer the advantage of home use, streamlining the monitoring process in hospital settings and mitigating the "first night bias" phenomenon. Data collected by these devices can be seamlessly transmitted to cloud platforms, enhancing their accessibility, convenience, and cost-effectiveness (Butpheng et al., 2020; Javaid et al., 2022). Wristwatches are capable of capturing various sleep parameters, including breathing volume, heart rate, snoring, sleep duration, hours of deep and light sleep, and duration of awakenings (Razjouyan et al., 2017; Yin et al., 2023).

Furthermore, wearables incorporate photoplethysmography, which can monitor pulse wave data, while lightweight sleep monitors aid in assessing sleep apnea and improving quality of sleep. Additionally, headband-like devices enable researchers to understand the activity of the brain through electroencephalography (De Fazio et al., 2022; Hayano et al., 2020; Imtiaz, 2021). Despite these advancements, the accuracy of these devices remains somewhat limited, with tendencies to overestimate sleep hours.

16.2.3 Internet of Things

Connected via the Internet, IoT facilitates the exchange of information between physical objects and the Internet. Evolving from and integrating various technologies such as ML, real-time analysis, sensors, and embedded systems, the IoT concept encompasses the notion of smart hospitals and other Internet-controlled devices, whether fixed or wireless (Nasajpour et al., 2020; Shin &

Hwang, 2017). These intelligent devices are capable of capturing and sharing data in everyday life to fulfil specific tasks. IoT applications span across smart cities, vehicles, entertainment systems, households, and interconnected healthcare settings (Hassebo & Tealab, 2023; Ullah et al., 2024). Central to IoT implementation in the medical domain are a multitude of medical devices, sensors, AI, diagnostic tools, and advanced imaging devices, all of which contribute to enhancing productivity and improving the quality of life across diverse industries and societies (Kelly et al., 2020; Mathkor et al., 2024).

Facilitating data transfer across computing, mechanical, and digital platforms via the Internet, IoT operates autonomously without human intervention. This technology has seen a surge in adoption for monitoring healthcare during the COVID-19 pandemic, addressing the challenges of inadequate and untimely health information dissemination, which has led to numerous fatalities (Mukati et al., 2023; Osama et al., 2023). Leveraging sensors, IoT enables prompt notification of health-related issues, with patient data securely stored in the cloud, facilitating efficient healthcare delivery. Additionally, IoT technology monitors individuals' daily activities, generating alerts for potential health concerns (Abdulmalek et al., 2022; Rejeb et al., 2023).

16.2.4 Telehealth

Utilizing electronic information and telecommunication technologies, telehealth enhances the remote delivery of clinical healthcare services, supports patient and healthcare provider education, and aids public health and health administration. Telehealth, also known as telemedicine, encompasses a broader scope, with telemedicine and eHealth falling under its umbrella (Gajarawala & Pelkowski, 2021). Defined by the World Health Organization as the application of information and communication technologies in healthcare, telehealth encompasses various technologies including telecommunication, telemonitoring, teleconsultation, tele-education, and telerehabilitation (Shaw, 2009; Tahsin et al., 2023). Originally aimed at providing medical expertise to remote areas, telehealth has evolved with technological advancements, enabling healthcare providers to offer virtual support to patients with chronic illnesses. Presently, telemedicine enables swift interventions and rapid responses, proving particularly beneficial in scenarios of limited healthcare access or when constant monitoring is crucial for managing unstable conditions such as post-hospitalization care for COPD exacerbation or chronic respiratory failure (Donner et al., 2021; Gaveikaite et al., 2020). Telemedicine and telehealth interventions have demonstrated efficacy in increasing patient engagement in health, promoting healthier lifestyles, and patient adherence (Bingham et al., 2021; Minuz et al., 2023; Vicente et al., 2022). This heightened involvement fosters a sense of security between patients and their families.

16.2.5 Mobile Health

mHealth, utilizing cellphone and wireless technology in healthcare, spans from disease monitoring to treatment aid, chronic illness management, and epidemic tracking (Devi et al., 2023; Stampe et al., 2021). Integrating mHealth enriches patient involvement, surveillance, and healthcare provision, while gamification, social features, and incentives amplify its effectiveness. Patient feedback indicates mHealth tools' ease of use facilitates disease progression monitoring and sharing with healthcare providers (Mustafa et al., 2023; Sardi et al., 2017).

The surge in fitness and medical app usage for health data monitoring has surged, escalating the demand for mHealth services (Said, 2023). Valued at $38.2 billion USD in 2021, the global mHealth apps market is anticipated to exhibit an 11.8% compound annual growth rate from 2022 to 2030 (Fortune Business Insights, n.d.). As mHealth technology advances and health awareness rises, its widespread adoption for chronic disease management is expected to burgeon further.

16.3 CURRENT CHALLENGES IN COPD TREATMENT

COPD management strategies resembling integrated care, though narrower in scope, collaborative self-management, self-care, and the patient-centered medical home model. Self-management, a cornerstone of integrated care, aims to empower patients by providing education on various topics including smoking cessation, early detection, and appropriate treatment of respiratory exacerbations, nutritional guidance, and stress management (Holland, 2014; Levy et al., 2023). While care coordination is essential for achieving optimal COPD outcomes, its efficacy is enhanced when coupled with self-management practices (Nici & ZuWallack, 2018; Schrijver et al., 2022). The patient-centered medical home model emphasizes the pivotal role of primary care providers in patient care, potentially underestimating the involvement of specialist physicians in clinical management (Jackson et al., 2013; Starfield, 2011). Implementing these interventions within the current healthcare framework can be resource-intensive (Deng et al., 2021). Pulmonary rehabilitation, integrating exercise training and collaborative self-management approaches, offers prolonged periods (often weeks) of healthcare provider engagement and care coordination, thus facilitating the integrated services crucial for optimizing care and improving outcomes (Bourbeau, 2009; Jing et al., 2023; Perumal, 2021).

16.4 CAN DIGITAL HEALTH IMPROVE COPD CARE?

As of now, utilizing a reactive approach in COPD management often leads to delays in recognizing treatment options, exacerbating the quality gap in care. Various interventions, such as pulmonary rehabilitation, smoking cessation, personalized inhalation therapy, home oxygen therapy, and home non-invasive ventilation, constitute COPD management (Figueiredo et al., 2022; Khan et al., n.d.). However, despite the existence of these treatments, significant challenges hinder their adoption and administration. Progress in digital health, encompassing telemedicine, wearable technology, AI, and ML, offers promise in overcoming these barriers and broadening access to a wider array of patients (Sun & Zhou, 2023; Yeung et al., 2023).

16.4.1 Self-management in COPD

Tailoring self-management plans to a patient's specific needs, disease severity, and comorbidities is recognized as beneficial in managing COPD. These plans equip patients with tools to effectively utilize resources, problem-solve, and recognize symptoms, fostering therapeutic relationships with healthcare professionals (Kermelly & Bourbeau, 2022; Welch et al., 2020). Digital health apps have emerged as promising complements to traditional paper-based plans, offering customizable content, remote access for clinicians, and the ability to capture patient data (Javaid et al., 2023; Senbekov et al., 2020). Early studies, such as one involving a web-based self-management system, have shown significant improvements in COPD impact scores and inhaler technique, leading to the development of comprehensive digital tools like "myCOPD" endorsed by the National Institute for Health and Care Excellence (NICE), which have demonstrated efficacy in improving COPD outcomes (Davies et al., 2023).

Alternate digital platforms, like the EDGE digital health tablet system, offer personalized self-management approaches with features such as adaptive symptom diaries and wireless pulse oximetry (Joo et al., 2024; Velardo et al., 2017). While such interventions have shown engagement and some improvements in health status, challenges remain regarding internal validity and generalizability. Issues like selection bias in trials and the representativeness of study populations raise questions about the broader applicability of these findings (Edmond et al., 2018; Hill et al., 2013). Larger, cross-comparable randomized controlled trials (RCTs) are needed to confirm the effectiveness of digital health apps in improving clinical outcomes and empowering specific patient groups for self-management (Iribarren et al., 2021; Ossenbrink et al., 2023).

Conventional telehealth interventions also show promise in COPD management, allowing for remote patient monitoring and early intervention. Despite benefits in reducing hospitalization and emergency visits, concerns persist regarding false positives and conflicting outcomes in studies (Donner et al., 2018; Janjua et al., 2021). Greater standardization is required in traditional telehealth approaches to produce more comparable evidence. However, the evolving landscape of digital innovations, designed with patient-centricity and customization in mind, may hold greater potential in engaging and empowering patients for effective self-management compared to traditional telehealth methods focused solely on symptom monitoring (Burton et al., 2022).

16.4.2 AI in COPD

The integration of AI methodologies within the realm of COPD diagnosis and classification represents a promising avenue in contemporary medical research (Estépar, 2020). Particularly, deep learning algorithms have been leveraged to analyze various diagnostic modalities such as lung auscultation, pulmonary function tests (PFTs), and imaging techniques like computed tomography (CT). For instance, a study conducted in Turkey demonstrated the efficacy of deep learning algorithms in analyzing lung sounds among participants exhibiting varying degrees of COPD severity, showcasing high accuracy rates above 94% and robust classification performance with an area under the curve (AUC) value of 0.9659. This underscores the potential utility of AI in facilitating COPD identification, particularly within resource-limited healthcare settings. Furthermore, collaborative efforts between pulmonologists and explainable AI platforms have yielded improved diagnostic accuracy in COPD identification, although certain challenges such as the necessity for comprehensive clinician training and the potential for AI supplementation to detrimentally affect diagnostic accuracy for some practitioners have been noted (Huang et al., 2023).

In addition to auscultatory analyses, AI-driven approaches have been extensively explored in the realm of COPD imaging, particularly through the integration of CT imaging data (Saad et al., 2023). Notably, studies have showcased the comparability of CT-based AI models with traditional spirometry in COPD staging. Furthermore, AI analyses of CT images have enabled the prediction of COPD-related outcomes and the stratification of patients into distinct risk categories (Hasenstab et al., 2021; Sugimori et al., 2021). Nonetheless, challenges surrounding data quality and quantity persist, necessitating the inclusion of larger, more diverse patient cohorts to ensure the robustness and generalizability of AI models.

Beyond individual modalities, there exists a burgeoning interest in adopting a multimodal diagnostic approach, amalgamating various data streams such as CT imaging, spirometry, demographic data, and biomarkers to augment COPD diagnostic capabilities (Myc et al., 2019; Y. Yang et al., 2022). Recent research has endeavored to integrate blood RNA-seq gene expression analyses with CT imaging, elucidating inflammatory patterns associated with specific COPD phenotypes (Ghosh et al., 2022; Morrow et al., 2019). This multimodal approach holds immense promise in elucidating COPD pathophysiology while concurrently enhancing diagnostic accuracy.

Despite the strides made, several challenges persist within the domain of AI-driven COPD diagnostics. Cost considerations and the imperative for further research to validate the efficacy of AI in COPD management underscore the complexity of integrating AI methodologies into clinical practice (Rajput et al., n.d.; Robertson et al., 2024). Nonetheless, AI represents a potent tool in augmenting COPD diagnosis, risk stratification, and prognosis prediction, potentially revolutionizing personalized treatment paradigms in the foreseeable future (Hasnain et al., 2023).

16.4.3 AECOPD

Pairing digital health applications with wearable physiological sensors to offer nuanced insights into COPD symptoms and facilitate feedback from both patients and healthcare providers proves beneficial in identifying AECOPD and directing management strategies to enhance patient quality of life (Coutu et al., 2023). In an RCT comparing the efficacy of the EDGE digital health system for enhancing COPD outcomes with conventional care involving a paper-based self-management plan, it was found that data derived from pulse oximetry, oxygen saturation, pulse rate, and respiratory rate, were predictive of exacerbations, yielding an AUC value of 0.682. Additionally, monitoring overnight pulse oximetry data demonstrated promise in discerning COPD phases, distinguishing between stable and exacerbation phases, with a positive value of 70% and a negative predictor value of 78% (Farmer et al., 2017).

Other sensor examples, such as accelerometers, offer portability, versatility, and accuracy. A meta-analysis encompassing 38 studies showcased a correlation between accelerometer-derived steps per day and forced expiratory volume in one second (FEV1), while a separate study involving 21 COPD patients unveiled an inverse relationship with hyperinflation. Further investigations by Moy et al., utilized ankle-worn accelerometers to establish a heightened risk of AECOPD and COPD hospitalization with reduced activity levels. Accelerometer data have also proven valuable in predicting readmissions post-AECOPD discharge and assessing mortality risk, underscoring their role in remote data monitoring and patient risk stratification (Hodkinson et al., 2019).

The integration of data streams from smart inhalers, physiological monitors, and external sources such as weather and pollution holds promise in furnishing essential insights for both self-management and clinical decision-making (Himes et al., 2019; Iribarren et al., 2021). This integration not only facilitates the identification of digital biomarkers for patient risk stratification but also aids in averting severe exacerbations. Overall, the amalgamation of digital health technologies with wearable sensors presents a compelling avenue for optimizing COPD management and enhancing patient outcomes.

16.5 TELEHEALTH IN COPD

Telehealth, encompassing remote monitoring and consultation, emerges as a pivotal tool across various facets of COPD management, including diagnosis, treatment, monitoring, education, and prevention (Rezende et al., n.d.). Illustrative clinical trials showcase its potential, such as an asynchronous remote monitoring system that significantly reduced the risk of hospitalization in COPD patients compared to standard care (Kermelly & Bourbeau, 2022; Pépin et al., 2022). Additionally, multicomponent interventions integrating telehealth elements demonstrated improvements in quality of life, though questions regarding study size and generalizability remain (Bashshur et al., 2016; Vegesna et al., 2017).

A systematic review of 29 clinical trials, predating the COVID-19 era, examined telehealth interventions in COPD management. While remote monitoring, consultation, or multicomponent interventions showed promise, the evidence quality varied widely, with inconclusive findings regarding their impact on exacerbations, quality of life, hospitalizations, or mortality. Insufficient data hindered subgroup analyses, and inconsistent outcome measurement and reporting further complicated assessment. Standardized outcome metrics and qualitative evaluations may elucidate contexts where telehealth best serves COPD patients, offering a pathway for more targeted and effective interventions (Kuan et al., 2022).

Before the COVID-19 pandemic, telehealth interventions were widely employed in areas like stroke care, psychiatry, and rural healthcare, but not extensively utilized for COPD management (Bhaskar et al., 2020; Takahashi et al., 2022). However, with the onset of the pandemic, healthcare systems faced challenges in providing in-person care due to shortages of personal protective equipment and concerns about virus transmission (Filip et al., 2022). As a result, routine one-on-one consultations, testing such as spirometry, and pulmonary rehabilitation were reduced, potentially leading to delays in COPD care (Cross et al., 2022). Additionally, limited social interactions during the pandemic may exacerbate stress and mental health issues among COPD patients (Madawala et al., 2023; Yohannes, 2021). Despite COPD patients being at higher risk for severe COVID-19, there was a notable decrease in COPD-related hospitalizations during the pandemic, possibly attributed to infection control measures and increased attention to COPD care, alongside the rise of telehealth as an alternative to in-person visits (Awatade et al., 2023; Sculley et al., 2022).

Healthcare providers quickly adapted to the changing landscape by shifting toward telehealth, with many consultations transitioning to telephone or video platforms (Amjad et al., 2023; Wosik et al., 2020). Surveys conducted during the early days of the pandemic revealed a significant shift toward telemedicine, with the majority of healthcare professionals adopting telehealth for COPD care. To accommodate this shift, reimbursement policies were adjusted, with Medicare and some private payers aligning telemedicine reimbursement rates with those for in-person visits. This enabled healthcare systems to maintain essential services and revenue streams, particularly in outpatient settings, amidst the challenges posed by the pandemic (Pinnock et al., 2022; Wu et al., 2021).

Implementation science elucidates methodologies aimed at seamless integration of evidence-based interventions into healthcare practices and policies, focusing on mechanisms, contexts, and determinants influencing intervention effectiveness (Chinman et al., 2017; Wensing, 2015). Conversely, human-centered design, synonymous with "design thinking," embodies an iterative approach to solution development tailored to end-user needs, encompassing patients, providers, systems, and stakeholders (Bengtson et al., 2020; Lorusso et al., 2021). Health systems are increasingly leveraging human-centered design methodologies informed by implementation science insights to optimize care delivery. While telephone and video telehealth modalities currently bridge gaps in face-to-face consultations, their potential extends beyond traditional healthcare models (Burrell, 2023; Ye et al., 2023). Advanced monitoring technologies provide continuous vital sign assessment, yielding insights unattainable through episodic measurements (Leenen et al., 2020; Selvaraju et al., 2022).

Telehealth's evolution necessitates transcending its role as a substitute for in-person care toward augmenting existing models or conceiving novel ones to enhance accessibility, especially for marginalized populations (Atkinson et al., 2023; Speedie et al., 2008). A human-centered design approach, emphasizing appropriateness, acceptability, and adoptability, is pivotal in crafting telehealth services tailored to both patient and healthcare system needs, thus realizing telehealth's transformative potential.

16.6 mHEALTH IN COPD

The existing literature indicates the potential for smartphone integration in COPD management, with mHealth playing a pivotal role in addressing modifiable risk factors, as evidenced by several studies (M. Alwashmi et al., 2016; Alwashmi et al., 2020). One intervention, incorporating COPD information, dose reminders, audio-visual materials, motivational components, and inhalation technique training, demonstrated enhanced therapeutic adherence among COPD patients (Rogliani et al., 2017). Similarly, a mobile-phone-based system facilitated a home endurance exercise training program, resulting in improved exercise capacity, strengthened limb muscles, and reduced systemic inflammation (Angulo et al., 2020; Qiu et al., 2023). Smartphone-enabled collection of COPD symptom diaries enabled early identification of exacerbation symptoms, facilitating timely intervention (Johnston et al., 2013; Konstantinidis et al., 2022). Moreover, remote delivery of COPD management interventions such as pulmonary rehabilitation and physical activity was feasible. However, medical devices like spirometers and pulse oximeters provide objective data unattainable through smartphones alone (Cox et al., 2021; Tomasic et al., 2018). Recent technological advancements have enabled seamless integration between smartphones and medical devices, but studies primarily focused on technical effectiveness, with limited patient involvement in intervention design and inadequate attention to patient perceptions, usability, and satisfaction (Rundo et al., 2020; Wagan et al., 2022).

Currently, numerous mobile apps for COPD exist in the market, but they often possess limited features or specific focuses. Importantly, there is generally an absence of clinical evidence regarding their efficacy and effectiveness (Silva et al., 2023; Sobnath et al., 2017). Recently, several RCTs

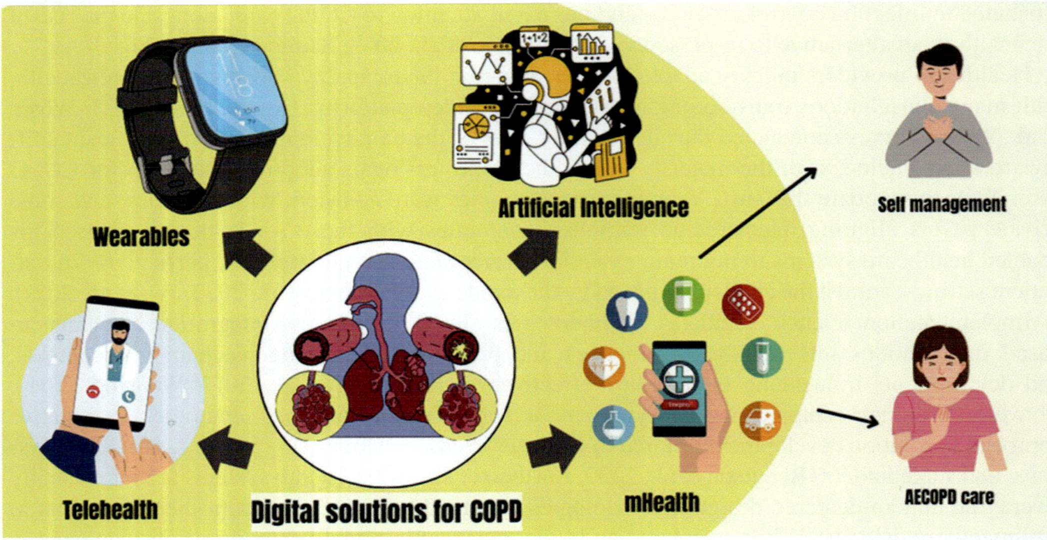

FIGURE 16.1 Digital solutions seem to be wide enough to address chronic obstructive pulmonary disease. In terms of telehealth, mHealth, wearables, and artificial intelligence.

have assessed the potential benefits of utilizing mHealth for COPD; however, they are typically constrained by small sample sizes (n ≤100) and narrow scopes, such as intervention for worsening COPD symptoms, physical activities, or impacts of climate change (Bischoff et al., 2023; Chong et al., 2023). Furthermore, many mHealth studies for COPD encounter numerous issues, including high drop-out rates (ranging from 20% to 36%), large variations in patient adherence rates (ranging from 20% to 98%), and inconsistent user experiences (mixed with technical challenges and varying levels of satisfaction) (Alghamdi et al., 2021; Cucciniello et al., 2021). Therefore, further clinical validation is essential to utilize mHealth effectively for improving COPD care.

In contrast to many extant studies, the MH-COPD program was formulated to encompass all fundamental components delineated by evidence-based guidelines in Australia and to integrate within COPD clinics (R. Smith et al., 2020). Additionally, the program is oriented toward augmenting patient self-management. Even though nurses and COPD physicians within the program review patient data via the clinician portal, their interventions predominantly center on providing support and motivation to patients for program adherence. Furthermore, a multitude of automated messages and alerts are disseminated through the mHealth system to reinforce the intervention, thereby minimizing the workload on clinicians in the program. As a result, the MH-COPD study presents a distinctive opportunity to investigate the potential of leveraging mHealth innovations to amplify patient self-management and health outcomes, thereby furnishing evidence for the efficacy of mHealth in advancing COPD care within the community (Figure 16.1) (Jing et al., 2023).

16.7 POLICIES AND REGULATIONS FOR DIGITAL HEALTH

Amidst the global expansion of digital health, numerous stakeholders are engaged in a broad array of service provisions. Initially centered on platforms and healthcare providers, the focus has shifted to encompass various responsibilities assigned to end-users, particularly patients (Landers et al., 2023; Wong et al., 2021). This chapter delves into the regulations governing the digital health ecosystem, which vary in their applicability across different regions. While some regulations are universally recognized, others are still evolving within nascent and diverse healthcare systems, such as that of India (Jain, 2023; Stephanie & Sharma, 2019).

Associated with several potential risks, digital health poses challenges such as telehealth possibly falling short of equipment or system failures, standard of care, susceptibility to abuse and manipulation of electronic data, and inadequate data protection in terms of security, authenticity, confidentiality, and privacy (Camara et al., 2015; Murdoch, 2021). Clarifying the responsibilities of each stakeholder is essential from a legal perspective, necessitating comprehensive documentation through policies, contracts, and agreements (Freeman & Mcvea, 2001). However, there is currently no universally applicable solution to the legal challenges entwined with digital health.

Ensuring patient privacy in digital zones is paramount, although mere compliance with regulatory frameworks does not suffice; proactive establishment of privacy protections is imperative to avert potential breaches (Giuffrè & Shung, 2023; McGraw & Mandl, 2021). Despite being a nascent and evolving field, privacy-related complexities must be addressed at all stages of development, underscoring the importance of treating privacy in digital health with due seriousness (Abernethy et al., n.d.; Filkins et al., 2016).

Critical to digital health is effective data management, requiring regulation to prevent potential mismanagement by healthcare providers and companies while promoting patient access to EHRs (Basil et al., n.d.; Keshta & Odeh, 2021). Transitioning toward a model where individuals retain control over their health data, rather than large commercial entities, may aid in mitigating the risk of data mismanagement (Abouelmehdi et al., 2018; Siala & Wang, 2022). Both the United States and the European Union have initiated steps to improve patient access to their health data; however, further legislative measures are essential to keep a balance between facilitating patient data access and safeguarding it from unauthorized or unintended access (Bradford et al., 2020; Tertulino et al., 2024).

16.8 FUTURE PROSPECTS

With a wide range of digital transformation approaches for COPD care, this chapter has demonstrated the predominant application of telemedicine, including telemonitoring and mHealth, aimed at enhancing COPD self-management and AECOPD (Fernandez-Granero et al., 2015; Pimenta et al., 2023). These applications exhibit the potential to reduce hospitalizations and underscore the necessity for rigorous evaluation of their effects on mortality in future investigations (Souza et al., 2023; Yelne et al., n.d.). Despite the implementation of multiple digital technologies such as EMR, ICD codes, SNOMET, and computerized decision support systems in hospital care, their practical utilization for COPD remains limited (Papadopoulos et al., 2022; Velickovski et al., 2014). Furthermore, while air pollution in the ambient environment has been linked to increased risks of hospitalization and mortality, translating these findings into preventive measures remains elusive and warrants further exploration (Genc et al., 2012; Karimi et al., 2022). Similarly, the clinical evidence supporting digitalization in public health surveillance remains insufficient to draw definitive conclusions.

Remote assessments have been found to significantly increase care burdens on providers. These issues pose practical challenges to digital health transformation and necessitate careful consideration and rigorous evaluation in future studies (Hermes et al., 2020; Paul et al., 2023). Unlike conventional clinical trials, evaluating digital health interventions is complex. An assessment of these interventions should account for the maturity level of the digital system under scrutiny (Guo et al., 2020; McLean et al., 2023; Murray et al., 2016).

The detection of acute COPD exacerbations represents a crucial clinical objective in COPD care but remains elusive. Currently, diagnosis relies on subjective assessments of COPD symptoms, with no objective measures through digital transformation approaches reliably predicting acute exacerbations (Ko et al., 2016; Pauwels et al., 2013). Hence, innovative research in monitoring devices and predictive models is imperative to enhance telemonitoring interventions in the future.

Effects of significant and non-significant have been reported in telemedical studies for COPD self-management, potentially attributed to heterogeneous designs of telemonitoring interventions (Ding, Fatehi, et al., 2019; Koh et al., 2023). Therefore, identifying underlying care processes contributing to care improvements is essential. Additionally, given the significant variability in patients' health conditions and therapeutic responses, identifying ideal candidates and effective personalized intervention strategies is crucial for achieving reliable clinical outcomes in future studies (Fogel, 2018; Légaré et al., 2018).

Numerous studies have justified digital solutions to improve COPD self-management. However, the successful functionalization of these solutions has been hindered by technical challenges, low patient compliance, and lack of personalization (Long et al., 2023). To address these issues, an iterative approach with continuous improvement is paramount. Moreover, acknowledging user preferences often enhances uptake and adherence. Digital health systems designed with a patient-centered approach and allowing for personalization of services typically exhibit higher compliance rates between patients and professionals.

The complexity of digital health has intensified with recent technologies such as smartphones and wearable devices, as well as enhanced computing power and capabilities. Consequently, AI is gaining traction in the health sector. Pulmonary medicine has been the focus of numerous research studies and experiments exploring AI applications. Most experiments with AI for lung diseases have focused on pulmonary test diagnosis and CT scanning. Moreover, recent research trends show interest in assessing COPD symptoms, such as analyzing lung sounds using deep learning techniques. Evaluation of personalized data gathered through smartphones and wearable devices offers new avenues for early detection and treatment of COPD.

Expanding on the idea, digital health has the potential to exacerbate existing healthcare disparities, especially among the elderly demographic. While digital interventions can offer benefits to elderly individuals managing chronic conditions requiring long-term care, concerns persist regarding digital exclusion (Gulliford & Alageel, 2019). Current estimates indicate that approximately 37% of the global population, totaling around 7.8 billion, faces digital exclusion (Kunonga et al., 2021). Within China alone, roughly 200 million elderly individuals lack Internet access, limiting their ability to utilize online health services (Qi et al., 2023). This digital disparity significantly impacts the elderly population, necessitating urgent attention and intervention. Despite the challenges hindering digital equity among the elderly, viable solutions exist (Kokorelias et al., 2022; Wilson et al., 2021). Clear communication of the advantages and disadvantages of digital health, coupled with supplementary support and resources, can help alleviate access barriers (Frishammar et al., 2023; Jardim & Nascimento, 2019). Additionally, adapting health applications to accommodate their specific limitations and proactively educating them on smartphone usage and online medical services can bridge the gap (Haleem et al., 2021; Haleem, Javaid, Singh, et al., 2022). Government initiatives promoting the benefits of digital health, especially in rural areas with significant elderly populations, can further aid in closing the divide. Evidence suggests that encouraging elderly individuals to familiarize themselves with new technologies can narrow the digital equity gap (Campanozzi et al., 2023; Ma et al., 2022). As digital health continues to evolve, tailoring it to meet the unique needs of the elderly is imperative for achieving comprehensive access to healthcare.

Transitioning to the broader context, the digital health revolution represents a significant cultural shift, introducing disruptive technologies that reshape the traditional dynamic between patients and healthcare providers (Meskó et al., 2017). This transformation not only facilitates collaborative health management but also empowers individuals, particularly those with chronic conditions, to engage in self-management practices (Grady & Gough, 2014; Vainauskienė & Vaitkienė, 2021). By educating the patients about their health and behavior, digital health enhances patient understanding of their conditions and fosters a biopsychosocial-digital approach to healthcare (Ahmadvand et al., 2018; Long et al., 2023). Consequently, clinicians are evolving from mere conveyors of medical knowledge to facilitators, guiding patients through the intricate realm of health information. To adapt to these evolving dynamics, clinicians must possess a comprehensive understanding of digital technologies and embrace a collaborative approach to the doctor-patient relationship (Győrffy et al., 2020; Kaba & Sooriakumaran, 2007). Upholding compassionate healing as the cornerstone of healthcare remains paramount amidst these changes. Patients, on their part, should actively engage

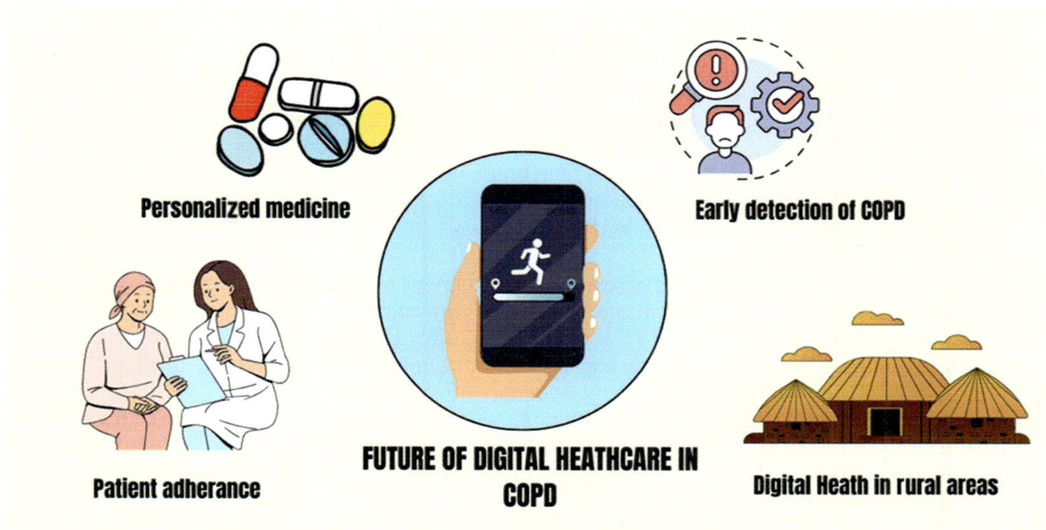

FIGURE 16.2 Certainly, the future of digital healthcare is bright, with applied activities in chronic diseases like chronic obstructive pulmonary disease.

with clinicians, discussing the information they receive and verifying its accuracy to strengthen the doctor–patient relationship and enhance medical services (Hsu et al., 2008; Sauerbrei et al., 2023). Effective communication between doctors and patients has the potential to improve clinical outcomes significantly. Figure 16.2 shows the perspective of the digital healthcare system.

16.9 CONCLUSION

Preventing and treating COPD involves providing effective care to reduce exacerbations and hospitalizations. Identifying individuals at high risk of early COPD presents a relevant way for preventive measures. However, many patients face difficulties in adhering to their treatment plans due to insufficient resources, limited knowledge about the disease, and inadequate clinical support. The evolution of digital health in the areas of mhealth, HIT, telehealth, AI, wearable technology, IoT, and digital therapeutics, holds potential for the early diagnosis and management of COPD.

The mainstream integration of digital health into medical practice is indeed an advancement, that is fostering the development of novel clinical models. However, the absence of real-time practice shows significant challenges. Although certain clinical studies explore the incorporation of digital aspects in COPD management, many others lack a comprehensive framework and encounter constraints in funding. For functional success, strategies for the adoption of new technologies must be developed immediately. This includes the development and evaluation of innovative technologies following established protocols to build a definite evidence base, thus supporting digital health. Additionally, the design of digital health devices should prioritize user-centric approaches, thereby promoting sustained utilization. Furthermore, establishing appropriate regulatory policies and adequate tools with security measures becomes essential for managing the substantial data volumes generated.

REFERENCES

Abdulmalek, S., Nasir, A., Jabbar, W. A., Almuhaya, M. A. M., Bairagi, A. K., Khan, Md. A.-M., & Kee, S.-H. (2022). IoT-based healthcare-monitoring system towards improving quality of life: A review. *Healthcare*, *10*(10), 1993. https://doi.org/10.3390/healthcare10101993

Abernethy, A., Adams, L., Barrett, M., Bechtel, C., Brennan, P., Butte, A., Faulkner, J., Fontaine, E., Friedhoff, S., Halamka, J., Howell, M., Johnson, K., Long, P., McGraw, D., Miller, R., Lee, P., Perlin, J., Rucker, D., Sandy, L., … Valdes, K. (n.d.). The promise of digital health: Then, now, and the future. *NAM Perspectives*, *2022*, 10.31478/202206e. https://doi.org/10.31478/202206e

Abouelmehdi, K., Beni-Hessane, A., & Khaloufi, H. (2018). Big healthcare data: Preserving security and privacy. *Journal of Big Data*, *5*(1), 1. https://doi.org/10.1186/s40537-017-0110-7

Abudiab, S., & Fuller-Thomson, E. (2022). Flourishing despite chronic obstructive pulmonary disease (COPD): Findings from a nationally representative survey of Canadians aged 50 and older. *International Journal of Environmental Research and Public Health*, *19*(23), 16337. https://doi.org/10.3390/ijerph192316337

Agrol, N. S. (2023). Comparative analysis of telemedicine programs in India and other countries, and to propose solutions for addressing the challenges associated with telemedicine implementation. *8*(8).

Ahmadvand, A., Gatchel, R., Brownstein, J., & Nissen, L. (2018). The biopsychosocial-digital approach to health and disease: Call for a paradigm expansion. *Journal of Medical Internet Research*, *20*(5), e189. https://doi.org/10.2196/jmir.9732

Alghamdi, S., Rajah, A., Aldabayan, Y., Aldhahir, A., Alqahtani, J., & Alzahrani, A. (2021). Chronic obstructive pulmonary disease patients' acceptance in e-health clinical trials. *International Journal of Environmental Research and Public Health*, *18*, 5230. https://doi.org/10.3390/ijerph18105230

Allioui, H., & Mourdi, Y. (2023). Exploring the full potentials of IoT for better financial growth and stability: A comprehensive survey. *Sensors*, *23*(19), 8015. https://doi.org/10.3390/s23198015

Alotaibi, Y. K., & Federico, F. (2017). The impact of health information technology on patient safety. *Saudi Medical Journal*, *38*(12), 1173–1180. https://doi.org/10.15537/smj.2017.12.20631

Alshamrani, M. (2022). IoT and artificial intelligence implementations for remote healthcare monitoring systems: A survey. *Journal of King Saud University - Computer and Information Sciences*, *34*(8, Part A), 4687–4701. https://doi.org/10.1016/j.jksuci.2021.06.005

Alwashmi, M., Hawboldt, J., Davis, E., Marra, C., Gamble, J.-M., & Abu Ashour, W. (2016). The effect of smartphone interventions on patients with chronic obstructive pulmonary disease exacerbations: A systematic review and meta-analysis. *JMIR mHealth and uHealth*, *4*(3), e105. https://doi.org/10.2196/mhealth.5921

Alwashmi, M. F., Fitzpatrick, B., Davis, E., Farrell, J., Gamble, J.-M., & Hawboldt, J. (2020). Features of a mobile health intervention to manage chronic obstructive pulmonary disease: A qualitative study. *Therapeutic Advances in Respiratory Disease*, *14*, 1753466620951044. https://doi.org/10.1177/1753466620951044

Amjad, A., Kordel, P., & Fernandes, G. (2023). A review on innovation in healthcare sector (telehealth) through artificial intelligence. *Sustainability*, *15*(8), Article 8. https://doi.org/10.3390/su15086655

Angulo, J., El Assar, M., Álvarez-Bustos, A., & Rodríguez-Mañas, L. (2020). Physical activity and exercise: Strategies to manage frailty. *Redox Biology*, *35*, 101513. https://doi.org/10.1016/j.redox.2020.101513

Atkinson, J., Hastie, R., Walker, S., Lindquist, A., & Tong, S. (2023). Telehealth in antenatal care: Recent insights and advances. *BMC Medicine*, *21*(1), 332. https://doi.org/10.1186/s12916-023-03042-y

Awatade, N. T., Wark, P. A. B., Chan, A. S. L., Mamun, S. A. A., Mohd Esa, N. Y., Matsunaga, K., Rhee, C. K., Hansbro, P. M., & Sohal, S. S. (2023). The complex association between COPD and COVID-19. *Journal of Clinical Medicine*, *12*(11), 3791. https://doi.org/10.3390/jcm12113791

Bashshur, R. L., Shannon, G. W., Bashshur, N., & Yellowlees, P. M. (2016). The empirical evidence for telemedicine interventions in mental disorders. *Telemedicine Journal and E-Health*, *22*(2), 87–113. https://doi.org/10.1089/tmj.2015.0206

Basil, N. N., Ambe, S., Ekhator, C., & Fonkem, E. (n.d). Health records database and inherent security concerns: A review of the literature. *Cureus*, *14*(10), e30168. https://doi.org/10.7759/cureus.30168

Bengtson, M., Bharadwaj, M., Bosch, A., Nyakundi, H., Matoke-Muhia, D., Dekker, C., & Diehl, J.-C. (2020). Matching development of point-of-care diagnostic tests to the local context: A case study of visceral leishmaniasis in Kenya and Uganda. *Global Health: Science and Practice*, *8*(3), 549–565. https://doi.org/10.9745/GHSP-D-20-00028

Benis, A., Tamburis, O., Chronaki, C., & Moen, A. (2021). One digital health: A unified framework for future health ecosystems. *Journal of Medical Internet Research*, *23*, 22189. https://doi.org/10.2196/22189

Bhaskar, S., Bradley, S., Chattu, V. K., Adisesh, A., Nurtazina, A., Kyrykbayeva, S., Sakhamuri, S., Moguilner, S., & Banach, M. (2020). Telemedicine as the new outpatient clinic gone digital: Position paper from the Pandemic Health System REsilience PROGRAM (REPROGRAM) International Consortium (Part 2). *Frontiers in Public Health*, *8*. https://doi.org/10.3389/fpubh.2020.00410

Bingham, J. M., Black, M., Anderson, E. J., Li, Y., Toselli, N., Fox, S., Martin, J. R., Axon, D. R., & Silva-Almodóvar, A. (2021). Impact of telehealth interventions on medication adherence for patients with type 2 diabetes, hypertension, and/or dyslipidemia: A systematic review. *Annals of Pharmacotherapy*, *55*(5), 637–649. https://doi.org/10.1177/1060028020950726

Bird, M., Li, L., Ouellette, C., Hopkins, K., McGillion, M. H., & Carter, N. (2019). Use of synchronous digital health technologies for the care of children with special health care needs and their families: Scoping review. *JMIR Pediatrics and Parenting*, *2*(2), e15106. https://doi.org/10.2196/15106

Bischoff, E. W., Ariens, N., Boer, L., Vercoulen, J., Akkermans, R. P., Bemt, L., van den, & Schermer, T. R. (2023). Effects of adherence to an mHealth tool for self-management of COPD exacerbations. *International Journal of Chronic Obstructive Pulmonary Disease*, *18*, 2381–2389. https://doi.org/10.2147/COPD.S431199

Bollmeier, S. G., & Hartmann, A. P. (2020). Management of chronic obstructive pulmonary disease: A review focusing on exacerbations. *American Journal of Health-System Pharmacy: AJHP*, *77*(4), 259–268. https://doi.org/10.1093/ajhp/zxz306

Bouabida, K., Lebouché, B., & Pomey, M.-P. (2022). Telehealth and COVID-19 pandemic: An overview of the telehealth use, advantages, challenges, and opportunities during COVID-19 pandemic. *Healthcare*, *10*(11), 2293. https://doi.org/10.3390/healthcare10112293

Bourbeau, J. (2009). The role of collaborative self-management in pulmonary rehabilitation. *Seminars in Respiratory and Critical Care Medicine*, *30*(6), 700–707. https://doi.org/10.1055/s-0029-1242639

Bourbeau, J., Nault, D., & Dang-Tan, T. (2004). Self-management and behavior modification in COPD. *Patient Education and Counseling*, *52*, 271–277. https://doi.org/10.1016/S0738-3991(03)00102-2

Bradford, L., Aboy, M., & Liddell, K. (2020). International transfers of health data between the EU and USA: A sector-specific approach for the USA to ensure an 'adequate' level of protection. *Journal of Law and the Biosciences*, *7*(1), lsaa055. https://doi.org/10.1093/jlb/lsaa055

Burrell, D. N. (2023). Dynamic evaluation approaches to telehealth technologies and artificial intelligence (AI) telemedicine applications in healthcare and biotechnology organizations. *Merits*, *3*(4), Article 4. https://doi.org/10.3390/merits3040042

Burton, L., Rush, K. L., Smith, M. A., Davis, S., Rodriguez Echeverria, P., Suazo Hidalgo, L., & Görges, M. (2022). Empowering patients through virtual care delivery: Qualitative study with micropractice clinic patients and health care providers. *JMIR Formative Research*, *6*(4), e32528. https://doi.org/10.2196/32528

Butpheng, C., Yeh, K.-H., & Xiong, H. (2020). Security and privacy in IoT-cloud-based e-health systems—A comprehensive review. *Symmetry*, *12*(7), Article 7. https://doi.org/10.3390/sym12071191

Camara, C., Peris-Lopez, P., & Tapiador, J. E. (2015). Security and privacy issues in implantable medical devices: A comprehensive survey. *Journal of Biomedical Informatics*, *55*, 272–289. https://doi.org/10.1016/j.jbi.2015.04.007

Campanozzi, L. L., Gibelli, F., Bailo, P., Nittari, G., Sirignano, A., & Ricci, G. (2023). The role of digital literacy in achieving health equity in the third millennium society: A literature review. *Frontiers in Public Health*, *11*, 1109323. https://doi.org/10.3389/fpubh.2023.1109323

Cano, I., Lluch-Ariet, M., Gomez-Cabrero, D., Maier, D., Kalko, S., Cascante, M., Tegnér, J., Miralles, F., Herrera, D., Roca, J., & Synergy-COPD consortium. (2014). Biomedical research in a digital health framework. *Journal of Translational Medicine*, *12*(2), S10. https://doi.org/10.1186/1479-5876-12-S2-S10

Chakrabarti, S. (2015). Usefulness of telepsychiatry: A critical evaluation of videoconferencing-based approaches. *World Journal of Psychiatry*, *5*(3), 286–304. https://doi.org/10.5498/wjp.v5.i3.286

Chinman, M., Woodward, E. N., Curran, G. M., & Hausmann, L. R. M. (2017). Harnessing implementation science to increase the impact of health disparity research. *Medical Care*, *55*(Suppl 9 2), S16–S23. https://doi.org/10.1097/MLR.0000000000000769

Chong, S. O. K., Pedron, S., Abdelmalak, N., Laxy, M., & Stephan, A.-J. (2023). An umbrella review of effectiveness and efficacy trials for app-based health interventions. *Npj Digital Medicine*, *6*(1), 1–11. https://doi.org/10.1038/s41746-023-00981-x

Coutu, F.-A., Iorio, O. C., & Ross, B. A. (2023). Remote patient monitoring strategies and wearable technology in chronic obstructive pulmonary disease. *Frontiers in Medicine*, *10*. https://doi.org/10.3389/fmed.2023.1236598

Cox, N. S., Dal Corso, S., Hansen, H., McDonald, C. F., Hill, C. J., Zanaboni, P., Alison, J. A., O'Halloran, P., Macdonald, H., & Holland, A. E. (2021). Telerehabilitation for chronic respiratory disease. *The Cochrane Database of Systematic Reviews*, *2021*(1), CD013040. https://doi.org/10.1002/14651858.CD013040.pub2

Cravo, A., Attar, D., Freeman, D., Holmes, S., Ip, L., & Singh, S. J. (2022). The importance of self-management in the context of personalized care in COPD. *International Journal of Chronic Obstructive Pulmonary Disease*, *17*, 231–243. https://doi.org/10.2147/COPD.S343108

Cross, A. J., Liang, J., Thomas, D., Zairina, E., Abramson, M. J., & George, J. (2022). Educational interventions for health professionals managing chronic obstructive pulmonary disease in primary care. *The Cochrane Database of Systematic Reviews*, *2022*(5), CD012652. https://doi.org/10.1002/14651858.CD012652.pub2

Cucciniello, M., Petracca, F., Ciani, O., & Tarricone, R. (2021). Development features and study characteristics of mobile health apps in the management of chronic conditions: A systematic review of randomised trials. *Npj Digital Medicine*, *4*(1), 1–11. https://doi.org/10.1038/s41746-021-00517-1

Davids, J., & Ashrafian, H. (2022). AIM and mHealth, Smartphones and Apps. In N. Lidströmer & H. Ashrafian (Eds.), *Artificial Intelligence in Medicine* (pp. 1229–1246). Springer International Publishing. https://doi.org/10.1007/978-3-030-64573-1_242

Davies, H., Chappell, M., Wang, Y., Phalguni, A., Wake, S., Arber, M., & Shore, J. (2023). myCOPD app for managing chronic obstructive pulmonary disease: A NICE medical technology guidance for a digital health technology. *Applied Health Economics and Health Policy*, *21*(5), 689–700. https://doi.org/10.1007/s40258-023-00811-x

De Fazio, R., Mattei, V., Al-Naami, B., De Vittorio, M., & Visconti, P. (2022). Methodologies and wearable devices to monitor biophysical parameters related to sleep dysfunctions: An overview. *Micromachines*, *13*(8), 1335. https://doi.org/10.3390/mi13081335

de Zambotti, M., Cellini, N., Goldstone, A., Colrain, I. M., & Baker, F. C. (2019). Wearable sleep technology in clinical and research settings. *Medicine and Science in Sports and Exercise*, *51*(7), 1538–1557. https://doi.org/10.1249/MSS.0000000000001947

Deng, N., Sheng, L., Jiang, W., Hao, Y., Wei, S., Wang, B., Duan, H., & Chen, J. (2021). A home-based pulmonary rehabilitation mHealth system to enhance the exercise capacity of patients with COPD: Development and evaluation. *BMC Medical Informatics and Decision Making*, *21*(1), 325. https://doi.org/10.1186/s12911-021-01694-5

Devi, D. H., Duraisamy, K., Armghan, A., Alsharari, M., Aliqab, K., Sorathiya, V., Das, S., & Rashid, N. (2023). 5G technology in healthcare and wearable devices: A review. *Sensors (Basel, Switzerland)*, *23*(5), 2519. https://doi.org/10.3390/s23052519

Ding, H., Fatehi, F., Maiorana, A., Bashi, N., Hu, W., & Edwards, I. (2019). Digital health for COPD care: The current state of play. *Journal of Thoracic Disease*, *11*(Suppl 17), S2210–S2220. https://doi.org/10.21037/jtd.2019.10.17

Ding, H., Karunanithi, M., Ireland, D., McCarthy, L., Hakim, R., Phillips, K., Pradhan, R., Seah, E.-H., Bowman, R. V., Fong, K., Masel, P., & Yang, I. A. (2019). Evaluation of an innovative mobile health programme for the self-management of chronic obstructive pulmonary disease (MH-COPD): Protocol of a randomised controlled trial. *BMJ Open*, *9*(4), e025381. https://doi.org/10.1136/bmjopen-2018-025381

Donner, C. F., Raskin, J., ZuWallack, R., Nici, L., Ambrosino, N., Balbi, B., Blackstock, F., Casaburi, R., Dreher, M., Effing, T., Goldstein, R., Krishnan, J., Lareau, S. C., Make, B. J., Maltais, F., Meek, P., Morgan, M., Pépin, J.-L., Rabbito, C., … Williams, L. (2018). Incorporating telemedicine into the integrated care of the COPD patient a summary of an interdisciplinary workshop held in Stresa, Italy, 7–8 September 2017. *Respiratory Medicine*, *143*, 91–102. https://doi.org/10.1016/j.rmed.2018.09.003

Donner, C. F., ZuWallack, R., & Nici, L. (2021). The role of telemedicine in extending and enhancing medical management of the patient with chronic obstructive pulmonary disease. *Medicina*, *57*(7), Article 7. https://doi.org/10.3390/medicina57070726

Edmond, S. N., Turk, D. C., Williams, D. A., & Kerns, R. D. (2018). Considerations of trial design and conduct in behavioral interventions for the management of chronic pain in adults. *Pain Reports*, *4*(3), e655. https://doi.org/10.1097/PR9.0000000000000655

Estépar, R. S. J. (2020). Artificial intelligence in COPD: New venues to study a complex disease. *Barcelona Respiratory Network Reviews*, *6*(2), 144–160. https://doi.org/10.23866/BRNRev:2019-0014

Farmer, A., Williams, V., Velardo, C., Shah, S. A., Yu, L.-M., Rutter, H., Jones, L., Williams, N., Heneghan, C., Price, J., Hardinge, M., & Tarassenko, L. (2017). Self-management support using a digital health system compared with usual care for chronic obstructive pulmonary disease: Randomized controlled trial. *Journal of Medical Internet Research*, *19*(5), e144. https://doi.org/10.2196/jmir.7116

Farzandipour, M., Nabovati, E., Sharif, R., Arani, M. H., & Anvari, S. (2017). Patient self-management of asthma using Mobile health applications: A systematic review of the functionalities and effects. *Applied Clinical Informatics*, *8*(4), 1068–1081. https://doi.org/10.4338/ACI-2017-07-R-0116

Fernandez-Granero, M. A., Sanchez-Morillo, D., & Leon-Jimenez, A. (2015). Computerised analysis of telemonitored respiratory sounds for predicting acute exacerbations of COPD. *Sensors*, *15*(10), Article 10. https://doi.org/10.3390/s151026978

Figueiredo, R. G., Laudano, C., Muniz, J., & de Bessa, J. (2022). Home-based respiratory care for COPD patients. *Sinusitis*, *6*(2), Article 2. https://doi.org/10.3390/sinusitis6020007

Filip, R., Gheorghita Puscaselu, R., Anchidin-Norocel, L., Dimian, M., & Savage, W. K. (2022). Global challenges to public health care systems during the COVID-19 pandemic: A review of pandemic measures and problems. *Journal of Personalized Medicine*, *12*(8), 1295. https://doi.org/10.3390/jpm12081295

Filkins, B. L., Kim, J. Y., Roberts, B., Armstrong, W., Miller, M. A., Hultner, M. L., Castillo, A. P., Ducom, J.-C., Topol, E. J., & Steinhubl, S. R. (2016). Privacy and security in the era of digital health: What should translational researchers know and do about it? *American Journal of Translational Research*, *8*(3), 1560–1580.

Fogel, D. B. (2018). Factors associated with clinical trials that fail and opportunities for improving the likelihood of success: A review. *Contemporary Clinical Trials Communications*, *11*, 156–164. https://doi.org/10.1016/j.conctc.2018.08.001

Fortune Business Insights. (n.d.). mHealth Apps Market Size, Share & Industry Analysis. Retrieved March 5, 2024, from https://www.fortunebusinessinsights.com/mhealth-apps-market-102020

Freeman, R., & Mcvea, J. (2001). A stakeholder approach to strategic management. *SSRN Electronic Journal*. https://doi.org/10.2139/ssrn.263511

Frishammar, J., Essén, A., Bergström, F., & Ekman, T. (2023). Digital health platforms for the elderly? Key adoption and usage barriers and ways to address them. *Technological Forecasting and Social Change*, *189*, 122319. https://doi.org/10.1016/j.techfore.2023.122319

Gajarawala, S. N., & Pelkowski, J. N. (2021). Telehealth benefits and barriers. *The Journal for Nurse Practitioners*, *17*(2), 218–221. https://doi.org/10.1016/j.nurpra.2020.09.013

Gaveikaite, V., Grundstrom, C., Winter, S., Schonenberg, H., Isomursu, M., Chouvarda, I., & Maglaveras, N. (2020). Challenges and opportunities for telehealth in the management of chronic obstructive pulmonary disease: A qualitative case study in Greece. *BMC Medical Informatics and Decision Making*, *20*(1), 216. https://doi.org/10.1186/s12911-020-01221-y

Genc, S., Zadeoglulari, Z., Fuss, S. H., & Genc, K. (2012). The adverse effects of air pollution on the nervous system. *Journal of Toxicology*, *2012*, 782462. https://doi.org/10.1155/2012/782462

Ghosh, A. J., Saferali, A., Lee, S., Chase, R. P., Moll, M., Morrow, J. D., Yun, J. H., Castaldi, P. J., & Hersh, C. P. (2022). Blood RNA sequencing shows overlapping gene expression across COPD phenotype domains. *Thorax*, *77*(2), 115–122. https://doi.org/10.1136/thoraxjnl-2020-216401

Giuffrè, M., & Shung, D. L. (2023). Harnessing the power of synthetic data in healthcare: Innovation, application, and privacy. *Npj Digital Medicine*, *6*(1), 1–8. https://doi.org/10.1038/s41746-023-00927-3

Grady, P. A., & Gough, L. L. (2014). Self-management: A comprehensive approach to management of chronic conditions. *American Journal of Public Health*, *104*(8), e25. https://doi.org/10.2105/AJPH.2014.302041

Guillodo, E., Lemey, C., Simonnet, M., Walter, M., Baca-García, E., Masetti, V., Moga, S., Larsen, M., Ropars, J., & Berrouiguet, S. (2020). Clinical applications of Mobile health Wearable–Based sleep monitoring: Systematic review. *JMIR mHealth and uHealth*, *8*(4), e10733. https://doi.org/10.2196/10733

Guk, K., Han, G., Lim, J., Jeong, K., Kang, T., Lim, E.-K., & Jung, J. (2019). Evolution of wearable devices with real-time disease monitoring for personalized healthcare. *Nanomaterials*, *9*(6). https://doi.org/10.3390/nano9060813

Gulliford, M., & Alageel, S. (2019). Digital health intervention at older ages. *The Lancet. Digital Health*, *1*(8), e382–e383. https://doi.org/10.1016/S2589-7500(19)30194-3

Guo, C., Ashrafian, H., Ghafur, S., Fontana, G., Gardner, C., & Prime, M. (2020). Challenges for the evaluation of digital health solutions—A call for innovative evidence generation approaches. *NPJ Digital Medicine*, *3*, 110. https://doi.org/10.1038/s41746-020-00314-2

Győrffy, Z., Radó, N., & Mesko, B. (2020). Digitally engaged physicians about the digital health transition. *PLoS ONE*, *15*(9), e0238658. https://doi.org/10.1371/journal.pone.0238658

Haleem, A., Javaid, M., Pratap Singh, R., & Suman, R. (2022). Medical 4.0 technologies for healthcare: Features, capabilities, and applications. *Internet of Things and Cyber-Physical Systems*, *2*, 12–30. https://doi.org/10.1016/j.iotcps.2022.04.001

Haleem, A., Javaid, M., Qadri, M. A., & Suman, R. (2022). Understanding the role of digital technologies in education: A review. *Sustainable Operations and Computers*, *3*, 275–285. https://doi.org/10.1016/j.susoc.2022.05.004

Haleem, A., Javaid, M., Singh, R. P., & Suman, R. (2021). Telemedicine for healthcare: Capabilities, features, barriers, and applications. *Sensors International*, *2*, 100117. https://doi.org/10.1016/j.sintl.2021.100117

Hasenstab, K. A., Yuan, N., Retson, T., Conrad, D. J., Kligerman, S., Lynch, D. A., & Hsiao, A. (2021). Automated CT staging of chronic obstructive pulmonary disease severity for predicting disease progression and mortality with a deep learning convolutional neural network. *Radiology: Cardiothoracic Imaging*, *3*(2), e200477. https://doi.org/10.1148/ryct.2021200477

Hasnain, M., Hayat, A., & Hussain, A. (2023). Revolutionizing chronic obstructive pulmonary disease care with the open AI application: ChatGPT. *Annals of Biomedical Engineering*, *51*(10), 2100–2102. https://doi.org/10.1007/s10439-023-03238-6

Hassebo, A., & Tealab, M. (2023). Global models of smart cities and potential IoT applications: A review. *IoT*, *4*(3), Article 3. https://doi.org/10.3390/iot4030017

Hayano, J., Yamamoto, H., Nonaka, I., Komazawa, M., Itao, K., Ueda, N., Tanaka, H., & Yuda, E. (2020). Quantitative detection of sleep apnea with wearable watch device. *PLoS ONE*, *15*(11), e0237279. https://doi.org/10.1371/journal.pone.0237279

Hermes, S., Riasanow, T., Clemons, E. K., Böhm, M., & Krcmar, H. (2020). The digital transformation of the healthcare industry: Exploring the rise of emerging platform ecosystems and their influence on the role of patients. *Business Research*, *13*(3), 1033–1069. https://doi.org/10.1007/s40685-020-00125-x

Hill, L. G., Rosenman, R., Tennekoon, V., & Mandal, B. (2013). Selection effects and prevention program outcomes. *Prevention Science : The Official Journal of the Society for Prevention Research*, *14*(6), 557–569. https://doi.org/10.1007/s11121-012-0342-x

Himes, B. E., Leszinsky, L., Walsh, R., Hepner, H., & Wu, A. C. (2019). Mobile Health and inhaler-based monitoring devices for asthma management. *The Journal of Allergy and Clinical Immunology. In Practice*, *7*(8), 2535–2543. https://doi.org/10.1016/j.jaip.2019.08.034

Hodkinson, A., Kontopantelis, E., Adeniji, C., van Marwijk, H., McMillan, B., Bower, P., & Panagioti, M. (2019). Accelerometer- and pedometer-based physical activity interventions among adults with cardiometabolic conditions. *JAMA Network Open*, *2*(10), e1912895. https://doi.org/10.1001/jamanetworkopen.2019.12895

Holland, A. E. (2014). Physiotherapy management of acute exacerbations of chronic obstructive pulmonary disease. *Journal of Physiotherapy*, *60*(4), 181–188. https://doi.org/10.1016/j.jphys.2014.08.018

Hsu, C., Phillips, W. R., Sherman, K. J., Hawkes, R., & Cherkin, D. C. (2008). Healing in primary care: A vision shared by patients, physicians, nurses, and clinical staff. *Annals of Family Medicine*, *6*(4), 307–314. https://doi.org/10.1370/afm.838

Huang, D.-M., Huang, J., Qiao, K., Zhong, N.-S., Lu, H.-Z., & Wang, W.-J. (2023). Deep learning-based lung sound analysis for intelligent stethoscope. *Military Medical Research*, *10*, 44. https://doi.org/10.1186/s40779-023-00479-3

Hudd, T. R. (2020). Emerging role of pharmacists in managing patients with chronic obstructive pulmonary disease. *American Journal of Health-System Pharmacy: AJHP*, *77*(19), 1625–1630. https://doi.org/10.1093/ajhp/zxaa216

Hurst, J. R., Buist, A. S., Gaga, M., Gianella, G. E., Kirenga, B., Khoo, E. M., Mendes, R. G., Mohan, A., Mortimer, K., Rylance, S., Siddharthan, T., Singh, S. J., van Boven, J. F. M., Williams, S., Zhang, J., & Checkley, W. (2021). Challenges in the implementation of chronic obstructive pulmonary disease guidelines in low- and middle-income countries: An official American thoracic society workshop report. *Annals of the American Thoracic Society*, *18*(8), 1269–1277. https://doi.org/10.1513/AnnalsATS.202103-284ST

Ignatowicz, A., Atherton, H., Bernstein, C. J., Bryce, C., Court, R., Sturt, J., & Griffiths, F. (2019). Internet videoconferencing for patient–clinician consultations in long-term conditions: A review of reviews and applications in line with guidelines and recommendations. *Digital Health*, *5*. https://doi.org/10.1177/2055207619845831

Imtiaz, S. A. (2021). A systematic review of sensing technologies for wearable sleep staging. *Sensors*, *21*(5), Article 5. https://doi.org/10.3390/s21051562

Iribarren, S. J., Akande, T. O., Kamp, K. J., Barry, D., Kader, Y. G., & Suelzer, E. (2021). Effectiveness of mobile apps to promote health and manage disease: Systematic review and meta-analysis of randomized controlled trials. *JMIR mHealth and uHealth*, *9*(1), e21563. https://doi.org/10.2196/21563

Jackson, G. L., Powers, B. J., Chatterjee, R., Bettger, J. P., Kemper, A. R., Hasselblad, V., Dolor, R. J., Irvine, R. J., Heidenfelder, B. L., Kendrick, A. S., Gray, R., & Williams, J. W. (2013). The patient centered medical home. A systematic review. *Annals of Internal Medicine*, *158*(3), 169–178. https://doi.org/10.7326/0003-4819-158-3-201302050-00579

Jain, D. (2023). Regulation of digital healthcare in India: Ethical and legal challenges. *Healthcare*, *11*(6), 911. https://doi.org/10.3390/healthcare11060911

Janjua, S., Carter, D., Threapleton, C. J., Prigmore, S., & Disler, R. T. (2021). Telehealth interventions: Remote monitoring and consultations for people with chronic obstructive pulmonary disease (COPD). *The Cochrane Database of Systematic Reviews*, *2021*(7), CD013196. https://doi.org/10.1002/14651858.CD013196.pub2

Jardim, J. R., & Nascimento, O. A. (2019). The importance of inhaler adherence to prevent COPD exacerbations. *Medical Sciences*, *7*(4), 54. https://doi.org/10.3390/medsci7040054

Javaid, M., Haleem, A., Singh, R. P., Rab, S., Suman, R., & Khan, I. H. (2022). Evolutionary trends in progressive cloud computing based healthcare: Ideas, enablers, and barriers. *International Journal of Cognitive Computing in Engineering*, *3*, 124–135. https://doi.org/10.1016/j.ijcce.2022.06.001

Javaid, M., Haleem, A., Singh, R. P., & Suman, R. (2023). 5G technology for healthcare: Features, serviceable pillars, and applications. *Intelligent Pharmacy*, *1*(1), 2–10. https://doi.org/10.1016/j.ipha.2023.04.001

Jen, M. Y., Kerndt, C. C., & Korvek, S. J. (2024). Health Information Technology. In *StatPearls*. StatPearls Publishing. http://www.ncbi.nlm.nih.gov/books/NBK470186/

Jing, Y., Ma, Y., Zhang, H., Wu, Z., Li, Y., Li, H., Huang, M., Lin, L., & Xu, Y. (2023). Pulmonary rehabilitation integrated coached exercise training for patients with COPD: A study protocol for a randomized controlled trial. *Trials*, *24*(1), 69. https://doi.org/10.1186/s13063-022-07058-2

Johnston, N. W., Lambert, K., Hussack, P., de Verdier, M. G., Higenbottam, T., Lewis, J., Newbold, P., Jenkins, M., Norman, G. R., Coyle, P. V., & McIvor, R. A. (2013). Detection of COPD exacerbations and compliance with patient-reported daily symptom diaries using a smart phone-based information system [corrected.]. *Chest*, *144*(2), 507–514. https://doi.org/10.1378/chest.12-2308

Joo, M.-I., Kang, D.-Y., Kang, M.-S., & Kim, H.-C. (2024). A patient management system using an edge computing-based IoT pulse oximeter. *Applied Sciences*, *14*(1), Article 1. https://doi.org/10.3390/app14010414

Kaba, R., & Sooriakumaran, P. (2007). The evolution of the doctor-patient relationship. *International Journal of Surgery (London, England)*, *5*(1), 57–65. https://doi.org/10.1016/j.ijsu.2006.01.005

Kang, H. S., & Exworthy, M. (2022). Wearing the future—Wearables to empower users to take greater responsibility for their health and care: Scoping review. *JMIR mHealth and uHealth*, *10*(7), e35684. https://doi.org/10.2196/35684

Kaplan, A., Boivin, M., Bouchard, J., Kim, J., Hayes, S., & Licskai, C. (2023). The emerging role of digital health in the management of asthma. *Therapeutic Advances in Chronic Disease*, *14*. https://doi.org/10.1177/20406223231209329

Karimi, B., Moradzadeh, R., & Samadi, S. (2022). Air pollution and COVID-19 mortality and hospitalization: An ecological study in Iran. *Atmospheric Pollution Research*, *13*(7), 101463. https://doi.org/10.1016/j.apr.2022.101463

Kasoju, N., Remya, N. S., Sasi, R., Sujesh, S., Soman, B., Kesavadas, C., Muraleedharan, C. V., Varma, P. R. H., & Behari, S. (2023). Digital health: Trends, opportunities and challenges in medical devices, pharma and bio-technology. *CSI Transactions on ICT*, *11*(1), 11–30. https://doi.org/10.1007/s40012-023-00380-3

Kelly, J. T., Campbell, K. L., Gong, E., & Scuffham, P. (2020). The internet of things: Impact and implications for health care delivery. *Journal of Medical Internet Research*, *22*(11), e20135. https://doi.org/10.2196/20135

Kermelly, S. B., & Bourbeau, J. (2022). eHealth in self-managing at a distance patients with COPD. *Life*, *12*(6), Article 6. https://doi.org/10.3390/life12060773

Keshta, I., & Odeh, A. (2021). Security and privacy of electronic health records: Concerns and challenges. *Egyptian Informatics Journal*, *22*(2), 177–183. https://doi.org/10.1016/j.eij.2020.07.003

Khan, K. S., Jawaid, S., Memon, U. A., Perera, T., Khan, U., Farwa, U. E., Jindal, U., Afzal, M. S., Razzaq, W., Abdin, Z. U., & Khawaja, U. A. (n.d). Management of chronic obstructive pulmonary disease (COPD) exacerbations in hospitalized patients from admission to discharge: A comprehensive review of therapeutic interventions. *Cureus*, *15*(8), e43694. https://doi.org/10.7759/cureus.43694

Kickbusch, I., Piselli, D., Agrawal, A., Balicer, R., Banner, O., Adelhardt, M., Capobianco, E., Fabian, C., Gill, A. S., Lupton, D., Medhora, R. P., Ndili, N., Ryś, A., Sambuli, N., Settle, D., Swaminathan, S., Morales, J. V., Wolpert, M., Wyckoff, A. W., … Wong, B. L. H. (2021). The Lancet and Financial Times Commission on governing health futures 2030: Growing up in a digital world. *The Lancet*, *398*(10312), 1727–1776. https://doi.org/10.1016/S0140-6736(21)01824-9

Ko, F. W., Chan, K. P., Hui, D. S., Goddard, J. R., Shaw, J. G., Reid, D. W., & Yang, I. A. (2016). Acute exacerbation of COPD. *Respirology (Carlton, Vic.)*, *21*(7), 1152–1165. https://doi.org/10.1111/resp.12780

Koh, J. H., Chong, L. C. Y., Koh, G. C. H., & Tyagi, S. (2023). Telemedical interventions for chronic obstructive pulmonary disease management: Umbrella review. *Journal of Medical Internet Research*, *25*, e33185. https://doi.org/10.2196/33185

Kokorelias, K. M., Nelson, M. L., Tang, T., Steele Gray, C., Ellen, M., Plett, D., Jarach, C. M., Xin Nie, J., Thavorn, K., & Singh, H. (2022). Inclusion of older adults in digital health technologies to support hospital-to-home transitions: Secondary analysis of a rapid review and equity-informed recommendations. *JMIR Aging*, *5*(2), e35925. https://doi.org/10.2196/35925

Konstantinidis, A., Kyriakopoulos, C., Ntritsos, G., Giannakeas, N., Gourgoulianis, K. I., Kostikas, K., & Gogali, A. (2022). The role of digital tools in the timely diagnosis and prevention of acute exacerbations of COPD: A comprehensive review of the literature. *Diagnostics*, *12*(2), Article 2. https://doi.org/10.3390/diagnostics12020269

Kuan, P. X., Chan, W. K., Ying, D. K. F., Rahman, M. A. A., Peariasamy, K. M., Lai, N. M., Mills, N. L., & Anand, A. (2022). Efficacy of telemedicine for the management of cardiovascular disease: A systematic review and meta-analysis. *The Lancet Digital Health*, *4*(9), e676–e691. https://doi.org/10.1016/S2589-7500(22)00124-8

Kunonga, T. P., Spiers, G. F., Beyer, F. R., Hanratty, B., Boulton, E., Hall, A., Bower, P., Todd, C., & Craig, D. (2021). Effects of digital technologies on older People's access to health and social care: Umbrella review. *Journal of Medical Internet Research*, *23*(11), e25887. https://doi.org/10.2196/25887

Landers, C., Vayena, E., Amann, J., & Blasimme, A. (2023). Stuck in translation: Stakeholder perspectives on impediments to responsible digital health. *Frontiers in Digital Health*, *5*, 1069410. https://doi.org/10.3389/fdgth.2023.1069410

Leenen, J. P. L., Leerentveld, C., van Dijk, J. D., van Westreenen, H. L., Schoonhoven, L., & Patijn, G. A. (2020). Current evidence for continuous vital signs monitoring by wearable wireless devices in hospitalized adults: Systematic review. *Journal of Medical Internet Research*, *22*(6), e18636. https://doi.org/10.2196/18636

Légaré, F., Adekpedjou, R., Stacey, D., Turcotte, S., Kryworuchko, J., Graham, I. D., Lyddiatt, A., Politi, M. C., Thomson, R., Elwyn, G., & Donner-Banzhoff, N. (2018). Interventions for increasing the use of shared decision making by healthcare professionals. *The Cochrane Database of Systematic Reviews*, *2018*(7), CD006732. https://doi.org/10.1002/14651858.CD006732.pub4

Levy, M. L., Bacharier, L. B., Bateman, E., Boulet, L.-P., Brightling, C., Buhl, R., Brusselle, G., Cruz, A. A., Drazen, J. M., Duijts, L., Fleming, L., Inoue, H., Ko, F. W. S., Krishnan, J. A., Mortimer, K., Pitrez, P. M., Sheikh, A., Yorgancıoğlu, A., & Reddel, H. K. (2023). Key recommendations for primary care from the 2022 Global Initiative for Asthma (GINA) update. *Npj Primary Care Respiratory Medicine*, *33*(1), 1–13. https://doi.org/10.1038/s41533-023-00330-1

Li, J.-P. O., Liu, H., Ting, D. S. J., Jeon, S., Chan, R. V. P., Kim, J. E., Sim, D. A., Thomas, P. B. M., Lin, H., Chen, Y., Sakomoto, T., Loewenstein, A., Lam, D. S. C., Pasquale, L. R., Wong, T. Y., Lam, L. A., & Ting, D. S. W. (2021). Digital technology, tele-medicine and artificial intelligence in ophthalmology: A global perspective. *Progress in Retinal and Eye Research*, *82*, 100900. https://doi.org/10.1016/j.preteyeres.2020.100900

Li, K. F. (2013). Smart home technology for telemedicine and emergency management. *Journal of Ambient Intelligence and Humanized Computing*, *4*(5), 535. https://doi.org/10.1007/s12652-012-0129-8

Long, H., Li, S., & Chen, Y. (2023). Digital health in chronic obstructive pulmonary disease. *Chronic Diseases and Translational Medicine*, *9*(2), 90–103. https://doi.org/10.1002/cdt3.68

Lorusso, L., Allied, A., Lee, J., & Worden, E. (2021). Design thinking for healthcare: Transliterating the creative problem-solving method into architectural practice. *HERD: Health Environments Research & Design Journal*, *14*, 193758672199422. https://doi.org/10.1177/1937586721994228

Lu, L., Zhang, J., Xie, Y., Gao, F., Xu, S., Wu, X., & Ye, Z. (2020). Wearable health devices in health care: Narrative systematic review. *JMIR mHealth and uHealth*, *8*(11). https://doi.org/10.2196/18907

Ma, J., Sun, X., Wang, X., Liu, B., & Shi, K. (2023). Factors affecting patient adherence to inhalation therapy: An application of SEIPS Model 2.0. *Patient Preference and Adherence*, *17*, 531–545. https://doi.org/10.2147/PPA.S395327

Ma, T., Zhang, S., Zhu, S., Ni, J., Wu, Q., & Liu, M. (2022). The new role of nursing in digital inclusion: Reflections on smartphone use and willingness to increase digital skills among Chinese older adults. *Geriatric Nursing*, *48*, 118–126. https://doi.org/10.1016/j.gerinurse.2022.09.004

Madawala, S., Quach, A., Lim, J. Y., Varatharaj, S., Perera, B., Osadnik, C., & Barton, C. (2023). Healthcare experience of adults with COPD during the COVID-19 pandemic: A rapid review of international literature. *BMJ Open Respiratory Research*, *10*(1), e001514. https://doi.org/10.1136/bmjresp-2022-001514

Masoumian Hosseini, M., Masoumian Hosseini, S. T., Qayumi, K., Hosseinzadeh, S., & Sajadi Tabar, S. S. (2023). Smartwatches in healthcare medicine: Assistance and monitoring: A scoping review. *BMC Medical Informatics and Decision Making*, *23*, 248. https://doi.org/10.1186/s12911-023-02350-w

Mathkor, D. M., Mathkor, N., Bassfar, Z., Bantun, F., Slama, P., Ahmad, F., & Haque, S. (2024). Multirole of the internet of medical things (IoMT) in biomedical systems for managing smart healthcare systems: An overview of current and future innovative trends. *Journal of Infection and Public Health*, *17*(4), 559–572. https://doi.org/10.1016/j.jiph.2024.01.013

McGraw, D., & Mandl, K. D. (2021). Privacy protections to encourage use of health-relevant digital data in a learning health system. *NPJ Digital Medicine*, *4*, 2. https://doi.org/10.1038/s41746-020-00362-8

McLean, K. A., Knight, S. R., Diehl, T. M., Varghese, C., Ng, N., Potter, M. A., Zafar, S. N., Bouamrane, M.-M., & Harrison, E. M. (2023). Readiness for implementation of novel digital health interventions for postoperative monitoring: A systematic review and clinical innovation network analysis. *The Lancet Digital Health*, *5*(5), e295–e315. https://doi.org/10.1016/S2589-7500(23)00026-2

Meskó, B., Drobni, Z., Bényei, É., Gergely, B., & Győrffy, Z. (2017). Digital health is a cultural transformation of traditional healthcare. *mHealth*, *3*, 38. https://doi.org/10.21037/mhealth.2017.08.07

Minuz, P., Albini, F. L., Imbalzano, E., Izzo, R., Masi, S., Pengo, M. F., Pucci, G., Scalise, F., Salvetti, M., Tocci, G., Cicero, A., Iaccarino, G., Savoia, C., Sechi, L., Parati, G., Borghi, C., Volpe, M., Ferri, C., Grassi, G. ... The Italian Society of Arterial Hypertension (SIIA). (2023). Telemedicine and digital medicine in the clinical management of hypertension and hypertension-related cardiovascular diseases: A position paper of the Italian Society of Arterial Hypertension (SIIA). *High Blood Pressure & Cardiovascular Prevention*, *30*(5), 387–399. https://doi.org/10.1007/s40292-023-00595-0

Mishra, P., & Singh, G. (2023). Internet of medical things healthcare for sustainable smart cities: Current status and future prospects. *Applied Sciences*, *13*(15), Article 15. https://doi.org/10.3390/app13158869

Monaco, A., Casteig Blanco, A., Cobain, M., Costa, E., Guldemond, N., Hancock, C., Onder, G., Pecorelli, S., Silva, M., Tournoy, J., Trevisan, C., Votta, M., Yfantopoulos, J., Yghemonos, S., Clay, V., Mondello Malvestiti, F., De Schaetzen, K., Sykara, G., & Donde, S. (2021). The role of collaborative, multistakeholder partnerships in reshaping the health management of patients with noncommunicable diseases during and after the COVID-19 pandemic. *Aging Clinical and Experimental Research*, *33*(10), 2899–2907. https://doi.org/10.1007/s40520-021-01922-y

Monaghesh, E., & Hajizadeh, A. (2020). The role of telehealth during COVID-19 outbreak: A systematic review based on current evidence. *BMC Public Health*, *20*(1), 1193. https://doi.org/10.1186/s12889-020-09301-4

Morrow, J. D., Chase, R. P., Parker, M. M., Glass, K., Seo, M., Divo, M., Owen, C. A., Castaldi, P., DeMeo, D. L., Silverman, E. K., & Hersh, C. P. (2019). RNA-sequencing across three matched tissues reveals shared and tissue-specific gene expression and pathway signatures of COPD. *Respiratory Research*, *20*(1), 65. https://doi.org/10.1186/s12931-019-1032-z

Mukati, N., Namdev, N., Dilip, R., Hemalatha, N., Dhiman, V., & Sahu, B. (2023). Healthcare assistance to COVID-19 patient using internet of things (IoT) enabled technologies. *Materials Today. Proceedings*, *80*, 3777–3781. https://doi.org/10.1016/j.matpr.2021.07.379

Mumtaz, H., Riaz, M. H., Wajid, H., Saqib, M., Zeeshan, M. H., Khan, S. E., Chauhan, Y. R., Sohail, H., & Vohra, L. I. (2023). Current challenges and potential solutions to the use of digital health technologies in evidence generation: A narrative review. *Frontiers in Digital Health*, *5*. https://www.frontiersin.org/journals/digital-health/articles/10.3389/fdgth.2023.1203945

Murdoch, B. (2021). Privacy and artificial intelligence: Challenges for protecting health information in a new era. *BMC Medical Ethics*, *22*(1), 122. https://doi.org/10.1186/s12910-021-00687-3

Murray, E., Hekler, E. B., Andersson, G., Collins, L. M., Doherty, A., Hollis, C., Rivera, D. E., West, R., & Wyatt, J. C. (2016). Evaluating digital health interventions: Key questions and approaches. *American Journal of Preventive Medicine*, *51*(5), 843–851. https://doi.org/10.1016/j.amepre.2016.06.008

Mustafa, A. S., Ali, N., Dhillon, J. S., & Sedera, D. (2023). An integrated model for evaluating the sustainability of gamified mobile health apps: An instrument development and validation. *Healthcare*, *11*(7), 1051. https://doi.org/10.3390/healthcare11071051

Myc, L. A., Shim, Y. M., Laubach, V. E., & Dimastromatteo, J. (2019). Role of medical and molecular imaging in COPD. *Clinical and Translational Medicine*, *8*, 12. https://doi.org/10.1186/s40169-019-0231-z

Najjar, R. (2023). Redefining radiology: A review of artificial intelligence integration in medical imaging. *Diagnostics*, *13*(17), 2760. https://doi.org/10.3390/diagnostics13172760

Nasajpour, M., Pouriyeh, S., Parizi, R. M., Dorodchi, M., Valero, M., & Arabnia, H. R. (2020). Internet of things for current COVID-19 and future pandemics: An exploratory study. *Journal of Healthcare Informatics Research*, *4*(4), 325–364. https://doi.org/10.1007/s41666-020-00080-6

National Academies of Sciences, Engineering, and Medicine; Health and Medicine Division; Board on Health Care Services; Committee on Implementing High-Quality Primary Care; Robinson, S. K., Meisnere, M., Robert L. Phillips, J., & McCauley, L. (2021). Digital Health and Primary Care. In *Implementing High-Quality Primary Care: Rebuilding the Foundation of Health Care.* National Academies Press (US). https://www.ncbi.nlm.nih.gov/books/NBK571817/

Neto, O. L., & Wyl, V. V. (2024). Digital transformation of public health for noncommunicable diseases: Narrative viewpoint of challenges and opportunities. *JMIR Public Health and Surveillance, 10*(1), e49575. https://doi.org/10.2196/49575

Nici, L., & ZuWallack, R. (2018). Integrated care in chronic obstructive pulmonary disease and rehabilitation. *COPD: Journal of Chronic Obstructive Pulmonary Disease, 15*(3), 223–230. https://doi.org/10.1080/15412555.2018.1501671

Ogunbayo, O. J., Russell, S., Newham, J. J., Heslop-Marshall, K., Netts, P., Hanratty, B., & Kaner, E. (2017). Understanding the factors affecting self-management of COPD from the perspectives of healthcare practitioners: A qualitative study. *Npj Primary Care Respiratory Medicine, 27*(1), 1–9. https://doi.org/10.1038/s41533-017-0054-6

Osama, M., Ateya, A. A., Sayed, M. S., Hammad, M., Pławiak, P., Abd El-Latif, A. A., & Elsayed, R. A. (2023). Internet of medical things and healthcare 4.0: Trends, requirements, challenges, and research directions. *Sensors, 23*(17), Article 17. https://doi.org/10.3390/s23177435

Ossenbrink, L., Haase, T., Timpel, P., Schoffer, O., Scheibe, M., Schmitt, J., Deckert, S., & Harst, L. (2023). Effectiveness of digital health interventions containing game components for the self-management of type 2 diabetes: Systematic review. *JMIR Serious Games, 11*, e44132. https://doi.org/10.2196/44132

Pandya, A., & Lodha, P. (2021). Social connectedness, excessive screen time during COVID-19 and mental health: A review of current evidence. *Frontiers in Human Dynamics, 3.* https://www.frontiersin.org/articles/10.3389/fhumd.2021.684137

Pang, T. Y., Lee, T.-K., & Murshed, M. (2023). Towards a new paradigm for digital health training and education in Australia: Exploring the implication of the fifth industrial revolution. *Applied Sciences, 13*(11), Article 11. https://doi.org/10.3390/app13116854

Papadopoulos, P., Soflano, M., Chaudy, Y., Adejo, W., & Connolly, T. M. (2022). A systematic review of technologies and standards used in the development of rule-based clinical decision support systems. *Health and Technology, 12*(4), 713–727. https://doi.org/10.1007/s12553-022-00672-9

Paul, M., Maglaras, L., Ferrag, M. A., & Almomani, I. (2023). Digitization of healthcare sector: A study on privacy and security concerns. *ICT Express, 9*(4), 571–588. https://doi.org/10.1016/j.icte.2023.02.007

Pauwels, E. K. J., Coumou, A. W., Kostkiewicz, M., & Kairemo, K. (2013). [18F]fluoro-2-deoxy-d-glucose positron emission tomography/computed tomography imaging in oncology: Initial staging and evaluation of cancer therapy. *Medical Principles and Practice, 22*(5), 427–437. https://doi.org/10.1159/000346303

Pépin, J.-L., Degano, B., Tamisier, R., & Viglino, D. (2022). Remote monitoring for prediction and management of acute exacerbations in chronic obstructive pulmonary disease (AECOPD). *Life, 12*(4), 499. https://doi.org/10.3390/life12040499

Perumal, S. D. (2021). Renewed vision on pulmonary rehabilitation service delivery for chronic obstructive pulmonary disease management beyond COVID-19. *Chronic Diseases and Translational Medicine, 7*(2), 107–116. https://doi.org/10.1016/j.cdtm.2021.01.003

Philpot, L. M., Dugani, S. B., Singla, A., DeZutter, M., & Ebbert, J. O. (2023). digital care horizon: A framework for extending health care through digital transformation. *Mayo Clinic Proceedings: Digital Health, 1*(3), 210–216. https://doi.org/10.1016/j.mcpdig.2023.05.005

Pimenta, S., Hansen, H., Demeyer, H., Slevin, P., & Cruz, J. (2023). Role of digital health in pulmonary rehabilitation and beyond: Shaping the future. *ERJ Open Research, 9*(2), 00212–02022. https://doi.org/10.1183/23120541.00212-2022

Pinnock, H., Murphie, P., Vogiatzis, I., & Poberezhets, V. (2022). Telemedicine and virtual respiratory care in the era of COVID-19. *ERJ Open Research, 8*(3), 00111–02022. https://doi.org/10.1183/23120541.00111-2022

Puah, S. H., Goh, C. Y., Chan, C. L., Teoh, A. K. J., Zhang, H., Shen, Z., & Neo, L. P. (2022). Mobile Device: A useful tool to teach inhaler devices to healthcare professionals. *BMC Medical Education, 22*(1), 238. https://doi.org/10.1186/s12909-022-03302-0

Qi, C., Wang, Y., Qi, X., Jiao, Y., Que, C., & Chen, Y. (2023). Barriers to providing internet-based home care services for urban older adults in China: A qualitative study of the service providers. *BMC Geriatrics, 23*(1), 320. https://doi.org/10.1186/s12877-023-04028-4

Qiu, Y., Fernández-García, B., Lehmann, H. I., Li, G., Kroemer, G., López-Otín, C., & Xiao, J. (2023). Exercise sustains the hallmarks of health. *Journal of Sport and Health Science, 12*(1), 8–35. https://doi.org/10.1016/j.jshs.2022.10.003

Rajput, S., Sharma, P. K., & Malviya, R. (n.d.). Artificial intelligence for emergency medical care. *Health Care Science*, *n/a*(n/a). https://doi.org/10.1002/hcs2.72

Razjouyan, J., Lee, H., Parthasarathy, S., Mohler, J., Sharafkhaneh, A., & Najafi, B. (2017). Improving sleep quality assessment using wearable sensors by including information from Postural/sleep position changes and body acceleration: A comparison of chest-worn sensors, wrist actigraphy, and polysomnography. *Journal of Clinical Sleep Medicine : JCSM : Official Publication of the American Academy of Sleep Medicine*, *13*(11), 1301–1310. https://doi.org/10.5664/jcsm.6802

Reisman, M. (2017). EHRs: The challenge of making electronic data usable and interoperable. *Pharmacy and Therapeutics*, *42*(9), 572–575.

Rejeb, A., Rejeb, K., Treiblmaier, H., Appolloni, A., Alghamdi, S., Alhasawi, Y., & Iranmanesh, M. (2023). The internet of things (IoT) in healthcare: Taking stock and moving forward. *Internet of Things*, *22*, 100721. https://doi.org/10.1016/j.iot.2023.100721

Renukappa, S., Mudiyi, P., Suresh, S., Abdalla, W., & Subbarao, C. (2022). Evaluation of challenges for adoption of smart healthcare strategies. *Smart Health*, *26*, 100330. https://doi.org/10.1016/j.smhl.2022.100330

Rezende, L. C., Ribeiro, E. G., Parreiras, L. C., Guimarães, R. A., dos Reis, G. M., Carajá, A. F., Franco, T. B., Mendes, L. P. de S., Augusto, V. M., & Silva, K. L. (n.d). Telehealth and telemedicine in the management of adult patients after hospitalization for COPD exacerbation: A scoping review. *Jornal Brasileiro de Pneumologia*, *49*(3), e20220067. https://doi.org/10.36416/1806-3756/e20220067

Robertson, N. M., Centner, C. S., & Siddharthan, T. (2024). Integrating artificial intelligence in the diagnosis of COPD globally: A way forward. *Chronic Obstructive Pulmonary Diseases (Miami, Fla.)*, *11*(1), 114–120. https://doi.org/10.15326/jcopdf.2023.0449

Rogliani, P., Ora, J., Puxeddu, E., Matera, M. G., & Cazzola, M. (2017). Adherence to COPD treatment: Myth and reality. *Respiratory Medicine*, *129*, 117–123. https://doi.org/10.1016/j.rmed.2017.06.007

Ronquillo, Y., Meyers, A., & Korvek, S. J. (2024). Digital Health. In *StatPearls*. StatPearls Publishing. http://www.ncbi.nlm.nih.gov/books/NBK470260/

Rundo, L., Pirrone, R., Vitabile, S., Sala, E., & Gambino, O. (2020). Recent advances of HCI in decision-making tasks for optimized clinical workflows and precision medicine. *Journal of Biomedical Informatics*, *108*, 103479. https://doi.org/10.1016/j.jbi.2020.103479

Saad, M. M., Bayoumy, A. A., EL-Nisr, M. M., Zaki, N. M., Khalil, T. H., & ELSerafi, A. F. (2023). Assessment of artificial intelligence-aided chest computed tomography in diagnosis of chronic obstructive airway disease: An observational study. *Egyptian Journal of Radiology and Nuclear Medicine*, *54*(1), 97. https://doi.org/10.1186/s43055-023-01043-8

Said, G. R. E. (2023). Factors affecting mHealth technology adoption in developing countries: The case of Egypt. *Computers*, *12*(1), Article 1. https://doi.org/10.3390/computers12010009

Sardi, L., Idri, A., & Fernández-Alemán, J. L. (2017). A systematic review of gamification in e-health. *Journal of Biomedical Informatics*, *71*, 31–48. https://doi.org/10.1016/j.jbi.2017.05.011

Sauerbrei, A., Kerasidou, A., Lucivero, F., & Hallowell, N. (2023). The impact of artificial intelligence on the person-centred, doctor-patient relationship: Some problems and solutions. *BMC Medical Informatics and Decision Making*, *23*(1), 73. https://doi.org/10.1186/s12911-023-02162-y

Schrijver, J., Lenferink, A., Brusse-Keizer, M., Zwerink, M., Valk, P. D., Palen, J., & Effing, T. W. (2022). Self-management interventions for people with chronic obstructive pulmonary disease. *The Cochrane Database of Systematic Reviews*, *2022*(1), CD002990. https://doi.org/10.1002/14651858.CD002990.pub4

Sculley, J. A., Musick, H., & Krishnan, J. A. (2022). Telehealth in chronic obstructive pulmonary disease: Before, during, and after the coronavirus disease 2019 pandemic. *Current Opinion in Pulmonary Medicine*, *28*(2), 93–98. https://doi.org/10.1097/MCP.0000000000000851

Selvaraju, V., Spicher, N., Wang, J., Ganapathy, N., Warnecke, J. M., Leonhardt, S., Swaminathan, R., & Deserno, T. M. (2022). Continuous monitoring of vital signs using cameras: A systematic review. *Sensors*, *22*(11), Article 11. https://doi.org/10.3390/s22114097

Senbekov, M., Saliev, T., Bukeyeva, Z., Almabayeva, A., Zhanaliyeva, M., Aitenova, N., Toishibekov, Y., & Fakhradiyev, I. (2020). The recent progress and applications of digital technologies in healthcare: A review. *International Journal of Telemedicine and Applications*, *2020*, e8830200. https://doi.org/10.1155/2020/8830200

Sharma, A., Singh, A., Gupta, V., & Arya, S. (2022). Advancements and future prospects of wearable sensing technology for healthcare applications. *Sensors & Diagnostics*, *1*(3), 387–404. https://doi.org/10.1039/D2SD00005A

Shaw, D. K. (2009). Overview of telehealth and its application to cardiopulmonary physical therapy. *Cardiopulmonary Physical Therapy Journal*, *20*(2), 13–18.

Shi, Y., Zhang, C., Pan, S., Chen, Y., Miao, X., He, G., Wu, Y., Ye, H., Weng, C., Zhang, H., Zhou, W., Yang, X., Liang, C., Chen, D., Hong, L., & Su, F. (2023). The diagnosis of tuberculous meningitis: Advancements in new technologies and machine learning algorithms. *Frontiers in Microbiology*, *14*, 1290746. https://doi.org/10.3389/fmicb.2023.1290746

Shin, D., & Hwang, Y. (2017). Integrated acceptance and sustainability evaluation of internet of medical things: A dual-level analysis. *Internet Research*, *27*(5), 1227–1254. https://doi.org/10.1108/IntR-07-2016-0200

Siala, H., & Wang, Y. (2022). SHIFTing artificial intelligence to be responsible in healthcare: A systematic review. *Social Science & Medicine*, *296*, 114782. https://doi.org/10.1016/j.socscimed.2022.114782

Silva, J., Hipólito, N., Machado, P., Flora, S., & Cruz, J. (2023). Technological features of smartphone apps for physical activity promotion in patients with COPD: A systematic review. *Pulmonology*. https://doi.org/10.1016/j.pulmoe.2023.06.005

Smith, A. A., Li, R., & Tse, Z. T. H. (2023). Reshaping healthcare with wearable biosensors. *Scientific Reports*, *13*(1), 4998. https://doi.org/10.1038/s41598-022-26951-z

Smith, R., Osadnik, C., & Lee, A. (2020). Educational topics and their rationale for inclusion within pulmonary rehabilitation – A systematic review. *Patient Education and Counseling*, *103*. https://doi.org/10.1016/j.pec.2020.04.009

Sobnath, D. D., Philip, N., Kayyali, R., Nabhani-Gebara, S., Pierscionek, B., Vaes, A. W., Spruit, M. A., & Kaimakamis, E. (2017). Features of a mobile support app for patients with chronic obstructive pulmonary disease: Literature review and current applications. *JMIR mHealth and uHealth*, *5*(2), e17. https://doi.org/10.2196/mhealth.4951

Souza, K. M., Carrasco, G., Rojas-Cortés, R., Michel Barbosa, M., Bambirra, E. H. F., Castro, J. L., & Alvares-Teodoro, J. (2023). Effectiveness of nirmatrelvir-ritonavir for the treatment of patients with mild to moderate COVID-19 and at high risk of hospitalization: Systematic review and meta-analyses of observational studies. *PLOS ONE*, *18*(10), e0284006. https://doi.org/10.1371/journal.pone.0284006

Speedie, S. M., Ferguson, A. S., Sanders, J., & Doarn, C. R. (2008). Telehealth: The promise of new care delivery models. *Telemedicine Journal and E-Health: The Official Journal of the American Telemedicine Association*, *14*(9), 964–967. https://doi.org/10.1089/tmj.2008.0114

Stampe, K., Kishik, S., & Müller, S. D. (2021). Mobile Health in chronic disease management and patient empowerment: Exploratory qualitative investigation into patient-physician consultations. *Journal of Medical Internet Research*, *23*(6), e26991. https://doi.org/10.2196/26991

Starfield, B. (2011). Is patient-centered care the same as person-focused care? *The Permanente Journal*, *15*(2), 63. https://doi.org/10.7812/tpp/10-148

Stephanie, L., & Sharma, R. (2019). Digital health eco-systems: An epochal review of practice-oriented research. *International Journal of Information Management*, *53*, 102032. https://doi.org/10.1016/j.ijinfomgt.2019.10.017

Stoumpos, A. I., Kitsios, F., & Talias, M. A. (2023). Digital transformation in healthcare: Technology acceptance and its applications. *International Journal of Environmental Research and Public Health*, *20*(4), 3407. https://doi.org/10.3390/ijerph20043407

Sugimori, H., Shimizu, K., Makita, H., Suzuki, M., & Konno, S. (2021). A comparative evaluation of computed tomography images for the classification of spirometric severity of the chronic obstructive pulmonary disease with deep learning. *Diagnostics*, *11*(6). https://doi.org/10.3390/diagnostics11060929

Sun, G., & Zhou, Y.-H. (2023). AI in healthcare: Navigating opportunities and challenges in digital communication. *Frontiers in Digital Health*, *5*. https://www.frontiersin.org/journals/digital-health/articles/10.3389/fdgth.2023.1291132

Tagde, P., Tagde, S., Bhattacharya, T., Tagde, P., Chopra, H., Akter, R., Kaushik, D., & Rahman, Md. H. (2021). Blockchain and artificial intelligence technology in e-health. *Environmental Science and Pollution Research International*, *28*(38), 52810–52831. https://doi.org/10.1007/s11356-021-16223-0

Tahsin, F., Armas, A., Kirakalaprathapan, A., Kadu, M., Sritharan, J., & Steele Gray, C. (2023). Information and communications technologies enabling integrated primary care for patients with complex care needs: Scoping review. *Journal of Medical Internet Research*, *25*, e44035. https://doi.org/10.2196/44035

Takahashi, E. A., Schwamm, L. H., Adeoye, O. M., Alabi, O., Jahangir, E., Misra, S., & Still, C. H., & null, null. (2022). An overview of telehealth in the management of cardiovascular disease: A scientific statement from the American Heart Association. *Circulation*, *146*(25), e558–e568. https://doi.org/10.1161/CIR.0000000000001107

Tertulino, R., Antunes, N., & Morais, H. (2024). Privacy in electronic health records: A systematic mapping study. *Journal of Public Health*, *32*(3), 435–454. https://doi.org/10.1007/s10389-022-01795-z

Thimbleby, H. (2013). Technology and the future of healthcare. *Journal of Public Health Research*, *2*(3), e28. https://doi.org/10.4081/jphr.2013.e28

Tomasic, I., Tomasic, N., Trobec, R., Krpan, M., & Kelava, T. (2018). Continuous remote monitoring of COPD patients—Justification and explanation of the requirements and a survey of the available technologies. *Medical & Biological Engineering & Computing*, *56*(4), 547–569. https://doi.org/10.1007/s11517-018-1798-z

Tonga, K. O., & Oliver, B. G. (2023). Effectiveness of pulmonary rehabilitation for chronic obstructive pulmonary disease therapy: Focusing on traditional medical practices. *Journal of Clinical Medicine*, *12*(14), Article 14. https://doi.org/10.3390/jcm12144815

Ullah, A., Anwar, S. M., Li, J., Nadeem, L., Mahmood, T., Rehman, A., & Saba, T. (2024). Smart cities: The role of internet of things and machine learning in realizing a data-centric smart environment. *Complex & Intelligent Systems*, *10*(1), 1607–1637. https://doi.org/10.1007/s40747-023-01175-4

Vainauskienė, V., & Vaitkienė, R. (2021). Enablers of patient knowledge empowerment for self-management of chronic disease: An integrative review. *International Journal of Environmental Research and Public Health*, *18*(5), 2247. https://doi.org/10.3390/ijerph18052247

van der Heijden, M., Lucas, P. J. F., Lijnse, B., Heijdra, Y. F., & Schermer, T. R. J. (2013). An autonomous mobile system for the management of COPD. *Journal of Biomedical Informatics*, *46*(3), 458–469. https://doi.org/10.1016/j.jbi.2013.03.003

Vegesna, A., Tran, M., Angelaccio, M., & Arcona, S. (2017). Remote patient monitoring via non-invasive digital technologies: A systematic review. *Telemedicine Journal and E-Health*, *23*(1), 3–17. https://doi.org/10.1089/tmj.2016.0051

Velardo, C., Shah, S. A., Gibson, O., Clifford, G., Heneghan, C., Rutter, H., Farmer, A., Tarassenko, L., & EDGE COPD Team. (2017). Digital health system for personalised COPD long-term management. *BMC Medical Informatics and Decision Making*, *17*(1), 19. https://doi.org/10.1186/s12911-017-0414-8

Velickovski, F., Ceccaroni, L., Roca, J., Burgos, F., Galdiz, J. B., Marina, N., & Lluch-Ariet, M. (2014). Clinical decision support systems (CDSS) for preventive management of COPD patients. *Journal of Translational Medicine*, *12*(Suppl 2), S9. https://doi.org/10.1186/1479-5876-12-S2-S9

Verma, A., Gudi, N., Yadav, U. N., Roy, M. P., Mahmood, A., Nagaraja, R., & Nayak, P. (n.d.). Prevalence of COPD among population above 30 years in India: A systematic review and meta-analysis. *Journal of Global Health*, *11*, 04038. https://doi.org/10.7189/jogh.11.04038

Vicente, M. A., Fernández, C., Guilabert, M., Carrillo, I., Martín-Delgado, J., & Mira, J. J. (2022). Patient engagement using telemedicine in primary care during COVID-19 pandemic: A trial study. *International Journal of Environmental Research and Public Health*, *19*(22), 14682. https://doi.org/10.3390/ijerph192214682

Wagan, S. A., Koo, J., Siddiqui, I. F., Attique, M., Shin, D. R., & Qureshi, N. M. F. (2022). Internet of medical things and trending converged technologies: A comprehensive review on real-time applications. *Journal of King Saud University - Computer and Information Sciences*, *34*(10), 9228–9251. https://doi.org/10.1016/j.jksuci.2022.09.005

Walker, D. M., Tarver, W. L., Jonnalagadda, P., Ranbom, L., Ford, E. W., & Rahurkar, S. (2023). Perspectives on challenges and opportunities for interoperability: Findings from key informant interviews with stakeholders in Ohio. *JMIR Medical Informatics*, *11*, e43848. https://doi.org/10.2196/43848

Wang, H., Naghavi, M., Allen, C., Barber, R. M., Bhutta, Z. A., Carter, A., Casey, D. C., Charlson, F. J., Chen, A. Z., Coates, M. M., Coggeshall, M., Dandona, L., Dicker, D. J., Erskine, H. E., Ferrari, A. J., Fitzmaurice, C., Foreman, K., Forouzanfar, M. H., Fraser, M. S., … Murray, C. J. L. (2016). Global, regional, and national life expectancy, all-cause mortality, and cause-specific mortality for 249 causes of death, 1980–2015: A systematic analysis for the Global Burden of Disease Study 2015. *The Lancet*, *388*(10053), 1459–1544. https://doi.org/10.1016/S0140-6736(16)31012-1

Watson, A., & Wilkinson, T. M. A. (2022). Digital healthcare in COPD management: A narrative review on the advantages, pitfalls, and need for further research. *Therapeutic Advances in Respiratory Disease*, *16*, 17534666221075493. https://doi.org/10.1177/17534666221075493

Wedzicha, J. A., & Seemungal, T. A. (2007). COPD exacerbations: Defining their cause and prevention. *Lancet (London, England)*, *370*(9589), 786–796. https://doi.org/10.1016/S0140-6736(07)61382-8

Welch, L., Orlando, R., Lin, S. X., Vassilev, I., & Rogers, A. (2020). Findings from a pilot randomised trial of a social network self-management intervention in COPD. *BMC Pulmonary Medicine*, *20*(1), 162. https://doi.org/10.1186/s12890-020-1130-1

Wensing, M. (2015). Implementation science in healthcare: Introduction and perspective. *Zeitschrift Fur Evidenz, Fortbildung Und Qualitat Im Gesundheitswesen*, *109*(2), 97–102. https://doi.org/10.1016/j.zefq.2015.02.014

Williamson, S. M., & Prybutok, V. (2024). Balancing privacy and progress: A review of privacy challenges, systemic oversight, and patient perceptions in AI-driven healthcare. *Applied Sciences*, *14*(2), Article 2. https://doi.org/10.3390/app14020675

Wilson, J., Heinsch, M., Betts, D., Booth, D., & Kay-Lambkin, F. (2021). Barriers and facilitators to the use of e-health by older adults: A scoping review. *BMC Public Health*, *21*, 1556. https://doi.org/10.1186/s12889-021-11623-w

Wong, B. L. H., Khurana, M. P., Smith, R. D., El-Omrani, O., Pold, A., Lotfi, A., O'Leary, C. A., & Saminarsih, D. S. (2021). Harnessing the digital potential of the next generation of health professionals. *Human Resources for Health*, *19*(1), 50. https://doi.org/10.1186/s12960-021-00591-2

Wosik, J., Fudim, M., Cameron, B., Gellad, Z. F., Cho, A., Phinney, D., Curtis, S., Roman, M., Poon, E. G., Ferranti, J., Katz, J. N., & Tcheng, J. (2020). Telehealth transformation: COVID-19 and the rise of virtual care. *Journal of the American Medical Informatics Association : JAMIA*, *27*(6), 957–962. https://doi.org/10.1093/jamia/ocaa067

Wu, F., Burt, J., Chowdhury, T., Fitzpatrick, R., Martin, G., Scheer, J. W. van der, & Hurst, J. R. (2021). Specialty COPD care during COVID-19: Patient and clinician perspectives on remote delivery. *BMJ Open Respiratory Research*, *8*(1), e000817. https://doi.org/10.1136/bmjresp-2020-000817

Yang, F., Wang, Y., Yang, C., Hu, H., & Xiong, Z. (2018). Mobile health applications in self-management of patients with chronic obstructive pulmonary disease: A systematic review and meta-analysis of their efficacy. *BMC Pulmonary Medicine*, *18*, 147. https://doi.org/10.1186/s12890-018-0671-z

Yang, Y., Chen, Z., Li, W., Zeng, N., Guo, Y., Wang, S., Duan, W., Liu, Y., Chen, H., Li, X., Chen, R., & Kang, Y. (2022). Multi-modal data combination strategy based on chest HRCT images and PFT parameters for intelligent dyspnea identification in COPD. *Frontiers in Medicine*, *9*, 980950. https://doi.org/10.3389/fmed.2022.980950

Ye, J., He, L., & Beestrum, M. (2023). Implications for implementation and adoption of telehealth in developing countries: A systematic review of China's practices and experiences. *NPJ Digital Medicine*, *6*, 174. https://doi.org/10.1038/s41746-023-00908-6

Yelne, S., Chaudhary, M., Dod, K., Sayyad, A., & Sharma, R. (n.d). Harnessing the power of AI: A comprehensive review of its impact and challenges in nursing science and healthcare. *Cureus*, *15*(11), e49252. https://doi.org/10.7759/cureus.49252

Yeung, A. W. K., Torkamani, A., Butte, A. J., Glicksberg, B. S., Schuller, B., Rodriguez, B., Ting, D. S. W., Bates, D., Schaden, E., Peng, H., Willschke, H., van der Laak, J., Car, J., Rahimi, K., Celi, L. A., Banach, M., Kletecka-Pulker, M., Kimberger, O., Eils, R., & Atanasov, A. G. (2023). The promise of digital healthcare technologies. *Frontiers in Public Health*, *11*. https://www.frontiersin.org/journals/public-health/articles/10.3389/fpubh.2023.1196596

Yin, J., Xu, J., & Ren, T.-L. (2023). Recent progress in long-term sleep monitoring technology. *Biosensors*, *13*(3), Article 3. https://doi.org/10.3390/bios13030395

Yohannes, A. M. (2021). COPD patients in a COVID-19 society: Depression and anxiety. *Expert Review of Respiratory Medicine*, *15*(1), 5–7. https://doi.org/10.1080/17476348.2020.1787835

Zwerink, M., Brusse-Keizer, M., van der Valk, P. D., Zielhuis, G. A., Monninkhof, E. M., van der Palen, J., Frith, P. A., & Effing, T. (2014). Self management for patients with chronic obstructive pulmonary disease. *The Cochrane Database of Systematic Reviews*, *2014*(3), CD002990. https://doi.org/10.1002/14651858.CD002990.pub3

17 Clinical Trials and Regulatory Issues of Nanocarriers Employed in the Treatment of Chronic Obstructive Pulmonary Disease

Lokesh Nagar, Annu Saini, Tarif Hussian, and Nisha Gulati
Faculty of Pharmaceutical Sciences, Maharshi Dayanand University
Rohtak, Haryana, India

Sachin Kumar Singh
School of Pharmaceutical Sciences, Lovely Professional University
Phagwara, Punjab, India
Faculty of Health, Australian Research Centre in Complementary and Integrative Medicine, University of Technology Sydney
Ultimo, NSW, Australia

Kamal Dua
Faculty of Pharmaceutical Sciences, Maharshi Dayanand University
Rohtak, Haryana, India
Faculty of Health, Australian Research Centre in Complementary and Integrative Medicine, University of Technology Sydney
Ultimo, NSW, Australia
Discipline of Pharmacy, Graduate School of Health,
Faculty of Health, University of Technology Sydney
Ultimo, NSW 2007, Australia

Harish Dureja
Faculty of Pharmaceutical Sciences, Maharshi Dayanand University
Rohtak, Haryana, India
Faculty of Health, Australian Research Centre in Complementary and Integrative Medicine, University of Technology Sydney
Ultimo, NSW, Australia

17.1 INTRODUCTION

Chronic obstructive pulmonary disease (COPD) is a progressive respiratory illness marked by restricted airflow, persistent coughing, increased mucus production, and difficulty breathing (WHO, 2023). It primarily stems from prolonged exposure to irritants like cigarette smoke, though

DOI: 10.1201/9781003498926-17

environmental pollutants, genetic factors, and respiratory infections also play roles (Gilmour et al., 2006). Managing COPD is challenging due to its irreversible nature and the presence of comorbidities such as cardiovascular disease and osteoporosis (Li et al., 2022). The treatment goals encompass alleviating symptoms, boosting exercise capacity, reducing exacerbation risks, and enhancing overall quality of life (Vogelmeier et al., 2020). Bronchodilators, including short-acting and long-acting beta-agonists and anticholinergics, form the cornerstone of pharmacotherapy, often supplemented with inhaled corticosteroids for patients prone to exacerbations (Almadhoun & Sharma, 2023). However, adherence to medication can be hindered by complex regimens and inhaler techniques. Exacerbations often necessitate systemic corticosteroids and antibiotics, further complicating the management of disease (Jardim & Nascimento, 2019). Non-pharmacological interventions like pulmonary rehabilitation and smoking cessation programs are crucial but underutilized aspects of COPD care (Wedzicha, 2008). Despite therapeutic advancements, COPD remains a significant global burden, emphasizing the ongoing necessity for comprehensive management strategies targeting its multifaceted challenges (Singh et al., 2023). COPD, characterized by progressive airflow limitation and chronic inflammation of the airways, poses significant challenges for conventional drug delivery methods due to the complex nature of the respiratory system and the need for targeted delivery to affected areas (Khan et al., 2023). Nanotechnology provides unique opportunities to overcome these challenges by designing drug delivery systems at the nanoscale level (Patra et al., 2018).

The landscape of COPD treatment has been transformed by nanotechnology, which has facilitated the creation of nano-sized drug carriers like liposomes, nanoparticles, dendrimers, and micelles, revolutionizing the approach to treatment (Ehsan et al., 2023). These carriers encapsulate therapeutic agents, enhancing their Pharmacokinetic (PK) and Pharmacodynamic (PD) properties. They offer various advantages including prolonged circulation time, improved drug stability, targeted delivery to specific lung regions, and reduced systemic side effects (Din et al., 2017). Their small size facilitates deep penetration into the lungs, reaching sites of inflammation and tissue damage more effectively than traditional methods (Ibarra-Sánchez et al., 2022). Furthermore, nanotechnology allows the incorporation of functionalities into drug delivery systems, such as stimuli-responsive properties, enabling controlled drug release in response to specific triggers present in the diseased lung microenvironment (Liu et al., 2024). This control over space and time improves the effectiveness of treatment while reducing unintended side effects. Surface modification of nanoparticles with ligands or antibodies facilitates specific targeting of diseased cells, thereby enhancing therapeutic outcomes in COPD patients (Li et al., 2023). Overall, the application of nanotechnology in COPD treatment shows significant promise for improving efficacy, patient compliance, and quality of life. However, it is necessary to optimize nanocarrier design, understand their interactions with the respiratory system, and evaluate long-term safety and efficacy in clinical settings (Jessamine et al., 2024).

The importance of establishing regulatory frameworks to ensure the safety and effectiveness of nano-based drug therapies for COPD cannot be overstated, given the potential benefits they offer in treating respiratory diseases (Passi et al., 2020). Nanomaterial-conjugated drugs, such as poly lactic-co-glycolic acid (plga) nanoparticles, have demonstrated considerable therapeutic potential in combating COPD by targeting inflammation and oxidative stress (Saxena et al., 2022). These nanoparticles can effectively deliver drugs like antioxidants and anti-inflammatory agents, offering a promising avenue for COPD treatment with minimized side effects. The progress and implementation of nanotherapeutics necessitate rigorous regulatory supervision to guarantee their safety, efficacy, and quality in clinical applications (Xu et al., 2020). Regulatory frameworks play a crucial role in ensuring the safety and efficacy of nano-based drug therapies intended for the treatment of COPD (Cheng et al., 2023). Stringent regulations are crucial for mitigating risks and maximizing therapeutic benefits in the realm of nano-sized drug delivery systems, given their complexity and potential impact on human health (Elumalai et al., 2024). These regulatory frameworks encompass stringent preclinical and clinical testing requirements, ensuring thorough evaluation of nanomedicines before market approval. Regulatory agencies like the U.S. Food and Drug Administration (FDA) and European Medicines Agency (EMA) require details on thorough toxicity assessments, PK studies, and efficacy evaluations for nano-based therapies targeting COPD (Ramos et al., 2022). These assessments are crucial to maintain stringent standards and ensure the safety and effectiveness of such treatments. Moreover, these frameworks facilitate transparent communication between developers, regulators, and healthcare providers, fostering a collaborative environment for innovation while prioritizing patient safety (Bhati et al., 2023). In addition, the regulatory guidelines ensure that nano-based drug therapies undergo meticulous scrutiny, addressing concerns related to biocompatibility, stability, and targeted delivery to the lungs (Mehta et al., 2023). Ultimately, stringent regulatory oversight not only safeguards patient well-being but also fosters public trust in the promising advancements of nanotechnology in COPD treatment (Nagar et al., 2024). This chapter addresses the regulatory landscape, along with future prospects for nanocarriers employed in COPD treatment.

17.2 REGULATORY CONSIDERATIONS FOR NANO-BASED DRUG FORMULATIONS

While numerous nanotechnology-based products have gained approval for medical use, global regulatory ethics and policies continue to lag behind, impeding the full realization of nanomedicines' potential (Đorđević et al., 2021). The primary obstacles in regulating nanomedicine stem from the unique characteristics of these products. Nanomedicines display unique PD and PK characteristics when contrasted with the individual materials and payloads they comprise (Figure 17.1) (Alghamdi et al., 2022).

The regulatory landscape for nano-based drug formulations targeting COPD is intricate and requires comprehensive scrutiny (Peng et al., 2024). The safety assessment of nanomaterials used in drug delivery systems necessitates rigorous evaluation. Regulatory bodies such as the FDA in the United States and the EMA in Europe prioritize thorough toxicological assessments to ensure

<table>
<tr><td>Characterization method
Nanomaterials often exhibit unique properties that challenge traditional characterization methods.</td><td>Toxicity assessment
Understanding nanomaterial exposure health risks is complex, demanding specialized testing protocols.</td><td>Regulatory frameworks
Existing regulations may not adequately address the specific risks and benefits of nanomedicines/nanomaterials.</td></tr>
<tr><td>Standardization
Inconsistent testing methods hinder nanomaterial safety and efficacy evaluation.</td><td rowspan="2">Key challenges in regulation of nanomaterials/ nanomedicines</td><td>International coordination
Global safety standards require harmonized regulatory approaches across jurisdictions.</td></tr>
<tr><td>Risk communication
Effectively conveying nanomedicine uncertainties and benefits is crucial for informed decisions.</td><td>Ethical considerations
Regulators face the challenge of balancing nanomedicine benefits with ethical concerns like privacy and equity.</td></tr>
<tr><td>Long-term effects
Understanding the potential long-term environmental and health impacts of nanomaterial exposure requires ongoing research.</td><td>Rapid technological advancements
Regulatory frameworks must adapt quickly to keep pace with evolving nanotechnology innovations.</td><td>Labeling and tracking
Labeling and tracing nanomaterials throughout their lifecycle posess logistical hurdles.</td></tr>
</table>

FIGURE 17.1 Challenges encountered in the regulation of nanomaterials (Foulkes et al., 2020).

that nanomaterials do not induce adverse effects on pulmonary tissues or systemic circulation (Halamoda-Kenzaoui et al., 2021). Furthermore, comprehending the PKs and biodistribution of nano-formulated drugs is vital for establishing dosing regimens and potential off-target effects (Onoue et al., 2014). Comprehensive preclinical studies investigating these aspects are essential for securing regulatory approval. Secondly, establishing the efficacy of nano-based formulations in treating COPD is contingent upon robust clinical trials (ElSaygh et al., 2021). Regulatory agencies mandate well-designed studies that demonstrate the superiority or at least non-inferiority of nano-formulated drugs in comparison to conventional therapies (Farjadian et al., 2019). These trials should encompass not only symptom relief but also the assessment of crucial clinical endpoints such as exacerbation rates, lung function enhancement, and quality of life. Moreover, given the heterogeneity of COPD patients, subgroup analyses may be necessary to discern populations most likely to derive benefit from nano-based treatments (Martinez et al., 2022).

Manufacturing processes for nano-drug formulations require rigorous quality control measures to ensure consistency, purity, and stability of the product (Desai, 2012). Regulatory authorities (RAs) require compliance with Good Manufacturing Practices (GMP) guidelines to guarantee the safety and effectiveness of products. Furthermore, precise documentation of formulation characteristics such as particle size distribution, surface chemistry, and drug loading is essential for regulatory submissions (Patel & Chotai, 2011). Any modifications in manufacturing processes or formulation composition after approval necessitate regulatory oversight to uphold product integrity. Ensuring compliance with regulatory requirements is essential for advancing the development and commercialization of safe, effective, and high-quality nano-based drug formulations tailored for managing COPD (Thapa & Kim, 2022).

In recent years, there has been a notable rise in the development of guidelines intended to assist medicinal product developers and clinical trial sponsors in crafting high-quality documentation, particularly in the realm of nanomedicines and nanocarriers (Dri et al., 2022). These guidelines are essential elements of the chemistry, manufacturing, and control (CMC) section of investigational medicinal product dossiers for assessing such innovative products (Chiodin et al., 2019). The growing number of guidelines underscores the regulatory focus and dedication towards addressing the challenges and nuances of nanotechnology applications in the pharmaceutical sector (Soares et al., 2018). Despite variations in regulatory approaches and response times among different RAs, there's a clear trend indicating heightened attention and efforts in this area (Bremer-Hoffmann et al., 2018). This trend indicates a rise in the number of clinical trial applications integrating nanotechnology-based products by sponsors (Allan et al., 2021). It also underscores the adaptability of the regulatory landscape to emerging innovations, especially as they advance into clinical trial phases (Lee et al., 2019). This trajectory indicates a proactive approach by regulators to ensure the safety, effectiveness, and quality of medicinal products based on nanotechnology, thereby promoting confidence in their development and eventual market approval (Shandilya et al., 2020).

17.3 REGULATORY AGENCIES RESPONSIBLE FOR DRUG APPROVAL AND OVERSIGHT

The regulatory landscape surrounding nano-based drugs is a multifaceted domain shaped by scientific advancements, safety considerations, and the evolving frameworks of regulatory bodies' worldwide (Malik et al., 2023). Notably, in regions like the United States and Europe, regulatory agencies such as the FDA and EMA have formulated guidelines specifically tailored to address the complexities of nanomedicine products (Halamoda-Kenzaoui et al., 2018). These guidelines primarily concentrate on delineating the distinctive attributes of nanomaterials, including size, surface area, and surface chemistry, and assessing their potential implications on safety and efficacy (Lin et al., 2014). Furthermore, RAs are keenly attentive to the possibility of nanomaterial accumulation in tissues or organs, which could engender unforeseen long-term effects (Gupta & Xie, 2018). Consequently, stringent preclinical and clinical investigations are typically mandated to scrutinize the safety, PKs, and PDs of nano-based drugs before they are deemed suitable for market approval (Kad et al., 2020).

Collaboration among regulatory agencies internationally is crucial for harmonizing standards and facilitating the global development and approval of nano-based drugs. The progress and implementation of nanotherapeutics necessitate rigorous regulatory supervision to guarantee their safety, efficacy, and quality in clinical applications (Gonin et al., 2023). However, challenges persist in aligning regulatory approaches due to variations in interpretation and priorities among countries. As nanotechnology continues to advance, regulatory agencies must adapt and update their guidelines to address emerging concerns and integrate new scientific knowledge (Amenta et al., 2015). Despite the potential benefits of nano-based drugs, significant hurdles remain regarding concerns about their safety, manufacturing consistency, and long-term effects, which pose challenges for regulatory approval. Regulators are tasked with balancing the imperative for innovation with ensuring patient safety, necessitating a cautious approach to evaluating nanomedicine products (Souto et al., 2024). Continuous research into the safety and efficacy of nano-based drugs, among regulators, industry stakeholders, and the scientific community, will be essential for navigating the evolving regulatory landscape and delivering promising nanomedicine therapies to patients worldwide (Yusuf et al., 2023).

17.3.1 United States Food and Drug Administration Regulations

The regulation of nanotechnology products, such as nanomedicines, falls under the oversight of the FDA, which employs its existing statutory and RAs in addition to product-specific standards within its jurisdiction (Fatehi et al., 2012). Over the years, the FDA has issued guidelines concerning nanomaterials in various sectors, such as food, cosmetics, and animal feed. However, there has been no officially documented guidance specifically tailored for nanomedicine (Fytianos et al., 2020). In 2017, the FDA issued initial draft guidance pertaining to drug products, specifically addressing biological products incorporating nanomaterials. Following this, in 2018, further instructions were issued for liposome drug products CMCs; human PKs and bioavailability; as well as labeling documentation, with the aim of assisting the industry (Dabas et al., 2023). These guidance documents are intended for industry use in the registration and approval of nanomedicine incorporating nanomaterials. These nanomaterials form a component of the FDA's continual adoption of the Good Guidance Practice (21 CFR 10.115) standards, incorporating suggestions from the Nanotechnology Task Force Report issued in 2007 (de Vlieger et al., 2019). The FDA does not aim to offer a definitive classification of safe and harmful nanomedicine products; rather, it employs a case-by-case approach to evaluate nano products (Mühlebach, 2018). The release of these guidance documents by the FDA signals a proactive move toward tackling the distinctive safety and efficacy considerations of this growing category of therapeutic products (Gutierrez et al., 2020). This Draft Guidance marks a significant step in recognizing the complexity of drug products containing nanomaterials, as well as the challenges associated with the approval of intricate generics, all aimed at safeguarding patient well-being (Emily et al., 2018). According to FDA guidelines, drug products containing nanomaterials are defined as those with dimensions, structures, or properties attributable to nanoscale dimensions (approximately 1 nm to 100 nm), or up to one micrometer (1,000 nm), regardless of whether these dimensions fall within the nanoscale range (Jeevanandam et al., 2018).

The FDA conducts safety information reviews and provides post-marketing safety guidelines to manufacturers, while also overseeing post-market scrutiny of products to safeguard consumer well-being (Suke et al., 2015). The FDA formed a nanotechnology task force, which concluded that the current safety assessment framework is generally applicable to nanomaterials. Numerous directives pertaining to the characterization and safety evaluation within nanotechnology have been issued, reflecting the prevailing positions of regulatory bodies or notified entities, albeit lacking legal binding (Paradise, 2019). The 505(b)(2) regulatory pathway allows for the submission of a New Drug Application (NDA) to the FDA. Nano-based drugs formulated with nanotechnology might be eligible to utilize this route provided they meet certain criteria. Factors such as, a new formulation or possessing distinguishing characteristics from previously approved drugs may qualify a nano-based drug for a 505(b)(2) submission. However, the FDA evaluates each case individually to determine compliance with pathway requirements (FDA, 2022).

In the clinical development of drug products following a 505(b)(2) approval pathway, particularly for medium- and high-risk products containing nanomaterials, simply demonstrating bioequivalence between the proposed and referenced products may not sufficiently ensure therapeutic equivalence (Klein et al., 2021). Additional evidence regarding the comparability of disposition and exposure-response relationship of the active ingredient across both products may be necessary. For medium- and high-risk drug products containing nanomaterials seeking bioequivalence with a referenced product, conducting a single-dose comparative absorption, distribution, metabolism, and excretion (ADME) study can offer further assurance regarding the similarity of disposition of the active ingredient compared to the referenced product (Bartlett et al., 2014). It's crucial to acknowledge that while ADME studies offer valuable insights, they might not always identify nuanced yet medically relevant distinctions in the speed and extent of active ingredient release into specific tissues (Tibbitts et al., 2015). To address this limitation, both single- and multiple-dose studies examining the PK and PD characteristics of the proposed product and the referenced product can provide a more comprehensive understanding. These studies allow for exploration of the exposure-response relationship, which is crucial for assessing therapeutic equivalence (Pereira et al., 2022). For medium- and high-risk nano-based products, a 505(b)(2) approval pathway, a combination of bioequivalence studies, single-dose comparative ADME studies, and PK/PD studies are recommended to ensure a thorough evaluation of therapeutic equivalence (Freije et al., 2020).

17.3.2 European Union: European Medicines Agency Regulations

In the European Union (EU), nanomedicines are governed by the General Medicinal Product legislation, overseen by the EMA. The European Commission (EC) published Recommendation 2011/696/EU, which offers a foundational definition of "nanomaterial" for regulatory purposes (European Medicines Agency, 2024). Despite its non-binding nature, this recommendation serves as a reference point for EU legislation and policy concerning nanomedicines (Hemmrich & McNeil, 2023). While the EMA has initiated preliminary guidelines for standardizing nanomedicine preparation, official regulatory guidelines are still awaited. Presently, nanomedicines in the EU are regulated based on risk/benefit-analysis principles. In line with the European Chemicals Legislation (REACH EC 1907/2006), the European Chemicals Agency (ECHA) focuses on ensuring the safety of nanomaterials (Pita et al., 2016). The 2018 revision of the REACH legislation brought about fresh obligations for downstream users and introduced additional safety assessment criteria for nanomaterials (Tschiche et al., 2022). The European Food Safety Authority (EFSA) holds the responsibility of evaluating the hazards linked with nanomaterials in food intended for humans and animals. The EFSA has provided significant guidance on conducting risk assessments and uncertainty analyses concerning nanomaterials in food, along with suggestions for additional research (More et al., 2021a, 2021b).

The EMA has introduced revised regulations specific to nanomedicines, focusing on nano-specific safety evaluations, data requirements, and the approval processes for their applications (Hardy et al., 2018). According to EU regulations, these guidelines encompass details like particle dimensions, chemical and physical characteristics, toxicological profiles, pertinent safety information, potential exposure scenarios, and the identification of nanomaterials (Vasileios Bampidis et al., 2021). The EMA requires the submission of comprehensive physicochemical characterization data for nanomaterials, encompassing parameters such as zeta potential, particle size, size distribution, and spectroscopic analyses investigations (Mudalige et al., 2019). Additionally, there is a highlighted necessity for data pertaining to degradation products, impurity profiling, and solid-state characterization, utilizing methods. A collectively formulated strategic document by the European Nano Safety Cluster (ENSC) outlines the research priorities for nano safety spanning from 2015 to 2025 (Piveteau et al., 2020). This document outlines numerous nano safety research endeavors supported by various EU member nations. Its strategic objective is to advance comprehension of engineered nanomaterial characteristics (ENMs), the hazard mechanisms associated with ENMs, exposure considerations, as well as the assessment and management of risks related to them (Afantitis et al., 2020).

The EMA has been involved in thorough deliberations concerning the regulatory standards for different categories of nano-medicinal products intended for human use (Dorbeck-Jung & Chowdhury, 2011). Reflection papers have comprehensively documented discussions on a range of topics, such as liposomal generics, polymeric micelle formulations, surface coatings of nanoscaled parenteral dosage forms, and diagnostic iron oxide nanoparticles. These studies provide invaluable insights into the specific regulatory considerations and guidelines applicable to these specialized pharmaceutical formulations (Abou Assi et al., 2021).

17.3.3 Japan: Pharmaceuticals and Medical Devices Agency Regulations

In Japan, the regulation of medicines is overseen by the Ministry of Health, Labor and Welfare (MHLW) and the Pharmaceuticals and Medical Devices Agency (PMDA) (Japan Health Authority [PMDA], 2024). Despite the absence of a formal definition or specific framework for nanomedicines in Japan, there have been significant regulatory developments. In 2016, a comprehensive guideline outlining the developmental procedures for liposome-based pharmaceutical products was introduced (Tenchov et al., 2021). The regulation of nanomedicines is primarily governed by the Pharmaceutical Affairs Law, which acts as the principal legislation overseeing medicinal products. Regulation is conducted individually for each case. It is important to highlight that regulatory bodies, including the PMDA and MHLW, are actively engaged in compiling and evaluating data related to nanomedicines (PMDA, 2023). The partnership between the PMDA and the MHLW has resulted in the development of joint publications, including the MHLW/EMA reflection paper addressing block copolymer micelle-based pharmaceuticals (Tanimoto, 2015). Furthermore, the PMDA has provided assistance to the MHLW in developing management guidance for specific nanotechnology-based medicines. Despite the publication of significant documents by the MHLW in 2016, such as a reflection paper on nucleic acids (siRNA)-loaded nanotechnology-based drug products and a guideline for the development of liposome drug products, there remains a deficiency in having a dedicated regulatory framework designed for evaluating all nanomedicines and nanocarriers during clinical trials. (Chopra et al., 2023).

17.3.4 Canada: Health Canada Regulations

Health Canada has formulated a working definition of nanomaterials, delineating them as any product, material, substance, ingredient, device, system, or structure that is artificially produced and falls within the nanoscale range (1–100 nm) in at least one spatial dimension, or demonstrates one or more nanoscale phenomena, regardless of whether its dimensions are smaller or larger than the nanoscale in all spatial aspects (Ferraris et al., 2021). Concerns regarding nanomedicines in Canada have been recognized since 2010. Presently, the assessment and authorization of nanomedicines occur within established legislative and regulatory structures without explicit mention of nanomaterials. Nonetheless, Health Canada has implemented a working definition of nanomaterials, enabling the solicitation of specific information to enhance understanding and assess potential risks and benefits, including those in clinical trial (CT) applications involving nanomaterials (Wasti et al., 2023). Despite the absence of a dedicated guideline for clinical trial submissions involving nanotechnology products, Health Canada requires various information for assessing nanomaterials in regulated products or substances, including those in clinical trial applications. This information includes the intended application, functionality, and purpose of the nanomaterial. It is also specifics about resultant products, manufacturing processes, data on toxicity, eco-toxicity, metabolism, and environmental fate (Canada, 2011). Furthermore, the request includes the need for risk assessment and risk management strategies, if they have been contemplated or put into action. Although Health Canada provides online resources detailing the characteristics and physico-chemical properties of nanomaterials, there is presently no organized or specialized methodology specifically designed for clinical trials (Giubilato et al., 2020).

17.3.5 India: Central Drugs Standard Control Organization Regulations

The Ministry of Health and Family Welfare (MHFW) plays a crucial role in supervising the regulation of nanotechnology applications, primarily through its Directorate General of Health Services, which includes the Central Drugs Standard Control Organization (CDSCO) (Srivastava & Chowdhury, 2008). The guideline titled "Guidelines for Evaluation of Nanopharmaceuticals in India" has been developed within this framework. It categorizes nanopharmaceuticals considering factors such as the biodegradability of nanomaterials, the characteristics of nanomaterials, the various nanoforms of the ingredient, and the approval status of both the drug and nanomaterial utilized (Jampilek & Kralova, 2020). Moreover, the guideline delineates criteria concerning toxicological and clinical trial data, furnishing vital information for the assessment of nanomaterials. A nanopharmaceutical is a type of pharmaceutical formulation incorporating nanomaterials, designed for therapeutic, diagnostic, or health-related applications, with constituent materials typically ranging in size from 1 bn to 100 nm in at least one dimension. Furthermore, particles ranging in size from greater than 100 nm to less than 1000 nm are also encompassed by this definition, given that they demonstrate modified or distinct pharmaceutical attributes in comparison to the active pharmaceutical ingredient (Yetisgin et al., 2020). The regulatory requirements specified in the Drugs and Cosmetics Rules, 1945, and other relevant regulations, apply equally to nanopharmaceuticals, ensuring that the general requirements and guidelines for approving the manufacture or importation of any new drug encompass nanotechnology-based pharmaceuticals as well. The evaluation process for nanotechnology involves a "case-by-case approach," demanding specialized expertise to effectively develop nanopharmaceuticals. This method ensures thorough consideration of the distinct traits and potential risks associated with nanomaterials during regulatory assessments (Souto et al., 2020).

17.4 CHALLENGES IN REGULATORY APPROVAL FOR NANOCARRIERS FOR COPD

Regulatory approval for nanocarriers intended for the treatment of COPD presents several unique challenges due to the absence of disease-specific guidelines. First, concerns arise regarding the safety profile of nanocarriers when administered via inhalation (Pontes & Grenha, 2020). Nanoparticles may provoke immune responses or induce inflammation in the respiratory tract, potentially worsening COPD symptoms in patients. Consequently, extensive preclinical studies are necessary to assess the biocompatibility and toxicity of nanocarriers in the lungs (Gulati et al., 2021). Regulatory bodies mandate the thorough characterization of nanocarriers, encompassing their physicochemical attributes, stability, and drug release kinetics. This is essential to guarantee uniform performance and mitigate potential hazards to patients. Second, demonstrating the efficacy of nanocarriers in treating COPD poses another significant challenge (Berbel Manaia et al., 2017). COPD is a complex disease with various underlying mechanisms, such as airway inflammation, mucous hypersecretion, and oxidative stress. Nanocarriers must effectively target specific cellular or molecular pathways implicated in COPD pathogenesis to provide therapeutic benefits. Designing nanocarriers with optimal size, surface properties, and drug-loading capacity to enhance pulmonary delivery and drug retention in the lungs is critical for achieving therapeutic efficacy (Alfahad et al., 2021). Furthermore, clinical trials for COPD treatments typically require large sample sizes and long-term follow-up to evaluate improvements in lung function, symptom control, and exacerbation rates, prolonging the regulatory approval process. Addressing these challenges requires close collaboration among researchers, clinicians, and RAs to ensure the development of safe and effective nanocarrier-based therapies for COPD (Parris et al., 2019).

As of the current date, there have been no nano-based drugs specifically approved for the treatment of COPD available in the market. Nonetheless, research in this area has demonstrated promising advancements (Kumar et al., 2023). Nanotechnology offers distinctive opportunities in drug delivery for respiratory diseases like COPD by enabling targeted delivery, enhanced drug stability, and potentially reduced side effects. Various experimental nano-based formulations have exhibited efficacy in preclinical studies, yet further research is necessary to translate these findings into clinically approved treatments (Luo et al., 2021). One notable avenue of exploration involves employing nanoparticles to encapsulate existing COPD medications, such as bronchodilators or corticosteroids. These nanoparticles can augment the drugs' effectiveness by facilitating their delivery directly to the lungs, the primary site of action (Sato & Kawakami, 2023). The drug targeting and minimizing systemic exposure can be refined as the nano-based formulations have the potential to mitigate adverse effects while maximizing therapeutic benefits. Clinical trials are essential for assessing the safety and effectiveness of treatments in patients with COPD (Holder et al., 2023). Clinical trials are instrumental in showcasing the safety and effectiveness of nano-based medications for COPD. The increase in the number of nanomedicine clinical trials, particularly for COPD treatment, underscores the growing interest and investment in this innovative therapeutic approach. With numerous nanomedicines already in the market or undergoing clinical trials, the focus on COPD treatment highlights the potential of nanotechnology to revolutionize the management of chronic lung diseases (Virmani et al., 2022). The development of multifunctional nanocarriers tailored for COPD treatment is essential to enhance drug delivery efficiency and therapeutic outcomes. Factors such as particle size, charge, and shape significantly influence the effectiveness of nanocarriers in penetrating lung tissues and delivering drugs to target sites within the respiratory system (Singh et al., 2019). Optimizing the design of nanocarriers to overcome mucociliary clearance and enhance drug retention in the lungs is crucial for improving the treatment outcomes of COPD patients. Current research endeavors concentrate on developing novel nanostructures specifically engineered to address the complex pathophysiology of COPD. These may include particles designed to target inflamed airway tissues, alleviate oxidative stress, or modulate immune responses (Yan & Sha, 2023). Furthermore, advancements in nanotechnology enable the integration of diagnostic or imaging agents into therapeutic nanoparticles, facilitating simultaneous treatment and monitoring of COPD progression. While these innovative approaches hold promise, their clinical translation necessitates comprehensive evaluation to ensure both safety and efficacy. Hence, sustained research efforts are indispensable to realize the full potential of nano-based therapies for COPD and enhance the management of this debilitating condition (Kia et al., 2023). A few nanoformulations have been patented by researchers for COPD treatment (see Table 17.1).

17.5 FUTURE DIRECTIONS AND CONCLUSION

In recent years, there has been notable attention toward utilizing nanocarrier technology to improve the management of COPD. Nanocarriers, including liposomes, polymeric nanoparticles, and dendrimers, offer several advantages such as targeted drug delivery, extended drug release, and reduced systemic side effects. Researchers are actively exploring the potential of nanocarrier-based formulations to deliver bronchodilators, corticosteroids, and anti-inflammatory agents directly to the lungs, thereby improving therapeutic outcomes while minimizing systemic exposure. However, the development of nanocarrier-based therapies for COPD is hindered

TABLE 17.1
List of Patents on Nanostructures for COPD

S. No.	Publication Number	Year	Inventor	Title	Ref.
1.	CN115227682B	2023	Li Nan; Li Xu	Preparation method of powder aerosol for targeted small-airway slow-release delivery of COPD therapeutic drug	https://patents.google.com/patent/CN115227682B/en?oq=CN115227682B
2.	CN105310986A	2016	Diao Yuanyuan; Zhang Ting	Olodaterol lung targeting nanoparticle and its preparation method	https://patents.google.com/patent/CN105310986A/en?oq=+CN105310986A
3.	US2022226500A1	2021	Peter Butzloff	Fullerene nanoparticle composition containing ellagic acid	https://patents.google.com/patent/US20220226500A1/en?oq=US2022226500A1
4.	CN106389387A	2017	Xin Wang; Linfang Lei	Roflumilast nanoparticle and preparation method thereof	https://patents.google.com/patent/CN106389387A/en?oq=+CN106389387A
5.	KR20200012587A	2020	Yang Se-ran, Jang Su-jin, Jang Sang-ho, Hwang Jeong-sun, Han Byeol	Composition for prevention or treatment of COPD comprising PTD-bFGF or increased exosome by PTD-bFGF	https://patents.google.com/patent/KR20200012587A/en?oq=KR20200012587A
6.	US20210308067A1	2021	Gerard Bernard McCauley	Extracellular vesicles for inhalation	https://patents.google.com/patent/US20210308067A1/en?oq=US2021308067A1
7.	WO2022023456A1	2022	Ida Christina Hovdal, Sandra Gracin, Annica Maarit Jarke, Mikael Johan Alvin Brülls, Noemi Gaglianone, Ankur Ajmera	Pharmaceutical compositions comprising nano-embedded microparticles and methods of use	https://patents.google.com/patent/WO2022023456A1/en?oq=+WO2022023456A1

by regulatory challenges, including safety concerns, efficacy assessment, and manufacturing standards. Collaboration among regulatory bodies, industry partners, and academia is crucial to establish standardized evaluation, approval, and commercialization processes for nanocarrier-based therapies globally. Furthermore, there is a growing focus on patient-centered approaches and personalized medicine in COPD management. COPD is a heterogeneous disease characterized by varying clinical presentations and treatment responses among individuals. Personalized medicine seeks to tailor treatment strategies to each patient's unique characteristics, including genetic predisposition, biomarker profile, disease severity, and comorbidities. Advances in omics technologies, such as genomics, proteomics, and metabolomics, enable the identification of biomarkers associated with COPD phenotypes and treatment response. Integrating these biomarkers into clinical practice can facilitate early disease detection, risk stratification, and treatment optimization, ultimately enhancing patient outcomes and quality of life. However, implementing personalized medicine in COPD faces challenges related to biomarker validation, data interpretation, and healthcare infrastructure, underscoring the importance of interdisciplinary collaborations and healthcare system improvements.

The potential of nanocarriers in treating COPD is promising, however a noticeable gap exists in comprehensive clinical trials and regulatory frameworks across different countries. Preclinical studies have demonstrated positive results regarding the efficacy and safety of nanocarriers, but the transition to clinical trials has been impeded by various challenges. These challenges include inadequate standardization of protocols, variability in patient populations, and inconsistent regulatory guidelines, all of which contribute to the lack of robust clinical evidence. Additionally, COPD's multifactorial nature requires extensive and rigorous clinical studies to assess the long-term safety, efficacy, and potential adverse effects of nanocarriers, aspects often overlooked in current research endeavors. The deficiency of clinical trials investigating nanocarriers in COPD treatment raises concerns regarding the translational potential of these innovative therapies. Without comprehensive clinical evidence, healthcare providers and regulatory agencies encounter uncertainties regarding the optimal use, dosing regimens, and safety profiles of nanocarrier-based therapies for COPD patients. Addressing these gaps necessitates collaborative efforts among researchers, clinicians, and regulatory bodies to design and execute well-structured clinical trials adhering to standardized protocols and regulatory requirements. Only through robust clinical investigations can the potential of nanocarrier-based therapies in COPD treatment be realized, ensuring improved patient outcomes and regulatory confidence in their widespread clinical implementation.

REFERENCES

Abou Assi, R., Abdulbaqi, I. M., & Siok Yee, C. (2021). The evaluation of drug delivery nanocarrier development and pharmacological briefing for metabolic-associated fatty liver disease (MAFLD): An update. *Pharmaceuticals*, *14*(3), 215. https://doi.org/10.3390/ph14030215

Afantitis, A., Melagraki, G., Isigonis, P., Tsoumanis, A., Varsou, D. D., Valsami-Jones, E., Papadiamantis, A., Ellis, L.-J. A., Sarimveis, H., Doganis, P., Karatzas, P., Tsiros, P., Liampa, I., Lobaskin, V., Greco, D., Serra, A., Kinaret, P. A. S., Saarimäki, L. A., Grafström, R., & Kohonen, P. (2020). NanoSolveIT project: Driving nanoinformatics research to develop innovative and integrated tools for in silico nanosafety assessment. *Computational and Structural Biotechnology Journal*, *18*, 583–602. https://doi.org/10.1016/j.csbj.2020.02.023

Alfahad, A. J., Alzaydi, M. M., Aldossary, A. M., Alshehri, A. A., Almughem, F. A., Zaidan, N. M., & Tawfik, E. A. (2021). Current views in chronic obstructive pulmonary disease pathogenesis and management. *Saudi Pharmaceutical Journal*, *29*(12), 1361–1373. https://doi.org/10.1016/j.jsps.2021.10.008

Alghamdi, M. A., Fallica, A. N., Virzì, N., Kesharwani, P., Pittalà, V., & Greish, K. (2022). The promise of nanotechnology in personalized medicine. *Journal of Personalized Medicine*, *12*(5), 673. https://doi.org/10.3390/jpm12050673

Allan, J., Belz, S., Hoeveler, A., Hugas, M., Okuda, H., Patri, A., Rauscher, H., Silva, P., Slikker, W., Sokull-Kluettgen, B., Tong, W., & Anklam, E. (2021). Regulatory landscape of nanotechnology and nanoplastics from a global perspective. *Regulatory Toxicology and Pharmacology*, *122*, 104885. https://doi.org/10.1016/j.yrtph.2021.104885

Almadhoun, K., & Sharma, S. (2023, April 28). *Bronchodilators*. StatPearls Publishing. https://www.ncbi.nlm.nih.gov/books/NBK519028/

Amenta, V., Aschberger, K., Arena, M., Bouwmeester, H., Botelho Moniz, F., Brandhoff, P., Gottardo, S., Marvin, H. J. P., Mech, A., Quiros Pesudo, L., Rauscher, H., Schoonjans, R., Vettori, M. V., Weigel, S., & Peters, R. J. (2015). Regulatory aspects of nanotechnology in the agri/feed/food sector in EU and non-EU countries. *Regulatory Toxicology and Pharmacology*, *73*(1), 463–476. https://doi.org/10.1016/j.yrtph.2015.06.016

Bartlett, J. A., Brewster, M. E., Brown, P. C., Cabral-Lilly, D., Cruz, C. N., David, R. M., Eickhoff, W., Haubenreisser, S., Jacobs, A., Malinoski, F., Morefield, E., Nalubola, R., Prud'homme, R. K., Sadrieh, N., Sayes, C. M., Shahbazian, H., Subbarao, N., Tamarkin, L., Tyner, K. M., & Uppoor, R. (2014). *Summary Report of PQRI Workshop on Nanomaterial in Drug Products: Current Experience and Management of Potential Risks*. *17*(1), 44–64. https://doi.org/10.1208/s12248-014-9701-9

Berbel Manaia, E., Paiva Abuçafy, M., Chiari-Andréo, B. G., Lallo Silva, B., Oshiro-Júnior, J. A., & Chiavacci, L. (2017). Physicochemical characterization of drug nanocarriers. *International Journal of Nanomedicine*, *12*, 4991–5011. https://doi.org/10.2147/ijn.s133832

Bhati, D., Deogade, M. S., & Kanyal, D. (2023). Improving patient outcomes through effective hospital administration: A comprehensive review. *Cureus*, *15*(10). https://doi.org/10.7759/cureus.47731

Bremer-Hoffmann, S., Halamoda-Kenzaoui, B., & Borgos, S. E. (2018). Identification of regulatory needs for nanomedicines. *Journal of Interdisciplinary Nanomedicine*, *3*(1), 4–15. https://doi.org/10.1002/jin2.34

Canada, H. (2011). *Nanotechnology-Based Health Products and Food - Canada.ca*. Canada.ca. https://www.canada.ca/en/health-canada/services/drugs-health-products/nanotechnology-based-health-products-food.html

Cheng, X., Xie, Q., & Sun, Y. (2023). Advances in nanomaterial-based targeted drug delivery systems. *Frontiers in Bioengineering and Biotechnology*, *11*. https://doi.org/10.3389/fbioe.2023.1177151

Chiodin, D., Cox, E. M., Edmund, A. V., Kratz, E., & Lockwood, S. H. (2019). Regulatory affairs 101: Introduction to investigational new drug applications and clinical trial applications. *Clinical and Translational Science*, *12*(4), 334–342. https://doi.org/10.1111/cts.12635

Chopra, H., Mohanta, Y. K., Rauta, P. R., Ahmed, R., Mahanta, S., Mishra, P. K., Panda, P., Rabaan, A. A., Alshehri, A. A., Othman, B., Alshahrani, M. A., Alqahtani, A. S., AL Basha, B. A., & Dhama, K. (2023). An insight into advances in developing nanotechnology based therapeutics, drug delivery, diagnostics and vaccines: Multidimensional applications in tuberculosis disease management. *Pharmaceuticals*, *16*(4), 581. https://doi.org/10.3390/ph16040581

Dabas, R., Musyuni, P., & Aggarwal, G. (2023). An overview of regulatory perspectives of nanoparticles and liposomal products in India and USA. *Journal of Generic Medicines*, *19*(4), 166–178. https://doi.org/10.1177/17411343231206469

de Vlieger, J. S. B., Crommelin, D. J. A., Tyner, K., Drummond, D. C., Jiang, W., McNeil, S. E., Neervannan, S., Crist, R. M., & Shah, V. P. (2019). Report of the AAPS Guidance Forum on the FDA Draft Guidance for industry: "Drug products, including biological products, that contain nanomaterials." *The AAPS Journal*, *21*(4). https://doi.org/10.1208/s12248-019-0329-7

Desai, N. (2012). Challenges in development of nanoparticle-based therapeutics. *The AAPS Journal*, *14*(2), 282–295. https://doi.org/10.1208/s12248-012-9339-4

Din, F. ud, Aman, W., Ullah, I., Qureshi, O. S., Mustapha, O., Shafique, S., & Zeb, A. (2017). Effective use of nanocarriers as drug delivery systems for the treatment of selected tumors. *International Journal of Nanomedicine*, *12*, 7291–7309. https://doi.org/10.2147/ijn.s146315

Dorbeck-Jung, B. R., & Chowdhury, N. (2011). Is the European medical products authorisation regulation equipped to cope with the challenges of nanomedicines? *Law & Policy*, *33*(2), 276–303. https://doi.org/10.1111/j.1467-9930.2011.00339.x

Đorđević, S., Gonzalez, M. M., Conejos-Sánchez, I., Carreira, B., Pozzi, S., Acúrcio, R. C., Satchi-Fainaro, R., Florindo, H. F., & Vicent, M. J. (2021). Current hurdles to the translation of nanomedicines from bench to the clinic. *Drug Delivery and Translational Research*, *12*(3), 500–525. https://doi.org/10.1007/s13346-021-01024-2

Dri, D. A., Rinaldi, F., Carafa, M., & Marianecci, C. (2022). Nanomedicines and nanocarriers in clinical trials: Surfing through regulatory requirements and physico-chemical critical quality attributes. *Drug Delivery and Translational Research*, *13*(3), 757–769. https://doi.org/10.1007/s13346-022-01262-y

Ehsan, M., Qayoomian, M., Beigoli, S., & Boskabady, M. H. (2023). Recent advances in nanoparticle applications in respiratory disorders: A review. *Frontiers in Pharmacology*, *14*. https://doi.org/10.3389/fphar.2023.1059343

ElSaygh, J., Zaher, A., Nathani, P., & Omballi, M. (2021). A review of clinical trials that contributed to chronic obstructive pulmonary disease treatment protocols. *Cureus*. https://doi.org/10.7759/cureus.14618

Elumalai, K., Srinivasan, S., & Shanmugam, A. (2024). Review of the efficacy of nanoparticle-based drug delivery systems for cancer treatment. *Biomedical Technology*, *5*, 109–122. https://doi.org/10.1016/j.bmt.2023.09.001

Emily, M., Ioanna, N., Scott, B., & Beat, F. (2018). Reflections on FDA draft guidance for products containing nanomaterials: Is the abbreviated new drug application (ANDA) a suitable pathway for nanomedicines? *The AAPS Journal*, *20*(5). https://doi.org/10.1208/s12248-018-0255-0

European Medicines Agency. (2024). *Multidisciplinary: nanomedicines | European Medicines Agency*. https://www.ema.europa.eu/en/human-regulatory-overview/research-and-development/scientific-guidelines/multidisciplinary-guidelines/multidisciplinary-nanomedicines

Farjadian, F., Ghasemi, A., Gohari, O., Roointan, A., Karimi, M., & Hamblin, M. R. (2019). Nanopharmaceuticals and nanomedicines currently on the market: Challenges and opportunities. *Nanomedicine*, *14*(1), 93–126. https://doi.org/10.2217/nnm-2018-0120

Fatehi, L., Wolf, S. M., McCullough, J., Hall, R., Lawrenz, F., Kahn, J. P., Jones, C., Campbell, S. A., Dresser, R. S., Erdman, A. G., Haynes, C. L., Hoerr, R. A., Hogle, L. F., Keane, M. A., Khushf, G., Nancy, E., Kokkoli, Marchant, G., Maynard, A. D., & Philbert, M. (2012). Recommendations for nanomedicine human subjects research oversight: An evolutionary approach for an emerging field. *Journal of Law, Medicine & Ethics*, *40*(4), 716–750. https://doi.org/10.1111/j.1748-720x.2012.00703.x

FDA, (2022). Abbreviated Approval Pathways for Drug Product: 505(b)(2) or ANDA *FDA*. https://www.fda.gov/drugs/cder-small-business-industry-assistance-sbia/abbreviated-approval-pathways-drug-product-505b2-or-anda

Ferraris, C., Rimicci, C., Garelli, S., Ugazio, E., & Battaglia, L. (2021). Nanosystems in cosmetic products: A brief overview of functional, market, regulatory and safety concerns. *Pharmaceutics*, *13*(9), 1408. https://doi.org/10.3390/pharmaceutics13091408

Foulkes, R., Man, E., Thind, J., Yeung, S., Joy, A., & Hoskins, C. (2020). The regulation of nanomaterials and nanomedicines for clinical application: Current and future perspectives. *Biomaterials Science*, *8*(17), 4653–4664. https://doi.org/10.1039/d0bm00558d

Freije, I., Lamouche, S., & Tanguay, M. (2020). Review of drugs approved via the 505(b)(2) pathway: Uncovering drug development trends and regulatory requirements. *Therapeutic Innovation & Regulatory Science*, *54*(1), 128–138. https://doi.org/10.1007/s43441-019-00036-y

Fytianos, G., Rahdar, A., & Kyzas, G. Z. (2020). Nanomaterials in cosmetics: Recent updates. *Nanomaterials*, *10*(5), 979. https://doi.org/10.3390/nano10050979

Gilmour, M. I., Jaakkola, M. S., London, S. J., Nel, A. E., & Rogers, C. A. (2006). How exposure to environmental tobacco smoke, outdoor air pollutants, and increased pollen burdens influences the incidence of asthma. *Environmental Health Perspectives*, *114*(4), 627–633. https://doi.org/10.1289/ehp.8380

Giubilato, E., Cazzagon, V., Amorim, M. J. B., Blosi, M., Bouillard, J., Bouwmeester, H., Costa, A. L., Fadeel, B., Fernandes, T. F., Fito, C., Hauser, M., Marcomini, A., Nowack, B., Pizzol, L., Powell, L., Prina-Mello, A., Sarimveis, H., Scott-Fordsmand, J. J., Semenzin, E., & Stahlmecke, B. (2020). Risk management framework for nano-biomaterials used in medical devices and advanced therapy medicinal products. *Materials*, *13*(20), 4532. https://doi.org/10.3390/ma13204532

Gonin, W., Ezan, V., Bastos, H., Bellissant, E., Maison, P., & Ankri, J. (2023). Regulatory assessment of nano-enabled health products in public health interest. Position of the scientific advisory board of the French national agency for the safety of medicines and health products. *Frontiers in Public Health*, *11*. https://doi.org/10.3389/fpubh.2023.1125577

Gulati, N., Chellappan, D. K., MacLoughlin, R., Dua, K., & Dureja, H. (2021). Inhaled nano-based therapeutics for inflammatory lung diseases: Recent advances and future prospects. *Life Sciences*, *285*, 119969. https://doi.org/10.1016/j.lfs.2021.119969

Gupta, R., & Xie, H. (2018). Nanoparticles in daily life: Applications, toxicity and regulations. *Journal of Environmental Pathology, Toxicology and Oncology: Official Organ of the International Society for Environmental Toxicology and Cancer*, *37*(3), 209–230. https://doi.org/10.1615/JEnvironPatholToxicolOncol.2018026009

Gutierrez, L., Cauchon, N. S., Christian, T. R., Giffin, M. J., & Abernathy, M. J. (2020). The confluence of innovation in therapeutics and regulation: Recent CMC considerations. *Journal of Pharmaceutical Sciences*, *109*(12), 3524–3534. https://doi.org/10.1016/j.xphs.2020.09.025

Halamoda-Kenzaoui, B., Holzwarth, U., Roebben, G., Bogni, A., & Bremer-Hoffmann, S. (2018). Mapping of the available standards against the regulatory needs for nanomedicines. *Wiley Interdisciplinary Reviews: Nanomedicine and Nanobiotechnology*, *11*(1), e1531. https://doi.org/10.1002/wnan.1531

Halamoda-Kenzaoui, B., Vandebriel, R. J., Howarth, A., Siccardi, M., David, C. A. W., Liptrott, N. J., Santin, M., Borgos, S. E., Bremer-Hoffmann, S., & Caputo, F. (2021). Methodological needs in the quality and safety characterisation of nanotechnology-based health products: Priorities for method development and standardisation. *Journal of Controlled Release*, *336*, 192–206. https://doi.org/10.1016/j.jconrel.2021.06.016

Hardy, A., Benford, D., Halldorsson, T., Jeger, M. J., Knutsen, H. K., More, S., Naegeli, H., Noteborn, H., Ockleford, C., Ricci, A., Rychen, G., Schlatter, J. R., Silano, V., Solecki, R., Turck, D., Younes, M., Chaudhry, Q., Cubadda, F., Gott, D., & Oomen, A. (2018). Guidance on risk assessment of the application of nanoscience and nanotechnologies in the food and feed chain: Part 1, human and animal health. *EFSA Journal*, *16*(7). https://doi.org/10.2903/j.efsa.2018.5327

Hemmrich, E., & McNeil, S. E. (2023). Active ingredient vs excipient debate for nanomedicines. *Nature Nanotechnology*, *18*(7), 692–695. https://doi.org/10.1038/s41565-023-01371-w

Holder, J. E., Oliveira, E., Lodeiro, C., Trim, C. M., Byrne, L. J., Bértolo, E., & Wilson, C. M. (2023). The use of nanoparticles for targeted drug delivery in non-small cell lung cancer. *Frontiers in Oncology, 13*. https://doi.org/10.3389/fonc.2023.1154318

Ibarra-Sánchez, L. Á., Gámez-Méndez, A., Martínez-Ruiz, M., Nájera-Martínez, E. F., Morales-Flores, B. A., Melchor-Martínez, E. M., Sosa-Hernández, J. E., Parra-Saldívar, R., & Iqbal, H. M. N. (2022). Nanostructures for drug delivery in respiratory diseases therapeutics: Revision of current trends and its comparative analysis. *Journal of Drug Delivery Science and Technology*, *70*, 103219. https://doi.org/10.1016/j.jddst.2022.103219

Jampilek, J., & Kralova, K. (2020). Potential of nanonutraceuticals in increasing immunity. *Nanomaterials*, *10*(11), 2224. https://doi.org/10.3390/nano10112224

Japan Health Authority (PMDA). (2024). *Japan Health Authority (PMDA) Regulations*. Global Regulatory Partners, Inc. https://globalregulatorypartners.com/resource-center/regulatory-intelligence-platform/japan-regulatory-intelligence/japan-health-authority-pmda-regulations/#:~:text=In%20Japan%2C%20The%20Pharmaceuticals%20and

Jardim, J. R., & Nascimento, O. A. (2019). The importance of inhaler adherence to prevent COPD exacerbations. *Medical Sciences*, *7*(4), 54. https://doi.org/10.3390/medsci7040054

Jeevanandam, J., Barhoum, A., Chan, Y. S., Dufresne, A., & Danquah, M. K. (2018). Review on nanoparticles and nanostructured materials: History, sources, toxicity and regulations. *Beilstein Journal of Nanotechnology*, *9*(1), 1050–1074. https://doi.org/10.3762/bjnano.9.98

Jessamine, V., Mehndiratta, S., De Rubis, G., Paudel, K. R., Shetty, S., Suares, D., Chellappan, D. K., Oliver, B. G., Hansbro, P. M., & Dua, K. (2024). The application of nanoparticles as advanced drug delivery systems in attenuating COPD. *Heliyon*, e25393–e25393. https://doi.org/10.1016/j.heliyon.2024.e25393

Kad, A., Pundir, A., Arya, S. K., Bhardwaj, N., & Khatri, M. (2020). An elucidative review to analytically sieve the viability of nanomedicine market. *Journal of Pharmaceutical Innovation*, *17*(1), 249–265. https://doi.org/10.1007/s12247-020-09495-5

Khan, K. S., Jawaid, S., Memon, U. A., Perera, T., Khan, U., Farwa, U. E., Jindal, U., Afzal, M. S., Razzaq, W., Abdin, Z. U., & Khawaja, U. A. (2023). Management of chronic obstructive pulmonary disease (COPD) exacerbations in hospitalized patients from admission to discharge: A comprehensive review of therapeutic interventions. *Cureus*, *15*(8). https://doi.org/10.7759/cureus.43694

Kia, P., Ruman, U., Pratiwi, A. R., & Hussein, M. Z. (2023). Innovative therapeutic approaches based on nanotechnology for the treatment and management of tuberculosis. *International Journal of Nanomedicine*, *18*, 1159–1191. https://doi.org/10.2147/IJN.S364634

Klein, K., Borchard, G., Shah, V. P., Flühmann, B., McNeil, S. E., & Vlieger, J. S. B. (2021). A pragmatic regulatory approach for complex generics through the U.S. FDA 505(j) or 505(b)(2) approval pathways. *Annals of the New York Academy of Sciences*, *1502*(1), 5–13. https://doi.org/10.1111/nyas.14662

Kumar, M., Hilles, A. R., Almurisi, S. H., Bhatia, A., & Mahmood, S. (2023). Micro and nano-carriers-based pulmonary drug delivery system: Their current updates, challenges, and limitations – A review. *JCIS Open*, *12*, 100095. https://doi.org/10.1016/j.jciso.2023.100095

Lee, M., Ly, H., Möller, C. C., & Ringel, M. (2019). Innovation in regulatory science is meeting evolution of clinical evidence generation. *Clinical Pharmacology & Therapeutics*, *105*(4), 886–898. https://doi.org/10.1002/cpt.1354

Li, J., Wang, Q., Xia, G., Adilijiang, N., Li, Y., Hou, Z., Fan, Z., & Li, J. (2023). Recent advances in targeted drug delivery strategy for enhancing oncotherapy. *Pharmaceutics*, *15*(9), 2233. https://doi.org/10.3390/pharmaceutics15092233

Li, Y., Gao, H., Zhao, L., & Wang, J. (2022). Osteoporosis in COPD patients: Risk factors and pulmonary rehabilitation. *The Clinical Respiratory Journal*, *16*(7), 487–496. https://doi.org/10.1111/crj.13514

Lin, P.-C., Lin, S., Wang, P. C., & Sridhar, R. (2014). Techniques for physicochemical characterization of nanomaterials. *Biotechnology Advances*, *32*(4), 711–726. https://doi.org/10.1016/j.biotechadv.2013.11.006

Liu, Y., Lin, Z., Wang, Y., Chen, L., Wang, Y., & Luo, C. (2024). Nanotechnology in inflammation: Cutting-edge advances in diagnostics, therapeutics and theranostics. *Theranostics*, *14*(6), 2490–2525. https://doi.org/10.7150/thno.91394

Luo, M.-X., Hua, S., & Shang, Q.-Y. (2021). Application of nanotechnology in drug delivery systems for respiratory diseases (Review). *Molecular Medicine Reports*, *23*(5). https://doi.org/10.3892/mmr.2021.11964

Malik, S., Muhammad, K., & Waheed, Y. (2023). Emerging applications of nanotechnology in healthcare and medicine. *Molecules*, *28*(18), 6624–6624. https://doi.org/10.3390/molecules28186624

Martinez, F. J., Agusti, A., Celli, B. R., Han, M. K., Allinson, J. P., Bhatt, S. P., Calverley, P., Chotirmall, S. H., Chowdhury, B., Darken, P., Da Silva, C. A., Donaldson, G., Dorinsky, P., Dransfield, M., Faner, R., Halpin, D. M., Jones, P., Krishnan, J. A., Locantore, N., & Martinez, F. D. (2022). Treatment trials in young patients with chronic obstructive pulmonary disease and Pre–Chronic obstructive pulmonary disease patients: Time to move forward. *American Journal of Respiratory and Critical Care Medicine*, *205*(3), 275–287. https://doi.org/10.1164/rccm.202107-1663so

Mehta, M., Thuy, A., Bui, Y., Yagiz Alp Aksoy, X., Goldys, E. M., & Deng, W. (2023). Lipid-based nanoparticles for Drug/Gene delivery: An overview of the production techniques and difficulties encountered in their industrial development. *ACS Materials Science Au*, *3*(6). https://doi.org/10.1021/acsmaterialsau.3c00032

More, S. J., Vasileios Bampidis, Benford, D., Bragard, C., Halldorsson, T. I., Hernandez-Jerez, A. F., Susanne Hougaard Bennekou, Kostas Koutsoumanis, Lambré, C., Kyriaki Machera, Naegeli, H., Nielsen, S., Josef Rudolf Schlatter, Schrenk, D., Silano, V., Turck, D., Younes, M., Castenmiller, J., Chaudhry, Q., & Cubadda, F. (2021a). Guidance on risk assessment of nanomaterials to be applied in the food and feed chain: Human and animal health. *EFSA Journal*, *19*(8). https://doi.org/10.2903/j.efsa.2021.6768

More, S. J., Vasileios Bampidis, Benford, D., Bragard, C., Halldorsson, T. I., Hernandez-Jerez, A. F., Susanne Hougaard Bennekou, Kostas Koutsoumanis, Lambré, C., Kyriaki Machera, Naegeli, H., Søren Saxmose Nielsen, Josef Rudolf Schlatter, Schrenk, D., Silano, V., Turck, D., Younes, M., Castenmiller, J., Chaudhry, Q., & Cubadda, F. (2021b). Guidance on technical requirements for regulated food and feed product applications to establish the presence of small particles including nanoparticles. *EFSA Journal*, *19*(8). https://doi.org/10.2903/j.efsa.2021.6769

Mudalige, T., Qu, H., Van Haute, D., Ansar, S. M., Paredes, A., & Ingle, T. (2019). Characterization of nanomaterials. *Nanomaterials for Food Applications*, 313–353. https://doi.org/10.1016/b978-0-12-814130-4.00011-7

Mühlebach, S. (2018). Regulatory challenges of nanomedicines and their follow-on versions: A generic or similar approach? *Advanced Drug Delivery Reviews*, *131*, 122–131. https://doi.org/10.1016/j.addr.2018.06.024

Nagar, L., Gulati, N., Saini, A., Singh, S., Gupta, G., MacLoughlin, R., Kumar Chellappan, D., Dua, K., & Dureja, H. (2024). Recent trends and applications of nanostructure-based drug delivery in alleviating chronic obstructive pulmonary disease (COPD). *Current Drug Delivery*, *21*. https://doi.org/10.2174/0115672018289883240226113353

Onoue, S., Yamada, S., & Chan, K. (2014). Nanodrugs: Pharmacokinetics and safety. *International Journal of Nanomedicine*, 1025. https://doi.org/10.2147/ijn.s38378

Paradise, J. (2019). Regulating nanomedicine at the food and drug administration. *AMA Journal of Ethics*, *21*(4), E347–355. https://doi.org/10.1001/amajethics.2019.347

Parris, B. A., O'Farrell, H. E., Fong, K. M., & Yang, I. A. (2019). Chronic obstructive pulmonary disease (COPD) and lung cancer: Common pathways for pathogenesis. *Journal of Thoracic Disease*, *11*(S17), S2155–S2172. https://doi.org/10.21037/jtd.2019.10.54

Passi, M., Shahid, S., Chockalingam, S., Sundar, I. K., & Packirisamy, G. (2020). Conventional and nanotechnology based approaches to combat chronic obstructive pulmonary disease: Implications for chronic airway diseases. *International Journal of Nanomedicine*, *15*, 3803–3826. https://doi.org/10.2147/IJN.S242516

Patel, K. T., & Chotai, N. P. (2011). Documentation and records: Harmonized GMP requirements. *Journal of Young Pharmacists*, *3*(2), 138–150. https://doi.org/10.4103/0975-1483.80303

Patra, J. K., Das, G., Fraceto, L. F., Campos, E. V. R., Rodriguez-Torres, M. del P., Acosta-Torres, L. S., Diaz-Torres, L. A., Grillo, R., Swamy, M. K., Sharma, S., Habtemariam, S., & Shin, H.-S. (2018). Nano based drug delivery systems: Recent developments and future prospects. *Journal of Nanobiotechnology*, *16*(1). https://doi.org/10.1186/s12951-018-0392-8

Peng, S., Wang, W., Zhang, R., Wu, C., Pan, X., & Huang, Z. (2024). Nano-formulations for pulmonary delivery: Past, present, and future perspectives. *Pharmaceutics*, *16*(2), 161. https://doi.org/10.3390/pharmaceutics16020161

Pereira, L. C., Fátima, M. A. de, Santos, V. V., Brandão, C. M., Alves, I. A., & Azeredo, F. J. (2022). Pharmacokinetic/Pharmacodynamic modeling and application in antibacterial and antifungal pharmacotherapy: A narrative review. *Antibiotics*, *11*(8), 986. https://doi.org/10.3390/antibiotics11080986

Pita, R., Ehmann, F., & Papaluca, M. (2016). Nanomedicines in the EU—Regulatory overview. *The AAPS Journal*, *18*(6), 1576–1582. https://doi.org/10.1208/s12248-016-9967-1

Piveteau, L., Morad, V., & Kovalenko, M. V. (2020). Solid-state NMR and NQR spectroscopy of lead-halide perovskite materials. *Journal of the American Chemical Society*, *142*(46), 19413–19437. https://doi.org/10.1021/jacs.0c07338

PMDA. (2023). *Regulations and Approval/Certification of Medical Devices | Pharmaceuticals and Medical Devices Agency*. Www.pmda.go.jp. https://www.pmda.go.jp/english/review-services/reviews/0004.html

Pontes, J. F., & Grenha, A. (2020). Multifunctional nanocarriers for lung drug delivery. *Nanomaterials*, *10*(2), 183. https://doi.org/10.3390/nano10020183

Ramos, T. I., Villacis-Aguirre, C. A., López-Aguilar, K. V., Santiago Padilla, L., Altamirano, C., Toledo, J. R., & Santiago Vispo, N. (2022). The Hitchhiker's guide to human therapeutic nanoparticle development. *Pharmaceutics*, *14*(2), 247. https://doi.org/10.3390/pharmaceutics14020247

Sato, K., & Kawakami, H. (2023). Mitochondrial dysfunction and nanocarrier-based treatments in chronic obstructive pulmonary disease (COPD). *Oxygen*, *3*(4), 394–406. https://doi.org/10.3390/oxygen3040026

Saxena, J., Bisen, M., Misra, A., Srivastava, V. K., Kaushik, S., Siddiqui, A. J., Mishra, N., Singh, A., & Jyoti, A. (2022). Targeting COPD with PLGA-based nanoparticles: Current status and prospects. *BioMed Research International*, *2022*, 1–13. https://doi.org/10.1155/2022/5058121

Shandilya, N., Marcoulaki, E., Barruetabeña, L., Llopis, I. R., Noorlander, C., Jiménez, A. S., Oudart, Y., Puelles, R. C., Pérez-Fernández, M., Falk, A., Resch, S., Sips, A., & Fransman, W. (2020). Perspective on a risk-based roadmap towards the implementation of the safe innovation approach for industry. *NanoImpact*, *20*, 100258. https://doi.org/10.1016/j.impact.2020.100258

Singh, A. P., Biswas, A., Shukla, A., & Maiti, P. (2019). Targeted therapy in chronic diseases using nanomaterial-based drug delivery vehicles. *Signal Transduction and Targeted Therapy*, *4*(1). https://doi.org/10.1038/s41392-019-0068-3

Singh, P. V., Singh, P., & Anjankar, A. (2023). Harnessing the therapeutic potential of stem cells in the management of chronic obstructive pulmonary disease: A comprehensive review. *Cureus*. https://doi.org/10.7759/cureus.44498

Soares, S., Sousa, J., Pais, A., & Vitorino, C. (2018). Nanomedicine: Principles, properties, and regulatory issues. *Frontiers in Chemistry*, *6*(360). https://doi.org/10.3389/fchem.2018.00360

Souto, E. B., Blanco-Llamero, C., Krambeck, K., Kiran, N. S., Yashaswini, C., Postwala, H., Severino, P., Priefer, R., Prajapati, B. G., & Maheshwari, R. (2024). Regulatory insights into nanomedicine and gene vaccine innovation: Safety assessment, challenges, and regulatory perspectives. *Acta Biomaterialia*. https://doi.org/10.1016/j.actbio.2024.04.010

Souto, E. B., Silva, G. F., Dias-Ferreira, J., Zielinska, A., Ventura, F., Durazzo, A., Lucarini, M., Novellino, E., & Santini, A. (2020). Nanopharmaceutics: Part I—Clinical trials legislation and good manufacturing practices (GMP) of nanotherapeutics in the EU. *Pharmaceutics*, *12*(2), 146. https://doi.org/10.3390/pharmaceutics12020146

Srivastava, N., & Chowdhury, N. (2008). Regulation of health related nano applications in India: Exploring the limitations of the current regulatory design. *SSRN Electronic Journal*. https://doi.org/10.2139/ssrn.1105685

Suke, S. G., Kosta, P., & Negi, H. (2015). Role of pharmacovigilance in India: An overview. *Online Journal of Public Health Informatics*, *7*(2). e61487. https://doi.org/10.5210/ojphi.v7i2.5595

Tanimoto, T. (2015). A perspective on the benefit-risk assessment for new and emerging pharmaceuticals in Japan. *Drug Design, Development and Therapy*, 1877. https://doi.org/10.2147/dddt.s62636

Tenchov, R., Bird, R., Curtze, A. E., & Zhou, Q. (2021). Lipid nanoparticles—From liposomes to mRNA vaccine delivery, a landscape of research diversity and advancement. *ACS Nano*, *15*(11). https://doi.org/10.1021/acsnano.1c04996

Thapa, R. K., & Kim, J. O. (2022). Nanomedicine-based commercial formulations: Current developments and future prospects. *Journal of Pharmaceutical Investigation*, *53*(1), 19–33. https://doi.org/10.1007/s40005-022-00607-6

Tibbitts, J., Canter, D., Graff, R., Smith, A., & Khawli, L. A. (2015). Key factors influencing ADME properties of therapeutic proteins: A need for ADME characterization in drug discovery and development. *MAbs*, *8*(2), 229–245. https://doi.org/10.1080/19420862.2015.1115937

Tobechukwu, C., Ugochukwu, S., Ufedo, L., Princess, C., Emmanuel, C., Juliet, O., Charles, C., Shedrach, C., Onyinyechi, L., Evaristus, C., Nwadike, U., John, O., Bravo, U., Ossai, E., & Bennett, C. (2023). Advances in drug delivery systems, challenges and future directions. *Heliyon*, *9*(6), e17488–e17488. https://doi.org/10.1016/j.heliyon.2023.e17488

Tschiche, H. R., Bierkandt, F. S., Creutzenberg, O., Fessard, V., Franz, R., Giese, B., Greiner, R., Haas, K., Haase, A., Hartwig, A., Hund-Rinke, K., Iden, P., Kromer, C., Loeschner, K., Mutz, D., Rakow, A., Rasmussen, K., Rauscher, H., Richter, H., & Schoon, J. (2022). Environmental considerations and current status of grouping and regulation of engineered nanomaterials. *Environmental Nanotechnology, Monitoring and Management*, *18*, 100707–100707. https://doi.org/10.1016/j.enmm.2022.100707

Virmani, T., Kumar, G., Virmani, R., Sharma, A., & Pathak, K. (2022). Nanocarrier-based approaches to combat chronic obstructive pulmonary disease. *Nanomedicine*. https://doi.org/10.2217/nnm-2021-0403

Vogelmeier, C., Román, M., Singh, D., Han, M. K., Roisin, R., & Ferguson, G. (2020). Goals of COPD treatment: Focus on symptoms and exacerbations. *Respiratory Medicine*, *166*(1), 105938. https://doi.org/10.1016/j.rmed.2020.105938

Wasti, S. T., Lee, I. H., Kim, S., Lee, J.-H., & Kim, H. (2023). Ethical and legal challenges in nanomedical innovations: A scoping review. *Ethical and Legal Challenges in Nanomedical Innovations: A Scoping Review*, *14*, 1163392–1163392. https://doi.org/10.3389/fgene.2023.1163392

Wedzicha, J. A. (2008). *Chronic Obstructive Pulmonary Disease Exacerbations*. Crc Press.

WHO. (2023, March 16). *Chronic obstructive pulmonary disease (COPD)*. Www.who.int. https://www.who.int/news-room/fact-sheets/detail/chronic-obstructive-pulmonary-disease-(copd)#:~:text=In%20people%20with%20COPD%2C%20the

Xu, You Lin, Fang, S., & Song, L. (2020). Novel drug delivery systems targeting oxidative stress in chronic obstructive pulmonary disease: A review. *Journal of Nanobiotechnology*, *18*(1). https://doi.org/10.1186/s12951-020-00703-5

Yan, X., & Sha, X. (2023). Nanoparticle-mediated strategies for enhanced drug penetration and retention in the airway mucosa. *Pharmaceutics*, *15*(10), 2457–2457. https://doi.org/10.3390/pharmaceutics15102457

Yetisgin, A. A., Cetinel, S., Zuvin, M., Kosar, A., & Kutlu, O. (2020). Therapeutic nanoparticles and their targeted delivery applications. *Molecules*, *25*(9), 2193. https://doi.org/10.3390/molecules25092193

Yusuf, A., Almotairy, A. R. Z., Henidi, H., Alshehri, O. Y., & Aldughaim, M. S. (2023). Nanoparticles as drug delivery systems: A review of the implication of Nanoparticles' physicochemical properties on responses in biological systems. *Polymers*, *15*(7), 1596. https://doi.org/10.3390/polym15071596

https://patents.google.com/patent/CN115227682B/en?oq=CN115227682B

https://patents.google.com/patent/CN105310986A/en?oq=+CN105310986A

https://patents.google.com/patent/US20220226500A1/en?oq=US2022226500A1

https://patents.google.com/patent/CN106389387A/en?oq=+CN106389387A

https://patents.google.com/patent/KR20200012587A/en?oq=KR20200012587A

https://patents.google.com/patent/US20210308067A1/en?oq=US2021308067A1

https://patents.google.com/patent/WO2022023456A1/en?oq=+WO2022023456A1

https://classic.clinicaltrials.gov/ct2/show/NCT03794583

https://classic.clinicaltrials.gov/ct2/show/NCT01168895

18 Prospects of Advanced Drug Delivery Systems in Combating Chronic Obstructive Pulmonary Disease

Tanisqa Mall
Department of Chemistry, University of Petroleum & Energy Studies
Dehradun, Uttarakhand, India

Mousmee Sharma
Department of Chemistry, Uttaranchal University
Dehradun, Uttarakhand, India

Parteek Prasher
Department of Chemistry, University of Petroleum & Energy Studies
Dehradun, Uttarakhand, India

18.1 INTRODUCTION

Chronic obstructive pulmonary disease (COPD) is a respiratory disorder that worsens over time and causes difficulty in breathing due to persistent airflow limitations and chronic inflammation in the lungs [1]. It is a significant issue in public health and a primary contributor to illness and death on a global scale. The primary factor responsible for developing COPD is the exposure to harmful particles or gases, primarily through smoking cigarettes, but also as a result of occupational exposure and air pollution. The pathogenesis of COPD is intricate and multifaceted, encompassing several pathogenic mechanisms, including inflammation of the airways, excessive production of mucus, oxidative stress, and structural alterations in the lungs, such as emphysema and remodeling of the airways. These pathological processes contribute to the gradual deterioration of lung function, resulting in symptoms such as persistent cough, production of mucus, and difficulty breathing (shortness of breath). Ultimately, these processes affect the survival of patients and quality of life. The management of COPD depends on a combination of pharmacological and non-pharmacological interventions, with the main goal of relieving symptoms, enhancing lung function, and preventing exacerbations. Bronchodilators, specifically long-acting muscarinic antagonists (LAMAs) and long-acting β2-agonists (LABAs), are the primary drugs used in the treatment of respiratory conditions. They effectively alleviate symptoms and enhance lung function. In patients with frequent or severe exacerbations, inhaled corticosteroids (ICS) are commonly administered alongside bronchodilators to mitigate airway inflammation and avoid exacerbations [2, 3].

Although there are multiple therapy alternatives available, effectively administering medications to the lungs continues to be a major difficulty in treating COPD. Traditional methods of administering drugs, such as dry powder inhalers (DPIs), and metered-dose inhalers (MDIs) have drawbacks including inadequate drug distribution in the lungs, suboptimal drug absorption, and the inability to deliver drugs directly to the intended site of action [4].

DOI: 10.1201/9781003498926-18

These constraints can result in diminished treatment effectiveness and heightened systemic adverse effects. The difficulties linked to traditional drug delivery techniques in managing COPD have led to the investigation of more sophisticated drug delivery technologies. These innovative methodologies strive to address the constraints of traditional systems by boosting medication targeting, optimizing drug bioavailability, and minimizing off-target effects. Given these difficulties, it is increasingly important to investigate sophisticated methods of delivering drugs that are specifically designed for managing COPD. These innovative approaches strive to improve the effectiveness, safety, and patient compliance with treatment while also tackling the drawbacks of traditional formulations. Researchers want to transform the treatment of COPD by utilizing advanced technologies like nanotechnology, controlled-release formulations, and targeted delivery systems. These developments offer the potential for accurate medication targeting, extended drug release kinetics, and enhanced therapeutic indices, ultimately increasing treatment outcomes and improving the quality of life for persons with COPD.

Nanotechnology-based drug delivery systems, pulmonary delivery devices, and targeted formulations are among the promising strategies being investigated to address the unmet needs in COPD treatment. This chapter will present a comprehensive overview of COPD, including its pathogenesis, as well as the current difficulties encountered in traditional drug administration.

18.2 NANOTECHNOLOGY-BASED DRUG DELIVERY SYSTEMS FOR COPD

Nanotechnology has created opportunities for the advancement of sophisticated drug delivery systems in different medical fields, such as COPD. Nanotechnology-enabled medication delivery methods utilize the distinct characteristics of materials at the nanoscale, providing numerous benefits compared to traditional formulations. Nanotechnology-based drug delivery systems in the context of COPD provide the potential to overcome several significant obstacles associated with conventional drug delivery. These obstacles include inadequate drug solubility, limited bioavailability, quick elimination from the lungs, and absence of targeted distribution [5]. By encapsulating or conjugating drugs with nanocarriers, these systems can enhance drug stability, improve pulmonary deposition, prolong drug retention in the lungs, and facilitate targeted delivery to specific cells or tissues.

Nanoparticulate drug delivery systems, such as polymeric nanoparticles, lipid-based nanocarriers (such as liposomes and solid lipid nanoparticles), and inorganic nanoparticles (such as mesoporous silica nanoparticles and gold nanoparticles), have been extensively studied for delivering drugs to the lungs in patients with COPD.

These nanocarriers can be engineered to encapsulate or adsorb therapeutic agents, such as small molecule drugs, proteins, peptides, or nucleic acids, and can be tailored to achieve desired physicochemical properties, surface characteristics, and release profiles. Furthermore, nanotechnology-based drug delivery systems offer opportunities for targeted delivery to specific cells or tissues involved in the pathogenesis of COPD [6]. By conjugating nanocarriers with specific ligands or antibodies, these systems can selectively target receptors or antigens expressed on inflammatory cells, epithelial cells, or other relevant cell types, potentially enhancing therapeutic efficacy and minimizing off-target effects. Nanocarriers can also be leveraged for gene therapy approaches in COPD, serving as vehicles for the delivery of therapeutic nucleic acids, such as small interfering RNA (siRNA) or plasmid DNA, to modulate gene expression or replace defective genes involved in the disease pathogenesis. In addition to nanoparticulate systems, nanoemulsions and nanomicellar formulations have emerged as promising strategies for improving the solubility and bioavailability of hydrophobic drugs used in COPD treatment, enhancing their pulmonary delivery and therapeutic efficacy [7]. This chapter will provide an overview of various nanotechnology-based drug delivery systems, including nanoparticles, nanocarriers for gene therapy, nanoemulsions, and nanomicellar formulations, and their potential applications in the management of COPD. The advantages, challenges, and recent developments in these nanotechnology-based approaches will be discussed, highlighting their prospects in combating this chronic respiratory disease.

Nanoparticles offer a potential approach for delivering drugs directly to the lungs, specifically by utilizing the intracellular cell adhesion molecule and platelet endothelial cell adhesion molecule receptors found on pulmonary endothelial cells. These nanoparticles function as carriers for drugs such as nicotinamide adenine dinucleotide phosphate H+ (NADPH) oxidase inhibitors, catalase (CAT), and superoxide dismutase (SOD) with the goal of reducing oxidative stress in the respiratory system [8].

18.3 DIFFERENT TYPES OF NANOPARTICLES USED FOR DRUG DELIVERY IN COPD

18.3.1 Polymeric Nanoparticles

Polymeric nanoparticles are colloidal carriers composed of biocompatible and biodegradable polymers, such as poly (lactic-co-glycolic acid) (PLGA), chitosan, or polyethylene glycol (PEG). They can encapsulate or adsorb therapeutic agents, protecting them from degradation and enabling sustained release kinetics. Polymeric nanoparticles can be tailored for pulmonary delivery by optimizing particle size, surface charge, and aerodynamic properties. Their versatility allows for surface modifications to enhance mucoadhesion, improve cellular uptake, or enable targeted delivery [9].

18.3.2 Liposomes

Liposomes are composed of an external hydrophobic layer and an internal hydrophilic layer, with sizes typically ranging from 50 nm to 100 nm, which is dictated by the combination of cholesterol and phospholipids [10]. Liposomes, which include phospholipids, are ideal for enclosing hydrophilic medications like paclitaxel and doxorubicin. As a result, the U.S. Food and Drug Administration (FDA) has approved this drug delivery system based on nanoparticles [11]. The early utilization of liposome-based products entailed employing liposomes generated from bovine surfactant to treat acute respiratory distress syndrome in babies through the administration of the substance directly into the lungs. Nevertheless, there are still difficulties in preserving the structural integrity of liposomes after inhalation, which requires continuous endeavors to enhance the effectiveness of medicines based on liposomes. At present, clinical development is being conducted on dry inhaled powder formulations such as Arikace and Pulmaquin. Arikace is a medication that contains liposomal amikacin combined with dipalmitoyl-phosphatidylcholine and cholesterol. It is specifically designed to target lung infections caused by *Pseudomonas aeruginosa* in individuals with cystic fibrosis. On the other hand, Pulmaquin is a medication that uses liposomal ciprofloxacin to treat lung infections. Additionally, ongoing research is being conducted on liposome-based formulations that include antioxidants to manage acute lung damage caused by oxidants and hypoxia [12, 13].

18.3.3 Lipid-based Nanoparticles

Lipid-based nanoparticles, including liposomes, solid lipid nanoparticles (SLNs), and nanostructured lipid carriers (NLCs), are composed of biocompatible lipids and can encapsulate both hydrophilic and hydrophobic drugs. Liposomes are spherical vesicles consisting of one or more phospholipid bilayers surrounding an aqueous core, while SLNs and NLCs are solid lipid matrices stabilized by surfactants. These systems can prolong drug residence time in the lungs, improve drug solubility, and facilitate controlled release. Surface modifications with targeting ligands can enhance their specificity toward diseased cells or tissues [14].

18.3.4 Inorganic Nanoparticles

Inorganic nanoparticles, such as mesoporous silica nanoparticles (MSNs) and gold nanoparticles (AuNPs), have unique physicochemical properties suitable for drug delivery applications in COPD. MSNs possess a high surface area, tunable pore size, and the ability to accommodate various therapeutic agents. Their surface can be functionalized to improve biocompatibility, control drug release, or enable targeted delivery. AuNPs exhibit excellent biocompatibility, facile surface modification, and photothermal and photodynamic properties, making them attractive for theranostic applications in COPD.

These nanoparticulate systems offer advantages over conventional formulations, including improved drug solubility and stability, controlled and sustained release kinetics, enhanced pulmonary deposition, and the potential for targeted delivery to specific cells or tissues involved in COPD pathogenesis. However, factors such as particle size, surface charge, and aerodynamic properties need to be carefully optimized to ensure effective pulmonary delivery [15].

18.3.5 Dendrimers

Dendrimers are highly branched polymeric macromolecules characterized by well-defined radical branching. Dendrimers with small size, spherical shape, and lipophilicity exhibit superior penetration into cell membranes compared to linear polymers. The loading capacity of dendrimers depends on factors such as the type and number of active sites, loading capacity, external functional groups, and lipophilicity. Formulations based on dendrimers have demonstrated utility in addressing endothelial dysfunction, inflammation, and metastasis in cancer. PEGylation, the attachment of PEG chains, enhances the therapeutic potential of dendrimers in the treatment of pulmonary diseases following inhalation [16].

18.4 MUCOADHESIVE DRUG DELIVERY SYSTEMS

Mucoadhesive drug delivery systems have gained significant attention in the treatment of COPD due to their potential to improve drug delivery to the lungs and enhance therapeutic efficacy. COPD is a progressive respiratory disease characterized by airflow limitation and chronic inflammation in the lungs, making targeted and sustained drug delivery essential for effective management. The concept of mucoadhesion involves the adherence of a drug delivery system to the mucosal surfaces, such as the respiratory tract, due to attractive forces between the system and the mucus layer. This adherence can prolong the residence time of the drug at the site of action, leading to improved drug absorption and enhanced therapeutic effects. Mucoadhesive drug delivery systems for COPD can be designed in various forms, including:

18.4.1 Mucoadhesive Nanoparticles and Microparticles

Mucoadhesive polymers are being increasingly recognized as effective carriers for delivering drugs to the lungs for the treatment of COPD. These polymers have certain characteristics and chemical groups that have a strong attraction to the mucosal surfaces of the respiratory system. This allows the drugs to stay in the body for a longer time and enhances their effectiveness in treating diseases. Cationic polymers such as chitosan engage in electrostatic interactions with the negatively charged groups of mucus, whereas anionic polymers like alginate can establish hydrogen bonds with the mucin layer of the mucosa, thereby enhancing their ability to adhere to mucus [17]. Solid lipid microparticles (SLMs) have been investigated as mucoadhesive carriers due to their capacity to establish hydrogen bonds with the mucin layer, enabling regulated medication release and extended retention in the lungs. Microparticles loaded with fluticasone propionate (FP) have demonstrated significant promise for the treatment of COPD. SLMs made from alginate and chitosan have been created to efficiently transfer FP to the secondary bronchi through pulmonary release. These SLMs have a desirable mass median aerodynamic diameter (MMAD) of 3.5–4.0 μm [18]. The mucoadhesive microparticles containing hyaluronic acid, which were loaded with budesonide, had a MMAD that fell within the range suitable for inhalation (3.12–5.35 µm). Additionally, they had a fine particle fraction (FPF) of 35.6%, which has the potential for further enhancement. The microparticles in animal models showed a longer time to reach maximum concentration (Tmax), slower absorption, and higher bioavailability because of the mucoadhesive properties of hyaluronic acid. Nevertheless, it is crucial to thoroughly assess parameters such as the timing of drug release, the efficacy of drug encapsulation, and the capacity to redistribute the drug to optimize these adhesive systems for the purpose of delivering drugs effectively and precisely in the management of COPD [19].

Nanocomposite microparticles (NCMPs) are a highly promising method for delivering antioxidants to the lungs in the treatment of COPD. NCMPs offer the benefits of nanoparticles while also providing the convenience of microparticles for distribution. This is achieved by enclosing nanoparticles within respirable microparticles using the spray-drying method. Under normal physiological settings, NCMPs undergo dissociation into their original nanoparticles, hence preserving their drug release and delivery characteristics. These systems have enhanced antioxidant activity in laboratory tests compared to formulations that are not encapsulated, making them appropriate for treating oxidative stress in COPD. NCMPs can be created by combining antioxidants such as N-acetylcysteine and therapeutic compounds such microRNAs, while ensuring that they have appropriate sizes and fractions for effective delivery to the lungs [20]. To achieve a continuous release of medication, nanoparticles containing drugs can be placed into microparticles that can expand in size. This method has the potential to specifically target the alveolar region. Important factors to consider include choosing suitable excipients and drying settings to preserve the characteristics of nanoparticles, assessing the release patterns of the nanoparticles in laboratory and animal studies, and refining the formulations to achieve the required release rate and clearance by macrophages.

18.4.2 Mucoadhesive Liposomes and Lipid-Based Carriers

Liposomes and lipid-based carriers provide substantial benefits for delivering drugs in different lung diseases. Liposomes are lipid bilayer structures that can transport both hydrophilic and hydrophobic medicines. They enable regulated drug release by allowing diffusion through the bilayer. They can be customized for precise distribution and maintain stability and safety during administration. For example, liposomes that have both a photothermal agent and a folate ligand, known as theragnostic liposomes, have been shown to effectively prevent tumor growth while causing minimal adverse effects. These liposomes employ thermal stimulation to produce bubbles, which in turn create imperfections in the lipid bilayer and improve the efficiency of drug release. Lipid-based nanoparticles, such as NLCs, consist of solid and lipid phases dispersed in an aqueous surfactant. They provide benefits such as the convenient and safe delivery of unstable medications [18, 21, 22]. Nanoparticle-lipid complexes have demonstrated potential in specifically addressing a range of pulmonary illnesses, such as pulmonary aspergillosis, lung cancer, tuberculosis, acute lung damage, and cystic fibrosis. For example, the pH-sensitive NLC formulations that included β-elemene and doxorubicin showed a combined anticancer activity and better suppression of tumors compared to formulations that contained only one of the drugs. Prostaglandin E2-containing inhalable NLCs have been developed for idiopathic pulmonary fibrosis. These NLCs have shown effectiveness in decreasing animal death and lowering lung injury.

In addition, the use of lipid and carbohydrate carriers in the formulation of proliposomes shows great potential for delivering drugs to the lungs. Proliposomes that have been spray-dried and contain medications such as salbutamol sulfate have demonstrated the ability to form liposomes when they come into contact with water. The inhalable proliposome microparticles can be customized to achieve precise medication-release control and improve efficiency in targeting the lungs. Liposomes and lipid-based carriers are highly adaptable and efficient systems for delivering drugs to the lungs [23]. They provide precise control over drug release, targeted effects, and enhanced therapeutic results in a range of respiratory conditions.

18.4.3 Mucoadhesive Dry Powder Inhalers

The advancements in DPI technology have created opportunities for the creation of novel mucoadhesive formulations and devices, which have the potential to significantly transform the administration of drugs to the lungs for the management of COPD. One important area of concentration has been the improvement of powder compositions to increase the ability to produce aerosols, target specific areas, and maintain a consistent release of medication over time. This involves the addition of mucoadhesive polymers or the inclusion of nanoparticles and nanocomposites that contain mucoadhesive components into dry powder formulations that can be inhaled. When these mucoadhesive particles are deposited in the lungs, they stick to the respiratory mucosa, which helps them stay in place for a longer time and allows for a slow release of the medicine. As a result, this method has the potential to enhance the effectiveness of treatment, decrease the frequency of dosage, and potentially alleviate the adverse effects linked to widespread distribution throughout the body [24, 25]. The formulation methodologies have been further enhanced by the parallel improvements in DPI device design. Breath-actuated devices, which include visual and audible cues, have been developed to improve the accuracy and effectiveness of medicine delivery to specific areas of the respiratory tract. Furthermore, the advancement of low-resistance DPIs that can be activated at lower inhalation flows (≤30 L/min) and "active" or power-assisted DPIs that utilize external energy sources (such as battery-driven impellers or vibrating piezoelectric crystals) could effectively tackle the difficulties experienced by COPD patients with impaired lung function and limited ability to inhale deeply. This would guarantee consistent and reliable drug delivery [26, 27]. Through the integration of these advancements in formulation and device design, mucoadhesive DPI devices have the potential to surpass the constraints of conventional methods for delivering drugs to the lungs. The potential for precise and consistent drug administration at the specific site of action, along with enhanced patient adherence and less dependence on patient inhalation effort, offers significant potential for optimizing the therapy of COPD over a broad range of disease severities. In the end, these innovations have the potential to create individualized and efficient treatment plans, enhancing the well-being of individuals with COPD and solving the currently unmet medical requirements of this debilitating respiratory ailment. The addition of mucoadhesive polymers or excipients to the dry powder formulation can improve the ability of the particles to stick to the respiratory mucosa, resulting in longer drug retention. Mucoadhesive drug delivery systems offer several benefits in the treatment of COPD, including enhanced drug absorption, prolonged drug release, decreased frequency of dose, and specific targeting of the lungs. Nevertheless, it is crucial to address some obstacles, like mucus clearance processes, compatibility with inhalation devices, and the possibility of local discomfort, while developing these systems. In summary, mucoadhesive drug delivery devices present a favorable method for improving the efficacy of COPD treatment by optimizing drug administration to the respiratory tract and enhancing patient adherence through decreased dose frequency and increased therapeutic results [28].

18.5 ADVANCED FORMULATION APPROACHES

18.5.1 Protein and Peptide Delivery Systems

The pulmonary route presents a favorable method for administering therapeutic proteins and peptides due to its several advantages compared to traditional administration routes [28]. Directly targeting the lungs for local respiratory disorders like asthma, COPD, and cystic fibrosis can lead to increased medication accumulation at the specific site of action, which may boost the effectiveness of the treatment. In addition, the lungs possess a substantial surface area (about 70–140 m^2) and a well-developed network of blood vessels, which allows for the efficient absorption of proteins and peptides into the bloodstream. This property makes them suitable for the treatment of disorders that affect the entire body. In addition, pulmonary distribution avoids the initial breakdown of medications by the liver, known as first-pass hepatic metabolism, which can greatly decrease the amount of orally delivered drugs that actually enters the bloodstream. Inhalation provides a less intrusive method compared to parenteral injections, which are frequently used for administering various protein and peptide therapies. However, the lungs have natural defense systems that can impede the efficient delivery and absorption of proteins and peptides, notwithstanding their advantages. Mucociliary clearance is a fundamental mechanism in the airways that involves the physical movement of foreign particles through the mucous layer into the oropharynx, where they are then either eaten or expelled. In addition, coughing functions as a mechanism to expel bigger particles that settle in and cause irritation to the throat. Macrophages in the alveolar region have the ability to absorb and break down particulate contaminants [29].

In addition, the lungs contain a diverse array of peptidases, such as elastase, chymotrypsin, prolyl endopeptidase, collagenase, carboxypeptidases, angiotensin-converting enzyme (ACE), neutral endopeptidase, aminopeptidase P, cathepsin B, mast cell proteases, and neutrophil proteases [30–32]. These enzymes have a substantial impact on the transport and bioavailability of therapeutic proteins and peptides by accelerating their proteolytic breakdown. Peptides that have less than 30 amino acids often have low bioavailability because they are easily broken down by peptidases in the lungs. However, peptides with molecular weights ranging from 6 kDa to 50 kDa are less vulnerable to degradation and can be absorbed more effectively.

The primary mechanisms by which proteins and peptides are taken up from the lungs are transcytosis and paracellular transport. Transcytosis is the process by which macromolecules are transported from the external environment into the interior of cells. This can be facilitated by receptors or can occur through non-specific mechanisms [33]. Proteins and peptides can be absorbed directly into epithelial cells by membrane transporters, such as the epithelial transferrin receptor or hypothesized β-lactam-transporting peptide transporters. This process facilitates receptor-mediated delivery [34]. Conversely, paracellular transport refers to the transportation of molecules through the intercellular connections and gaps created by the shedding of dead cells. The paracellular transport of cationic peptides has been observed to have an inverse relationship with their molecular size. According to current hypotheses, peptides with molecular weights below 40 kDa, including insulin and calcitonin, are mainly transported through the paracellular pathway. On the other hand, bigger molecules (>40 kDa) are primarily absorbed through transcytosis. Furthermore, the process of diffusional absorption, which is often associated with tiny molecules, has also been documented in specific peptides, including thyrotropin-releasing hormone [35].

Different methods have been investigated to address the obstacles in delivering therapeutic proteins and peptides to the lungs. The purpose of utilizing protease inhibitors is to decrease enzymatic breakdown by pulmonary peptidases, thus enhancing bioavailability. Absorption enhancers, such as surfactants, liposomes, and cyclodextrins, have demonstrated the ability to increase the movement of substances across cells and through the spaces between cells. Various formulation tactics, such as DPIs, nebulizers, and controlled-release systems like nanoparticles or microparticles, have been studied to improve the efficiency of delivering medication, maintain its stability, and ensure targeted release in the respiratory tract. Although these techniques have shown encouraging outcomes in preclinical investigations, their effective use in clinical settings depends on overcoming safety issues. Immunogenicity continues to be a possible concern, as proteins and peptides from outside sources have the potential to induce immunological responses either in specific areas or across the body [36]. Moreover, the potential negative consequences of extended inhalation of these large compounds are unclear and demand additional assessment. Furthermore, it is crucial to conduct comprehensive evaluations of the safety profiles of various new excipients, including absorption enhancers. This is because certain excipients have demonstrated hazardous effects in preclinical tests, while others have displayed positive outcomes. Notwithstanding these difficulties, the continuous research endeavors in this domain strive to create secure and efficient compositions for administering proteins and peptides through the lungs, thereby unleashing their therapeutic capabilities for both respiratory ailments and systemic disorders. Further advancements in this field have the potential to result in groundbreaking and user-oriented therapeutic alternatives, providing enhanced results and quality of life for persons impacted by various illnesses [33, 37, 38].

18.5.2 Polymer-based Drug Delivery Systems

Polymer-based medication delivery systems offer an advanced method for tackling the intricacies of chronic lung illnesses. These systems utilize the characteristics of nanoparticles, namely those made of polymers, to overcome obstacles and efficiently transport medicines to the lungs. A significant benefit of polymer-based drug delivery systems is their capacity to improve the pharmacokinetics of medications and achieve accurate targeting inside the lung tissues. By employing meticulous design and engineering techniques, nanoparticles made from polymers can effectively enclose medications and shield them from deterioration, enabling gradual and regulated release over an extended period. This focused administration method reduces the occurrence of side effects across the body and maximizes the effectiveness of the treatment. Polymer-based nanoparticles hold significant promise in the field of chronic lung illnesses, including COPD, asthma, and cystic fibrosis [39]. These disorders are distinguished by the presence of inflammation and excessive production of mucus in the airways, which presents considerable difficulties in delivering drugs effectively. Preclinical investigations have shown that inhalable nanoparticles, such as liposomes and polymeric nanoparticles, have displayed exceptional effectiveness. These nanoparticles have been found to provide persistent therapeutic effects and enhance lung function in animal models. In addition, nanoparticles made from polymers show potential for delivering genes, allowing for the correction of genetic abnormalities that cause certain lung disorders [40]. Polymer-based nanoparticles provide a revolutionary method for gene therapy by enclosing genetic material and aiding its transportation to specific cells. This presents new opportunities for treating illnesses such as cystic fibrosis and α1-antitrypsin deficiency, in which gene abnormalities are a key factor in the development of the disease. Advancements in drug delivery systems based on polymers have allowed for the precise distribution of new medicines. Medications such as mepolizumab and CXCR-2 antagonists, which specifically target inflammatory pathways, demonstrate potential in decreasing inflammation and preventing worsening of symptoms in people with asthma and COPD. Using polymer-based nanoparticles, researchers can optimize the distribution of these medications to the lungs and boost their therapeutic effects [8, 41, 42].

Different synthetic polymers, including polylactic acid (PLA), PLGA, polyacrylates, and polyanhydrides, as well as natural polymers including albumin, chitosan, gelatin, alginate, and collagen, are used to treat pulmonary disorders. Artificial polymers are favored over organic polymers because of their capacity to efficiently maintain medication release. Therefore, synthetic polymers are frequently employed in the production of nanoparticles for medication delivery systems that specifically target lung illnesses. Therapeutic chemicals are either absorbed or enclosed within the core of PLGA nanoparticles, providing safeguard against degradation. This encapsulation enables the traversal of cellular barriers and effective medication release.

PLGA nanoparticles have the ability to simultaneously transport chemotherapeutic drugs and molecular agents, and they are taken up by cells by mechanisms such as clathrin-coated endocytosis or fluid-phase pinocytosis [43]. Adsorption of target molecules is enhanced by modifying the surface charge using cationic polymers such as chitosan. Nevertheless, the presence of persistent airway inflammation and excessive mucosa secretion presents difficulties for nano-based drug delivery systems, as they must overcome inherent defense mechanisms and barriers to target affected cells. While there is increasing interest in developing targeted nanoparticles for chronic inflammatory and obstructive airway disorders such as COPD, there is relatively limited research on new medication delivery systems in this field. Currently, preclinical investigations are being conducted on various formulations and methods to synthesize nanoparticles laden with drugs that target specific cell types in the airways [44].

PLGA nanoparticles are efficient carriers for the treatment of pulmonary inflammation and its related diseases. For example, drug-loaded PLGA nanoparticles such as PLGA-PEGPS-341 provide gradual and continuous release of medication for the treatment of cystic fibrosis lung illness in mice models. PLGA nanoparticles that have been altered and enclosed with tuftsin demonstrate notable intracellular effectiveness in *in vitro* models of lung infections associated with *Mycobacterium tuberculosis*. The addition of hydrophilic PEG polymers increases the movement of polymeric nanoparticles in human lung mucus and sputum. Although nanomedicines coupled with PLGA offer numerous benefits, they encounter obstacles such as drug toxicity and limited therapeutic effectiveness [44, 45]. Therefore, it is imperative to create new medications and strengthen PLGA-drug conjugates to obtain improved effectiveness with low harm for the treatment of COPD. Additional investigation is crucial for monitoring the drug release kinetics of medications loaded in PLGA nanoparticles in the respiratory tract and comprehending their impact on immune cells at the molecular level [43]. Moreover, nanoparticles made from polymers present possibilities for molecular imaging of chronic lung disorders, allowing for a better understanding of how the diseases develop and how they respond to treatment. Through the process of conjugating imaging agents to nanoparticles, scientists can observe disease-specific indicators and track the progress of therapeutic treatments in real time. Despite the difficulties related to pulmonary medication administration, such as the risk of organ toxicity and the presence of the mucus barrier, current studies on polymer-based drug delivery methods are making progress [39].

18.6 EMERGING TRENDS IN COPD DRUG DELIVERY

18.6.1 3D-printed Drug Delivery Devices

Asthma and COPD are commonly treated with medications that are breathed. COPD is a major global health issue that causes millions of deaths each year and puts tremendous strain on healthcare systems worldwide. COPD includes chronic bronchitis and emphysema, which are characterized by inflammation of the airways and damage to lung tissues. This results in symptoms such as coughing, difficulty breathing, and wheezing. Although smoking is a significant contributing factor to COPD, the disease can also affect individuals who do not smoke, and it is more common among older age groups. However, the effectiveness of treatment depends on the precise use of inhalation equipment. Optimal drug delivery while using devices like MDIs and DPIs requires the use of proper technique. DPIs rely on the patient's inhalation rate and the size of the drug particles to ensure successful administration. However, elderly patients and individuals with impaired lung function may face challenges in achieving the necessary flow rates, leading to suboptimal medicine administration. To address this issue, it was proposed to provide an additional device for DPIs that would enhance airflow and improve the reliability of drug administration [46].

The add-on device was conceived using 3D printing technology, which offers flexibility and precision in fabricating complex structures for pharmaceutical applications. The prototype add-on device was fabricated using fused deposition modeling (FDM). The device utilizes a centrifugal fan to improve the movement and erosion of medication particles, leading to the creation of smaller particles that can be deposited more efficiently in the lungs. The fan's speed can be adjusted to accommodate patients with varying inspiratory forces. Computational fluid dynamics were used to assess the airflow patterns and distribution of medicine particles in the modified DPI. A two-factor central composite design was used to investigate the effects of inhalation flow rate and device input power on the emitted dosages of salmeterol xinafoate and FP, which are the active ingredients in the chosen DPI (Accuhaler). The drugs were measured using high performance liquid chromatography (HPLC) tests, and a particle least square model was developed to verify the appropriateness of the add-on device for different patient populations. To summarize, the work demonstrates the feasibility of using 3D printing to produce an extra device for DPIs that could improve the delivery of medication and better treatment outcomes for persons with lung diseases. Through the optimization of formulation characteristics and the utilization of modern characterization techniques, researchers have the ability to create tailored drugs that offer enhanced effectiveness, safety, and patient adherence. This eventually leads to an improved quality of life for patients who suffer from COPD.

Novel methods of administering medication, such as the combination of 3D printing and hot-melt extrusion (HME) [47], show potential for enhancing the effectiveness of treatment in patients with COPD. These techniques allow for the creation of customized medicine formulations designed specifically for the unique needs of each patient. This provides controlled release of the drug and improves the effectiveness of the treatment. Research has investigated the utilization of different polymers and additives in 3D-printed tablets to attain certain drug release patterns, such as prolonged release or expedited release for fast symptom relief. For instance, theophylline, a widely used bronchodilator, has been included into formulations by the utilization of 3D printing and HME processes. These formulations utilize a blend of polymers such as Eudragit RL, Eudragit RS, Eudragit E, hydroxypropyl cellulose (HPC), and polyethylene oxides (PEO) to attain precise rates of drug release. Researchers have shown that by modifying the content and structure of the tablets, they may achieve continuous-release features that last for up to 10 hours. Additionally, they can also achieve accelerated release, which allows for a rapid commencement of effect [47, 48]. Moreover, research has examined the impact of various plasticizers, fillers, and polymer ratios on the rate at which drugs are released and the properties of tablets. The physical form, surface morphology, and thermal properties of the drug-loaded tablets were analyzed using techniques such as X-ray diffraction (XRD), scanning electron microscopy (SEM), and differential scanning calorimetry (DSC). These investigations offer vital insights into the mechanisms that govern drug release from 3D-printed formulations and contribute to the development of improved drug delivery methods for managing COPD. In summary, the integration of 3D printing with HME provides a flexible framework for creating innovative drug delivery systems specifically designed to address the distinct obstacles associated with COPD treatment [49].

18.6.2 Theranostic Nanocarriers

Theranostics, the combination of diagnostic testing and therapeutic intervention, has great potential for enhancing personalized medicine, especially in the treatment of chronic obstructive airway illnesses such as cystic fibrosis and COPD [50]. Nanotechnology provides a robust framework for creating theranostic strategies specifically designed to address the distinct difficulties associated with these illnesses. Nano-based theranostics provide numerous benefits for individuals with cystic fibrosis and COPD, including the ability to diagnose lung inflammation in real time and deliver treatment directly to the affected area. This approach ensures that each patient receives the most effective therapy while reducing any negative side effects from the medication. To facilitate the use of theranostics for cystic fibrosis and COPD, it is crucial to create nanocarriers that can target the airways and perform several functions. The purpose of these nanocarriers is to transport both diagnostic and therapeutic substances at the same time, combining many functionalities including cell targeting, imaging sensitivity, and therapy into a unified system.

Although multifunctional nanosystems have demonstrated potential in the treatment and detection of cancer, it is crucial to continue developing new nanosystems that are specifically designed to address the unique problems of cystic fibrosis and COPD [51, 52]. The use of metal oxide nanoparticles in cancer theranostics is restricted in chronic airway disorders due to their toxicity and tendency to cause persistent inflammation. Amphiphilic block copolymers provide a flexible option for creating several types of nanostructures, such as micelles and vesicles, that may effectively transport both hydrophilic and hydrophobic substances. Presently, the primary focus of research is on the development of multifunctional polymeric vesicles, namely those created by combining PEG and PLGA [53]. These vesicles are intended for the simultaneous delivery of medicines for COPD/CF and molecular probes. These nanocarriers have the capability to use non-invasive imaging techniques to track the distribution of drugs and the effectiveness of anti-inflammatory treatments in real-time. This allows for the development of theranostics for patients with cystic fibrosis and COPD.

PEG, renowned for its superb water solubility, minimal toxicity, and absence of accumulation in reticuloendothelial system (RES) cells, is frequently employed as the hydrophilic polymer component in amphiphilic block copolymers. On the contrary, PLGA is preferred due to its capacity to degrade naturally and its compatibility with living organisms, which makes it appropriate for therapeutic uses. The advancement of cutting-edge real-time imaging technologies, such as single-photon-emission computed tomography and positron-emission tomography/CT, has the potential to greatly enhance the assessment of drug development stages and decrease the expenses associated with clinical procedures. Non-invasive imaging techniques are essential for evaluating medication action in lung epithelia and inflammatory cells, enabling more efficient drug development and assessment of chronic lung disease conditions in a cost-effective manner. In summary, the development of theranostic nanocarriers has great potential in improving the diagnosis and treatment of cystic fibrosis and COPD, leading to better outcomes and quality of life for those affected [54].

18.7 FUTURE PERSPECTIVE AND CONCLUSION

Advancements in medicine and nanobiotechnology have recently resulted in the creation of several nanoparticles that show great potential for targeted therapy in chronic lung disorders. Nevertheless, numerous obstacles impede their practical implementation in clinical settings. One of the main concerns is the possibility of organ toxicity, especially with inorganic nanoparticles. These nanoparticles may not break down readily or be removed from the area where they are delivered, which can lead to risks of inflammation and scarring in the lungs. Furthermore, the interactions between nanoparticles and the immune system might lead to lung injury, making it crucial to carefully choose and design them to reduce these dangers.

Another major hindrance is the mucous barrier present in the respiratory tract, which is worsened by excessive mucus production in chronic lung disorders. This barrier hinders the entry of inhaled nanoparticles into the underlying respiratory epithelium. Diverse approaches have been investigated to address this obstacle, such as modifying the surface with PEG, enclosing mucolytic agents, and utilizing mannitol to generate osmotic gradients. These strategies all focus on augmenting the ability of nanoparticles to penetrate mucus and enhance their delivery to specific lung tissues. Although the use of nanoparticle-based medication delivery in respiratory medicine is now in its early phases, advancements in novel nanomaterials provide optimism for overcoming these challenges.

Biologics and gene therapies are progressing, providing innovative methods to regulate inflammation and tissue remodeling in the lungs of individuals with COPD. Personalized medicine, facilitated by the field of pharmacogenomics and the identification of biomarkers, permits the customization of treatment plans according to the unique characteristics of each patient. Smart inhaler technologies offer immediate monitoring and optimization of drug adherence and inhalation method. Drug delivery efficiency is improved by employing targeted drug delivery techniques, such as liposomes and dendrimers. Combining medicines and utilizing regenerative medicine technologies provide a holistic strategy to managing and perhaps repairing damaged tissues. Artificial intelligence and machine learning algorithms are employed to examine clinical data to optimize individualized treatment.

REFERENCES

1. D. S. Dhanjal *et al.*, "Novel controlled release pulmonary drug delivery systems: Current updates and challenges," in *Modeling and control of drug delivery systems*, Elsevier, 2021, pp. 253–272.
2. B. Mishra and J. Singh, "Novel drug delivery systems and significance in respiratory diseases," in *Targeting chronic inflammatory lung diseases using advanced drug delivery systems*, Elsevier, 2020, pp. 57–95.
3. S. D. Shukla *et al.*, "Chronic respiratory diseases: An introduction and need for novel drug delivery approaches," in *Targeting chronic inflammatory lung diseases using advanced drug delivery systems*, Elsevier, 2020, pp. 1–31.
4. R. A. Pleasants and J. F. Donohue, "Current Perspectives of Pharmacotherapies for COPD," *Respir Care*, vol. 68, no. 7, pp. 927–938, 2023.
5. M. E. Taghavizadeh Yazdi, M. Qayoomian, S. Beigoli and M. H. Boskabady, "Recent Advances in Nanoparticle Applications in Respiratory Disorders: A Review," *Front Pharmacol*, vol. 14, p. 1059343, 2023.

6. Y. Xu, H. Liu and L. Song, "Novel Drug Delivery Systems Targeting Oxidative Stress in Chronic Obstructive Pulmonary Disease: A Review," *J Nanobiotechnol*, vol. 18, pp. 1–25, 2020.
7. T. Virmani, G. Kumar, R. Virmani, A. Sharma and K. Pathak, "Nanocarrier-Based Approaches to Combat Chronic Obstructive Pulmonary Disease," *Nanomedicine*, vol. 17, no. 24, pp. 1833–1854, 2022.
8. J. Saxena *et al.*, "Targeting COPD With PLGA-Based Nanoparticles: Current Status and Prospects," *Biomed Res Int*, vol. 2022, 5058121, 13 pages, 2022.
9. M. Elmowafy *et al.*, "Polymeric Nanoparticles for Delivery of Natural Bioactive Agents: Recent Advances and Challenges," *Polymers (Basel)*, vol. 15, no. 5, p. 1123, 2023.
10. M. Paranjpe and C. C. Müller-Goymann, "Nanoparticle-Mediated Pulmonary Drug Delivery: A Review," *Int J Mol Sci*, vol. 15, no. 4, pp. 5852–5873, 2014.
11. E. D. Hood, M. Chorny, C. F. Greineder, I. S. Alferiev, R. J. Levy and V. R. Muzykantov, "Endothelial Targeting of Nanocarriers Loaded With Antioxidant Enzymes for Protection Against Vascular Oxidative Stress and Inflammation," *Biomaterials*, vol. 35, no. 11, pp. 3708–3715, 2014.
12. S. Chennakesavulu, A. Mishra, A. Sudheer, C. Sowmya, C. S. Reddy and E. Bhargav, "Pulmonary Delivery of Liposomal Dry Powder Inhaler Formulation for Effective Treatment of Idiopathic Pulmonary Fibrosis," *Asian J Pharm Sci*, vol. 13, no. 1, pp. 91–100, 2018.
13. L. Willis, D. Hayes and H. M. Mansour, "Therapeutic Liposomal Dry Powder Inhalation Aerosols for Targeted Lung Delivery," *Lung*, vol. 190, pp. 251–262, 2012.
14. C. Heneweer, S. E. M. Gendy and O. Peñate-Medina, "Liposomes and Inorganic Nanoparticles for Drug Delivery and Cancer Imaging," *Ther Deliv*, vol. 3, no. 5, pp. 645–656, 2012.
15. W. Paul and C. P. Sharma, "Inorganic Nanoparticles for Targeted Drug Delivery," in *Biointegration of Medical Implant Materials, second edition*, 2020, pp. 333–373.
16. A. K. Mandal, "Dendrimers in Targeted Drug Delivery Applications: A Review of Diseases and Cancer," *Int J Polym Mater Polym Biomaterials*, vol. 70, no. 4, pp. 287–297, 2021.
17. S. Mansuri, P. Kesharwani, K. Jain, R. K. Tekade and N. K. Jain, "Mucoadhesion: A Promising Approach in Drug Delivery System," *React Funct Polym*, vol. 100, pp. 151–172, 2016.
18. E. Amore *et al.*, "Mucoadhesive Solid Lipid Microparticles for Controlled Release of a Corticosteroid in the Chronic Obstructive Pulmonary Disease Treatment," *Nanomedicine*, vol. 12, no. 19, pp. 2287–2302, 2017.
19. T. Liu, M. Han, F. Tian, D. Cun, J. Rantanen and M. Yang, "Budesonide Nanocrystal-Loaded Hyaluronic Acid Microparticles for Inhalation: In Vitro and in Vivo Evaluation," *Carbohydr Polym*, vol. 181, pp. 1143–1152, 2018.
20. A. F. Ourique, P. dos Santos Chaves, G. D. Souto, A. R. Pohlmann, S. S. Guterres and R. C. R. Beck, "Redispersible Liposomal-N-Acetylcysteine Powder for Pulmonary Administration: Development, in Vitro Characterization and Antioxidant Activity," *Eur J Pharm Sci*, vol. 65, pp. 174–182, 2014.
21. T. S. Patil and A. S. Deshpande, "Nanostructured Lipid Carriers-Based Drug Delivery for Treating Various Lung Diseases: A State-of-the-Art Review," *Int J Pharm*, vol. 547, no. 1, pp. 209–225, 2018.
22. V. De Leo *et al.*, "Preparation of Drug-Loaded Small Unilamellar Liposomes and Evaluation of Their Potential for the Treatment of Chronic Respiratory Diseases," *Int J Pharm*, vol. 545, no. 1–2, pp. 378–388, 2018.
23. N. Dhiman, J. Sarvaiya and P. Mohindroo, "A Drift on Liposomes to Proliposomes: Recent Advances and Promising Approaches," *J Liposome Res*, vol. 32, no. 4, pp. 317–331, 2022.
24. M. Bonini and O. S. Usmani, "The Importance of Inhaler Devices in the Treatment of COPD," *COPD Research and Practice*, vol. 1, pp. 1–9, 2015.
25. F. Lavorini, M. Pistolesi and O. S. Usmani, "Recent Advances in Capsule-Based Dry Powder Inhaler Technology," *Multidiscip Respir Med*, vol. 12, pp. 1–7, 2017.
26. P. Colthorpe, T. Voshaar, T. Kieckbusch, E. Cuoghi and J. Jauernig, "Delivery Characteristics of a Low-Resistance Dry-Powder Inhaler Used to Deliver the Long-Acting Muscarinic Antagonist Glycopyrronium," *J Drug Assess*, vol. 2, no. 1, pp. 11–16, 2013.
27. M. Corradi *et al.*, "NEXThaler, an Innovative Dry Powder Inhaler Delivering an Extrafine Fixed Combination of Beclometasone and Formoterol to Treat Large and Small Airways in Asthma," *Expert Opin Drug Deliv*, vol. 11, no. 9, pp. 1497–1506, 2014.
28. M. Cazzola, F. Cavalli, O. S. Usmani and P. Rogliani, "Advances in Pulmonary Drug Delivery Devices for the Treatment of Chronic Obstructive Pulmonary Disease," *Expert Opin Drug Deliv*, vol. 17, no. 5, pp. 635–646, May 2020. doi: 10.1080/17425247.2020.1739021.
29. S.-A. Cryan, "Carrier-Based Strategies for Targeting Protein and Peptide Drugs to the Lungs," *AAPS J*, vol. 7, pp. E20–E41, 2005.

30. J. Fujita *et al.*, "Evaluation of Elastase and Antielastase Balance in Patients With Chronic Bronchitis and Pulmonary Emphysema," *Am Rev Respir Dis*, vol. 142, no. 1, pp. 57–62, 1990.
31. M. Orlowski, J. Orlowski, M. Lesser and K. H. Kilburn, "Proteolytic Enzymes in Bronchopulmonary Lavage Fluids: Cathepsin B-Like Activity and Prolyl Endopeptidase," *J Lab Clin Med*, vol. 97, no. 4, pp. 467–476, 1981.
32. R. E. Stratford Jr and V. H. L. Lee, "Aminopeptidase Activity in Homogenates of Various Absorptive Mucosae m the Albino Rabbit: Implications in Peptide Delivery," *Int J Pharm*, vol. 30, no. 1, pp. 73–82, 1986.
33. P. C. L. Kwok and H.-K. Chan, "Pulmonary delivery of peptides and proteins," in *Peptide and protein delivery*, Elsevier, 2011, pp. 23–46.
34. D. A. Groneberg, M. Nickolaus, J. Springer, F. Döring, H. Daniel and A. Fischer, "Localization of the Peptide Transporter PEPT2 in the Lung: Implications for Pulmonary Oligopeptide Uptake," *Am J Pathol*, vol. 158, no. 2, pp. 707–714, 2001.
35. J. S. Patton, "Mechanisms of Macromolecule Absorption by the Lungs," *Adv Drug Deliv Rev*, Vol, vol. 19, no. 1, pp. 3–36, 1996.
36. R. K. Wolff, "Safety of Inhaled Proteins for Therapeutic Use," *J Aerosol Med*, vol. 11, no. 4, pp. 197–219, 1998.
37. D. K. Chellappan *et al.*, "Protein and Peptide Delivery to Lungs by Using Advanced Targeted Drug Delivery," *Chem Biol Interact*, vol. 351, p. 109706, 2022.
38. L. Gomes dos Reis and D. Traini, "Advances in the Use of Cell Penetrating Peptides for Respiratory Drug Delivery," *Expert Opin Drug Deliv*, vol. 17, no. 5, pp. 647–664, 2020.
39. F. Emami, S. J. Mostafavi Yazdi and D. H. Na, "Poly (Lactic Acid)/poly (Lactic-Co-Glycolic Acid) Particulate Carriers for Pulmonary Drug Delivery," *J Pharm Investig*, vol. 49, pp. 427–442, 2019.
40. X. Tong, W. Pan, T. Su, M. Zhang, W. Dong and X. Qi, "Recent Advances in Natural Polymer-Based Drug Delivery Systems," *React Funct Polym*, vol. 148, p. 104501, 2020.
41. J. Y. Yhee, J. Im and R. S. Nho, "Advanced Therapeutic Strategies for Chronic Lung Disease Using Nanoparticle-Based Drug Delivery," *J Clin Med*, vol. 5, no. 9, p. 82, 2016.
42. E. Elmowafy, R. Osman and R. A. H. Ishak, "Polymer-Based Novel Lung Targeted Delivery Systems," *Curr Pharm Des*, vol. 23, no. 3, pp. 373–392, 2017.
43. S. Rezvantalab *et al.*, "PLGA-Based Nanoparticles in Cancer Treatment," *Front Pharmacol*, vol. 9, p. 1260, 2018.
44. N. Vij, "Synthesis and Evaluation of Airway Targeted PLGA Nanoparticles for Drug Delivery in Obstructive Lung Diseases," *Methods Mol Biol*, vol. 906, pp. 303–310, 2012.
45. N. Vij *et al.*, "Development of PEGylated PLGA Nanoparticle for Controlled and Sustained Drug Delivery in Cystic Fibrosis," *J Nanobiotechnology*, vol. 8, pp. 1–18, 2010.
46. M. Palo, J. Holländer, J. Suominen, J. Yliruusi and N. Sandler, "3D Printed Drug Delivery Devices: Perspectives and Technical Challenges," *Expert Rev Med Devices*, vol. 14, no. 9, pp. 685–696, 2017.
47. Q. Li *et al.*, "Preparation and Investigation of Controlled-Release Glipizide Novel Oral Device With Three-Dimensional Printing," *Int J Pharm*, vol. 525, no. 1, pp. 5–11, 2017, doi: https://doi.org/10.1016/j.ijpharm.2017.03.066.
48. A. Isreb, K. Baj, M. Wojsz, M. Isreb, M. Peak and M. A. Alhnan, "3D Printed Oral Theophylline Doses With Innovative 'Radiator-Like' Design: Impact of Polyethylene Oxide (PEO) Molecular Weight," *Int J Pharm*, vol. 564, pp. 98–105, 2019, doi: https://doi.org/10.1016/j.ijpharm.2019.04.017.
49. R. Varghese, S. Salvi, P. Sood, J. Karsiya and D. Kumar, "3D Printed Medicine for the Management of Chronic Diseases: The Road Less Travelled," *Ann 3D Print Med*, vol. 5, p. 100043, 2022.
50. B. D. Kurmi, J. Kayat, V. Gajbhiye, R. K. Tekade and N. K. Jain, "Micro- and Nanocarrier-Mediated Lung Targeting," *Expert Opin Drug Deliv*, vol. 7, no. 7, pp. 781–794, 2010. doi: 10.1517/17425247.2010.492212.
51. K. C. Weng *et al.*, "Targeted Tumor Cell Internalization and Imaging of Multifunctional Quantum Dot-Conjugated Immunoliposomes in Vitro and in Vivo," *Nano Lett*, vol. 8, no. 9, pp. 2851–2857, 2008.
52. J.-M. Verbavatz, T. Ma, R. Gobin and A. S. Verkman, "Absence of Orthogonal Arrays in Kidney, Brain and Muscle from Transgenic Knockout Mice Lacking Water Channel Aquaporin-4," *J Cell Sci*, vol. 110, no. 22, pp. 2855–2860, 1997.
53. S. S. Davis, "Biomedical Applications of Nanotechnology—implications for Drug Targeting and Gene Therapy," *Trends Biotechnol*, vol. 15, no. 6, pp. 217–224, 1997.
54. N. Vij, "Nano-Based Theranostics for Chronic Obstructive Lung Diseases: Challenges and Therapeutic Potential," *Expert Opin Drug Deliv*, vol. 8, no. 9, pp. 1105–1109, 2011.

Index

Note: Locators in *italics* represent figures and **bold** indicate tables in the text.

D

E

For Product Safety Concerns and Information please contact our EU representative GPSR@taylorandfrancis.com Taylor & Francis Verlag GmbH, Kaufingerstraße 24, 80331 München, Germany